Proceedings of Symposia in
PURE MATHEMATICS

Volume 87

Spectral Analysis, Differential Equations and Mathematical Physics: A Festschrift in Honor of Fritz Gesztesy's 60th Birthday

Helge Holden
Barry Simon
Gerald Teschl
Editors

American Mathematical Society
Providence, Rhode Island

2010 *Mathematics Subject Classification.* Primary 34N05, 34L05, 35P05, 35L45, 46E35, 47A05, 47B36, 81T08, 81Q10, 93E03.

Photographs in preface courtesy of Gerald Teschl and F. Gesztesy, respectively.

Library of Congress Cataloging-in-Publication Data

Spectral analysis, differential equations and mathematical physics : a festschrift in honor of Fritz Gesztesy's 60th birthday / Helge Holden, Barry Simon, Gerald Teschl, editors.
p. cm – (Proceedings of symposia in pure mathematics ; volume 87)
Includes bibliographical references.
ISBN 978-0-8218-7574-2 (alk. paper)
1. Differential equations. 2. Mathematical physics. I. Gesztesy, Fritz, 1953– honouree. II. Holden, H. (Helge), 1956– editor of compilation. III. Simon, Barry, 1946– editor of compilation. IV. Teschl, Gerald, 1970– editor of compilation.

QC20.7.D47S64 2013
510–dc23 2012002392

10 9 8 7 6 5 4 3 2 1 18 17 16 15 14 13

Contents

Preface

A room without books is like a body without a soul.
— Attributed to Cicero (106 BC – 43 BC)

Fritz (Friedrich) was born to parents Friederike and Franz Gesztesy on November 5, 1953, in Leibnitz, Austria. He was raised there together with his younger sister, Doris.

Fritz attended the local Realgymnasium from 1964 to 1972 and, soon after the age of twelve, developed his passion for physics and mathematics. From this period onwards, he spent large parts of his free time, on one hand, in his electronics workshop (repairing and reassembling vacuum tube radios and TVs, just before the transistor revolution took place) and, on the other hand, studying B. Baule's seven-volume textbook "Die Mathematik des Naturforschers und Ingenieurs", known as "Der Baule" (developed at the Technical University of Graz, Austria).

Erwin Schödinger Institute, Vienna, July 2011.

Given his strong interests in physics and mathematics, the study of Theoretical Physics seemed the most natural choice to him and so he enrolled at the University of Graz in the fall of 1972. After studying seven semesters, he presented his dissertation on a topic in quantum field theory in early 1976. His Ph.D. advisors were Heimo Latal (University of Graz) and Ludwig Streit (University of Bielefeld,

Germany). At this point he had become disillusioned with Theoretical Physics *per se.* Strongly influenced by the monograph of T. Kato and the four-volume treatise by M. Reed and B. Simon, and especially under the guiding influence of Ludwig Pittner (University of Graz), Harald Grosse and Walter Thirring (both at the University of Vienna), and the four-volume course on Mathematical Physics by the latter, Fritz decided to devote his future energies to areas in Mathematical Physics.

Fritz was an instructor at the Institute of Theoretical Physics of the University of Graz from 1975, became Assistant Professor there in 1977, and Associate Professor (Docent) in 1982, a position he held until 1988, with several interruptions: The academic years 1980–81 and 1983–84 were spent at the University of Bielefeld as an Alexander von Humboldt fellow. Around 1986 the idea of a possible switch of continents was raised in conversations with Evans Harrell (Georgia Tech, Atlanta), and this idea slowly, but steadily, took more concrete form. After a Max Kade fellowship for the academic year 1987–88 at the California Institute of Technology, Pasadena, he assumed the position of Full Professor at the Department of Mathematics at the University of Missouri, Columbia, in the fall of 1988, his current affiliation. From 2002 he has held the M. & R. Houchins Distinguished Professorship.

Just a few days before his move to Columbia, Missouri, Fritz and Gloria Benoit were married in Bakersfield, California, in August 1988.

Fritz and Gloria on Maui, Hawaii, June 2008.

Fritz credits Ludwig Streit (Bielefeld), Sergio Albeverio (Bochum and Bonn), Raphael Høegh-Krohn (Oslo), and especially Barry Simon (Caltech) as having had the most influence on him over the years. In addition to his two years at Bielefeld and the year at Caltech, he spent time at various research institutions, including Leuven; CNRS, Luminy, Marseille; LPTHE, Orsay; BiBoS, Bielefeld; IMA, Minneapolis, Minnesota; CCM, Madeira; University of Vienna; Center for Advanced

Study (CAS) at the Norwegian Academy of Science and Letters. Fritz spent many summer months since 1990 collaborating with Helge Holden at the Norwegian University of Science and Technology, Trondheim, Norway, and with Barry Simon at Caltech.

Fritz has received a number of honors, including the Theodor Körner Award in the Natural Sciences, Vienna (1983), the Ludwig Boltzmann Award of the Austrian Physical Society (1987), and election to the Royal Norwegian Society of Sciences and Letters, Trondheim, Norway (2002). He was elected Fellow of the American Mathematical Society, inaugural class of 2013.

He has supervised or co-supervised three Ph.D. students at the University of Graz, one at the Technical University of Graz, one at the University of Louvain-La-Neuve, and nine at the University of Missouri. He takes great pride in the fact that some have become very successful in their own careers and now have successful students of their own. According to the Mathematics Genealogy Project, Fritz has 26 mathematical descendants.

Fritz's editorial responsibilities have included Mathematische Nachrichten, Journal of Mathematical Analysis and Applications, Operators and Matrices, and Journal of Spectral Theory.

Fritz's research interests developed from spectral and scattering theory for Schrödinger and Dirac-type operators in his early years until about 1988, to integrable systems and their connections with spectral theory (via trace formulas, etc.) from about 1988 to 2006. Since then his interests have primarily returned to various aspects of spectral theory for elliptic partial differential operators of relevance in mathematical physics.

Fritz is an exceptionally generous collaborator, sharing ideas and never saying no to immense calculations. He prefers to write the final version of the paper himself, securing precise statements, consistent notation and accurate bibliographies. No reference is too obscure to be checked carefully! Thus, it is no surprise that Fritz at the time this was written, has 95 co-authors and he lists over 240 publications. The author citation data base of MathSciNet shows that Fritz is cited 2295 times by 917 authors. His 1988 Springer monograph "Solvable Models in Quantum Mechanics", written jointly with S. Albeverio, R. Høegh-Krohn, and H. Holden, was translated into Russian and appeared with Mir Publishers in 1991. Its second edition, supplemented with an appendix by P. Exner, appeared in 2005 in the AMS-Chelsea series. It continues to be the authoritative treatise on solvable point interaction models and to this day remains an inspiration for research in this area.

As an avid collector of books (his personal library has approximately 5000 titles), Fritz preferred to have a volume of mathematical contributions instead of a conference in his honor. "Books are for life," he likes to say. Hence, this collection is primarily devoted to contributions in areas dear to his heart: Spectral Theory, Differential Equations, and Mathematical Physics. We are grateful to Sergei Gelfand,

Christine Thivierge, and the staff at AMS for their support throughout the preparations of this volume. We also thank all the authors for their contributions and the referees for their invaluable assistance.

Happy Birthday, Fritz!

Helge Holden
Barry Simon
Gerald Teschl

January, 2013

A mathematical descendants list

Fritz, Universität Graz, 1976

- Vladimir Batchenko, University of Missouri - Columbia, 2005
- Ronald Dickson, University of Missouri - Columbia, 1998
- Georg Karner, Universität Graz, 1986
- Miroslaw Mystkowski, University of Missouri - Columbia, 1997
- Charlotte Nessmann, Universität Graz, 1984
- Manfred Perusch, Universität Graz, 1982
- Ratnam Ratnaseelan, University of Missouri - Columbia, 1996
- Walter Renger, University of Missouri - Columbia, 1996
- Juma Shabani, Université Catholique de Louvain, 1986
  - Mathias Hounkpe, Université d'Abomey-Calavi, 1996
  - Alfred Vyabandi, Université d'Abomey-Calavi, 2001
- Wilhelm Sticka, University of Missouri - Columbia, 1995
- Gerald Teschl, University of Missouri - Columbia, 1995
  - Kerstin Ammann, Universität Wien, 2013
  - Jonathan Eckhardt, Universität Wien, 2012
  - Katrin Grunert, Universität Wien, 2010
  - Johanna Michor, Universität Wien, 2005
  - Alice Mikikits-Leitner, Universität Wien, 2009
- Mehmet Unal, University of Missouri - Columbia, 1995
- Karl Unterkofler, Technische Universität Graz, 1989
  - Julian King, Universität Innsbruck, 2010
  - Helin Koc Rauchenwald, Universität Wien, 2011
  - Klaus Rheinberger, Universität Innsbruck, 2006
  - Konrad Schwarz, Universität Innsbruck, 2009
  - Robert Tratnig, Technische Universität Graz, 2005
- Maxim Zinchenko, University of Missouri - Columbia, 2006

The most current information can be found at:
`http://genealogy.math.ndsu.nodak.edu/id.php?id=11336`

Publications of Friedrich (Fritz) Gesztesy[1]

1976:

[1] "Energiedichten und Renormierung im Modell einer Feldtheorie mit quadratischer Wechselwirkung". Dissertation, University of Graz, Austria, 1976.

[2] F. Gesztesy and H. G. Latal, *Renormalization, Nelson's symmetry and energy densities in a field theory with quadratic interaction*, Rep. Math. Phys. **14** (1978), no. 2, 215–224, DOI 10.1016/0034-4877(78)90044-7. MR527600 (80e:81077)

[3] F. Gesztesy and L. Pittner, *Electrons in logarithmic potentials. I. Solution of the Schrödinger equation*, J. Phys. A **11** (1978), no. 4, 679–686. MR0475458 (57 #15064a)

[4] F. Gesztesy and L. Pittner, *Electrons in logarithmic potentials. II. Solution of the Dirac equation*, J. Phys. A **11** (1978), no. 4, 687–695. MR0475459 (57 #15064b)

[5] F. Gesztesy and L. Pittner, *On the commutation relation* $[A, B] = -icI$, Lett. Nuovo Cimento (2) **22** (1978), no. 8, 332–335. MR502163 (82d:81059)

[6] F. Gesztesy and L. Pittner, *Uncertainty relations and quadratic forms*, J. Phys. A **11** (1978), no. 9, 1765–1770. MR506828 (81a:81028)

[7] F. Gesztesy and L. Pittner, *On the Friedrichs extension of ordinary differential operators with strongly singular potentials*, Acta Phys. Austriaca **51** (1979), no. 3-4, 259–268. MR553603 (81j:47035)

[8] F. Gesztesy and L. Pittner, *Diffraction of non-relativistic electron waves by a cylindrical capacitor*, J. Phys. A **12** (1979), no. 7, 1091–1104. MR534257 (80d:78005)

1978:

[9] "Diffraction of relativistic electron waves by a cylindrical capacitor"; with L. Pittner. J. Phys. A **12**, 2247–2254 (1979).

[10] W. Becker, F. Gesztesy, and H. Mitter, *On systems of periodic differential equations*, Lett. Math. Phys. **3** (1979), no. 4, 249–253, DOI 10.1007/BF01821842. MR545400 (81m:34063)

[11] F. Gesztesy and L. Pittner, *A generalization of the virial theorem for strongly singular potentials*, Rep. Math. Phys. **18** (1980), no. 2, 149–162 (1983), DOI 10.1016/0034-4877(80)90082-8. MR730744 (85e:81022)

[1]Updated: January 4, 2013

[12] F. Gesztesy, *On the one-dimensional Coulomb Hamiltonian*, J. Phys. A **13** (1980), no. 3, 867–875. MR560542 (80m:81023)
[13] "An efficient method for the summation of partial wave amplitudes for long-range potentials"; with C. B. Lang. Phys. Lett. **79A**, 295–297 (1980).
[14] F. Gesztesy, W. Plessas, and B. Thaller, *On the high-energy behaviour of scattering phase shifts for Coulomb-like potentials*, J. Phys. A **13** (1980), no. 8, 2659–2671. MR582916 (81k:81083)
[15] F. Gesztesy and C. B. Lang, *On the Abel summability of partial wave amplitudes for Coulomb-type interactions*, J. Math. Phys. **22** (1981), no. 2, 312–319, DOI 10.1063/1.524880. MR609622 (83f:40003)
[16] F. Gesztesy and B. Thaller, *Born expansions for Coulomb-type interactions*, J. Phys. A **14** (1981), no. 3, 639–657. MR605262 (83d:81088)
[17] "A note on quasiperiodic states"; with H. Mitter. J. Phys. **A14**, L79–L83 (1981).
[18] "On the universal low energy limit in nonrelativistic scattering theory"; with S. Albeverio and R. Høegh-Krohn. Acta Phys. Austriaca Suppl. **23**, 577–585 (1981).
[19] F. Gesztesy, *On the structure of Coulomb-type scattering amplitudes*, J. Math. Phys. **23** (1982), no. 1, 74–82, DOI 10.1063/1.525209. MR640373 (83c:81103)
[20] S. Albeverio, F. Gesztesy, and R. Høegh-Krohn, *The low energy expansion in nonrelativistic scattering theory*, Ann. Inst. H. Poincaré Sect. A (N.S.) **37** (1982), no. 1, 1–28 (English, with French summary). MR667880 (83k:81093)

1982:

[21] "Spectral concentration in the nonrelativistic limit"; with H. Grosse and B. Thaller. Phys. Lett. **116B**, 155–157 (1982).
[22] E. Brüning and F. Gesztesy, *Continuity of wave and scattering operators with respect to interactions*, J. Math. Phys. **24** (1983), no. 6, 1516–1528, DOI 10.1063/1.525890. MR708672 (85g:81179)
[23] D. Bollé, F. Gesztesy, and H. Grosse, *Time delay for long-range interactions*, J. Math. Phys. **24** (1983), no. 6, 1529–1541, DOI 10.1063/1.525891. MR708673 (85c:81041)
[24] S. Albeverio, F. Gesztesy, R. Høegh-Krohn, and L. Streit, *Charged particles with short range interactions*, Ann. Inst. H. Poincaré Sect. A (N.S.) **38** (1983), no. 3, 263–293 (English, with French summary). MR708965 (84j:81107)
[25] S. Albeverio, D. Bollé, F. Gesztesy, and R. Høegh-Krohn, *Low-energy parameters in nonrelativistic scattering theory*, Ann. Physics **148** (1983), no. 2, 308–326, DOI 10.1016/0003-4916(83)90242-7. MR714194 (84j:81108)

1983:

[26] "Efficient method for calculating relativistic corrections for spin-1/2 particles"; with H. Grosse and B. Thaller. Phys. Rev. Lett. **50**, 625–628 (1983).
[27] D. Bollé and F. Gesztesy, *On averaged angular time delay for two-body scattering*, Helv. Phys. Acta **56** (1983), no. 5, 1064–1069. MR728114 (85e:81114)
[28] D. Bollé, F. Gesztesy, and S. F. J. Wilk, *New results for scattering on the line*, Phys. Lett. A **97** (1983), no. 1-2, 30–34, DOI 10.1016/0375-9601(83)90094-4. MR720677 (85h:81067)

[29] "On non-degenerate ground states for Schrödinger operators". Rep. Math. Phys. **20**, 93–109 (1984).

[30] F. Gesztesy and L. Pittner, *Two-body scattering for Schrödinger operators involving zero-range interactions*, Rep. Math. Phys. **19** (1984), no. 2, 143–154, DOI 10.1016/0034-4877(84)90012-0. MR740351 (86d:81075)

[31] S. Albeverio, L. S. Ferreira, F. Gesztesy, R. Høegh-Krohn, and L. Streit, *Model dependence of Coulomb-corrected scattering lengths*, Phys. Rev. C (3) **29** (1984), no. 2, 680–683, DOI 10.1103/PhysRevC.29.680. MR734284 (85b:81236)

[32] S. Albeverio, F. Gesztesy, R. Høegh-Krohn, and W. Kirsch, *On point interactions in one dimension*, J. Operator Theory **12** (1984), no. 1, 101–126. MR757115 (86e:81037)

[33] F. Gesztesy, H. Grosse, and B. Thaller, *A rigorous approach to relativistic corrections of bound state energies for spin-$\frac{1}{2}$ particles*, Ann. Inst. H. Poincaré Phys. Théor. **40** (1984), no. 2, 159–174 (English, with French summary). MR747200 (86m:81042)

[34] "On relativistic corrections to bound state energies for two-fermion systems"; with H. Grosse and B. Thaller. Phys. Rev. **D30**, 2189–2193 (1984).

[35] "Low-energy parametrization of scattering observables in n-dimensional quantum systems"; with D. Bollé. Phys. Rev. Lett. **52**, 1469–1472 (1984).

[36] S. Albeverio, R. Høegh-Krohn, F. Gesztesy, and H. Holden, *Some exactly solvable models in quantum mechanics and the low energy expansions*, Proceedings of the second international conference on operator algebras, ideals, and their applications in theoretical physics (Leipzig, 1983), Teubner-Texte Math., vol. 67, Teubner, Leipzig, 1984, pp. 12–28. MR763518 (85i:81015)

[37] "Scattering observables in arbitrary dimension $n \geq 2$"; with D. Bollé. Phys. Rev. **A30**, 1279–1293 (1984).

[38] F. Gesztesy, *Perturbation theory for resonances in terms of Fredholm determinants*, Resonances—models and phenomena (Bielefeld, 1984), Lecture Notes in Phys., vol. 211, Springer, Berlin, 1984, pp. 78–104, DOI 10.1007/3-540-13880-3_67. MR777333 (86f:81145)

[39] "Scattering lengths in nonrelativistic three-body systems"; with G. Karner. In *Few-Body Problems in Physics*, Vol. II, B. Zeitnitz (ed.), Elsevier Science Publishers B. V., 1984, pp. 375–376.

[40] F. Gesztesy, H. Grosse, and B. Thaller, *First-order relativistic corrections and spectral concentration*, Adv. in Appl. Math. **6** (1985), no. 2, 159–176, DOI 10.1016/0196-8858(85)90009-0. MR789851 (86j:81026)

[41] D. Bollé, F. Gesztesy, and S. F. J. Wilk, *A complete treatment of low-energy scattering in one dimension*, J. Operator Theory **13** (1985), no. 1, 3–31. MR768299 (86f:34047)

[42] F. Gesztesy, H. Mitter, and M. Perusch, *Scattering theory for time-dependent Hamiltonians asymptotically constant in time*, J. Math. Anal. Appl. **110** (1985), no. 1, 265–282, DOI 10.1016/0022-247X(85)90349-X. MR803434 (86i:35114)

[43] F. Gesztesy and W. Kirsch, *One-dimensional Schrödinger operators with interactions singular on a discrete set*, J. Reine Angew. Math. **362** (1985), 28–50. MR809964 (87e:34034)

1985:

[44] "On the self-adjointness of Dirac operators with anomalous magnetic moment"; with B. Simon and B. Thaller. Proc. Amer. Math. Soc. **94**, 115–118 (1985).

[45] S. Albeverio, F. Gesztesy, W. Karwowski, and L. Streit, *On the connection between Schrödinger and Dirichlet forms*, J. Math. Phys. **26** (1985), no. 10, 2546–2553, DOI 10.1063/1.526771. MR803798 (87c:81041)

[46] D. Bollé, F. Gesztesy, and W. Schweiger, *Scattering theory for long-range systems at threshold*, J. Math. Phys. **26** (1985), no. 7, 1661–1674, DOI 10.1063/1.526963. MR793308 (87a:81145)

[47] S. Albeverio, Ph. Blanchard, F. Gesztesy, and L. Streit, *Quantum mechanical low energy scattering in terms of diffusion processes*, Stochastic aspects of classical and quantum systems (Marseille, 1983), Lecture Notes in Math., vol. 1109, Springer, Berlin, 1985, pp. 207–227, DOI 10.1007/BFb0101546. MR805999 (87c:81164)

[48] "On essential spectra of hard core type Schrödinger operators"; with E. Brüning. J. Phys. **A18**, L7–L11 (1985).

[49] W. Bulla and F. Gesztesy, *Deficiency indices and singular boundary conditions in quantum mechanics*, J. Math. Phys. **26** (1985), no. 10, 2520–2528, DOI 10.1063/1.526768. MR803795 (87d:35097)

[50] "An exactly solvable periodic Schrödinger operator"; with C. Macedo and L. Streit. J. Phys. **A18**, L503–L507 (1985).

1986:

[51] "Threshold behavior and Levinson's theorem for two-dimensional scattering systems: A surprise"; with D. Bollé, C. Danneels, and S. F. J. Wilk. Phys. Rev. Lett. **56**, 900–903 (1986).

[52] D. Bollé, F. Gesztesy, C. Nessmann, and L. Streit, *Scattering theory for general, nonlocal interactions: threshold behavior and sum rules*, Rep. Math. Phys. **23** (1986), no. 3, 373–408, DOI 10.1016/0034-4877(86)90032-7. MR913481 (88m:81131)

[53] "Scattering observables in arbitrary dimensions $n \geq 2$: An Addendum"; with D. Bollé. Phys. Rev. **A33**, 3517–3518 (1986).

[54] F. Gesztesy, *Scattering theory for one-dimensional systems with nontrivial spatial asymptotics*, Schrödinger operators, Aarhus 1985, Lecture Notes in Math., vol. 1218, Springer, Berlin, 1986, pp. 93–122, DOI 10.1007/BFb0073045. MR869597 (88a:81197)

[55] S. Albeverio, F. Gesztesy, R. Høegh-Krohn, H. Holden, and W. Kirsch, *The Schrödinger operator for a particle in a solid with deterministic and stochastic point interactions*, Schrödinger operators, Aarhus 1985, Lecture Notes in Math., vol. 1218, Springer, Berlin, 1986, pp. 1–38, DOI 10.1007/BFb0073042. MR869594 (88a:81030)

[56] F. Gesztesy, G. Karner, and L. Streit, *Charged particles with a short-range force: perturbation theory with respect to the range and to additional effects*, J. Math. Phys. **27** (1986), no. 1, 249–261, DOI 10.1063/1.527369. MR816440 (87g:81118)

[57] F. Gesztesy and H. Holden, *A unified approach to eigenvalues and resonances of Schrödinger operators using Fredholm determinants*, J. Math. Anal. Appl. **123** (1987), no. 1, 181–198, DOI 10.1016/0022-247X(87)90303-9. MR881540 (88c:35117)

[58] D. Bollé, F. Gesztesy, and M. Klaus, *Scattering theory for one-dimensional systems with $\int dx\, V(x) = 0$*, J. Math. Anal. Appl. **122** (1987), no. 2, 496–518, DOI 10.1016/0022-247X(87)90281-2. MR877834 (89k:34028a)

[59] F. Gesztesy and G. Karner, *On three-body scattering near thresholds*, SIAM J. Math. Anal. **18** (1987), no. 4, 1064–1086, DOI 10.1137/0518079. MR892489 (88j:81076)

[60] "Point interactions in two dimensions. Basic properties and applications to solid state physics"; with S. Albeverio, R. Høegh-Krohn, and H. Holden. J. reine angew. Math. **308**, 87–107 (1987).

[61] D. Bollé, F. Gesztesy, H. Grosse, and B. Simon, *Kreĭn's spectral shift function and Fredholm determinants as efficient methods to study supersymmetric quantum mechanics*, Lett. Math. Phys. **13** (1987), no. 2, 127–133, DOI 10.1007/BF00955200. MR886147 (88f:81043)

[62] D. Bollé, F. Gesztesy, H. Grosse, W. Schweiger, and B. Simon, *Witten index, axial anomaly, and Kreĭn's spectral shift function in supersymmetric quantum mechanics*, J. Math. Phys. **28** (1987), no. 7, 1512–1525, DOI 10.1063/1.527508. MR894842 (88j:81022)

[63] E. Brüning, M. Demuth, and F. Gesztesy, *Invariance of the essential spectra for perturbations with unbounded hard cores*, Lett. Math. Phys. **13** (1987), no. 1, 69–77, DOI 10.1007/BF00570770. MR878664 (88c:35115)

[64] J.-P. Antoine, F. Gesztesy, and J. Shabani, *Exactly solvable models of sphere interactions in quantum mechanics*, J. Phys. A **20** (1987), no. 12, 3687–3712. MR913638 (89d:81027)

[65] F. Gesztesy and P. Šeba, *New analytically solvable models of relativistic point interactions*, Lett. Math. Phys. **13** (1987), no. 4, 345–358, DOI 10.1007/BF00401163. MR895297 (89b:81034)

[66] F. Gesztesy and H. Holden, *A new class of solvable models in quantum mechanics describing point interactions on the line*, J. Phys. A **20** (1987), no. 15, 5157–5177. MR914699 (89a:81016)

[67] F. Gesztesy, *On stationary two-body scattering theory in two dimensions*, Models and methods in few-body physics (Lisbon, 1986), Lecture Notes in Phys., vol. 273, Springer, Berlin, 1987, pp. 609–629, DOI 10.1007/3-540-17647-0_42. MR899147

[68] F. Gesztesy, H. Holden, and W. Kirsch, *On energy gaps in a new type of analytically solvable model in quantum mechanics*, J. Math. Anal. Appl. **134** (1988), no. 1, 9–29, DOI 10.1016/0022-247X(88)90003-0. MR958850 (90c:81032)

[69] F. Gesztesy and B. Simon, *Topological invariance of the Witten index*, J. Funct. Anal. **79** (1988), no. 1, 91–102, DOI 10.1016/0022-1236(88)90031-6. MR950085 (90a:47032)

[70] D. Bollé, F. Gesztesy, and C. Danneels, *Threshold scattering in two dimensions*, Ann. Inst. H. Poincaré Phys. Théor. **48** (1988), no. 2, 175–204 (English, with French summary). MR952661 (89k:81184)

[71] W. Bulla, F. Gesztesy, and K. Unterkofler, *On relativistic energy band corrections in the presence of periodic potentials*, Lett. Math. Phys. **15** (1988), no. 4, 313–324, DOI 10.1007/BF00419589. MR952454 (90c:81038)
[72] F. Gesztesy and B. Simon, *On a theorem of Deift and Hempel*, Comm. Math. Phys. **116** (1988), no. 3, 503–505. MR937772 (89g:35080)
[73] F. Gesztesy, D. Gurarie, H. Holden, M. Klaus, L. Sadun, B. Simon, and P. Vogl, *Trapping and cascading of eigenvalues in the large coupling limit*, Comm. Math. Phys. **118** (1988), no. 4, 597–634. MR962490 (89m:81049)
[74] S. Albeverio, R. Figari, F. Gesztesy, R. Høegh-Krohn, H. Holden, and W. Kirsch, *Point interaction Hamiltonians for crystals with random defects*, Applications of selfadjoint extensions in quantum physics (Dubna, 1987), Lecture Notes in Phys., vol. 324, Springer, Berlin, 1989, pp. 87–99, DOI 10.1007/BFb0022960. MR1009843
[75] F. Gesztesy, H. Holden, and P. Šeba, *On point interactions in magnetic field systems*, Schrödinger operators, standard and nonstandard (Dubna, 1988), World Sci. Publ., Teaneck, NJ, 1989, pp. 146–164. MR1091996 (91m:81044)
[76] F. Gesztesy, *Some applications of commutation methods*, Schrödinger operators (Sønderborg, 1988), Lecture Notes in Phys., vol. 345, Springer, Berlin, 1989, pp. 93–117, DOI 10.1007/3-540-51783-9_18. MR1037318 (91g:58246)
[77] F. Gesztesy and B. Simon, *Constructing solutions of the mKdV-equation*, J. Funct. Anal. **89** (1990), no. 1, 53–60, DOI 10.1016/0022-1236(90)90003-4. MR1040955 (91e:35183)
[78] F. Gesztesy, W. Schweiger, and B. Simon, *Commutation methods applied to the mKdV-equation*, Trans. Amer. Math. Soc. **324** (1991), no. 2, 465–525, DOI 10.2307/2001730. MR1029000 (92b:35132)
[79] F. Gesztesy and Z. Zhao, *On critical and subcritical Sturm-Liouville operators*, J. Funct. Anal. **98** (1991), no. 2, 311–345, DOI 10.1016/0022-1236(91)90081-F. MR1111572 (93f:34146)
[80] F. Gesztesy, H. Holden, E. Saab, and B. Simon, *Explicit construction of solutions of the modified Kadomtsev-Petviashvili equation*, J. Funct. Anal. **98** (1991), no. 1, 211–228, DOI 10.1016/0022-1236(91)90096-N. MR1111199 (92h:35206)
[81] F. Gesztesy, *On the modified Korteweg-de Vries equation*, engineering (Leibnitz, 1989), Lecture Notes in Pure and Appl. Math., vol. 133, Dekker, New York, 1991, pp. 139–183. MR1171468 (93i:35123)
[82] F. Gesztesy and W. Schweiger, *Rational KP and mKP-solutions in Wronskian form*, Rep. Math. Phys. **30** (1991), no. 2, 205–222 (1992), DOI 10.1016/0034-4877(91)90025-I. MR1188396 (94a:58089)
[83] F. Gesztesy, *(m)KdV-soliton solutions on quasi-periodic finite-gap backgrounds*, Nonlinear fields: classical, random, semiclassical (Karpacz, 1991), World Sci. Publ., River Edge, NJ, 1991, pp. 171–194. MR1146003 (93e:58082)
[84] W. Bulla, F. Gesztesy, and K. Unterkofler, *Holomorphy of the scattering matrix with respect to c^{-2} for Dirac operators and an explicit treatment of relativistic corrections*, Comm. Math. Phys. **144** (1992), no. 2, 391–416. MR1152379 (93b:81302)
[85] F. Gesztesy, *Quasi-periodic, finite-gap solutions of the modified Korteweg-de Vries equation*, applications (Oslo, 1988), Cambridge Univ. Press, Cambridge, 1992, pp. 428–471. MR1190516 (94d:35144)

[86] F. Gesztesy and H. Holden, *A new representation of soliton solutions of the Kadomtsev-Petviashvili equation*, applications (Oslo, 1988), Cambridge Univ. Press, Cambridge, 1992, pp. 472–479. MR1190517 (93i:35124)

[87] F. Gesztesy and K. Unterkofler, *Isospectral deformations for Sturm-Liouville and Dirac-type operators and associated nonlinear evolution equations*, Rep. Math. Phys. **31** (1992), no. 2, 113–137, DOI 10.1016/0034-4877(92)90008-O. MR1227036 (94f:35124)

[88] F. Gesztesy, W. Karwowski, and Z. Zhao, *New types of soliton solutions*, Bull. Amer. Math. Soc. (N.S.) **27** (1992), no. 2, 266–272, DOI 10.1090/S0273-0979-1992-00309-9. MR1152159 (93c:35138)

[89] F. Gesztesy, W. Karwowski, and Z. Zhao, *Limits of soliton solutions*, Duke Math. J. **68** (1992), no. 1, 101–150, DOI 10.1215/S0012-7094-92-06805-0. MR1185820 (94b:35242)

[90] F. Gesztesy, G. M. Graf, and B. Simon, *The ground state energy of Schrödinger operators*, Comm. Math. Phys. **150** (1992), no. 2, 375–384. MR1194022 (93j:47070)

[91] F. Gesztesy and Z. Zhao, *Critical and subcritical Jacobi operators defined as Friedrichs extensions*, J. Differential Equations **103** (1993), no. 1, 68–93, DOI 10.1006/jdeq.1993.1042. MR1218739 (94m:47065)

[92] F. Gesztesy, H. Holden, B. Simon, and Z. Zhao, *On the Toda and Kac-van Moerbeke systems*, Trans. Amer. Math. Soc. **339** (1993), no. 2, 849–868, DOI 10.2307/2154302. MR1153014 (93m:58050)

[93] F. Gesztesy and B. Simon, *A short proof of Zheludev's theorem*, Trans. Amer. Math. Soc. **335** (1993), no. 1, 329–340, DOI 10.2307/2154271. MR1096260 (93c:34162)

[94] Friedrich Gesztesy, David Race, and Rudi Weikard, *On (modified) Boussinesq-type systems and factorizations of associated linear differential expressions*, J. London Math. Soc. (2) **47** (1993), no. 2, 321–340, DOI 10.1112/jlms/s2-47.2.321. MR1207952 (95c:35212)

[95] F. Gesztesy and R. Weikard, *Spectral deformations and soliton equations*, Differential equations with applications to mathematical physics, Math. Sci. Engrg., vol. 192, Academic Press, Boston, MA, 1993, pp. 101–139, DOI 10.1016/S0076-5392(08)62376-0. MR1207152 (93m:34138)

[96] F. Gesztesy, *A complete spectral characterization of the double commutation method*, J. Funct. Anal. **117** (1993), no. 2, 401–446, DOI 10.1006/jfan.1993.1132. MR1244942 (94m:47093)

[97] F. Gesztesy, H. Holden, B. Simon, and Z. Zhao, *Trace formulae and inverse spectral theory for Schrödinger operators*, Bull. Amer. Math. Soc. (N.S.) **29** (1993), no. 2, 250–255, DOI 10.1090/S0273-0979-1993-00431-2. MR1215308 (94c:34127)

[98] F. Gesztesy, D. Race, K. Unterkofler, and R. Weikard, *On Gelfand-Dickey and Drinfeld-Sokolov systems*, Rev. Math. Phys. **6** (1994), no. 2, 227–276, DOI 10.1142/S0129055X94000122. MR1269299 (95g:58104)

[99] F. Gesztesy and Z. Zhao, *Domain perturbations, Brownian motion, capacities, and ground states of Dirichlet Schrödinger operators*, Math. Z. **215** (1994), no. 1, 143–150, DOI 10.1007/BF02571703. MR1254817 (95g:60098)

[100] F. Gesztesy and H. Holden, *Trace formulas and conservation laws for nonlinear evolution equations*, Rev. Math. Phys. **6** (1994), no. 1, 51–95, DOI 10.1142/S0129055X94000055. MR1263198 (95h:35198a)

[101] "New trace formulas for Schrödinger operators". In *Evolution Equations*, G. Ferreyra, G. Goldstein, and F. Neubrander (eds.), Marcel Dekker, 1995, pp. 201–221.

[102] F. Gesztesy and R. Weikard, *Picard and finite-gap potentials*, Evolution equations (Baton Rouge, LA, 1992), Lecture Notes in Pure and Appl. Math., vol. 168, Dekker, New York, 1995, pp. 223–233. MR1300431 (95h:35191)

[103] F. Gesztesy and K. Unterkofler, *On the (modified) Kadomtsev-Petviashvili hierarchy*, Differential Integral Equations **8** (1995), no. 4, 797–812. MR1306592 (95h:35199)

[104] F. Gesztesy and Z. Zhao, *On positive solutions of critical Schrödinger operators in two dimensions*, J. Funct. Anal. **127** (1995), no. 1, 235–256, DOI 10.1006/jfan.1995.1010. MR1308624 (96a:35037)

[105] F. Gesztesy and R. Weikard, *On Picard potentials*, Differential Integral Equations **8** (1995), no. 6, 1453–1476. MR1329850 (96e:34141)

[106] Fritz Gesztesy and Roman Svirsky, *(m)KdV solitons on the background of quasi-periodic finite-gap solutions*, Mem. Amer. Math. Soc. **118** (1995), no. 563, iv+88. MR1303091 (96c:35162)

[107] F. Gesztesy and R. Weikard, *Treibich-Verdier potentials and the stationary (m)KdV hierarchy*, Math. Z. **219** (1995), no. 3, 451–476, DOI 10.1007/BF02572375. MR1339715 (96e:14030)

[108] F. Gesztesy and B. Simon, *Rank-one perturbations at infinite coupling*, J. Funct. Anal. **128** (1995), no. 1, 245–252, DOI 10.1006/jfan.1995.1030. MR1317717 (95m:47014)

[109] F. Gesztesy, H. Holden, and B. Simon, *Absolute summability of the trace relation for certain Schrödinger operators*, Comm. Math. Phys. **168** (1995), no. 1, 137–161. MR1324393 (96b:34110)

[110] F. Gesztesy and R. Weikard, *Lamé potentials and the stationary (m)KdV hierarchy*, Math. Nachr. **176** (1995), 73–91, DOI 10.1002/mana.19951760107. MR1361127 (98a:58086)

[111] F. Gesztesy, H. Holden, B. Simon, and Z. Zhao, *Higher order trace relations for Schrödinger operators*, Rev. Math. Phys. **7** (1995), no. 6, 893–922, DOI 10.1142/S0129055X95000347. MR1348829 (97d:34094)

[112] F. Gesztesy and H. Holden, *On new trace formulae for Schrödinger operators*, Acta Appl. Math. **39** (1995), no. 1-3, 315–333, DOI 10.1007/BF00994640. KdV '95 (Amsterdam, 1995). MR1329568 (96f:35126)

[113] M. Demuth, F. Gesztesy, J. van Casteren, and Z. Zhao, *Finite capacities in spectral theory*, Partial differential operators and mathematical physics (Holzhau, 1994), Oper. Theory Adv. Appl., vol. 78, Birkhäuser, Basel, 1995, pp. 89–97. MR1365320 (97c:47052)

[114] F. Gesztesy and R. Weikard, *Floquet theory revisited*, Differential equations and mathematical physics (Birmingham, AL, 1994), Int. Press, Boston, MA, 1995, pp. 67–84. MR1703573 (2000i:34163)

[115] Fritz Gesztesy and Rudi Weikard, *A characterization of elliptic finite-gap potentials*, C. R. Acad. Sci. Paris Sér. I Math. **321** (1995), no. 7, 837–841 (English, with English and French summaries). MR1355838 (96k:58112)

[116] Fritz Gesztesy and Barry Simon, *The xi function*, Acta Math. **176** (1996), no. 1, 49–71, DOI 10.1007/BF02547335. MR1395669 (97e:47078)

[117] Fritz Gesztesy and Rudi Weikard, *Picard potentials and Hill's equation on a torus*, Acta Math. **176** (1996), no. 1, 73–107, DOI 10.1007/BF02547336. MR1395670 (97f:14046)

[118] F. Gesztesy and G. Teschl, *On the double commutation method*, Proc. Amer. Math. Soc. **124** (1996), no. 6, 1831–1840, DOI 10.1090/S0002-9939-96-03299-6. MR1322925 (96h:34171)

[119] F. Gesztesy and G. Teschl, *Commutation methods for Jacobi operators*, J. Differential Equations **128** (1996), no. 1, 252–299, DOI 10.1006/jdeq.1996.0095. MR1392402 (97i:47079)

[120] F. Gesztesy and B. Simon, *Uniqueness theorems in inverse spectral theory for one-dimensional Schrödinger operators*, Trans. Amer. Math. Soc. **348** (1996), no. 1, 349–373, DOI 10.1090/S0002-9947-96-01525-5. MR1329533 (96e:34030)

[121] F. Gesztesy, B. Simon, and G. Teschl, *Zeros of the Wronskian and renormalized oscillation theory*, Amer. J. Math. **118** (1996), no. 3, 571–594. MR1393260 (97g:34105)

[122] F. Gesztesy, M. Krishna, and G. Teschl, *On isospectral sets of Jacobi operators*, Comm. Math. Phys. **181** (1996), no. 3, 631–645. MR1414303 (97i:47048)

[123] F. Gesztesy, B. Simon, and G. Teschl, *Spectral deformations of one-dimensional Schrödinger operators*, J. Anal. Math. **70** (1996), 267–324, DOI 10.1007/BF02820446. MR1444263 (98m:34171)

[124] F. Gesztesy, R. Ratnaseelan, and G. Teschl, *The KdV hierarchy and associated trace formulas*, Recent developments in operator theory and its applications (Winnipeg, MB, 1994), Oper. Theory Adv. Appl., vol. 87, Birkhäuser, Basel, 1996, pp. 125–163. MR1399359 (97m:58095)

[125] F. Gesztesy, H. Holden, B. Simon, and Z. Zhao, *A trace formula for multidimensional Schrödinger operators*, J. Funct. Anal. **141** (1996), no. 2, 449–465, DOI 10.1006/jfan.1996.0137. MR1418515 (97i:47098)

[126] F. Gesztesy and H. Holden, *On trace formulas for Schrödinger-type operators*, and molecular physics (Minneapolis, MN, 1995), IMA Vol. Math. Appl., vol. 89, Springer, New York, 1997, pp. 121–145, DOI 10.1007/978-1-4612-1870-8_5. MR1487920 (98m:34172)

[127] W. Bulla, F. Gesztesy, W. Renger, and B. Simon, *Weakly coupled bound states in quantum waveguides*, Proc. Amer. Math. Soc. **125** (1997), no. 5, 1487–1495, DOI 10.1090/S0002-9939-97-03726-X. MR1371117 (97g:81009)

[128] F. Gesztesy, R. Nowell, and W. Pötz, *One-dimensional scattering theory for quantum systems with nontrivial spatial asymptotics*, Differential Integral Equations **10** (1997), no. 3, 521–546. MR1744860 (2000k:81392)

[129] F. Gesztesy and W. Renger, *New classes of Toda soliton solutions*, Comm. Math. Phys. **184** (1997), no. 1, 27–50, DOI 10.1007/s002200050051. MR1462498 (99f:58097)

[130] Fritz Gesztesy and Barry Simon, *Inverse spectral analysis with partial information on the potential. I. The case of an a.c. component in the spectrum*, Helv. Phys. Acta **70** (1997), no. 1-2, 66–71. Papers honouring the 60th birthday of Klaus Hepp and of Walter Hunziker, Part II (Zürich, 1995). MR1441597 (98f:81347)

[131] Rafael del Rio, Fritz Gesztesy, and Barry Simon, *Inverse spectral analysis with partial information on the potential. III. Updating boundary conditions*, Internat. Math. Res. Notices **15** (1997), 751–758, DOI 10.1155/S1073792897000494. MR1470376 (99a:34032)

[132] Fritz Gesztesy and Barry Simon, *m-functions and inverse spectral analysis for finite and semi-infinite Jacobi matrices*, J. Anal. Math. **73** (1997), 267–297, DOI 10.1007/BF02788147. MR1616422 (99c:47039)

[133] F. Gesztesy and M. Ünal, *Perturbative oscillation criteria and Hardy-type inequalities*, Math. Nachr. **189** (1998), 121–144, DOI 10.1002/mana.19981890108. MR1492926 (99a:34069)

[134] W. Bulla, F. Gesztesy, H. Holden, and G. Teschl, *Algebro-geometric quasi-periodic finite-gap solutions of the Toda and Kac-van Moerbeke hierarchies*, Mem. Amer. Math. Soc. **135** (1998), no. 641, x+79. MR1432141 (99b:58109)

[135] F. Gesztesy and R. Ratnaseelan, *An alternative approach to algebro-geometric solutions of the AKNS hierarchy*, Rev. Math. Phys. **10** (1998), no. 3, 345–391, DOI 10.1142/S0129055X98000112. MR1626836 (99d:58079)

[136] F. Gesztesy and W. Sticka, *On a theorem of Picard*, Proc. Amer. Math. Soc. **126** (1998), no. 4, 1089–1099, DOI 10.1090/S0002-9939-98-04668-1. MR1476130 (98m:34012)

[137] Fritz Gesztesy and Rudi Weikard, *A characterization of all elliptic algebro-geometric solutions of the AKNS hierarchy*, Acta Math. **181** (1998), no. 1, 63–108, DOI 10.1007/BF02392748. MR1654775 (99k:14052)

[138] Fritz Gesztesy, Konstantin A. Makarov, and Eduard Tsekanovskii, *An addendum to Krein's formula*, J. Math. Anal. Appl. **222** (1998), no. 2, 594–606, DOI 10.1006/jmaa.1998.5948. MR1628437 (99g:47047)

[139] Fritz Gesztesy and Rudi Weikard, *Elliptic algebro-geometric solutions of the KdV and AKNS hierarchies—an analytic approach*, Bull. Amer. Math. Soc. (N.S.) **35** (1998), no. 4, 271–317, DOI 10.1090/S0273-0979-98-00765-4. MR1638298 (99i:58075)

[140] F. Gesztesy and R. Weikard, *Toward a characterization of elliptic solutions of hierarchies of soliton equations*, Applied analysis (Baton Rouge, LA, 1996), Contemp. Math., vol. 221, Amer. Math. Soc., Providence, RI, 1999, pp. 133–161, DOI 10.1090/conm/221/03120. MR1647205 (99k:58090)

[141] Ronnie Dickson, Fritz Gesztesy, and Karl Unterkofler, *A new approach to the Boussinesq hierarchy*, Math. Nachr. **198** (1999), 51–108, DOI 10.1002/mana.19991980105. MR1670365 (99m:35204)

[142] Fritz Gesztesy and Barry Simon, *On the determination of a potential from three spectra*, Differential operators and spectral theory, Amer. Math. Soc. Transl. Ser. 2, vol. 189, Amer. Math. Soc., Providence, RI, 1999, pp. 85–92. MR1730505 (2000i:34026)

[143] Fritz Gesztesy, Konstantin A. Makarov, and Serguei N. Naboko, *The spectral shift operator*, Mathematical results in quantum mechanics (Prague, 1998), Oper. Theory Adv. Appl., vol. 108, Birkhäuser, Basel, 1999, pp. 59–90. MR1708788 (2000k:47012)

[144] "Corrections and addendum to *Inverse spectral analysis with partial information on the potential, III. Updating boundary conditions*"; with R. del Rio and B. Simon. Int. Math. Res. Notices **1999, No. 11**, 623–625.

[145] R. Dickson, F. Gesztesy, and K. Unterkofler, *Algebro-geometric solutions of the Boussinesq hierarchy*, Rev. Math. Phys. **11** (1999), no. 7, 823–879, DOI 10.1142/S0129055X9900026X. MR1702719 (2000d:14040)

[146] Fritz Gesztesy and Barry Simon, *Inverse spectral analysis with partial information on the potential. II. The case of discrete spectrum*, Trans. Amer. Math. Soc. **352** (2000), no. 6, 2765–2787, DOI 10.1090/S0002-9947-99-02544-1. MR1694291 (2000j:34019)

[147] Fritz Gesztesy and Helge Holden, *A combined sine-Gordon and modified Korteweg-de Vries hierarchy and its algebro-geometric solutions*, Differential equations and mathematical physics (Birmingham, AL, 1999), AMS/IP Stud. Adv. Math., vol. 16, Amer. Math. Soc., Providence, RI, 2000, pp. 133–173. MR1764748 (2001f:37114)

[148] Fritz Gesztesy and Eduard Tsekanovskii, *On matrix-valued Herglotz functions*, Math. Nachr. **218** (2000), 61–138, DOI 10.1002/1522-2616(200010)218:1¡61::AID-MANA61¿3.3.CO;2-4. MR1784638 (2001j:47018)

[149] Steve Clark, Fritz Gesztesy, Helge Holden, and Boris M. Levitan, *Borg-type theorems for matrix-valued Schrödinger operators*, J. Differential Equations **167** (2000), no. 1, 181–210, DOI 10.1006/jdeq.1999.3758. MR1785118 (2002d:34019)

[150] Fritz Gesztesy and Helge Holden, *The classical Boussinesq hierarchy revisited*, Skr. K. Nor. Vidensk. Selsk. **1** (2000), 15. MR1828737 (2002b:35178)

[151] Fritz Gesztesy and Helge Holden, *Darboux-type transformations and hyperelliptic curves*, J. Reine Angew. Math. **527** (2000), 151–183, DOI 10.1515/crll.2000.080. MR1794021 (2002b:37108)

[152] Fritz Gesztesy and Konstantin A. Makarov, *Some applications of the spectral shift operator*, Operator theory and its applications (Winnipeg, MB, 1998), Fields Inst. Commun., vol. 25, Amer. Math. Soc., Providence, RI, 2000, pp. 267–292. MR1759548 (2001f:47018)

[153] Fritz Gesztesy and Konstantin A. Makarov, *The Ξ operator and its relation to Krein's spectral shift function*, J. Anal. Math. **81** (2000), 139–183, DOI 10.1007/BF02788988. MR1785280 (2001i:47016)

[154] F. Gesztesy, C. K. R. T. Jones, Y. Latushkin, and M. Stanislavova, *A spectral mapping theorem and invariant manifolds for nonlinear Schrödinger equations*, Indiana Univ. Math. J. **49** (2000), no. 1, 221–243, DOI 10.1512/iumj.2000.49.1838. MR1777032 (2001g:37144)

[155] "The classical massive Thirring system revisited"; with V. Z. Enolskii and H. Holden. In *Stochastic Processes, Physics and Geometry: New Interplays. I. A Volume in Honor of Sergio Albeverio*, F. Gesztesy, H. Holden, J. Jost, S. Paycha, M. Röckner, and S. Scarlatti (eds.), Canadian Mathematical Society Conference Proceedings, Vol. 28, Amer. Math. Soc., Providence, RI, 2000, pp. 163–200.

[156] *Stochastic processes, physics and geometry: new interplays. II*, CMS Conference Proceedings, vol. 29, American Mathematical Society, Providence, RI, 2000. A volume in honor of Sergio Albeverio; Edited by Fritz Gesztesy, Helge Holden, Jürgen Jost, Sylvie Paycha, Michael Röckner and Sergio Scarlatti. MR1803398 (2001f:00037)

[157] Fritz Gesztesy and Alexander G. Ramm, *An inverse problem for point inhomogeneities*, Methods Funct. Anal. Topology **6** (2000), no. 2, 1–12. MR1783771 (2001h:47032)

[158] Fritz Gesztesy and Barry Simon, *On local Borg-Marchenko uniqueness results*, Comm. Math. Phys. **211** (2000), no. 2, 273–287, DOI 10.1007/s002200050812. MR1754515 (2001b:34020)

[159] F. Gesztesy, K. Unterkofler, and R. Weikard, *On a theorem of Halphen and its application to integrable systems*, J. Math. Anal. Appl. **251** (2000), no. 2, 504–526, DOI 10.1006/jmaa.2000.7026. MR1794755 (2001i:37108)

[160] Fritz Gesztesy and Helge Holden, *The Cole-Hopf and Miura transformations revisited*, Mathematical physics and stochastic analysis (Lisbon, 1998), World Sci. Publ., River Edge, NJ, 2000, pp. 198–214. MR1893107 (2003a:37107)

[161] Fritz Gesztesy and Barry Simon, *A new approach to inverse spectral theory. II. General real potentials and the connection to the spectral measure*, Ann. of Math. (2) **152** (2000), no. 2, 593–643, DOI 10.2307/2661393. MR1804532 (2001m:34185b)

[162] Fritz Gesztesy, *Integrable systems in the infinite genus limit*, Differential Integral Equations **14** (2001), no. 6, 671–700. MR1826956 (2002f:37124)

[163] Steve Clark and Fritz Gesztesy, *Weyl-Titchmarsh M-function asymptotics for matrix-valued Schrödinger operators*, Proc. London Math. Soc. (3) **82** (2001), no. 3, 701–724, DOI 10.1112/plms/82.3.701. MR1816694 (2002c:34144)

[164] Fritz Gesztesy, Nigel J. Kalton, Konstantin A. Makarov, and Eduard Tsekanovskii, *Some applications of operator-valued Herglotz functions*, Operator theory, system theory and related topics (Beer-Sheva/Rehovot, 1997), Oper. Theory Adv. Appl., vol. 123, Birkhäuser, Basel, 2001, pp. 271–321. MR1821917 (2002f:47049)

[165] Fritz Gesztesy and Helge Holden, *Dubrovin equations and integrable systems on hyperelliptic curves*, Math. Scand. **91** (2002), no. 1, 91–126. MR1917684 (2003d:37120)

[166] Fritz Gesztesy, Alexander Kiselev, and Konstantin A. Makarov, *Uniqueness results for matrix-valued Schrödinger, Jacobi, and Dirac-type operators*, Math. Nachr. **239/240** (2002), 103–145, DOI 10.1002/1522-2616(200206)239:1¡103::AID-MANA103¿3.0.CO;2-F. MR1905666 (2003i:47047)

[167] Steve Clark and Fritz Gesztesy, *Weyl-Titchmarsh M-function asymptotics, local uniqueness results, trace formulas, and Borg-type theorems for Dirac operators*, Trans. Amer. Math. Soc. **354** (2002), no. 9, 3475–3534 (electronic), DOI 10.1090/S0002-9947-02-03025-8. MR1911509 (2003i:34191)

[168] Fritz Gesztesy and Helge Holden, *Algebro-geometric solutions of the Camassa-Holm hierarchy*, Rev. Mat. Iberoamericana **19** (2003), no. 1, 73–142, DOI 10.4171/RMI/339. MR1993416 (2004e:37113)

[169] Fritz Gesztesy and Lev A. Sakhnovich, *A class of matrix-valued Schrödinger operators with prescribed finite-band spectra*, Reproducing kernel spaces and applications, Oper. Theory Adv. Appl., vol. 143, Birkhäuser, Basel, 2003, pp. 213–253. MR2019352 (2005g:47091)

[170] Eugene D. Belokolos, Fritz Gesztesy, Konstantin A. Makarov, and Lev A. Sakhnovich, *Matrix-valued generalizations of the theorems of Borg and*

Hochstadt, Evolution equations, Lecture Notes in Pure and Appl. Math., vol. 234, Dekker, New York, 2003, pp. 1–34. MR2073733 (2005j:47046)

[171] Steve Clark and Fritz Gesztesy, *On Povzner-Wienholtz-type self-adjointness results for matrix-valued Sturm-Liouville operators*, Proc. Roy. Soc. Edinburgh Sect. A **133** (2003), no. 4, 747–758, DOI 10.1017/S0308210500002651. MR2006200 (2004h:47067)

[172] Fritz Gesztesy and Konstantin A. Makarov, *(Modified) Fredholm determinants for operators with matrix-valued semi-separable integral kernels revisited*, Integral Equations Operator Theory **47** (2003), no. 4, 457–497, DOI 10.1007/s00020-003-1170-y. MR2021969 (2006g:47076)

[173] Vladimir Batchenko and Fritz Gesztesy, *The spectrum of Schrödinger operators with quasi-periodic algebro-geometric KdV potentials*, Mat. Fiz. Anal. Geom. **10** (2003), no. 4, 447–468. MR2020819 (2004i:37140)

[174] F. Gestezi and K. A. Makarov, $SL_2(\mathbf{R})$, *exponential representation of Herglotz functions, and spectral averaging*, Algebra i Analiz **15** (2003), no. 3, 104–144 (Russian, with Russian summary); English transl., St. Petersburg Math. J. **15** (2004), no. 3, 393–418. MR2052165 (2006f:47004)

[175] Radu C. Cascaval, Fritz Gesztesy, Helge Holden, and Yuri Latushkin, *Spectral analysis of Darboux transformations for the focusing NLS hierarchy*, J. Anal. Math. **93** (2004), 139–197, DOI 10.1007/BF02789306. MR2110327 (2006e:37128)

[176] Steve Clark and Fritz Gesztesy, *On Weyl-Titchmarsh theory for singular finite difference Hamiltonian systems*, J. Comput. Appl. Math. **171** (2004), no. 1-2, 151–184, DOI 10.1016/j.cam.2004.01.011. MR2077203 (2006i:39033)

[177] Fritz Gesztesy and Barry Simon, *Connectedness of the isospectral manifold for one-dimensional half-line Schrödinger operators*, J. Statist. Phys. **116** (2004), no. 1-4, 361–365, DOI 10.1023/B:JOSS.0000037217.89500.b3. MR2083146 (2005e:81057)

[178] Radu Cascaval and Fritz Gesztesy, *$\mathcal{I}$-self-adjointness of a class of Dirac-type operators*, J. Math. Anal. Appl. **294** (2004), no. 1, 113–121, DOI 10.1016/j.jmaa.2004.02.002. MR2059793 (2005d:47079)

[179] "Algebro-geometric solutions of the KdV and Camassa–Holm equation"; with H. Holden. Oberwolfach Workshop on *Wave Motion*, A. Constantin and J. Escher (organizers), Oberwolfach Report **5**, 275–280 (2004).

[180] "Evans Functions and Modified Fredholm Determinants"; with Y. Latushkin and K. A. Makarov. Oberwolfach Workshop on *Spectral Theory in Banach Spaces and Harmonic Analysis*, N. Kalton, A. G. R. McIntosh, and L. Weis (organizers), Oberwolfach Report **36**, 1950–1953 (2004).

[181] Jeffrey S. Geronimo, Fritz Gesztesy, and Helge Holden, *Algebro-geometric solutions of the Baxter-Szegő difference equation*, Comm. Math. Phys. **258** (2005), no. 1, 149–177, DOI 10.1007/s00220-005-1305-x. MR2166844 (2006e:37129)

[182] Volodymyr Batchenko and Fritz Gesztesy, *On the spectrum of Schrödinger operators with quasi-periodic algebro-geometric KdV potentials*, J. Anal. Math. **95** (2005), 333–387, DOI 10.1007/BF02791507. MR2145569 (2006a:34236)

[183] Steve Clark, Fritz Gesztesy, and Walter Renger, *Trace formulas and Borg-type theorems for matrix-valued Jacobi and Dirac finite difference*

operators, J. Differential Equations **219** (2005), no. 1, 144–182, DOI 10.1016/j.jde.2005.04.013. MR2181033 (2006e:47068)

2005:

[184] "Non-self-adjoint operators, infinite determinants, and some applications"; with Y. Latushkin, M. Mitrea, and M. Zinchenko. Russ. J. Math. Phys. **12**, 443–471 (2005).

[185] "On the spectrum of Jacobi operators with quasi-periodic algebro-geometric coefficients"; with V. Batchenko. Int. Math. Res. Papers **No. 10**, 511–563 (2005).

[186] Fritz Gesztesy, Karl Unterkofler, and Rudi Weikard, *An explicit characterization of Calogero-Moser systems*, Trans. Amer. Math. Soc. **358** (2006), no. 2, 603–656 (electronic), DOI 10.1090/S0002-9947-05-03886-9. MR2177033 (2006h:35229)

[187] Fritz Gesztesy and Maxim Zinchenko, *Weyl-Titchmarsh theory for CMV operators associated with orthogonal polynomials on the unit circle*, J. Approx. Theory **139** (2006), no. 1-2, 172–213, DOI 10.1016/j.jat.2005.08.002. MR2220038 (2007f:47027)

[188] Fritz Gesztesy and Maxim Zinchenko, *On spectral theory for Schrödinger operators with strongly singular potentials*, Math. Nachr. **279** (2006), no. 9-10, 1041–1082, DOI 10.1002/mana.200510410. MR2242965 (2007h:47076)

[189] Steve Clark and Fritz Gesztesy, *On self-adjoint and J-self-adjoint Dirac-type operators: a case study*, Recent advances in differential equations and mathematical physics, Contemp. Math., vol. 412, Amer. Math. Soc., Providence, RI, 2006, pp. 103–140, DOI 10.1090/conm/412/07770. MR2259103 (2009d:47045)

[190] Fritz Gesztesy and Vadim Tkachenko, *When is a non-self-adjoint Hill operator a spectral operator of scalar type?*, C. R. Math. Acad. Sci. Paris **343** (2006), no. 4, 239–242, DOI 10.1016/j.crma.2006.06.014 (English, with English and French summaries). MR2245385 (2007b:34211)

2006:

[191] "Local conservation laws and the Hamiltonian formalism for the Toda hierarchy revisited"; with H. Holden. Trans. Roy. Norwegian Soc. Sci. Lett. (2006) (3), 1–30.

[192] Fritz Gesztesy and Peter Yuditskii, *Spectral properties of a class of reflectionless Schrödinger operators*, J. Funct. Anal. **241** (2006), no. 2, 486–527, DOI 10.1016/j.jfa.2006.08.006. MR2271928 (2008a:34209)

[193] Fritz Gesztesy and Maxim Zinchenko, *A Borg-type theorem associated with orthogonal polynomials on the unit circle*, J. London Math. Soc. (2) **74** (2006), no. 3, 757–777, DOI 10.1112/S0024610706023167. MR2286444 (2007m:47071)

[194] Fritz Gesztesy, Yuri Latushkin, and Konstantin A. Makarov, *Evans functions, Jost functions, and Fredholm determinants*, Arch. Ration. Mech. Anal. **186** (2007), no. 3, 361–421, DOI 10.1007/s00205-007-0071-7. MR2350362 (2008k:34209)

[195] F. Gesztesy, M. Mitrea, and M. Zinchenko, *Multi-dimensional versions of a determinant formula due to Jost and Pais*, Rep. Math. Phys. **59** (2007), no. 3, 365–377, DOI 10.1016/S0034-4877(07)80072-3. MR2347795 (2009f:47020)

[196] *Spectral theory and mathematical physics: a Festschrift in honor of Barry Simon's 60th birthday*, Proceedings of Symposia in Pure Mathematics, vol. 76, American Mathematical Society, Providence, RI, 2007. Ergodic Schrödinger operators, singular spectrum, orthogonal polynomials, and inverse spectral theory; Papers from the conference held at the California Institute of Technology, Pasadena, CA, March 27–31, 2006; Edited by Fritz Gesztesy, Percy Deift, Cherie Galvez, Peter Perry and Wilhelm Schlag. MR2307744 (2007m:00018)

[197] "Algebro-geometric finite-band solutions of the Ablowitz–Ladik hierarchy"; with H. Holden, J. Michor, and G. Teschl. Int. Math. Res. Notices **2007**, rnm082, 1–55.

[198] Fritz Gesztesy, Marius Mitrea, and Maxim Zinchenko, *Variations on a theme of Jost and Pais*, J. Funct. Anal. **253** (2007), no. 2, 399–448, DOI 10.1016/j.jfa.2007.05.009. MR2370084 (2008k:35081)

[199] Stephen Clark, Fritz Gesztesy, and Maxim Zinchenko, *Weyl-Titchmarsh theory and Borg-Marchenko-type uniqueness results for CMV operators with matrix-valued Verblunsky coefficients*, Oper. Matrices **1** (2007), no. 4, 535–592, DOI 10.7153/oam-01-31. MR2363977 (2008h:34085)

[200] Fritz Gesztesy, Helge Holden, Johanna Michor, and Gerald Teschl, *The Ablowitz-Ladik hierarchy revisited*, Methods of spectral analysis in mathematical physics, Oper. Theory Adv. Appl., vol. 186, Birkhäuser Verlag, Basel, 2009, pp. 139–190, DOI 10.1007/978-3-7643-8755-6_8. MR2732077 (2012b:37174)

[201] Fritz Gesztesy, Marius Mitrea, and Maxim Zinchenko, *On Dirichlet-to-Neumann maps and some applications to modified Fredholm determinants*, Methods of spectral analysis in mathematical physics, Oper. Theory Adv. Appl., vol. 186, Birkhäuser Verlag, Basel, 2009, pp. 191–215, DOI 10.1007/978-3-7643-8755-6_9. MR2732078 (2012b:47053)

[202] Fritz Gesztesy, Helge Holden, and Gerald Teschl, *The algebro-geometric Toda hierarchy initial value problem for complex-valued initial data*, Rev. Mat. Iberoam. **24** (2008), no. 1, 117–182, DOI 10.4171/RMI/532. MR2435969 (2010a:37133)

[203] Fritz Gesztesy and Helge Holden, *Real-valued algebro-geometric solutions of the Camassa-Holm hierarchy*, Philos. Trans. R. Soc. Lond. Ser. A Math. Phys. Eng. Sci. **366** (2008), no. 1867, 1025–1054, DOI 10.1098/rsta.2007.2060. MR2377684 (2009b:37128)

[204] Stephen Clark, Fritz Gesztesy, and Maxim Zinchenko, *Borg-Marchenko-type uniqueness results for CMV operators*, Skr. K. Nor. Vidensk. Selsk. **1** (2008), 1–18. MR2517327 (2010g:34008)

[205] Fritz Gesztesy, Helge Holden, Johanna Michor, and Gerald Teschl, *Local conservation laws and the Hamiltonian formalism for the Ablowitz-Ladik hierarchy*, Stud. Appl. Math. **120** (2008), no. 4, 361–423, DOI 10.1111/j.1467-9590.2008.00405.x. MR2416645 (2009i:37156)

[206] Fritz Gesztesy, Konstantin A. Makarov, and Maxim Zinchenko, *Essential closures and AC spectra for reflectionless CMV, Jacobi, and Schrödinger operators revisited*, Acta Appl. Math. **103** (2008), no. 3, 315–339, DOI 10.1007/s10440-008-9238-y. MR2430447 (2010b:47118)

[207] Fritz Gesztesy, Yuri Latushkin, and Kevin Zumbrun, *Derivatives of (modified) Fredholm determinants and stability of standing and traveling waves*, J. Math.

Pures Appl. (9) **90** (2008), no. 2, 160–200, DOI 10.1016/j.matpur.2008.04.001 (English, with English and French summaries). MR2437809 (2012b:47035)

[208] Fritz Gesztesy and Marius Mitrea, *Generalized Robin boundary conditions, Robin-to-Dirichlet maps, and Krein-type resolvent formulas for Schrödinger operators on bounded Lipschitz domains*, Perspectives in partial differential equations, harmonic analysis and applications, Proc. Sympos. Pure Math., vol. 79, Amer. Math. Soc., Providence, RI, 2008, pp. 105–173. MR2500491 (2010k:35087)

[209] F. Gesztesy, A. Pushnitski, and B. Simon, *On the Koplienko spectral shift function. I. Basics*, Zh. Mat. Fiz. Anal. Geom. **4** (2008), no. 1, 63–107, 202 (English, with English and Ukrainian summaries). MR2404174 (2009k:47042)

[210] Fritz Gesztesy and Vadim Tkachenko, *A criterion for Hill operators to be spectral operators of scalar type*, J. Anal. Math. **107** (2009), 287–353, DOI 10.1007/s11854-009-0012-5. MR2496408 (2010d:47063)

[211] Fritz Gesztesy and Marius Mitrea, *Robin-to-Robin maps and Krein-type resolvent formulas for Schrödinger operators on bounded Lipschitz domains*, Modern analysis and applications. The Mark Krein Centenary Conference. Vol. 2: Differential operators and mechanics, Oper. Theory Adv. Appl., vol. 191, Birkhäuser Verlag, Basel, 2009, pp. 81–113, DOI 10.1007/978-3-7643-9921-4_6. MR2569392 (2011a:35088)

[212] Fritz Gesztesy and Maxim Zinchenko, *Local spectral properties of reflectionless Jacobi, CMV, and Schrödinger operators*, J. Differential Equations **246** (2009), no. 1, 78–107, DOI 10.1016/j.jde.2008.05.006. MR2467016 (2009k:47085)

[213] Fritz Gesztesy, Mark Malamud, Marius Mitrea, and Serguei Naboko, *Generalized polar decompositions for closed operators in Hilbert spaces and some applications*, Integral Equations Operator Theory **64** (2009), no. 1, 83–113, DOI 10.1007/s00020-009-1678-x. MR2501173 (2010k:47007)

[214] Fritz Gesztesy and Marius Mitrea, *Nonlocal Robin Laplacians and some remarks on a paper by Filonov on eigenvalue inequalities*, J. Differential Equations **247** (2009), no. 10, 2871–2896, DOI 10.1016/j.jde.2009.07.007. MR2568160 (2010k:35079)

[215] Stephen Clark, Fritz Gesztesy, and Maxim Zinchenko, *Minimal rank decoupling of full-lattice CMV operators with scalar- and matrix-valued Verblunsky coefficients*, Difference equations and applications, Uğur-Bahçeşehir Univ. Publ. Co., Istanbul, 2009, pp. 19–59. MR2664173 (2011i:47040)

[216] Fritz Gesztesy, Helge Holden, Johanna Michor, and Gerald Teschl, *The algebro-geometric initial value problem for the Ablowitz-Ladik hierarchy*, Discrete Contin. Dyn. Syst. **26** (2010), no. 1, 151–196, DOI 10.3934/dcds.2010.26.151. MR2552783 (2010m:37126)

[217] Sergei Avdonin, Fritz Gesztesy, and Konstantin A. Makarov, *Spectral estimation and inverse initial boundary value problems*, Inverse Probl. Imaging **4** (2010), no. 1, 1–9, DOI 10.3934/ipi.2010.4.1. MR2592779 (2011b:93038)

[218] "On Dirichlet-to-Neumann maps, nonlocal Interactions, and some applications to Fredholm determinants"; with M. Mitrea and M. Zinchenko. Few Body Systems **47**, 49–64 (2010).

[219] Mark S. Ashbaugh, Fritz Gesztesy, Marius Mitrea, Roman Shterenberg, and Gerald Teschl, *The Krein-von Neumann extension and its connection to an*

abstract buckling problem, Math. Nachr. **283** (2010), no. 2, 165–179, DOI 10.1002/mana.200910067. MR2604115 (2011f:47078)

[220] Mark S. Ashbaugh, Fritz Gesztesy, Marius Mitrea, and Gerald Teschl, *Spectral theory for perturbed Krein Laplacians in nonsmooth domains*, Adv. Math. **223** (2010), no. 4, 1372–1467, DOI 10.1016/j.aim.2009.10.006. MR2581375 (2011d:47052)

[221] S. Clark, F. Gesztesy, and M. Mitrea, *Boundary data maps for Schrödinger operators on a compact interval*, Math. Model. Nat. Phenom. **5** (2010), no. 4, 73–121, DOI 10.1051/mmnp/20105404. MR2662451 (2011e:34187)

[222] Fritz Gesztesy and Marius Mitrea, *A description of all self-adjoint extensions of the Laplacian and Kreĭn-type resolvent formulas on non-smooth domains*, J. Anal. Math. **113** (2011), 53–172, DOI 10.1007/s11854-011-0002-2. MR2788354 (2012d:47126)

[223] F. Gesztesy, I. Mitrea, D. Mitrea, and M. Mitrea, *On the nature of the Laplace-Beltrami operator on Lipschitz manifolds*, J. Math. Sci. (N. Y.) **172** (2011), no. 3, 279–346, DOI 10.1007/s10958-010-0199-0. Problems in mathematical analysis. No. 52. MR2839866

[224] Fritz Gesztesy and Helge Holden, *The damped string problem revisited*, J. Differential Equations **251** (2011), no. 4-5, 1086–1127, DOI 10.1016/j.jde.2011.04.025. MR2812583 (2012e:35137)

[225] Fritz Gesztesy, Yuri Latushkin, Konstantin A. Makarov, Fedor Sukochev, and Yuri Tomilov, *The index formula and the spectral shift function for relatively trace class perturbations*, Adv. Math. **227** (2011), no. 1, 319–420, DOI 10.1016/j.aim.2011.01.022. MR2782197 (2012c:47039)

[226] Fritz Gesztesy and Maxim Zinchenko, *Symmetrized perturbation determinants and applications to boundary data maps and Krein-type resolvent formulas*, Proc. Lond. Math. Soc. (3) **104** (2012), no. 3, 577–612, DOI 10.1112/plms/pdr024. MR2900237

[227] Fritz Gesztesy, Alexander Gomilko, Fedor Sukochev, and Yuri Tomilov, *On a question of A. E. Nussbaum on measurability of families of closed linear operators in a Hilbert space*, Israel J. Math. **188** (2012), 195–219, DOI 10.1007/s11856-011-0120-7. MR2897729

[228] Fritz Gesztesy, Jerome A. Goldstein, Helge Holden, and Gerald Teschl, *Abstract wave equations and associated Dirac-type operators*, Ann. Mat. Pura Appl. (4) **191** (2012), no. 4, 631–676, DOI 10.1007/s10231-011-0200-7. MR2993967

[229] "Weak convergence of spectral shift functions for one-dimensional Schrödinger operators"; with R. Nichols. Math. Nachrichten **285**, 1799–1838 (2012).

[230] Fritz Gesztesy and Roger Nichols, *An abstract approach to weak convergence of spectral shift functions and applications to multi-dimensional Schrödinger operators*, J. Spectr. Theory **2** (2012), no. 3, 225–266. MR2947287

[231] Fritz Gesztesy and Vadim Tkachenko, *A Schauder and Riesz basis criterion for non-self-adjoint Schrödinger operators with periodic and antiperiodic boundary conditions*, J. Differential Equations **253** (2012), no. 2, 400–437, DOI 10.1016/j.jde.2012.04.002. MR2921200

[232] "Initial value problems and Weyl–Titchmarsh theory for Schrödinger operators with operator-valued potentials"; with R. Weikard and M. Zinchenko. Operators and Matrices **7**, 241–283 (2013).

[233] "On a class of model Hilbert spaces"; with R. Weikard and M. Zinchenko. Discrete and Continuous Dynamical Systems (to appear).
[234] "Heat kernel bounds for elliptic partial differential operators in divergence form with Robin-type boundary conditions"; with M. Mitrea and R. Nichols. J. Analyse Math. (to appear).
[235] "Boundary data maps and Krein's resolvent formula for Sturm–Liouville operators on a finite interval"; with S. Clark, R. Nichols, and M. Zinchenko. Preprint 2012.
[236] "Supersymmetry and Schrödinger-type operators with distributional matrix-valued potentials"; with J. Eckhardt, R. Nichols, and G. Teschl. Preprint 2012.
[237] "Weyl–Titchmarsh theory for Sturm–Liouville operators with distributional coefficients"; with J. Eckhardt, R. Nichols, and G. Teschl. Opuscula Math. (to appear).
[238] "A survey on the Krein–von Neumann extension, the corresponding abstract buckling Problem, and Weyl-type spectral asymptotics for perturbed Krein Laplacians in nonsmooth domains"; with M. Ashbaugh, M. Mitrea, R. Shterenberg, and G. Teschl. Advances in Partial Differential Equations, M. Demuth and W. Kirsch (eds.), Birkhäuser, Basel, (to appear).
[239] "Inverse spectral theory for Sturm–Liouville operators with distributional coefficients"; with J. Eckhardt, R. Nichols, and G. Teschl. Preprint 2012.
[240] "On stability of square root domains for non-self-adjoint operators under additive perturbations"; with S. Hofmann and R. Nichols. Preprint 2012.
[241] "On spectral theory for Schrödinger operators with operator-valued potentials"; with R. Weikard and M. Zinchenko. Preprint 2013.
[242] "Stability of square root domains for one-dimensional non-self-adjoint second-order linear differential operators"; with S. Hofmann and R. Nichols. Preprint 2013.
[243] "Some remarks on the spectral problem underlying the Camassa–Holm hierarchy"; with R. Weikard. Preprint 2013.
[244] "The Birman–Schwinger principle and eigenvalue multiplicity questions revisited"; with H. Holden. Preprint 2013.

MONOGRAPHS

Solvable Models in Quantum Mechanics; with S. Albeverio, R. Høegh-Krohn, and H. Holden. Texts and Monographs in Physics, Springer-Verlag, Heidelberg–New York, 1988, 452 pages. (Translated into Russian by Yu. A. Kuperin, K. A. Makarov, and V. A. Geiler, Mir Publishers, Moscow, 1991.) MR0926273 (90a:81021)

The second and expanded edition of this monograph appeared as:
Solvable Models in Quantum Mechanics, 2nd edition; with S. Albeverio, R. Høegh-Krohn, and H. Holden. AMS–Chelsea Series, Amer. Math. Soc., 2005, 488 pages. With an appendix by P. Exner. MR2105735

Soliton Equations and Their Algebro-Geometric Solutions. Vol. I: (1 + 1)-*Dimensional Continuous Models*; with H. Holden. Cambridge Studies in Advanced Mathematics, Vol. 79, Cambridge Univ. Press, Cambridge, 2003, 505 pages. MR1992536

Soliton Equations and Their Algebro-Geometric Solutions. Vol. II: (1 + 1)-*Dimensional Discrete Models*; with H. Holden, J. Michor, and G. Teschl. Cambridge Studies in Advanced Mathematics, Vol. 114, Cambridge Univ. Press, Cambridge, 2008, 438 pages. MR2446594

VOLUMES CO-EDITED

Continued Fractions: From Analytic Number Theory to Constructive Approximation, B. C. Berndt and F. Gesztesy (eds.), Contemporary Mathematics **236**, Amer. Math. Soc., Providence, RI, 1999, 379 pages. MR1665358 (2000d:00018)

Stochastic Processes, Physics and Geometry: New Interplays. I. A Volume in Honor of Sergio Albeverio, F. Gesztesy, H. Holden, J. Jost, S. Paycha, M. Röckner, and S. Scarlatti (eds.), Canadian Mathematical Society Conference Proceedings, Vol. 28, Amer. Math. Soc., Providence, RI, 2000, 333 pages. MR1803374

Stochastic Processes, Physics and Geometry: New Interplays. II. A Volume in Honor of Sergio Albeverio, F. Gesztesy, H. Holden, J. Jost, S. Paycha, M. Röckner, and S. Scarlatti (eds.), Canadian Mathematical Society Conference Proceedings, Vol. 29, Amer. Math. Soc., Providence, RI, 2000, 647 pages. MR1803398

Spectral Theory and Mathematical Physics: A Festschrift in Honor of Barry Simon's 60th Birthday: Quantum Field Theory, Statistical Mechanics, and Nonrelativistic Quantum Systems, F. Gesztesy, Managing Editor, P. Deift, C. Galvez, P. Perry, and W. Schlag (eds.), Proceedings of Symposia in Pure Mathematics, Vol. 76.1, Amer. Math. Soc., Providence, RI, 2007, 496 pages. MR2310192

Spectral Theory and Mathematical Physics: A Festschrift in Honor of Barry Simon's 60th Birthday: Ergodic Schrödinger Operators, Singular Spectrum, Orthogonal Polynomials, and Inverse Spectral Theory, F. Gesztesy, Managing Editor, P. Deift, C. Galvez, P. Perry, and W. Schlag (eds.), Proceedings of Symposia in Pure Mathematics, Vol. 76.2, Amer. Math. Soc., Providence, RI, 2007, 464 pages. MR2307744

JOURNAL ISSUES CO-EDITED

Mathematische Nachrichten, **283**, Nos. 1–3 (2010), Erhard Schmidt Memorial Issue, Parts I–III, A. Böttcher, F. Gesztesy, and R. Mennicken (eds.), Wiley-VCH, pp. 1–159, 161–329, and 331–499.

The Mathematical Modelling of Natural Phenomena (MMNP), **5**, No. 4 (2010), Spectral problems. Issue dedicated to the memory of M. Birman, N. Apreutesei, D. Damanik, Yu. Egorov, F. Gesztesy, P. Kurasov, A. Laptev, S. Naboko, V. Volpert, V. Voulgalter (eds.), Cambridge University Press and EDP Sciences, pp. 1–469.

Mathematische Nachrichten, **285**, No. 14–15 (2012), Eduard R. Tsekanovskiĭ Special Issue on the Occasion of his Seventy-Fifth Birthday, F. Gesztesy, H. Langer, M. Malamud, and R. Mennicken (eds.), Wiley-VCH, pp. 1671–1931.

Proceedings of Symposia in Pure Mathematics
Volume **87**, 2013
http://dx.doi.org/10.1090/pspum/087/01431

Invariant measures for stochastic differential equations on networks

Sergio Albeverio, Luca Di Persio, and Elisa Mastrogiacomo

This paper is dedicated to Fritz Gesztesy on the occasion of his 60th birthday, with great admiration and gratefulness

Abstract. We study existence and uniqueness of an invariant measure for infinite dimensional stochastic differential equations with dissipative polynomially bounded nonlinear terms. We also exhibit the existence of a density with respect to a Gaussian measure. Moreover, we decompose the solution process into a stationary component and a component which vanishes asymptotically in the L^2-sense. Applications are given to neurobiological networks where the signals propagation is modelled by a system of coupled stochastic FitzHugh-Nagumo equations.

1. Introduction

In this paper we are concerned with the asymptotic properties of nonlinear stochastic differential equations on a finite network. In particular, the network is identified with a topological graph G and we study existence, uniqueness and regularity of the invariant measure associated to the transition semigroup of a semilinear diffusion process on G in presence of a Gaussian perturbation. We allow for a rather general nonlinear drift term, including dissipative functions of FitzHugh-Nagumo type (i.e. $f(u) = -u(u-1)(u-\xi)$, $u \in \mathbb{R}$, $0 < \xi < 1$) arising in various models of neurophysiology (see, for example, Keener and Sneyd [**41**] for details).

Neural networks have attracted a large amount of mathematical research, in particular due to their application to mathematical modeling in neuroscience. The phenomenology of neural electrical activity and the understanding of the ultimate mechanism responsible for it have been the object of a steadily increasing number of investigations by neurophysiologists, biologists, physicists and mathematicians during the last four decades. Part of the theoretical investigations have been first carried through using deterministic models for the mathematical analysis of electric potential spread inside neurons. We recall the celebrated axonal model due to A. L. Hodgkin and A. Huxley (see Hodgkin-Huxley [**39**]) for the diffusion of the transmembrane electrical potential in neural cells introduced since the 50s. This model consists of a system of four deterministic coupled differential equations describing

2010 *Mathematics Subject Classification.* Primary 45D05,93E20; Secondary 60H30.
Key words and phrases. Invariant measures, infinite dimensional stochastic differential equations, dissipative systems, neurobiological networks, stochastic FitzHugh-Nagumo systems.

the time evolution of the electrical potential and the influence of various ion channels. Successive simplifications of the model, trying to capture the key phenomena of the Hodgkin-Huxley system, lead to the FitzHugh-Nagumo equation, which is an equation with two stable states (see Georgescu [**34**]). We also mention the cable theory of electrotonic potential first developed by W. Rall in 1959 [**63**] Further deterministic models can be found, for instance, in the monograph by Keener and Sneyd (2009). Since the late seventies modifications of the deterministic models trying to capture stochastic effects (both intrinsic and extrinsic) start to appear, see Ricciardi (1977), Kallianpur and Wolpert (1984) [**40, 65**] and the references therein. Over the last few decades many more stochastic models of brain activities have been investigated, see, e.g., [**20, 32, 72–74**].

Our paper is devoted to an approach involving stochastic partial differential equations, both for single neurons and neuronal networks. This uses in particular methods of the theory of evolution equations, like semigroups (see, for example [**59**]) coupled with methods of stochastic PDE's as developed, for example, by Da Prato and Zabczyck [**28, 29**]. First works in this direction are [**18**], [**19**]. We describe a network modeled by a graph with space dependent diffusion processes propagating along the edges, hence described by a system of SPDE's. Within this setting, our problem is reformulated into a semilinear abstract evolution equation perturbed by a noise. We prove existence and uniqueness of solutions essentially by semigroup techniques and infinite dimensional stochastic analysis. In this perspective we continue the work we initiated in [**6**].

Moreover, we are concerned with the study of the asymptotic properties of the system. We shall analyse the dependence of models on the spatial structure, linking the abstract properties of well-posedness of the differential equation with the graph-theoretical properties of the underlying network. We look also for invariant measures, extending in particular the method implemented in Marcus [**51**] to the case of non globally Lipschitz coefficients. The existence of the invariant measure is obtained by proving that the solution of our equation can be split into the sum of a stationary process and a process vanishing at infinity: the required invariant measure is then the stationary distribution. Concerning uniqueness of the invariant measure, we adapt to our problem arguments used in finite dimensional settings, proving first the strong Feller property and the irreducibility. Finally, we are able to prove that the unique invariant measure associated with the system has a Radon-Nikodym derivative with respect to a properly chosen Gaussian measure μ, associated with the linear part of the SPDE, and we exhibit it explicitly.

Several works have been devoted to the study of invariant measures for stochastic differential equations in infinite dimensions with different types of noise (for early results see, e.g., [**21, 37, 38, 68, 75, 77**]). Recent contributions are given, both in the case of important finite dimensional equations and infinite dimensional systems (like porous media equation, Navier-Stokes and reaction diffusion systems). As an example we cite Da Prato and Barbu [**17**], where the authors treat the existence of invariant measures and the characterization of the Kolmogorov operator for the stochastic fast diffusion equation on a bounded domain of $\mathbb{R}^d$ by an approximation procedure. For other recent work on invariant measures for stochastic nonlinear equations see also, e.g., [**33, 54, 62**]. Further, we mention the work by Albeverio, Brezniak, Wu [**2**], concerning invariant measures for stochastic differential equations driven by Poisson type noise; for other types of non Gaussian noise with non-Lipschitz coefficients, see [**12**], [**19**], [**61**]. In particular in [**19**] the authors analyze reaction-diffusion problems on finite connected networks woth non-local Kirchhoff boundary conditions, the PDE along each edge being of the FitzHugh-Nagumo type perturbed by jump noise. The approach developed in all these works takes inspiration from the classical Krylov-Bogoliubov method (see, e.g., [**21**]). Finally, we mention the paper by Da Prato and Debussche [**26**], where the study of the invariant measure for stochastic PDEs with random coefficients is treated through infinite dimensional extensions of some concepts concerning structural properties of finite dimensional Markov process like ergodicity, the strong Feller property and irreducibility.

Before going over to the description of the organization of the present paper, let us briefly mention two types of direct relations with Fritz Gesztesy's own work. One concerns the discussion of perturbations of classical dynamical systems, in particular integrable ones. The other concerns the area of study of singular interactions in quantum mechanics and related systems.The latter comes from the study of certain nonlinear S(P)DE's on graphs as limits of corresponding equations on thin tubes around graphs. A special case of this describes quantum mechanical particles moving in thin tubes. The latter problem leads to relations with the problems of studying resonances in thin tubes and waves guides, the interesting fine properties of which are obtained in terms of point interactions, see e.g., [**4, 22**], and references therein.

The structure of the present paper is as follows. In Section 2 we present the setting of our problem. In Section 3 we give a reformulation of the system of evolution equations as an abstract evolution problem in Hilbert space. In Section 4 we prove basic and preparatory results. In Section 5 we prove a priori estimates leading to the existence and uniqueness results for solutions of our basic evolution equation. In Section 6 we show that the solution can be splitted into the sum of a stationary process and a process vanishing at infinity. Finally, in Section 7 we provide the explicit construction of the density of the invariant measure.

2. Setting of the problem

This paper is devoted to the study of existence, uniqueness and regularity of the invariant measure associated to the transition semigroup of a large class of semilinear diffusion processes on topological graphs. We treat the case where the nonlinear term is Gâteaux differentiable has at most polynomial growth together with all its derivatives, and fulfills appropriate dissipativity conditions. More precisely, we are concerned with a finite connected network, represented by a finite graph G with m edges $\mathsf{e}_1, \dots, \mathsf{e}_m$ and n vertices $\mathsf{v}_1, \dots, \mathsf{v}_n$. We assume that G is simple, that is, it has no multiple edges or loops. Following, e.g., [**57**], we normalize and parametrize the edges to be identified with the interval $[0,1]$ and we denote again by $\mathsf{e}_j(\cdot)$ the parametrization of the edge e_j (for more details about the above notation we refer to Mugnolo [**55**]). The valency of each vertex is denoted by $\Gamma(\mathsf{v}_i)$; precisely:

$$\Gamma(\mathsf{v}_i) := \#\{j \in \{1, \dots, m\} : \ \mathsf{e}_j(\mathsf{v}_i) = 1 \ \mathit{or} \ \mathsf{e}_j(\mathsf{v}_i) = 0\}. \tag{2.1}$$

The graph is described by the *incidence matrix* $\Phi = \Phi^+ - \Phi^-$, where $\Phi^+ := (\phi^+_{ij})_{n\times m}$ and $\Phi^- := (\phi^-_{ij})_{n\times m}$ are given by

$$\phi^+_{ij} := \begin{cases} 1, & \text{if } \mathsf{e}_j(0) = \mathsf{v}_i; \\ 0, & \text{otherwise,} \end{cases} \qquad \text{and} \qquad \phi^-_{ij} := \begin{cases} 1, & \text{if } \mathsf{e}_j(1) = \mathsf{v}_i; \\ 0, & \text{otherwise.} \end{cases}$$

For functions $u : G \to \mathbb{R}$, with an abuse of notation, we use the abbreviations $u_j(\mathsf{v}_i)$ to denote their values at 0 or 1, if $\mathsf{e}_j(0) = \mathsf{v}_i$ or $\mathsf{e}_j(1) = \mathsf{v}_i$, respectively. Moreover, when convenient, we shall write u in vector form, i.e. $u = (u_1, \dots, u_m)^t$ (^{t} standing for traspose). On G we consider the following system of stochastic diffusion equations:

$$\text{(E)} \qquad \begin{aligned} \partial_t u_j(t,x) = \partial_x(c_j(x)\partial_x u_j(x)) + p_j(x)u_j(x) + f_j(u_j(t,x)) + \dot{\beta}_j(t), \\ t > 0, x \in (0,1),\ j = 1, \dots, m. \end{aligned}$$

In the above equation, c_j and p_j ($j = 1, \dots, m$) are suitable smooth function, while $f_j, j = 1, \dots, m$ are suitable polynomials with odd degree (see Hypothesis 2.1 below for details). Moreover, $(\beta_j(t))_{t\geq 0}, j = 1, \dots, m$ represents the stochastic perturbation acting on the respective edges due to external sorrounding, and $\dot{\beta}_j(t)$ is the distributional derivative of the process β_j. Biological motivations lead to model this term by a Gaussian process. In fact, the evolution of the electrical potential on the molecular level can be perturbed by different types of random terms, each modeling the influence, at different time scales, of the surrounding medium. At a slow time scale, vesicles of neurotransmitters released by external neurons cause electrical impulses which arrive randomly at the soma (here modeled by the vertices of the graph) causing a sudden change in the membrane voltage potential of an amount, either positive or negative, depending on the composition of the vesicle and possibly even on the state of the neuron. The generality of the above diffusion is motivated by the discussion in the biological literature, see for example [**41**], where it is stressed that the basic cable properties are not constant throughout the dendritic tree. The above equation shall be endowed with suitable boundary and initial conditions. Since we are dealing with a diffusion in a network, we require, following for example [**55**], a continuity assumption and a Kirchhoff law on every node:

$$\text{(C)} \qquad u_j(\mathsf{v}_i) = u_k(\mathsf{v}_i), \qquad j,k \in \Gamma(\mathsf{v}_i),\ i = 1, \dots, n$$

$$\text{(K)} \qquad \sum_{j=1}^m \phi_{ij}\mu_j c_j(\mathsf{v}_i)\partial_x u_j(\mathsf{v}_i) = 0, \qquad i = 1, \dots, n.$$

Initial conditions at time $t = 0$ are taken for simplicity to be of the form

$$\text{(I)} \qquad u_j(0,x) = u_{j0}(x) \in C([0,1])\ j = 1, \dots, m.$$

Let us state the main assumptions on the data of the problem. Concerning the coefficients in equations (E), (C), (K), (I), we assume the following.

HYPOTHESIS 2.1.

(1) *The functions $f_j, j = 1, \dots, m$ satisfy the following conditions:*

(a) *f_j, $j = 1, \dots, m$ are continuous and differentiable on $\mathbb{R}$;*

(b) f_j, $j = 1, \dots, m$ *have at most subpolynomial growth at infinity, that is: there exist* $k \geq 1$ *and* $C > 0$ *such that*

$$|f_j(\xi)| \leq C(1 + |\xi|^{2k+1})$$
$$|f_j'(\xi)| \leq C(1 + |\xi|^{2k}), \qquad \xi \in \mathbb{R};$$

(c) f_j', $j = 1, \dots, m$ *are one sided bounded from above, in the sense that there exists* $\eta > 0$ *such that*

$$f_j'(\xi) \leq \eta, \qquad \xi \in \mathbb{R}.$$

(2) *The functions* $c_j(\cdot)$, $j = 1, \dots, m$ *belong to* $C^1([0,1])$, *and* $c_j(x) > 0$ *for every* $x \in [0,1]$. *In the following we will denote by* c *the minimum of* $c_j(\cdot), j = 1, \dots,$ *i.e.*

$$c := \min_{j=1,\dots,m} \min_{x\in[0,1]} |c_j(x)|.$$

We assume $c > 0$.

(3) *The functions* $p_j(\cdot)$, $j = 1, \dots, m$ *belong to* $L^\infty(0,1)$ *and* $p_j(x) > 0$ *for (Lebesgue) a.e.* $x \in (0,1)$. *In the following we will denote by* p *the essential infimum of* $p_j(\cdot), j = 1, \dots, m$, *i.e.*

$$\omega := \min_{j=1,\dots,m} \text{ess.inf}_{x\in[0,1]} |p_j(x)|.$$

We assume $p > 0$.

(4) μ_j, $j = 1, \dots, m$ *are real strictly positive numbers.*

(5) *The constants* ω *and* η *satisfy the following condition:*

$$\omega - \eta \geq 0.$$

Now, for $j = 1, \dots, m$, we set

$$A_j := \frac{\partial}{\partial x} c_j \frac{\partial}{\partial x} - p_j, \tag{2.2}$$

on the domain $D(A_j) = H^2(0,1;\mu_j \mathrm{d}x)$.

Concerning the stochastic term we assume the following.

HYPOTHESIS 2.2. *The processes* β_j, $j = 1, \dots, m$ *are space-dependent Wiener processes on a filtered probability space* $(\Omega, \mathcal{F}, (\mathcal{F}_t)_{t\geq 0}, \mathbb{P})$ *with continuous trajectories on* $[0,T]$ *for any* $T > 0$ *and values in* $L^2(0,1)$ *(as functions of* $x \in (0,1)$*); this means that, for any* $j = 1, \dots, m$,

$$\beta_j \in C([0,T]; L^2(\Omega; L^2(0,1)))$$

with $\mathcal{L}(\beta_j(t,\cdot)) \equiv \mathcal{N}(0, tQ_j)$ *for linear non-negative bounded trace class operators* Q_j, $j = 1, \dots, m$ *on* $L^2(0,1)$ *(with* $\mathcal{L}$ *the law and* $\mathcal{N}$ *the normal distribution of mean* 0 *and variance* tQ_j*). Finally, we assume that* Q_j *commutes with the linear operator* A_j *introduced in* (2.2).

REMARK 2.3. *Let* $W(t) := (\beta_1(t), \dots, \beta_m(t))$. *Then* $W(t)$ *is a stochastic perturbation acting on the Hilbert space* $\mathbf{X} := \Pi_{j=1}^m L^2(0,1;\mu_j \mathrm{d}x)$ *having covariance operator* $Q := \mathrm{diag}(Q_1, \dots, Q_m)$. *We notice that Hypothesis* 2.2 *implies that* $\mathrm{Tr}(Q) < \infty$. *More specific conditions will be imposed on the noise in the following (cf. Hypothesis* 4.5*).*

3. The abstract reformulation

We consider the product space

$$\mathbf{X} = \Pi_{j=1}^m L^2(0,1;\mu_j dx)$$

endowed with the natural inner product

$$\langle u, v\rangle_{\mathbf{X}} = \sum_{j=1}^m \int_0^1 u_j(x)v_j(x)\mu_j\, dx, \qquad u, v \in \mathbf{X}$$

and corresponding norm $|\cdot|_{\mathbf{X}}$. A general vector $u \in \mathbf{X}$ is a collection of functions $(u_j(\cdot))_{j=1,\dots,m}$, $u_j : [0,1] \to \mathbb{R}$ which represents the electrical potential inside the network. We introduce the *weighted incidence matrices* Φ_ω^+ and Φ_ω^- with entries

$$\omega_{ij}^+ := \begin{cases} \mu_j c_j(\mathsf{v}_i), & \text{if } \mathsf{e}_j(0) = \mathsf{v}_i; \\ 0, & \text{otherwise,} \end{cases} \qquad \omega_{ij}^- := \begin{cases} \mu_j c_j(\mathsf{v}_i), & \text{if } \mathsf{e}_j(1) = \mathsf{v}_i; \\ 0, & \text{otherwise.} \end{cases}$$

Notice that with the above notation, continuity and Kirchoff conditions for a given smooth function $w : G \to \mathbb{R}$, can be written as

$$(\mathrm{C}') \qquad \exists d \in \mathbb{R}^n \ s.t.\ (\Phi^+)^t d = w(0) \ and \ (\Phi^-)^t d = w(1),$$

$$(\mathrm{K}') \qquad \Phi_\omega^+ w'(0) = \Phi_\omega^- w'(1).$$

On the space $\mathbf{X}$ we introduce the linear operator $(A, D(A))$ defined by

$$(3.1) \qquad \begin{aligned} D(A) &:= \left\{u \in \Pi_{i=1}^m H^1(0,1;\mu_j \mathrm{d}x) \ : \ condition\ (\mathrm{C}')\ and\ (\mathrm{K}')\ hold\right\} \\ (Au)(x) &:= \left(\frac{\partial}{\partial x}\left(c_j(x)\frac{\partial}{\partial x}u_j(x)\right) - p_j(x)u_j(x)\right)_{j=1,\dots,m}. \end{aligned}$$

Moreover, we introduce the nonlinear operator $F : D(F) \subset \mathbf{X} \to \mathbf{X}$

$$\begin{aligned} D(F) &:= \Pi_{j=1}^m L^{4k+2}(0,1;\mu_j \mathrm{d}x)^m, \\ F(u) &:= (f_j(u_j))_{j=1,\dots,m} \end{aligned}$$

where k is the constant introduced in Hypothesis 2.1 - (1b). Finally, we introduce the stochastic process $W(t) = (W_j(t))_{j=1,\dots,m}$ which is a $\mathbf{X}$-valued cylindrical Wiener process with linear covariance operator

$$(3.2) \qquad Q := \operatorname{diag}(Q_j)_{j=1,\dots,m}.$$

With the above notation, equations (E),(C),(K), (I) can be rewritten in the following form:

$$(\mathrm{SDE}) \qquad \begin{aligned} du(t) &= [Au(t) + F(u(t))]dt + \sqrt{Q}dW(t), t \in \mathbb{R} \\ u(0) &= u_0 \end{aligned}$$

where $u_0 = u_0(\cdot) = (u_{01}(\cdot), \dots, u_{0m}(\cdot)) \in \mathbf{X}$.

Now, let us recall the definition of mild solution for the stochastic Cauchy problem SDE.

DEFINITION 3.1. *Given an $\mathcal{F}_t$-adapted cylindrical Wiener process on $(\Omega, \mathcal{F}, (\mathcal{F}_t)_{t\geq 0}, \mathbb{P})$, a process $(u(t))_{t\geq 0}$ is a mild solution of* (SDE) *if it satisfies $\mathbb{P}$-a.s. the following integral equation*

$$(3.3) \qquad u(t) = e^{tA}u_0 + \int_0^t e^{(t-s)A}F(u(s))\mathrm{d}s + \int_0^t e^{(t-s)A}\sqrt{Q}\mathrm{d}W(s), \qquad t \geq 0.$$

Equation (3.3) implies that the integrals on the right-member are well defined. In particular, the second integral, which we shall refer to as *stochastic convolution*, is understood as a mean- square continuous Gaussian process with values in H. For the analysis of the stochastic convolution and its properties, we refer to Section 4.3.

4. Preliminary results

4.1. Generation properties. Before proceeding with the analysis of the stochastic convolution, it is necessary to study the properties of the linear operator A and the nonlinear term F. As discussed in Mugnolo [**55**], the diffusion operator A on a network, endowed with continuity and Kirchhoff conditions at the nodes, can be studied efficiently by means of sesquilinear forms.

PROPOSITION 4.1. *Under Hypothesis* 2.1 - (2),(3) *the operator* $(A, D(A))$ *is self-adjoint, dissipative and has a compact resolvent. In particular, it generates an analytic* C_0*-semigroup of contractions.*

PROOF. For the sake of completeness, we provide a sketch of the proof following [**57**]. The idea is simply to relate the operator $(A, D(A))$ to a suitable form $a(u, v)$ having dense domain $\mathbf{Y} \subset \mathbf{X}$. The space $\mathbf{Y}$ is defined as

$$\mathbf{Y} := \left\{u \in \Pi_{j=1}^m H^1(0,1;\mu_j \mathrm{d}x) \ : \ (\mathrm{C}') \textit{ and } (\mathrm{K}') \textit{ hold}\right\}$$

while the form a is given by

$$a(u,v) := \sum_{j=1}^{m} \int_0^1 (\mu_j c_j(x) u_j'(x) v_j'(x) - p_j(x) u_j(x) v_j(x)) \mathrm{d}x, \qquad u, v \in \mathbf{Y}.$$

The form a is clearly positive and symmetric; furthermore, it is closed and continuous. Then a little computation shows that the operator uniquely associated with the form a is $(A, D(A))$ as defined above (see Mugnolo [**57**]). Classical results in Dirichlet forms theory, see for instance Ouhabaz [**58**], [**1, 36, 50**]lead then to the desired result.

□

The following result concerns the asymptotic behaviour of the semigroup $e^{tA}, t \geq 0$, corresponding to A.

COROLLARY 4.2. *Under Hypothesis* 2.1 - (2),(3), *the operator* A *is invertible and it generates a semigroup* $e^{tA}, t \geq 0$, *which is exponentially bounded, with growth bound given by the spectral bound* ω *of the operator* A.

PROOF. The dissipativity of A follows from the positivity and symmetry of the form corresponding with A. Moreover, it is straightforward to prove that

$$\langle Au, u \rangle \leq -\omega |u|^2_{\mathbf{X}},$$

where $\omega > 0$ is the quantity defined in 2.1 - (3). This implies that

$$\|e^{tA}\|_{L(\mathbf{X})} \leq e^{-\omega t} \qquad t \geq 0,$$

(where $\|\cdot\|_{\mathcal{L}(\mathbf{X})}$ is the norm in the space $\mathcal{L}(\mathbf{X})$ of the bounded linear operators in $\mathbf{X}$); thus $e^{tA}, t \geq 0$, is exponentially bounded with the claimed growth bound.

□

4.2. The nonlinear term. Concerning the nonlinear term F we have, under Hypothesis 2.1, the following result.

PROPOSITION 4.3.

(1) *The mapping $F : D(F) \subset \mathbf{X} \to \mathbf{X}$ is continuous, nonlinear, Fréchet differentiable and quasi-m-dissipative in the sense of* [**29**]*; in particular we have*

$$\langle F(u) - F(v) - \eta(u-v), u-v \rangle \leq 0, \qquad \textit{for all } u, v \in D(F),$$

where η is the constant introduced in Hypothesis 2.1 - (5).

(2) *Let K be the Banach space*

$$K := \Pi_{j=1}^{m} C([0,1]) \tag{4.1}$$

endowed with the norm

$$\|u\|_K := \sup_{j=1,\dots,m} |u_j|_\infty .$$

If $F_K^{(j)}$, $j = 0,1$ denotes the part of $F^{(j)}$ in K, that is

$$D(F_K^{(j)}) := \left\{ x \in D(F^{(j)}) \cap K ; F_K^{(j)}(x) \in K \right\}, \quad F_K^{(j)}(x) = F^{(j)}(x),\ x \in D(F_K^{(j)}),$$

then the following hold:

(a) *K is invariant under F_K, i.e. $F_K(K) \subset K$;*

(b) *there exists a positive real number γ such that*

$$\left\| F_K^{(j)}(u) \right\|_K \leq \gamma \left(1 + \|u\|_K^{2k+1-j} \right), \quad u \in K,\ j = 0,1 \tag{4.2}$$

and k being as in Hypothesis 2.1 - 1b.

PROOF. *Point 1.* Let $u, v \in \mathbf{X}$. By definition and assumption (1c) in Hypothesis 2.1,

$$\langle F(u) - F(v) - \eta(u-v), u-v \rangle_{\mathbf{X}} \leq \max_{j=1,\dots,m} \sup_{\xi \in \mathbb{R}} (f_j'(\xi) - \eta) |u-v|_{\mathbf{X}}^2 \leq 0.$$

Point 2.

(a) Since $f_j, j = 1, \dots, m$ are assumed to be continuous on $\mathbb{R}$, then, for any $u \in K$ we have $f_j(u_j) \in C[0,1]$ and, consequently, $F(u) \in K$ and, clearly, K is invariant under F.

(b) We have

$$\begin{aligned} \|F(u)\|_K &= \max_{j=1,\dots,m} |f_j(u_j)|_\infty \\ &\leq \max_{j=1,\dots,m} \sup_{\xi \in [0,1]} C(1 + |u_j(\xi)|^{2k+1}) \\ &\leq C(1 + \|u\|_K^{2k+1}), \end{aligned}$$

so that (4.2) holds with $j = 0$. Moreover, since $f_j, j = 1, \dots, m$ are assumed to be differentiable, F is Fréchet differentiable in K and for any $u, h \in K$ and $\xi \in [0,1]$ we have

$$\begin{aligned} \nabla F(u)[h](\xi) : &= \lim_{\varepsilon \to 0} \frac{F(u(\xi) + \varepsilon h(\xi)) - F(u(\xi))}{\varepsilon} \\ &= (f_j'(u_j(\xi)) h_j(\xi))_{j=1,\dots,m}. \end{aligned}$$

This implies that

$$\|\nabla F(u)[h]\|_{L(K)} \leq c(1 + \|u\|_K^{2k}), \quad u \in K, \text{ for some } c > 0,$$

where $L(K)$ denotes the space of linear bounded mappings from K into itself.

□

REMARK 4.4.

(1) For further use we will denote by F_η the mapping $F - \eta I$, which, to each $x = (x_j)_{j=1,\cdots m} \in D(F)$ assigns the function

$$((F - \eta I)(x))(\xi) = (f_j(x_j(\xi)) - \eta x_j(\xi)_{j=1,\ldots,m}, \ \xi \in [0,1].$$

We notice that, from Hypothesis 2.1- (1a),(1c), the functions

$$f_j(\xi)) - \eta\xi, \qquad j = 1, \ldots, m$$

are continuous, differentiable and non-increasing. Therefore it follows that F_η is dissipative (in the sense of [**29**]).

(2) Combining Hypothesis 2.1 - (5), Corollary 4.2 and Proposition 4.3 we get that

$$\langle A(x-y) - (F(x) - F(y)), x - y\rangle \leq -(\omega - \eta)|x - y|_{\mathbf{X}}^2 \leq 0, \quad x, y \in D(F),$$

that is, the operator $A + F$ is dissipative.

4.3. The stochastic convolution. In our case the stochastic convolution can be written as

$$W_A(t) := \int_0^t e^{(t-s)A}\sqrt{Q}\mathrm{d}W(s), \qquad t \geq 0.$$

It can be shown that (see [**28**, Theorem 5.2]), from Hypothesis 2.1 - 2.2, Remark 2.3 and by using Corollary 4.2, $(W_A(t))_{t\geq 0}$ is a $\mathbf{X}$-valued Gaussian process with mean 0 and covariance operator

$$Q_t := \int_0^t e^{sA}Qe^{sA^*}\mathrm{d}s, \qquad t \geq 0.$$

Moreover, it is $\mathbb{P}$-almost surely continuous on $[0, \infty)$ and it verifies the following estimate:

$$\sup_{t\geq 0} \mathbb{E}|W_A(t)|^{2a} \leq C$$

for any $a \in \mathbb{N}$ and some positive constant $C > 0$.

In the sequel we will need $(W_A(t))_{t\geq 0}$ to be more regular.

To this end, we introduce the interpolation spaces (see, e.g., [**70**]) and denote by $\mathbf{X}_\alpha$ with $\alpha \in (0,1)$ the index α space relative to $\mathbf{X}$ and A. We shall assume that the operator A satisfies the following additional conditions.

HYPOTHESIS 4.5.

(1) *There exists a complete orthonormal system* $\{\mathbf{e}_k\}_{k\in\mathbb{N}}$ *on* $\mathbf{X}$ *which diagonalizes* A. *We denote by* $\{-\alpha_k\}_{k\in\mathbb{N}}$, $\alpha_k \geq 0$, *the corresponding set of eigenvalues. Concerning the eigenfunctions* $\{\mathbf{e}_k\}_{k\in\mathbb{N}}$ *we shall assume that for any* $k \in \mathbb{N}$ *we have* $\mathbf{e}_k \in \mathbf{X}_{1/2}$ *and*

$$\sup_{k\in\mathbb{N}} \|\mathbf{e}_k\|_K < 1; \quad \sup_{k\in\mathbb{N}} \|\mathbf{e}_k\|_{1/2} \leq c\sqrt{\alpha_k}$$

for some positive constant c.

(2) *The bounded linear operator* $Q : \mathbf{X} \to \mathbf{X}$ *commutes with the linear operator* A *introduced in (2.2), thus diagonalizes on the same basis* $\{\mathbf{e}_k\}_{k\in\mathbb{N}}$ *which diagonalizes* A. *Moreover, if* $\lambda_k \in \mathbb{R}, k \in \mathbb{N}$ *is the corresponding set of eigenvalues, we have*

$$\mathrm{Tr}(A^{-(1-\iota)}Q) = \sum_{k=1}^{\infty} \frac{\lambda_k}{\alpha_k^{1-\iota}} < \infty,$$

for some $\iota \in (0,1)$.

Now let $\{\beta_k(t)\}_{k\in\mathbb{N}}$ for $j = 1, \ldots, m$ be a sequence of mutually independent real-valued Brownian motions defined on a stochastic basis $(\Omega, \mathcal{F}, (\mathcal{F}_t)_{t\geq 0}, \mathbb{P})$ and adapted to the nonanticipative filtration $\mathcal{F}_t$, $t \geq 0$. We define the cylindrical Wiener process $W(t)$ as

$$W(t) := \sum_{i=1}^{\infty} \mathbf{e}_k \beta_k(t),$$

where $\{\mathbf{e}_k\}_{k\in\mathbb{N}}$ is the complete orthonormal system of $\mathbf{X}$ introduced before. The series above defining $W(t)$ does not converge in $\mathbf{X}$, but it is convergent in any Hilbert space $\mathbf{U}$ such that the embedding of $\mathbf{U}$ into $\mathbf{X}$ is Hilbert-Schmidt (see [**28**, Chapter 4]).

Hypothesis 4.5 guarantees a stronger regularity of the stochastic convolution corresponding with our system. In particular, it turns out that the stochastic convolution is continuous not in a Hilbert space $\mathbf{X}$, but in the smaller Banach space K.

First, we notice that Hypothesis 4.5 implies that $e^{tA}K \subset K$. In fact, for any $x \in K$ we have

$$\|e^{tA}x\|_K = \left\| \sum_{k=1}^{\infty} e^{-\lambda_k t} x_k \mathbf{e}_k \right\|_K \leq \sum_{k=1}^{\infty} |x_k| \|\mathbf{e}_k\|_K \leq C\|x\|_K < \infty.$$

Then, reasoning as in [**28**, Theorem 5.20] we can prove the following

LEMMA 4.6. *The process* $W_A(t)$ *has a* K*-valued version with* α*-Hölder continuous paths for any* $\alpha \in (0, \iota/2)$.

Moreover, the following estimate holds:

$$\sup_{t\geq 0} \mathbb{E}\, \|W_A(t)\|_K^2 < \infty. \tag{4.3}$$

PROOF. We have the following representation for $W_A(t)$:

$$\begin{aligned} W_A(t) &= \int_0^t e^{(t-s)A} \sqrt{Q} \mathrm{d}W(s) \\ &= \int_0^t e^{(t-s)A} \sqrt{Q} \mathrm{d} \left(\sum_{k=1}^{\infty} \beta_k(s) \mathbf{e}_k \right) \\ &= \sum_{k=1}^{\infty} \sqrt{\lambda_k} \mathbf{e}_k \int_0^t e^{-\alpha_k (t-s)} \mathrm{d}\beta_k(s), \qquad t \geq 0, \xi \in (0,1), \end{aligned}$$

with the series converging in $L^2(\Omega, \mathcal{F}, \mathbb{P}; K)$ for all $t \geq 0$ and $\xi \in (0,1)$. This is because we have, for arbitrary $n, p \in \mathbb{N}$,

$$\begin{aligned}
\mathbb{E}\left\| \sum_{k=n+1}^{n+p} \sqrt{\lambda_k}\mathbf{e}_k \int_0^t e^{-\alpha_k(t-s)} \mathrm{d}\beta_k(s) \right\|_K^2
&= \sum_{k=n+1}^{n+p} \frac{(1-e^{-2\alpha_k t})}{2\alpha_k} \|\mathbf{e}_k\|_K^2 \\
&\leq \frac{C^2}{2} \sum_{k=n+1}^{n+p} \frac{\lambda_k}{\alpha_k} < \tilde{C},
\end{aligned}$$

with $\tilde{C}$ depending only on $\{\|\mathbf{e}_k\|_K, \alpha_k, \lambda_k\}_{k\in\mathbb{N}}$ Hence W_A is well-defined as a mean-square integrable process in K and estimate (4.3) holds thanks to the independence of $\tilde{C}$ on t. Moreover, for $s \leq t$,

$$W_A(t) - W_A(s) = \int_s^t e^{(t-r)A}\sqrt{Q}\mathrm{d}W(r) + \int_0^s \left(e^{(t-r)A} - e^{(s-r)A}\right)\sqrt{Q}\mathrm{d}W(r)$$

and since the integrals are independent it follows that

$$\begin{aligned}
\mathbb{E}\,\|W_A(t) - W_A(s)\|_K^2 &= \sum_{k=1}^{\infty} \lambda_k \|\mathbf{e}_k\|_K^2 \int_0^{t-s} e^{-2\alpha_k r}\mathrm{d}r \\
&\quad + \sum_{k=1}^{\infty} \lambda_k \|\mathbf{e}_k\|_K^2 \int_0^s \left| e^{-(t-s+r)\alpha_k} - e^{-r\alpha_k} \right|^2 \\
&= I_1(t;s) + I_2(t;s).
\end{aligned}$$

Next, by Hypothesis 4.5 we have

$$I_1(t;s) \leq \sum_{k=1}^{\infty} \lambda_k \int_0^{t-s} e^{-2\alpha_k r}\mathrm{d}r \leq \frac{1}{2}C^2 \sum_{k=1}^{\infty} \frac{\lambda_k}{\alpha_k^{1-\iota}} |t-s|^{\iota}.$$

Similarly,

$$\begin{aligned}
I_2(t;s) &\leq \frac{1}{2}C^2 \sum_{k=1}^{\infty} \frac{\lambda_k}{\alpha_k} \left[2(1-e^{-(t-s)\alpha_k}) - (1-e^{-2(t-s)\alpha_k}) - (e^{-t\alpha_k} - e^{-s\alpha_k})^2 \right] \\
&\leq C^2 \sum_{k=1}^{\infty} \frac{\lambda_k}{\alpha_k^{1-\iota}} (1-e^{-(t-s)\alpha_k}) \\
&\leq C^2 c_\iota \sum_{k=1}^{\infty} \frac{\lambda_k}{\alpha_k^{1-\iota}} |t-s|^{\iota}.
\end{aligned}$$

Collecting the above estimates and applying Kolmogorov's test the desired result follows.

□

REMARK 4.7. In the following (see Section 6) it will be convenient to consider the evolution equation (SDE) for arbitrary time $t \in \mathbb{R}$. Therefore we extend the process $(W(t))_{t\geq 0}$ for $t \leq 0$ by choosing a process $(\tilde{W}(t))_{t\geq 0}$ with the same law, but independent of it, and setting

$$W(t) = \tilde{W}(-t), \qquad t \leq 0. \tag{4.4}$$

It turns out from the estimate proved in the previous Lemma that, if we set

$$W_{A,\infty} := \sup_{t\geq 0} W_A(t) = \lim_{t\to\infty} \int_0^t e^{(t-s)A}\sqrt{Q}\mathrm{d}W(s),$$

where the limit above is considered in $L^2(\Omega)$, then $W_{A,\infty}$ is a well defined square summable Gaussian random variable with mean 0 and covariance operator

$$Q_\infty := \int_0^\infty e^{sA}Qe^{sA^\star}\mathrm{d}s.$$

5. A priori estimates. Existence and uniqueness

The arguments of this sections proceeds, basically, in the same spirit as Da Prato and Zabcyck [**29**, Chapter 5]. We prove an existence and uniqueness result for equation (SDE) in Sec. 3 through an approximating procedure on finite dimensional spaces and with a Lipschitz continuous nonlinearity.

We set $P_n : \mathbf{X} \to \mathbf{X}_n$ to be the projection of $\mathbf{X}$ onto the vector space generated by the first n eigenvectors $\mathbf{e}_1, \ldots, \mathbf{e}_n$ of A. We let $\{K_n\}_{n\in\mathbb{N}}$ be a sequence of finite dimensional subspaces of K and $\{\Pi_n\}$ a sequence of self-adjoint operators from $\mathbf{X}$ onto K_n such that $\Pi_n x \to x$ in K, for arbitrary $x \in K$.

REMARK 5.1. *We notice that such finite dimensional spaces and operators exist. An explicit construction is given, for example, in Zabczyck* [**77**, Proposition 3].

Moreover, for any $m \in \mathbb{N}$, we set F_m to be the m-th Yosida approximation of F, (i.e. $F_m := m(mI - F)^{-1}$). We know that F_m is Lipschitz continuous and it satisfies the following estimates:

$$|F_m(x) - F(x)|_{\mathbf{X}} \to 0, \qquad m \to \infty, x \in D(F)$$
$$|F_m(x)|_{\mathbf{X}} \leq |F(x)|_{\mathbf{X}}, \qquad x \in D(F), \forall m \in \mathbb{N}$$

and

$$\langle F_m(x) - F_m(y), x - y\rangle \leq m|x-y|^2_{\mathbf{X}}, \quad \forall m \in \mathbb{N}.$$

For any $n, m \in N$ we consider the following family of equations

$$\begin{cases} \mathrm{d}u_m^{(n)}(t) = Au_m^{(n)}(t)\mathrm{d}t + \Pi_n F_m(\Pi_n u_m^{(n)}(t))\mathrm{d}t + \sqrt{Q}\mathrm{d}W(t), \\ u_m^{(n)}(0) = P_n u_0 \in \mathbf{X}, \end{cases} \tag{5.1}$$

which can be seen as an approximating problem relative to (SDE). There exists a well-established theory on stochastic evolution equations in Hilbert spaces, see, for example, Da Prato and Zabcyck [**28**], that we shall apply in order to show that, for any $n, m \in N$, equation (5.1), admits a unique solution $u_m^{(n)}$. To this end, we first investigate the properties of the nonlinear term of the above approximating equations. We have the following result.

PROPOSITION 5.2. *For any $x \in \overline{D(F)}$, where $\overline{D(F)}$ means the closure of $D(F)$ in $\mathbf{X}$ (if $D(F)$ is not dense in $\mathbf{X}$, otherwise $\overline{D(F)} = \mathbf{X}$), we have*

$$\lim_{n,m\to\infty} \Pi_n F_m \Pi_n(x) = F(x), \qquad \textit{strongly, i.e.}$$
$$|\Pi_n F_m \Pi_n(x) - F(x)|_{\mathbf{X}} \to 0, \qquad \textit{as } n, m \to \infty.$$

PROOF. Let $x \in \overline{D(F)}$ and let $n, m \in \mathbb{N}$. Then

(5.2)
$$\begin{aligned}|\Pi_n F_m \Pi_n(x) - F(x)|^2_{\mathbf{X}} &= |\Pi_n F_m \Pi_n(x) - \Pi_n F \Pi_n(x) + \Pi_n F \Pi_n(x) - F(x)|^2_{\mathbf{X}} \\ &\leq |\Pi_n F_m \Pi_n(x) - \Pi_n F \Pi_n(x)|^2_{\mathbf{X}} + |\Pi_n F \Pi_n(x) - \Pi_n F(x) + \Pi_n F(x) - F(x)|^2_{\mathbf{X}} \\ &\leq |F_m \Pi_n(x) - F\Pi_n(x)|^2_{\mathbf{X}} + |F(\Pi_n x) - F(x)|^2_{\mathbf{X}} + |\Pi_n F(x) - F(x)|^2_{\mathbf{X}},\end{aligned}$$

where in the last step we used that $\|\Pi_n\|_{L(\mathbf{X};K)} \leq 1$. Now let $\varepsilon > 0$. We notice the following facts. First, since for any $y \in \mathbf{X}$ $\Pi_n y \to y$ as $n \to \infty$, we have that there exists $n_1 = n_1(\varepsilon, x)$ such that

$$|\Pi_n F(x) - F(x)|_{\mathbf{X}} \leq \varepsilon/3, \qquad \text{for all } n \geq n_1(\varepsilon, x).$$

Second, since F is continuous, there exists $\delta = \delta(x, \varepsilon)$ with the property that for any $y \in \mathbf{X}$ such that $|y - x|_{\mathbf{X}} < \delta$ then

$$|F(y) - F(x)|_{\mathbf{X}} < \varepsilon/3.$$

Using again the convergence of Π_n, we have in particular that there exists $n_2 = n_2(\delta, (x, \varepsilon)) \in \mathbb{N}$ such that $|\Pi_n x - x|_{\mathbf{X}} < \delta$ for any $n \geq n_2$ and, consequently that,

$$|F(\Pi_n x) - F(x)|_{\mathbf{X}} < \varepsilon/3, \qquad \text{for all } n \geq n_2.$$

Third, we notice that, since $F_m \to_{m\to\infty} F$, strongly on $\overline{D(F)}$, for any $n \in \mathbb{N}$ there exists $m_1 = m_1(n, x, \varepsilon) \in \mathbb{N}$ such that

$$|F_m(\Pi_n x) - F(\Pi_n x)|^2_{\mathbf{X}} < \frac{\varepsilon}{3}, \qquad \text{for all } m \geq m_1. \tag{5.3}$$

Now fixing $\bar{n} = \min\{n_1, n_2\}$, $\bar{m} = \bar{m}(\bar{n}, x)$ and taking into account the above claims, we conclude that

$$|\Pi_n F_m(\Pi_n x) - F(x)|_{\mathbf{X}} < \varepsilon, \qquad \text{for all } m > \bar{m}, n > \bar{n}.$$

This proves the proposition.

□

REMARK 5.3. We claim that if, as it happens in our case, F is an m-quasi-dissipative mapping, then also the mapping $\Pi_n F \Pi_n$ is m-quasi-dissipative. In fact, we first notice that for any $n \in \mathbb{N}$ the projection Π_n is symmetric, that is $\langle \Pi_n x, y\rangle = \langle x, \Pi_n y\rangle$ for any $x, y \in \mathbf{X}$; consequently we get from Hypothesis 2.1:

$$\begin{aligned}\langle \Pi_n F \Pi_n(x) - \Pi_n F \Pi_n(y), x - y\rangle &= \langle F\Pi_n(x) - F\Pi_n(y), \Pi_n x - \Pi_n y\rangle \\ &\leq \eta |\Pi_n x - \Pi_n y|^2_{\mathbf{X}} \\ &\leq \eta |x - y|^2_{\mathbf{X}},\end{aligned}$$

so that the claim follows.

PROPOSITION 5.4. *Let $n \in \mathbb{N}$ be fixed. For any $m \in \mathbb{N}$ let $(u_m^{(n)}(t))_{t\geq 0}$ denote the solution of the SDE*

$$\begin{cases} \mathrm{d}u_m^{(n)}(t) = Au_m^{(n)}(t)\mathrm{d}t + \Pi_n F_m \Pi_n(u_m^{(n)}(t))\mathrm{d}t + \sqrt{Q}\mathrm{d}W(t) \\ u_m^{(n)}(0) = P_n u_0 \in \mathbf{X}. \end{cases} \tag{Enm}$$

Moreover, let $u^{(n)}(t))_{t\geq 0}$ be the mild solution of the SDE

$$\begin{cases} \mathrm{d}u^{(n)}(t) = Au^{(n)}(t)\mathrm{d}t + \Pi_n F \Pi_n(u^{(n)}(t))\mathrm{d}t + \sqrt{Q}\mathrm{d}W(t) \\ u^{(n)}(0) = P_n u_0 \in \mathbf{X}. \end{cases} \tag{En}$$

Then $(u_m^{(n)}(t))_{m\in\mathbb{N}}$ converges to $u^{(n)}(t)$ for $m\to\infty$, uniformly on $[0,\infty)$, in the sense that

$$\mathbb{E}\sup_{t\geq 0}|u_m^{(n)}(t)-u^{(n)}(t)|^2_{\mathbf{X}}\to 0,\qquad m\to\infty,$$

for any fixed $n\in\mathbb{N}$.

PROOF. Existence and uniqueness of mild solutions for semilinear SDEs on Hilbert spaces under Lipschitz or dissipativity conditions on the nonlinear term is already known in the literature (see for example Da Prato-Zabzcyk [**28**]). Moreover, for any $T>0$ and $p\geq 2$, the following bounds hold:

$$\mathbb{E}\left(\sup_{t\in[0,T]}|u_m^{(n)}(t)|^p_{\mathbf{X}}\right)^{1/p}\leq C(1+|P_nu_0|^p_{\mathbf{X}})$$

$$\mathbb{E}\left(\sup_{t\in[0,T]}|u^{(n)}(t)|^p_{\mathbf{X}}\right)^{1/p}\leq C(1+|P_nu_0|^p_{\mathbf{X}}),\qquad\text{for some } c>0.$$

Hence problems (Enm) and (En) admit unique mild solutions which belong to $L^p(\Omega;C(0,T;\mathbf{X}))$ for any $p\geq 2$. In the following we are going to prove that

$$\mathbb{E}\sup_{t\geq 0}|u_m^{(n)}(t)-u^{(n)}(t)|^2_{\mathbf{X}}\ \to\ 0,\qquad m\ \to\ \infty,\quad\text{for all } n\in\mathbb{N}.$$

In fact, from (Enm) and (En) we have that the difference $u_m^{(n)}(t)-u^{(n)}(t)$ satisfies the following equality

$$\begin{aligned}\frac{\mathrm{d}}{\mathrm{d}t}|u_m^{(n)}(t)-u^{(n)}(t)|^2_{\mathbf{X}}&=2\langle A(u_m^{(n)}(t)-u^{(n)}(t)),u_m^{(n)}(t)-u^{(n)}(t)\rangle+\\&\quad 2\langle\Pi_nF_m\Pi_n(u_m^{(n)}(t))-\Pi_nF\Pi_n(u^{(n)}(t)),u_m^{(n)}(t)-u^{(n)}(t)\rangle.\end{aligned}$$

But Π_n is symmetric; hence the above equality can also be written in the form

$$\begin{aligned}\frac{\mathrm{d}}{\mathrm{d}t}|u_m^{(n)}(t)-u^{(n)}(t)|^2_{\mathbf{X}}&=2\langle A(u_m^{(n)}(t)-u^{(n)}(t)),u_m^{(n)}(t)-u^{(n)}(t)\rangle+\\&\quad 2\langle F_m\Pi_n(u_m^{(n)}(t))-F\Pi_n(u^{(n)}(t)),\Pi_nu_m^{(n)}(t)-\Pi_nu^{(n)}(t)\rangle.\end{aligned}\tag{5.4}$$

We observe that for any $x,y\in\overline{D(F)}$ and any $\xi>0$ we have (using the inequality $\langle a,b\rangle\leq\xi|a|^2_{\mathbf{X}}+1/\xi|b|^2_{\mathbf{X}}$, for $a,b\in\mathbf{X}$):

$$\langle F_m\Pi_n(x)-F\Pi_n(y),\Pi_nx-\Pi_ny\rangle\leq\frac{1}{\xi}|F_m\Pi_n(x)-F\Pi_n(x)|^2_{\mathbf{X}}+\xi|x-y|^2_{\mathbf{X}}.$$

Inserting this into (5.4) and using also the dissipativity condition stated in Hypothesis 2.1 we obtain

$$\begin{aligned}\frac{\mathrm{d}}{\mathrm{d}t}|u_m^{(n)}(t)-u^{(n)}(t)|^2_{\mathbf{X}}&\leq-2(\omega-\eta-\xi)|u_m^{(n)}(t)-u^{(n)}(t)|^2_{\mathbf{X}}\\&\quad+\frac{2}{\xi}|F_m\Pi_n(u_m^{(n)}(t))-F\Pi_n(u_m^{(n)}(t))|^2_{\mathbf{X}}.\end{aligned}$$

Choosing ξ sufficiently small and using Gronwall's lemma we get

$$|u_m^{(n)}(t)-u^{(n)}(t)|^2_{\mathbf{X}}\leq\frac{1}{\xi}|F_m\Pi_n(u_m^{(n)}(t))-F\Pi_n(u_m^{(n)}(t))|^2_{\mathbf{X}},\qquad t\geq 0;$$

hence letting $m \to \infty$ we see that $u_m^{(n)}(t) \to u^{(n)}(t)$, $t \geq 0$, uniformly, by the continuity and uniform boundedness of $u_m^{(n)}(t)$ and $u^{(n)}(t)$ in the $L^2(\Omega, \mathbb{P}) \times \mathbf{X}$-norm.

□

PROPOSITION 5.5. *For any $n \in \mathbb{N}$ let $(u^{(n)}(t))_{t \geq 0}$ denote the mild solution of the SDE (En), i.e.,*

$$\begin{cases} \mathrm{d}u^{(n)}(t) = Au^{(n)}(t)\mathrm{d}t + \Pi_n F \Pi_n(u^{(n)}(t))\mathrm{d}t + \sqrt{Q}\mathrm{d}W(t) \\ u^{(n)}(0) = P_n u_0 \in \mathbf{X}. \end{cases} \tag{5.5}$$

Then $(u^{(n)}(t))_{n \in \mathbb{N}}$ converges to the solution $u(t)$ of the stochastic differential equation

$$\begin{cases} \mathrm{d}u(t) = Au(t)\mathrm{d}t + F(u(t))\mathrm{d}t + \sqrt{Q}\mathrm{d}W(t) \\ u(0) = u_0 \in \mathbf{X}, \end{cases} \tag{5.6}$$

uniformly on $[0, \infty)$, in the sense that

$$\mathbb{E} \sup_{t \geq 0} |u^{(n)}(t) - u(t)|_{\mathbf{X}}^2 \to 0, \quad n \to \infty.$$

PROOF. Existence and uniqueness of solutions for problems (5.5) and (5.6) has been already proved. In the following we are going to prove that

$$\mathbb{E} \sup_{t \geq 0} |u^{(n)}(t) - u(t)|_{\mathbf{X}}^2 \to 0, \qquad n \to \infty.$$

In fact, proceeding as in the previous Proposition we have

$$\begin{aligned} \frac{\mathrm{d}}{\mathrm{d}t}|u^{(n)}(t) - u(t)|_{\mathbf{X}}^2 &= \langle A(u^{(n)}(t) - u(t)), u^{(n)}(t) - u(t)\rangle + \\ &\qquad \langle \Pi_n F \Pi_n(u^{(n)}(t)) - F(u(t)), u^{(n)}(t) - u(t)\rangle. \end{aligned} \tag{5.7}$$

Now let us focus on the second term on the right hand side of the previous equality. Using the dissipativity of F and $||\Pi_n|| \leq 1$, for any $x, y \in D(F)$ and $\xi > 0$ we get

$$\begin{aligned} &\langle \Pi_n F \Pi_n(x) - F(y), x - y\rangle \\ &\quad = \langle \Pi_n F \Pi_n(x) - \Pi_n F \Pi_n(y), x - y\rangle + \langle \Pi_n F \Pi_n(y) - F(y), x - y\rangle \\ &\quad \leq \frac{1}{\xi}|\Pi_n F \Pi_n(y) - F(y)|_{\mathbf{X}}^2 + (\eta + \xi)|x - y|_{\mathbf{X}}^2 \\ &\quad \leq \frac{1}{\xi}|\Pi_n F \Pi_n(y) - \Pi_n F(y)|_{\mathbf{X}}^2 + \frac{1}{\xi}|\Pi_n F(y) - F(y)|_{\mathbf{X}}^2 + (\eta + \xi)|x - y|_{\mathbf{X}}^2 \\ &\quad \leq \frac{1}{\xi}|F \Pi_n(y) - F(y)|_{\mathbf{X}}^2 + \frac{1}{\xi}|\Pi_n F(y) - F(y)|_{\mathbf{X}}^2 + (\eta + \xi)|x - y|_{\mathbf{X}}^2 \end{aligned} \tag{5.8}$$

Taking into account the above inequality and equality (5.7) we get from (5.8)

$$\begin{aligned} \frac{\mathrm{d}}{\mathrm{d}t}|u^{(n)}(t) - u(t)|_{\mathbf{X}}^2 &\leq -2(\omega - \eta - \xi)|u^{(n)}(t) - u(t)|_{\mathbf{X}}^2 + \\ &\frac{1}{\xi}|F\Pi_n(u(t)) - F(u(t))|_{\mathbf{X}}^2 + \frac{1}{\xi}|\Pi_n F(u(t)) - F(u(t))|_{\mathbf{X}}^2, \qquad t \geq 0, \end{aligned}$$

so that, for ξ sufficiently small and using Gronwall's lemma, we have

$$|u^{(n)}(t) - u(t)|^2_{\mathbf{X}} \leq |P_n u_0 - u_0|^2_{\mathbf{X}} + \frac{1}{\xi}|F\Pi_n(u(t)) - F(u(t))|^2_{\mathbf{X}} + \frac{1}{\xi}|\Pi_n F(u(t)) - F(u(t))|^2_{\mathbf{X}},$$

for $t \geq 0$. Now the thesis follows by the continuity and uniform boundedness of $u_n(t)$ and $u(t)$ in the $L^2(\Omega, \mathbb{P}) \times \mathbf{X}$-norm, the strong convergence of P_n to the identity and the continuity of F.

□

REMARK 5.6. (1) *Taking into account Hypothesis* 4.5 *and Lemma* 4.6 *we can conclude that the processes* $u_m^{(n)}, u_m$ *and* u *possess a* K*-continuous version when the initial datum* x *belongs to* K.

(2) *It can be proved (see, e.g.* [**77**, Proposition 1]*) that the approximating stochastic convolutions*

$$\int_0^t e^{(t-s)A}\sqrt{Q}\Pi_n \mathrm{d}W(t)$$

converge almost surely and uniformly on compact sets to the K*-continuous version* $W_A(t)$*. This result can be used to prove that* $u_m(t)$ *converges to* $u(t)$ *almost surely and uniformly on compact sets. The proof requires a generalization of* [**77**, Proposition 5].

Both properties will be crucial for the study of the invariant measure of the problem. We refer to Subsection 7.2 *for more details.*

6. The stationary process

In this section we are concerned with the asymptotic properties of the mild solution of the problem

$$\text{(SDE)} \qquad \begin{cases} \mathrm{d}u(t) = (Au(t) + F(u))\mathrm{d}t + \sqrt{Q}\mathrm{d}W \\ \mathrm{d}u(0) = u_0 \in D(F). \end{cases}$$

In particular, we are going to prove that its unique solution admits a characterization in terms of a stationary process r and a process v which vanishes at $t \to \infty$. To this end, we proceed by splitting the solution of the approximating problems into the sum of a stationary process $r_m^{(n)}$ and a vanishing process $v_m^{(n)}$ satisfying suitable properties. Heuristically, let us suppose that we are given two processes $r_m^{(n)}$ and $v_m^{(n)}$ defined, respectively, on the whole real line and on $\mathbb{R}^+$ by:

$$(r_{nm}) \qquad r_m^{(n)}(t) = \int_{-\infty}^t e^{(t-s)A}\Pi_n F_m \Pi_n(r_m^{(n)}(s))\mathrm{d}t + \int_{-\infty}^t e^{(t-s)A}\sqrt{Q}\mathrm{d}W(s)$$

and

(v_{nm})

$$v_m^{(n)}(t) = e^{tA}x - \int_{-\infty}^0 e^{(t-s)A}\Pi_n F_m \Pi_n(r_m^{(n)}(s))\mathrm{d}s - \int_{-\infty}^0 e^{(t-s)A}\sqrt{Q}\mathrm{d}W(s) + \int_0^t e^{(t-s)A}[\Pi_n F_m(\Pi_n u_m^{(n)}(s))\mathrm{d}s - \Pi_n F_m(\Pi_n r_m^{(n)}(s))]\mathrm{d}s.$$

Then, it is easy to prove that for any $t \geq 0$, $r_m^{(n)}(t) + v_m^{(n)}(t) = u_m^{(n)}(t)$.

In the following we prove that the processes $r_m^{(n)}, v_m^{(n)}$ defined by (r_{nm}) and (v_{nm}) give the required decomposition of $u_m^{(n)}(t)$. In particular, we prove that $r_m^{(n)}$ can be obtained as the limit of processes solving suitable differential equations and it turns out that this implies the existence and uniqueness of the mild solutions for the problems (r_{nm}) and (v_{nm}) with $r_m^{(n)}(t)$ being a stationary process and $\lim_{t\to\infty}\mathbb{E}|v_m^{(n)}(t)|^p = 0, p \geq 2$, for all $n, m \in \mathbb{N}$.

Before giving the results, we introduce the stochastic differential equation corresponding to r_{nm} and we give the notion of mild solution. Let us consider the problem

$$(6.1)\quad \begin{cases} \mathrm{d}r_m^{(n)}(t) = Ar_m^{(n)}(t)\mathrm{d}t + \Pi_n F_m \Pi_n(r_m^{(n)}(t))\mathrm{d}t + \Pi_n\sqrt{Q}\mathrm{d}W(t), & t \in \mathbb{R},\\ \lim_{t\to-\infty} r_m^{(n)}(t) = 0. \end{cases}$$

DEFINITION 6.1. *An $\mathcal{F}_t$-adapted process $r_m^{(n)}$ is said to be a* mild solution *to (6.1) if it satisfies the integral equation (r_{nm}), for any $t \in \mathbb{R}$.*

THEOREM 6.2. *For any $n, m \in \mathbb{N}$, there exists a unique mild solution $r_m^{(n)}$ to the equation*

$$(r_{nm})\quad \begin{cases} \mathrm{d}r_m^{(n)}(t) = Ar_m^{(n)}(t)\mathrm{d}t + \Pi_n F_m \Pi_n(r_m^{(n)}(t))\mathrm{d}t + \Pi_n\sqrt{Q}\mathrm{d}W(t), & t \in \mathbb{R},\\ \lim_{t\to-\infty} r_m^{(n)}(t) = 0 \end{cases}$$

such that

$$(6.2)\quad \sup_{t\in\mathbb{R}} \mathbb{E}|r_m^{(n)}(t)|^p_{\mathbf{X}} \leq C_p$$

for every $p \geq 2$ and for some positive constant C_p (independent of n and m). Further, $r_m^{(n)}$ is a stationary process, that is, for every $h \in \mathbb{R}^+$, $k \in \mathbb{N}$, any $0 \leq t_1 \cdots \leq t_k$ and any $A_1, \ldots, A_k \in \mathcal{B}(\mathbf{X})$ we have:

$$\mathbb{P}(r_m^{(n)}(t_1+h) \in A_1, \ldots, r_m^{(n)}(t_k+h)) = \mathbb{P}(r_m^{(n)}(t_1) \in A_1, \ldots, r_m^{(n)}(t_n) \in A_k).$$

PROOF. First, we prove uniqueness: let u, v be two solutions of (r_{nm}). Dissipativity of $A + \Pi_n F_m(\Pi_n \cdot))$ implies

$$\begin{aligned} \mathrm{d}|u(t)-v(t)|^2 &= \langle A(u(t)-v(t)) + \Pi_n F_m(\Pi_n u(t)) - \Pi_n F_m(\Pi_n v(t)), u(t)-v(t)\rangle \mathrm{d}t \\ &\leq -2(\omega-\eta)|u(t)-v(t)|^2\mathrm{d}t, \end{aligned}$$

and, by using Gronwall's lemma, we deduce that for any $\lambda > 0$ and $t \geq -\lambda$ the following inequality holds

$$|u(t)-v(t)|^2 \leq |u(-\lambda)-v(-\lambda)|^2 e^{-2(\omega-\eta)(t+\lambda)}.$$

Letting $\lambda \to \infty$ we conclude that $u(t) = v(t)$ for any $t \in \mathbb{R}$.

Existence can be proved again by an approximating procedure. For any $\lambda > 0$, we denote by $r_m^{(n)}(t;\lambda)$ the unique solution of

$$(6.3)\quad \begin{cases} \mathrm{d}r_m^{(n)}(t) = Ar_m^{(n)}(t)\mathrm{d}t + \Pi_n F_m \Pi_n(r_m^{(n)}(t))\mathrm{d}t + \Pi_n\sqrt{Q}\mathrm{d}W(t), & t \geq -\lambda,\\ r_m^{(n)}(-\lambda) = e^{-\lambda A}x. \end{cases}$$

Thanks to the smoothness properties of the stochastic convolution, we can assume that $r_m^{(n)}(\cdot;-\lambda)$ admits a K-continuous version, which we still denote by $r_m^{(n)}(\cdot;-\lambda)$

(see Remark 4.6 for more details). Here we prove that, for any $t \geq -\lambda$ and $p \geq 2$ the following estimate holds:

$$\sup_{-\lambda \leq t \leq 0} \mathbb{E}\|r_m^{(n)}(t;-\lambda)\|_K^p < C_p,$$

where C_p is a positive constant independent on n, m and λ, possibly depending on p. Consequently, for any $t \in \mathbb{R}$, we can define $r_m^{(n)}(t)$ as the limit of $r_m^{(n)}(t;-\lambda)$ for $\lambda \to \infty$ and this turns out to be the solution to (r_{nm}). For simplicity we consider the case $p = 2a, a \in \mathbb{N}$. We want to apply Itô's formula to the process $|r_m^{(n)}(t)|_{\mathbf{X}}^{2a}$. To this end, we recall the expressions for the first and second derivatives of the function $H(x) := |x|^{2a}$.

We have

$$\nabla F(x) = 2a|x|^{2(a-1)}x$$
$$\frac{1}{2}\mathrm{Tr}(Q\nabla F^2(x)) = a\mathrm{Tr}(Q)|x|^{2(a-1)} + (a-1)a|x|^{2(a-2)}|\sqrt{Q}x|^2.$$

Hence

$$\mathrm{d}|r_m^{(n)}(t;-\lambda)|^{2a} = \langle 2a|r_m^{(n)}(t;-\lambda)|^{2(a-1)}r_m^{(n)}(t;-\lambda), \mathrm{d}r_m^{(n)}(t;-\lambda)\rangle$$
$$+ a\mathrm{Tr}(Q)|r_m^{(n)}(t;-\lambda)|^{2(a-1)}\mathrm{d}t + (a-1)a|r_m^{(n)}(t;-\lambda)|^{2(a-2)}|\sqrt{Q}r_m^{(n)}(t;-\lambda)|^2\mathrm{d}t.$$

Now using the dissipativity of $A+F$, for sufficiently small $\xi > 0$, we get

$$\begin{aligned}
&\mathrm{d}|r_m^{(n)}(t;-\lambda)|^{2a} \\
&\quad = 2a|r_m^{(n)}(t;-\lambda)|^{2(a-1)}\langle Ar_m^{(n)}(t;-\lambda) + \Pi_n F_m \Pi_n(r_m^{(n)}(t;-\lambda)), r_m^{(n)}(t;-\lambda)\rangle \mathrm{d}t \\
&\qquad + 2a|r_m^{(n)}(t;-\lambda)|^{2(a-1)}\langle \Pi_n\sqrt{Q}\mathrm{d}W(t), r_m^{(n)}(t;-\lambda)\rangle \\
&\qquad + a\mathrm{Tr}(Q)|r_m^{(n)}(t;-\lambda)|^{2(a-1)}\mathrm{d}t + (a-1)a|r_m^{(n)}(t;-\lambda)|^{2(a-2)}|\sqrt{Q}r_m^{(n)}(t;-\lambda)|^2\mathrm{d}t \\
&\quad \leq -2(\omega-\eta)|r_m^{(n)}(t;-\lambda)|_{\mathbf{X}}^{2a} + 2a|r_m^{(n)}(t;-\lambda)|^{2(a-1)}\langle F(0), r_m^{(n)}(t;-\lambda)\rangle \mathrm{d}t \\
&\qquad + C_{a,Q}|r_m^{(n)}(t;-\lambda)|^{2(a-1)} + 2a|r_m^{(n)}(t;-\lambda)|^{2(a-1)}\langle \Pi_n\sqrt{Q}\mathrm{d}W(t), r_m^{(n)}(t;-\lambda)\rangle \\
&\quad \leq -2(\omega-\eta)|r_m^{(n)}(t;-\lambda)|_{\mathbf{X}}^{2a} + 2a|r_m^{(n)}(t;-\lambda)|^{2a-1}|F(0)|\mathrm{d}t \\
&\qquad\quad + C_{a,Q}|r_m^{(n)}(t;-\lambda)|^{2(a-1)} + 2a|r_m^{(n)}(t;-\lambda)|^{2(a-1)}\langle \Pi_n\sqrt{Q}\mathrm{d}W(t), r_m^{(n)}(t;-\lambda)\rangle \\
&\quad \leq -2(\omega-\eta+\xi)|r_m^{(n)}(t;-\lambda)|_{\mathbf{X}}^{2a} \\
&\qquad + \frac{1}{\xi}C_{a,Q,F(0)}\mathrm{d}t + 2a|r_m^{(n)}(t;-\lambda)|^{2(a-1)}\langle \Pi_n\sqrt{Q}\mathrm{d}W(t), r_m^{(n)}(t;-\lambda)\rangle.
\end{aligned}$$

and, integrating over $[-\lambda, t]$, $t \geq -\lambda$ we obtain:

$$|r_m^{(n)}(t;-\lambda)|^{2a} \leq e^{-2a\lambda\omega}|x|^{2a} - 2(\omega-\eta+\xi)\int_{-\lambda}^t |r_m^{(n)}(s;-\lambda)|_{\mathbf{X}}^{2a}\mathrm{d}s$$
$$+ \int_{-\lambda}^t |r_m^{(n)}(s;-\lambda)|^{2(a-1)}\langle \Pi_n\sqrt{Q}\mathrm{d}W(s), r_m^{(n)}(s;-\lambda)\rangle + (\mathrm{Tr}Q + |F(0)|^2)(t+\lambda).$$

Notice that the term

$$\int_{-\lambda}^t \langle |r_m^{(n)}(s;-\lambda)|^{2(a-1)}\Pi_n\sqrt{Q}\mathrm{d}W(s), r_m^{(n)}(s;-\lambda)\rangle$$

is a square integrable martingale with mean 0, so that, taking the expectation of both members in the previous inequality we obtain

$$\mathbb{E}|r_m^{(n)}(s;-\lambda)|^{2a} \le e^{-2(\lambda+t)\omega}|x|^{2a} + (\mathrm{TrQ} + \xi|F(0)|^{2a})(\lambda+t) - 2\left(\omega - \eta + \frac{1}{\xi}\right)\int_{-\lambda}^{t} \mathbb{E}|r_m^{(n)}(r;-\lambda)|^2 \mathrm{d}r\,,$$

so that Gronwall's lemma implies

$$\begin{aligned}\mathbb{E}|r_m^{(n)}(t;-\lambda)|^{2a} &\le \left[e^{-2a(\lambda+t)\omega}|x|^{2a} + (\mathrm{TrQ} + \xi|F(0)|^{2a})(\lambda+t)\right] e^{-2a\left(\omega-\eta+\frac{1}{\xi}\right)\lambda}\\ &\le C\,,\end{aligned}$$

where C is a suitable constant independent on m, n and λ.

Moreover, we can prove that, for any fixed $t \in \mathbb{R}$, the sequence $\{r_m^{(n)}(t;-\lambda)\}_{-\lambda\le t}$ is a Cauchy sequence, uniformly in t.

Now let $0 \le \gamma \le \lambda$. We need to estimate the norm:

$$\sup_{t\ge-\gamma} \mathbb{E}|r_m^{(n)}(t;-\lambda) - r_m^{(n)}(t;-\gamma)|^{2a}.$$

To this end we notice that the process $y_{\lambda,\gamma}(t) := r_m^{(n)}(t;-\lambda) - r_m^{(n)}(t;-\gamma)$ can be written as:

$$\begin{aligned} y_{\lambda,\gamma}(t) &= (e^{-\lambda A} - e^{-\gamma A})x \\ &\quad + \int_{-\lambda}^{-\gamma} e^{(-\gamma-s)A}\Pi_n F_m(\Pi_n r_m^{(n)}(s;-\lambda))\mathrm{d}s + \int_{-\lambda}^{-\gamma} e^{(-\gamma-s)A}\sqrt{Q}\mathrm{d}W(s) \\ &\quad + \int_{-\gamma}^{t} e^{(t-s)A}[\Pi_n F_m(\Pi_n r_m^{(n)}(s;-\lambda)) - \Pi_n F_m(\Pi_n r_m^{(n)}(s;-\gamma))]\mathrm{d}s. \end{aligned} \tag{6.4}$$

Moreover, the stochastic differential of $y_{\lambda,\gamma}$ is given by

$$\mathrm{d}y_{\lambda,\gamma}(t) = Ay_{\lambda,\gamma}\mathrm{d}t + [F_{n,m}(r_m^{(n)}(t;-\lambda)) - F_{n,m}(r_m^{(n)}(t;-\gamma))]\mathrm{d}t,$$

since the three terms in (6.4) does not depend on t. By applying Itô's formula to $|y_{\lambda,\gamma}|^{2a}$ and reasoning as above we get

$$\mathbb{E}|y_{\lambda,\gamma}(t)|^{2a} \le |e^{-\lambda x}|^{2a} + \mathbb{E}|y_{\lambda,\gamma}(-\gamma)|^{2a} - 2(\omega-\eta)\int_{-\lambda}^{t} \mathbb{E}|y_{\lambda,\gamma}(r)|^{2a}\mathrm{d}r.$$

Gronwall's lemma then implies

$$\begin{aligned}\sup_{t\ge-\gamma} \mathbb{E}|y_{\lambda,\gamma}(t)|^{2a} &\le (\mathbb{E}|y_{\lambda,\gamma}(-\gamma)|^{2a} + |e^{tA}x|^{2a})e^{-2(\omega-\eta)(\lambda+t)}\\ &\le C,\end{aligned}$$

where C is a constant independent on λ. We then conclude that for any $t \in \mathbb{R}$ there exists the limit $r_m^{(n)}(t) := \lim_{\lambda\to\infty} r_m^{(n)}(t;-\lambda)$ in $L^2(\Omega,\mathbb{P};K)$ and moreover,

$$\sup_{t\ge-\lambda} |\mathbb{E}r_m^{(n)}(t)|^{2a} \le C.$$

In addition, by the initial condition in (6.3) we deduce that $\lim_{t\to-\infty} r_m^{(n)}(t) = 0$.

Finally, for any $n, m \in \mathbb{N}$ the process $r_m^{(n)}$ is stationary. In order to prove this statement, we adapt to our case the argument given in [**51**]. More precisely, we introduce the following

$$\begin{cases} r_m^{(n,0)}(t) = x \\ r_m^{(n,k+1)}(t) = \int_{-\infty}^t e^{(t-s)A}\Pi_n F_m(\Pi_n r_m^{(n,k)}(s))\mathrm{d}\, s + W_{A,-\infty}(t). \end{cases} \tag{6.5}$$

We notice that (see [**51**, Theorem 2]) the limit $\lim_{k\to\infty} r^{(n,k)}(t) = \tilde{r}_m^{(n)}(t)$ exists and it is a stationary process. The crucial point is that $\tilde{r}_m^{(n)}$ and $r_m^{(n)}$ coincide. In fact, if we pass to the limit in (6.5), we see that $\tilde{r}_m^{(n)}$ solves equation (r_{nm}) so that, by uniqueness, $\tilde{r}_m^{(n)} \equiv r_m^{(n)}$. Consequently, $r_m^{(n)}$ is stationary.

□

Now we turn to discussing existence and uniqueness of a mild solution for the equation for the process $v_m^{(n)}$ and we show that it vanishes as $t \to \infty$. In the next result we formalize the above statement and we prove useful bounds independent on n and m.

PROPOSITION 6.3. *For any $n, m \in \mathbb{N}$, there exists a unique mild solution $(v_m^{(n)}(t))_{t\geq 0}$ of equation (v_{nm}). Moreover, for any $p \geq 1$, we have the following bound:*

$$\sup_{t\geq 0} \mathbb{E}|v_m^{(n)}(t)|^p \leq C_p, , \tag{6.6}$$

where C_p is a positive constant independent of n and m. In addition, we have the following limit

$$\lim_{t\to\infty} \mathbb{E}|v_m^{(n)}(t)|^p = 0, . \tag{6.7}$$

PROOF. Since we are dealing with a finite dimensional ordinary differential equation with Lipschitz nonlinearity, existence and uniqueness of a mild solution for (v_{nm}) follows immediately by well-known results in finite dimensions. Concerning estimate (6.6) we notice that, by construction, we have $v_m^{(n)} = u_m^{(n)} - r_m^{(n)}$, where $r_m^{(n)}$ is the process considered in Proposition 6.3. Hence, by the dissipativity of the mapping $\Pi_n F_m \Pi_n$ we get

$$\begin{aligned} \mathrm{d}|v_m^{(n)}(t)|^{2a} &= \mathrm{d}|u_m^{(n)}(t) - r_m^{(n)}(t)|^{2a} \\ &= 2a\langle Av_m^{(n)}(t) + \Pi_n F_m(\Pi_n u_m^{(n)}(t)) \\ &\qquad - \Pi_n F_m(\Pi_n r_m^{(n)}(t)), v_m^{(n)}(t)\rangle |v_m^{(n)}(t)|^{2a-2}\mathrm{d}t \\ &\leq -2a\omega|v_m^{(n)}(t)|^{2a} + 2a\eta|v_m^{(n)}(t)|^{2a}, \end{aligned}$$

so that integrating on $[0, t]$ and applying Gronwall's lemma, we obtain

$$\begin{aligned} \sup_{t\in[0,T]} \mathbb{E}|v_m^{(n)}(t)|^{2a} &\leq e^{2a(\omega-\eta)T}\mathbb{E}|v_m^{(n)}(0)|^{2a} \\ &\leq c_a e^{2a(\omega-\eta)T}\left[|u_0|_{\mathbf{X}}^{2a} + \mathbb{E}\int_{-\infty}^0 |e^{-sA}\Pi_n F_m(\Pi_n R^{(n)}(s))|_{\mathbf{X}}^{2a}\mathrm{d}s \right. \\ &\qquad \left. + \left|\int_{-\infty}^0 e^{-sA}\Pi_n\sqrt{Q}\mathrm{d}W(s)\right|^{2a}\right], \end{aligned}$$

where c_a is a positive constant depending only on a. Now we estimate the first integral in the above inequality: recalling that by assumptions (see condition (1b) in Hypothesis 2.1)

$$|\Pi_n F_m(\Pi_n x)|_{\mathbf{X}} \leq |F(\Pi_n x)|_{\mathbf{X}} \leq C(1+|\Pi_n x|_{\mathbf{X}}^{2k+1}) \leq C(1+|x|_{\mathbf{X}}^{2k+1}),$$

where C is a suitable positive constant and $k \in \mathbb{N}$, we get

$$\begin{aligned}\mathbb{E}\int_{-\infty}^{0} |e^{-sA}\Pi_n F_m(\Pi_n r_m^{(n)}(s))|_{\mathbf{X}}^{2a} ds &\leq C\mathbb{E}\int_{-\infty}^{0} e^{-2a\omega s}\left(1+|r_m^{(n)}(s)|^{\gamma}\right)^{2a} ds \\ &\leq C_a\left(\mathbb{E}\sup_{t\in\mathbb{R}}|r_m^{(n)}(t)|^{\gamma}+1\right)^{2a},\end{aligned}$$

for a positive constant C_a independent on n, m. Taking into account the above inequality and the bound proved in Remark 4.7 we deduce that

$$\sup_{t\in[0,T]} \mathbb{E}|v_m^{(n)}(t)|^{2a} \leq C_a\, e^{2a(\omega-\eta)T}(1+\sup_{t\geq 0}\mathbb{E}|r_m^{(n)}(t)|_{\mathbf{X}}^{2a\gamma}),$$

so that estimates (6.6) and (6.7) follow letting $T \to \infty$.

□

The next result states that $r_m^{(n)}$ and $v_m^{(n)}$ converge respectively to stochastic processes r and v in $L^p(\Omega, C([0,T];\mathbf{X}))$, $p \geq 1$ and for any $T > 0$.

PROPOSITION 6.4. *There exist a stationary process r and a process v in $L^p(\Omega; C([0,T];\mathbf{X}))$ such that*

$$\lim_{n,m\to\infty} r_m^{(n)}(t) = r(t)$$

$$\lim_{n,m\to\infty} v_m^{(n)}(t) = v(t).$$

Further, for any $p \geq 1$, $\lim_{t\to\infty}\mathbb{E}|v(t)|_{\mathbf{X}}^p = 0$.

PROOF. Without loss of generality, we assume that $p = 2a, a \in \mathbb{N}$. The proof is by contradiction. Assume that there exists $\varepsilon > 0$ such that, for all m and n,

$$\sup_{k,k'>n,j,j'>m} \mathbb{E}|r_j^{(k)}(t) - r_{j'}^{(k')}(t)|_{\mathbf{X}}^{2a} > 2\varepsilon.$$

Since the difference of two stationary processes is stationary, the expression on the left hand side is independent of time t. By choosing t large enough, thus making (by Prop. 6.3) $\mathbb{E}|v_j^{(k)}(t)|_{\mathbf{X}}^{2a}$ and $\mathbb{E}|v_{j'}^{(k')}(t)|_{\mathbf{X}}^{2a}$ sufficiently small, it is easy to show that

$$\sup_{t\geq 0}\sup_{k,k'>n,j,j'>m} \mathbb{E}|u_j^{(k)}(t) - u_{j'}^{(k')}(t)|_{\mathbf{X}}^{2a} > \varepsilon.$$

But this contradicts the fact that the limit $\lim_{n,m\to\infty} u_m^{(n)}(t)$ exists (see Theorem 5.5). As a consequence of the convergence of the sequence $r_m^{(n)}$ and $u_m^{(n)}$ we obtain the convergence of $v_m^{(n)}$.

The final step of the proof is to show that

$$\lim_{t\to\infty}\mathbb{E}|u(t)-r(t)|_{\mathbf{X}}^{2a} = 0.$$

Using a standard inequality we obtain

$$\mathbb{E}|u(t)-r(t)|_{\mathbf{X}}^{2a} \leq c_a\mathbb{E}|u(t)-r_m^{(n)}(t)|_{\mathbf{X}}^{2a}+c_a\mathbb{E}|u_m^{(n)}(t)-r_m^{(n)}(t)|_{\mathbf{X}}^{2a}+c_a\mathbb{E}|r_m^{(n)}(t)-r(t)|_{\mathbf{X}}^{2a},$$

for some strictly positive constant c_a depending only on a. If n, m are large enough, then the first and third terms are less than $\frac{\varepsilon}{c_a}$, uniformly in $t \geq 1$. Hence, taking t

large enough, we have $\mathbb{E}|u_m^{(n)}(t) - r_m^{(n)}(t)|^{2a} < \varepsilon/c_a$. Combining the previous three terms we have shown that $\mathbb{E}|u(t)-r(t)|_{\mathbf{X}}^{2a} < \varepsilon$ for sufficiently large t. This completes the proof of the Proposition 6.4.

□

Now the existence of the invariant measure follows immediately from the results above and the proof is standard. In fact, it is easily proved that the invariant measure is given by the law of the stationary process r. For the reader's convenience, we give the details below.

THEOREM 6.5. *An invariant measure for the process* $(u(t))_{t\geq 0}$ *is given by the law (independent on t)* $p := \mathcal{L}(r(t))$, $t \geq 0$.

PROOF. We recall that any invariant measure μ for the process $(u(t))_{t\geq 0}$ should satisfy the following equality:

$$P(t)^*\mu = \mu, \qquad \text{for all } t \geq 0, \tag{6.8}$$

where $P(t)^*\mu$ is the measure on $\mathbf{X}$ given by

$$P(t)^\star\mu(\Gamma) := \int_{\mathbf{X}} p(t;x,\Gamma)\mu(\mathrm{d}x), \qquad t \geq 0,\ \Gamma \in \mathcal{B}(\mathbf{X})$$

and $p(t;x,\mathrm{d}y)$ is the transition probability density function of $(u(t))_{t\geq 0}$. Hence it suffices to prove that the measure $p := \mathcal{L}(r(t))$ satisfies the equality (6.8).

We notice that, by the stationarity of $(r(t))_{t\geq 0}$; we have $p(\mathrm{d}y) = p(t;x,\mathrm{d}y)$. Then

$$P(t)^\star p(\Gamma) = \int_{\mathbf{X}} p(t;x,\Gamma)p(\mathrm{d}x) = \int_{\mathbf{X}} p(\Gamma)p(\mathrm{d}x) = p(\Gamma),$$

since $p(\mathbf{X}) = 1$. This proves the claim.

□

7. Density of the invariant measure

7.1. On the Yosida approximation of F_η. We devote this section to give an explicit expression for the density of the invariant measure. To this end, we proceed in the following way: first, we show that the mapping $F_\eta := F - \eta I$ (see Remark 4.4) can be represented as the gradient of a real valued differentiable mapping $G_\eta : \mathbf{X} \to \mathbb{R}$. Then we prove that a similar representation holds for the Yosida approximations $\{F_{\eta,\kappa}\}_{\kappa\in\mathbb{N}}$ of F_η. Finally, we obtain the density p of the invariant measure of the system (SDE) as the limit of the densities of the invariant measures corresponding to the approximating problems (5.2).

Let $g_1, \dots, g_m$ be primitive functions of $f_1, \dots, f_m$, i.e. functions such that $g_j' = f_j$, $j = 1, \dots, m$. The first result concerns the mapping F.

LEMMA 7.1. *Set*

$$G(x) := \sum_{j=1}^{m} \int_0^1 g_j(x_j(\xi))\mathrm{d}\xi, \qquad x \in K. \quad G_\eta(x) = \sum_{j=1}^{m} \int_0^1 \left(g_j(x_j(\xi)) - \frac{\eta}{2}x_j^2(\xi)\right) \mathrm{d}\xi. \tag{7.1}$$

Then

(1) *$G : K \to \mathbb{R}$ is a Gâteaux differentiable mapping such that the directional derivative of G at any point x coincides with $F(x)$.*

(2) *$G_\eta : K \to \mathbb{R}$ is a Gâteaux differentiable mapping such that the directional derivative of G_η at any point x coincides with $F_\eta(x)$.*
Moreover, G_η is bounded from above.

PROOF. To prove the result, it suffices to verify that, for any $x \in K$ and $h \in \mathbf{X}$ the following equality holds:

$$\nabla G(x)[h] = \langle F(x), h \rangle,$$

where G is the mapping defined by (7.1). Thus, we compute explicitly the gradient of G. Let $x \in K, h \in \mathbf{X}$. Then, by definition, we have

$$\nabla G(x)[h] = \lim_{\varepsilon \to 0} \sum_{j=1}^{m} \frac{1}{\varepsilon} \int_0^1 g_j(x_j(\xi) + \varepsilon h_j(\xi)) - g_j(x_j(\xi))\mathrm{d}\xi,$$

and, applying Taylor expansion of g_j around $x_j(\xi)$, we get

$$\begin{aligned}\nabla G(x)[h] &= \sum_{j=1}^{m} \int_0^1 f_j(x_j(\xi))h_j(\xi)\mathrm{d}\xi \\ &= \langle F(x), h \rangle_{\mathbf{X}}.\end{aligned}$$

Repeating the same arguments of the proof above, we see that also the mapping F_η can be written in term of the gradient of a real valued function defined on K. In particular, a straightforward calculation leads to $F(x) = \nabla G_\eta(x)$, where G_η is the mapping defined in (7.1), that is

$$G_\eta(x) = \sum_{j=1}^{m} \int_0^1 \left(g_j(x_j(\xi)) - \frac{\eta}{2} x_j^2(\xi) \right) \mathrm{d}\xi \,, x \in K.$$

Now we prove that G_η is bounded from above. We recall that the functions $g_j, j = 1, \ldots, m$ have been defined as having to satisfy the property $g_j' = f_j$, $j = 1, \ldots, m$. We choose

$$g_j(\xi) = \int_0^\xi f_j(t)\mathrm{d}t, \qquad \text{for any } \ \xi \geq 0$$

and

$$g_j(\xi) = -\int_\xi^0 f_j(t)\mathrm{d}t, \qquad \text{for any } \ \xi < 0.$$

Now we claim that, for any $\xi \in \mathbb{R}, j = 1, \ldots, m$, we have $g_j(\xi) - \frac{\eta}{2}\xi^2 \leq g_j(0)$, so that there exists a constant $C > 0$ such that $G_\eta(x) \leq C$ for every $x \in K$. In fact, we notice that by Lagrange's Theorem, for any $t \in [0, \xi]$ we have,

$$f_j(t) = f_j'(r)t$$

where r is a suitable value in $[0, \xi]$. Hence we can write

$$\int_0^\xi f_j(t)\,\mathrm{d}t = \int_0^\xi f_j'(r_j)t\,\mathrm{d}t = \frac{f_j'(r_j)}{2}t^2, \qquad \text{for some } r_j \in [0, \xi] \text{ and every } j = 1, \ldots, m.$$

We recall that, by assumption 1(c) in Hypothesis 2.1, we have

$$\max_{j=1,\ldots,m} \sup_{r \in \mathbb{R}} f_j'(r) \leq \eta. \tag{7.2}$$

Hence using inequality (7.2) we get

$$\int_0^{\xi} f_j(t)\,\mathrm{d}t \leq \frac{\eta}{2}\xi^2.$$

Analogously, when $\xi < 0$

$$\int_{\xi}^{0} -f_j(t)\,\mathrm{d}t = \int_{\xi}^{0} -f_j'(r)\,t\,\mathrm{d}t = \int_0^{-\xi} f_j'(r)t\,\mathrm{d}t \leq \frac{\eta}{2}\xi^2.$$

We conclude that

$$g_j(\xi) - \frac{\eta}{2}\xi^2 \leq 0, \qquad \xi \in \mathbb{R},$$

and, summing up over $j = 1, \ldots, m$, we obtain $G_\eta(x) \leq -0, x \in K$.

□

The second result concerns the Yosida approximations of F_η.

LEMMA 7.2. *For any $\kappa \in \mathbb{N}$, let $F_{\kappa,\eta}$ denote the κ-th Yosida approximation of $F - \eta I$. Then, there exist Gâteaux differentiable maps $G_{\eta,\kappa} : K \to \mathbb{R}$, $\kappa \in \mathbb{N}$ such that, for any $x \in K$, the directional derivative of $G_{\eta,\kappa}$ at x coincides with $F_{\kappa,\eta}$, i.e. $\nabla G_{\eta,\kappa}(x)[h] = \langle F_{\eta,\kappa}(x), h\rangle, h \in \mathbf{X}$. Further, for any $x \in K$, we have*

$$G_{\eta,\kappa}(x) \leq G_{\eta,\ell}(x) \leq G_\eta(x), \qquad \kappa \leq \ell \tag{7.3}$$

and

$$|G_{\eta,\kappa}(x) - G_\eta(x)| \to 0, \qquad \text{as } m \to \infty. \tag{7.4}$$

PROOF. We divide the proof into several steps. The first one concerns the investigation of the Yosida approximations of F_η and the construction of the mappings $G_{\eta,\kappa}, \kappa \in \mathbb{N}$. We recall that, by definition, the Yosida approximations $F_{\eta,\kappa}$ of F_η are defined by

$$F_{\eta,\kappa}(x) = \kappa(J_{\kappa,\eta}(x) - x), \quad x \in \mathbf{X}, \kappa \in \mathbb{N},$$

where

$$J_{\eta,\kappa}(x) = \left(I - \frac{1}{\kappa}F_\eta\right)^{-1}(x), \qquad x \in \mathbf{X}.$$

Let us explore the mappings $J_{\eta,\kappa}$. We define the following functions $h_{\eta,j,\kappa} : \mathbb{R} \to \mathbb{R}$:

$$h_{\eta,j,\kappa}(\xi) := \xi - \frac{1}{\kappa}\left(f_j(\xi) - \eta\xi\right) \tag{7.5}$$

$$= \left(1 + \frac{\eta}{\kappa}\right)\xi - \frac{1}{\kappa}f_j(\xi), \qquad j = 1, \ldots, m \tag{7.6}$$

and we notice that, by Hypothesis 2.1 - (1a),(1c), the functions $h_{\eta,j,\kappa}$ are continuous and differentiable. Moreover, since

$$\sup_{\xi\in\mathbb{R}} h'_{\eta,j,\kappa}(\xi) = \sup_{\xi\in\mathbb{R}}\left(1 + \frac{\eta}{\kappa} - \frac{1}{\kappa}f_j'(\xi)\right) \geq 0$$

we have also that the $h_{\eta,j,\kappa}$ are increasing. Thus $h_{\eta,j,\kappa}$, $j = 1, \ldots, m$, $\kappa \in \mathbb{N}$ admit continuous inverses $h^{-1}_{\eta,j,\kappa}$. We claim that, by construction,

$$J_{\eta,\kappa}(x) = (h^{-1}_{\eta,j,\kappa}(x_j))_{j=1,\ldots,m}, \qquad x \in \mathbf{X}. \tag{7.7}$$

Hence $F_{\eta,\kappa}$ can be written explicitly in terms of the functions $h_{\eta,j,\kappa}$ as:

$$F_{\eta,\kappa}(x) = \kappa \left(h_{\eta,j,\kappa}^{-1}(x_j) - x_j\right)_{j=1,\dots,m}$$

with $F_{\eta,\kappa}(x) \to F(x)$, $x \in \overline{D(F)}$. Further, the functions

$$f_{\eta,j,\kappa}(\xi) := \kappa(h_{\eta,j,\kappa}^{-1}(\xi) - \xi), \quad \xi \in \mathbb{R}, \; j = 1, \dots, m; \; \kappa \in \mathbb{N} \tag{7.8}$$

can be seen as the Yosida approximations of the functions $f_{\eta,j}$ and hence $f_{\eta,j,\kappa} \to f_{\eta,j}$ pointwise in $\mathbb{R}$.

For further use, we notice that $f_{\eta,j,\kappa}$, $j = 1, \dots, m; \kappa \in \mathbb{N}$ can be written in the following, equivalent, forms:

$$f_{\eta,j,\kappa}(t) = f_{\eta,j}(h_{\eta,j,\kappa}^{-1}(t)), \tag{7.9}$$

or

$$f_{\eta,j,\kappa}(t) = \kappa \left(h_{\eta,j,\kappa}^{-1}(t) - h_{\eta,j,\kappa}^{-1}\left(t - \frac{1}{\kappa} f_{\eta,j}(t)\right)\right). \tag{7.10}$$

In fact,

$$h_{\eta,j,\kappa}^{-1}(t) = \left(I - \frac{1}{\kappa} f_{\eta,j}\right)^{-1}(t)$$

or, equivalently,

$$\left(I - \frac{1}{\kappa} f_{\eta,j}\right) \circ h_{\eta,j,\kappa}^{-1}(t) = t,$$

from which it follows that

$$\frac{1}{\kappa} f_{\eta,j}(h_{\eta,j,\kappa}^{-1}(t)) = h_{\eta,j,\kappa}^{-1}(t) - t,$$

and, consequently, we have equality (7.9). Moreover, we can write

$$t - \frac{1}{\kappa} f_{\eta,j}(t) = h_{\eta,j,\kappa}(t),$$

so that, by construction, we obtain

$$\kappa \left(h_{\eta,j,\kappa}^{-1}(t) - h_{\eta,j,\kappa}^{-1}\left(t - \frac{1}{\kappa} f_{\eta,j}(t)\right)\right) = \kappa \left(h_{\eta,j,\kappa}^{-1}(t) - t\right) = f_{\eta,j,\kappa}(t),$$

i.e. equality (7.10).

Now let $g_{\eta,j,\kappa}$ be primitive functions of $f_{\eta,j,\kappa}$, $j = 1, \dots, m$; $\kappa \in \mathbb{N}$ and let $g_{\eta,j}$ be primitive functions of $f_{\eta,j}$. In particular we choose

$$g_{\eta,j,\kappa}(t) := \int_0^t f_{\eta,j,\kappa}(\xi) \mathrm{d}\xi, \qquad t \geq 0$$

$$g_{\eta,j}(t) := \int_0^t f_{\eta,j}(\xi) \mathrm{d}\xi, \qquad t \geq 0$$

and

$$g_{\eta,j,\kappa}(t) := -\int_t^0 f_{\eta,j,\kappa}(\xi) \mathrm{d}\xi, \qquad t \leq 0$$

$$g_{\eta,j}(t) := -\int_t^0 f_{\eta,j}(\xi) \mathrm{d}\xi, \qquad t \leq 0,$$

so that $g_{\eta,j,\kappa} \to g_{\eta,j}$, as $\kappa \to \infty$. We define the mapping $G_{\eta,\kappa}$ by

$$G_{\eta,\kappa}(x) := \sum_{j=1,\dots,m} \int_0^1 g_{\eta,j,\kappa}(x_j(\xi))\,\mathrm{d}\xi, \quad x = (x_j)_{j=1,\dots,m} \in K, \ \kappa \in \mathbb{N}$$

Reasoning as in Lemma 7.1, it can be proved that for any $\kappa \in \mathbb{N}$ the Gâteaux derivative of $G_{\eta,\kappa}$ coincides with the mapping $F_{\eta,\kappa}$ and, similarly, the Gâteaux derivative of G_η coincides with F_η.

The second step is to prove estimate (7.3). To this end, we recall that

$$G_\eta(x) = \sum_{j=1,\dots,m} \int_0^1 \left(g_j(x_j(\xi)) - \frac{\eta}{2} x_j^2(\xi) \right) \mathrm{d}\xi = \sum_{j=1,\dots,m} \int_0^1 g_{\eta,j}(x_j(\xi))\,\mathrm{d}\xi.$$

We suppose, for simplicity, that $f_j(0) = 0, j = 1, \dots, m$. Since, by assumption 1(c) in Hypothesis 2.1, $f_j(t) - \eta t$ is decreasing, the above constraint implies that

$$\begin{cases} f_j(t) - \eta t > 0 & \text{for } t < 0 \text{ and} \\ f_j(t) - \eta t \leq 0 & \text{for } t \geq 0 \end{cases}.$$

The crucial point here is to prove that, for any $j = 1, \dots, m$ and $\kappa \in \mathbb{N}$, the functions $f_{\eta,j,\kappa}$ are decreasing and the following inequalities hold:

$$\begin{cases} 0 \leq f_{\eta,j,\kappa} < f_{\eta,j} & \text{on } \mathbb{R}_+ \\ f_{\eta,j} < f_{\eta,j,\kappa} < 0, & \text{on } \mathbb{R}^-. \end{cases} \tag{7.11}$$

In fact, let us suppose that (7.11) holds. Then, by integration over $[0, t]$, $t \geq 0$ and, similarly, by integration over $[-t, 0], t \geq 0$ we have

$$g_{\eta,j,\kappa}(t) \leq g_{\eta,j}(t).$$

Obviously, the above inequality implies estimate (7.3).

It remains to prove (7.11). As noted in (7.9) $f_{\eta,j,\kappa}$ can be written as

$$f_{\eta,j,\kappa}(t) = f_{\eta,j}(h_{\eta,j,\kappa}^{-1}(t)),$$

where $h_{\eta,j,\kappa}$ are the functions defined in (7.5). Now we notice that, since $f_{\eta,j}$ is decreasing,

$$f_{\eta,j,\kappa}(t) - f_{\eta,j,\kappa}(r) = f_{\eta,j}(h_{\eta,j,\kappa}^{-1}(t)) - f_{\eta,j}(h_{\eta,j,\kappa}^{-1}(r)) \leq 0$$

if and only if $h_{\eta,j,\kappa}^{-1}(t) - h_{\eta,j,\kappa}^{-1}(r) \geq 0$. But this happens if and only if $t \geq r$ (since $h_{\eta,j,\kappa}^{-1}$ are increasing). Hence,

$$f_{\eta,j,\kappa}(t) - f_{\eta,j,\kappa}(r) \leq 0 \quad \text{if and only if} \quad t \geq r,$$

that is, $f_{\eta,j,\kappa}$ is decreasing. Moreover, recalling equality (7.10) and recalling that, for any $j = 1, \dots, m;\ \kappa \in \mathbb{N}$, we have

$$f_{\eta,j,\kappa}(t) \leq t - \left(t - \frac{1}{\kappa} f_{\eta,j}(t) \right) = \frac{1}{\kappa} f_{\eta,j}(t)$$

if and only if $f_{\eta,j}(t) \geq 0$, that is, $t \geq 0$. Moreover, in this case, $0 = f_{\eta,j,\kappa}(0) \leq f_{\eta,j,\kappa}(t)$. Analogously,

$$f_{\eta,j,\kappa}(t) \geq t - \left(t - \frac{1}{\kappa} f_{\eta,j}(t) \right) = \frac{1}{\kappa} f_{\eta,j}(t)$$

if and only if $f_{\eta,j}(t) \leq 0$, that is, $t \leq 0$. Moreover, in this case, $0 = f_{\eta,j,\kappa}(0) \geq f_{\eta,j,\kappa}(t)$. Summing up the above claims, we obtain (7.11).

Moreover, we notice that, for any $t \in \mathbb{R}$, the sequences $\{f_{\eta,j,\kappa}(t)\}_{\kappa\in\mathbb{N}}$ are monotone. In particular,

$$\begin{cases} 0 \le f_{\eta,j,\kappa}(t) \le f_{\eta,j,\ell}(t) \le f_{\eta,j}(t), & \text{if } t \le 0, \kappa \le \ell; \\ f_{\eta,j}(t) \le f_{\eta,j,\ell}(t) \le f_{\eta,j,\kappa}(t) \le 0, & \text{if } t \ge 0, \kappa \le \ell. \end{cases} \tag{7.12}$$

The above inequalities follows from the fact that, for any $t \le 0$, $j = 1, \dots, m$ and $\kappa, \ell \in \mathbb{N}$ such that $\kappa \le \ell$, the quantities $h^{-1}_{\eta,j,\kappa}(t)$ and $h^{-1}_{\eta,j,\ell}(t)$ satisfy

$$h^{-1}_{\eta,j,\kappa}(t) \ge h^{-1}_{\eta,j,\ell}(t).$$

Then, recalling equality (7.9) and the monotonicity of $f_{\eta,j}$ we get

$$f_{\eta,j,\kappa}(t) = f_{\eta,j}(h^{-1}_{\eta,j,\kappa}(t)) \le f_{\eta,j}(h^{-1}_{\eta,j,\ell}(t)) = f_{\eta,j,\ell}(t), \qquad t \le 0 \ \text{ and } \ \kappa \le \ell.$$

Similarly, we obtain the second expression in (7.12). A consequence of (7.12) is that, for any $j = 1, \dots, m$, the sequence $\{g_{\eta,j,\kappa}\}_{\kappa\in\mathbb{N}}$ is monotone increasing (and bounded from above). Then, by an application of the dominated convergence theorem, we conclude that $G_{\eta,\kappa}(x) \to G_\eta(x)$ as $\kappa \to \infty$, for any $x \in K$.

This concludes the proof.

□

7.2. The main result. We consider the following abstract linear stochastic differential equation:

$$\begin{cases} \mathrm{d}Z(t) = A_\eta Z(t)\mathrm{d}t + \sqrt{Q}\mathrm{d}W(t), & t \ge 0 \\ Z(0) = x \in K, \end{cases} \tag{7.13}$$

where A_η, Q and K have been defined respectively in (3.1),(3.2) and in (4.1). We know that under our assumptions (see 2.1), for any $x \in K$ equation (7.13) has a unique mild solution $Z(t;x)$ (cf., for example, [**29**, 5.3.1]). In particular, we have that $Z(t;x) = e^{tA_\eta}x + W_{A_\eta}(t)$. Moreover, arguing as in Sec. 4, this solution admits a K-continuous version (which we still denote by Z) provided that $x \in K$. We shall denote by $R(t), t \ge 0$ the corresponding transition semigroup, that is

$$R(t)\phi(x) := \mathbb{E}[\phi(Z(t;x))], \quad t \ge 0, x \in K, \phi \in C_b(K).$$

By [**29**, Theorem 6.2.1], there exists a unique invariant measure μ_0 for the semigroup $R(t)$, being a Gaussian measure $\mu_0 \equiv \mathcal{N}(0, Q_\infty)$ concentrated on K, with mean vector 0 and covariance operator Q_∞:

$$Q_\infty x := \int_0^\infty e^{tA_{K,\eta}} Q e^{tA^\star_{K,\eta}} x \mathrm{d}s, \qquad x \in E,$$

where $e^{tA_{K,\eta}}$ denotes the semigroup generated by A_K. It turns out that, under Hypothesis 2.1-1b, there exists an invariant measure μ_{F_η} for (SDE) on K that is absolutely continuous with respect to the measure μ_0, which in the present situation is of the form

$$\mu_0 = \mathcal{N}(0, Q_\infty) = \mathcal{N}\left(0, -\frac{1}{2}(A_\eta Q)^{-1}\right).$$

In particular, we are going to show that the invariant measure μ_{F_η} corresponding to the above problem is described by the following formula:

(7.14)

$$\mu_{F_\eta}(\mathrm{d}x) = c_\eta e^{2G_\eta(x)}\mu_0(\mathrm{d}x), \qquad x \in D(F), \text{ with normalizing constant } c_\eta > 0.$$

Thus, the main result of this section reads as follows.

THEOREM 7.3. *Let $(u_m^n(t))_{t\geq 0}$ be the solution of the stochastic differential equation (5.1) (i.e. (Enm)). Then the invariant measure corresponding to $(u(t))_{t\geq 0}$ is the distribution described by (7.14). Moreover, the corresponding semigroup is μ_0-symmetric, i.e.*

$$\int_K \phi(x)P(t)\psi(x)\mu_{F_\eta}(\mathrm{d}x) = \int_K \psi(x)P(t)\phi(x)\mu_{F_\eta}(\mathrm{d}x), \tag{7.15}$$

for all $\phi, \psi \in B_b(D(F))$.

PROOF. It is enough to prove the identity (7.15), from which, in particular, we get

$$\begin{aligned}(P(t)^\star\mu_{F_\eta})(\Gamma) &= \int_K \mathbf{1}_K(x)P(t)\mathbf{1}_\Gamma(x)\mu_{F_\eta}(\mathrm{d}x)\\ &= \int_K P(t)\mathbf{1}_E(x)\mathbf{1}_\Gamma(x)\mu_{F_\eta}(\mathrm{d}x) = \mu_{F_\eta}(\Gamma) \qquad \Gamma\in\mathcal{B}(K),\end{aligned}$$

and, consequently, that μ is the invariant measure for $(u(t))_{t\geq 0}$.

For any $\kappa, n \in \mathbb{N}$, we consider the transition semigroup $P^{\kappa,n}(t)$ corresponding to the solution $u_\kappa^{(n)}$ of the equation

$$\begin{cases}\mathrm{d}u_\kappa^{(n)}(t) = A_\eta u_\kappa^{(n)}(t) + \Pi_n F_{\eta,\kappa}(\Pi_n u_\kappa^{(n)}(t))\mathrm{d}t + \Pi_n\sqrt{Q}\mathrm{d}W(t), \qquad t\geq 0.\\ u_\kappa^{(n)}(0) = P_n x.\end{cases}$$

By the finite dimensional theory the Markovian semigroup $P^{\kappa,n}$ is symmetric with respect to the measure $\mu^{(n)}_{F_{\eta,\kappa}}$:

$$\mu^{(n)}_{F_{\eta,\kappa}}(\mathrm{d}x) = C_\kappa^{(n)} e^{G_{\eta,\kappa}(\Pi_n x)}\mu_0^{(n)}(dx), \qquad n\in\mathbb{N},$$

($C_\kappa^{(n)}$ being the normalizing constants and $\mu_0^{(n)}$ the images of the projections P_n, $n\in\mathbb{N}$, see Section 5). Consequently, for arbitrary bounded and continuous functions ϕ and ψ on K as $n\to\infty$:

$$\int_K \phi(x)P^{\kappa,n}(t)\psi(x)e^{G_{\eta,\kappa}(\Pi_n)}\mu_0^{(n)}(\mathrm{d}x) = \int_K \psi(x)P^{\kappa,n}(t)\phi(x)e^{G_{\eta,\kappa}(\Pi_n)}\mu_0^{(n)}(\mathrm{d}x) \tag{7.16}$$

for any $t\geq 0, n\in\mathbb{N}$. Now notice that

$$\int_K \phi(x)P^{\kappa,n}(t)\psi(x)e^{G_{\eta,\kappa}(\Pi_n x)}\mu_0^{(n)}(\mathrm{d}x) = \int_K \psi(P_n x)P^{\kappa,n}(t)\phi(P_n x)e^{G_{\eta,\kappa}(\Pi_n P_n x)}\mu_0(\mathrm{d}x). \tag{7.17}$$

We also observe also that $\Pi_n x\to x$ in K for every $x\in K$ and that $P_n x\to x$ for μ_0-almost all $x\in K$. For the former fact we refer to Remark 5.1. The latter follows from the following fact: the random variables $Z^{(n)}\in L^2(\mu_0, K)$ given by

$$Z^{(n)}(x) = \sqrt{2\alpha_n}\langle x, \mathbf{e}_j\rangle, \qquad x\in K, \text{ (with } \alpha_n \text{ as in Sec. 4, Hyp. 4.5)}$$

are Gaussian, normalized and independent and therefore the sequence

$$\sum_{j=1}^n Z^{(n)}(x)\frac{\mathbf{e}_n}{\sqrt{2\alpha_n}}, \qquad x\in K, n\in\mathbb{N}$$

converges μ_0-almost surely in K norm to a random variable ζ such that $\mathcal{L}(\zeta) = \mu_0$, see G. Da Prato and J. Zabczyk [**29**, Section 2.2.3]. Since K is continuously embedded in $\mathbf{X}$, the limit has to be $\zeta(x) = x$. Moreover one can check, see J. Zabczyk [**77**, Theorem 4], that for any sequence $\{x_n\} \subset K$ convergent to x in K and for any function $\psi \in C_b(K)$ one has

$$\lim_{n\to\infty} P^{\kappa,n}(t)\psi(x_n) = P^{\kappa}(t)\psi(x). \tag{7.18}$$

Letting $n \to \infty$ in (7.16) and (7.17) we obtain, for any $\phi, \psi \in C_b(K)$

$$\int_K \phi(x)P^{\kappa}(t)\psi(x)e^{G_{\eta,\kappa}(x)}\mu_0(\mathrm{d}x) = \int_K \psi(x)P^{\kappa}(t)\phi(x)e^{G_{\eta,\kappa}(x)}\mu_0(\mathrm{d}x).$$

Similarly, we can see that $P^{\kappa}(t)\phi \to P(t)\phi$, as $n \to \infty$, uniformly on compact sets. The crucial point is to prove that $u_m(t; P_n x) \to u(t;x)$, almost surely and uniformly on every interval $[0,T]$. But this was done in the proof of Proposition 5.5 (see also Remark 5.6). Consequently, letting $\kappa \to \infty$ we conclude that $P(t)$ is μ_0 symmetric and equality (7.14) holds.

□

Acknowledgments

The first author first met Fritz in Bielefeld in the 70's, during Fritz' PhD studies. He was particularly struck by Fritz' sharpness of judgement, passionate devotion to mathematics and physics, his determination to understand and clarify the deep relations underlying their wonderful interactions. The care and precision of Fritz' approach to problems on the cutting edge of physics and mathematics, joined with his great technical skills and his straight and at the same time charming character makes him an ideal partner for cooperation. The first author would like to thank him for the joy of collaboration and friendship over so many years. All authors would like to congratulate Fritz most warmly for his 60th birthday, express him their sincere admiration and wish him many more years of good health and many satisfactions in all his endeavours.

The authors are very grateful to several persons and Institutions for warm hospitality, stimulating discussions and financial support at various stages of this work: Prof. S. Bonaccorsi and Prof. L. Tubaro, Department of Mathematics, and Prof. M. Andreatta and Prof. F. Catanese, CIRM, University of Trento; NEST-project of the Provincia Autonoma di Trento; Prof. T. Ratiu, Centre Interfacultaire Bernoulli, EPFL and Profs. R. Dalang, M. Dozzi, F. Flandoli and F. Russo for invitations to the semester Programme SPDE's at that Institute.

References

[1] Sergio Albeverio, *Theory of Dirichlet forms and applications*, Lectures on probability theory and statistics (Saint-Flour, 2000), Lecture Notes in Math., vol. 1816, Springer, Berlin, 2003, pp. 1–106, DOI 10.1007/3-540-44922-1. MR2009816 (2005a:60120)

[2] Sergio Albeverio, Zdzisław Brzeźniak, and Jiang-Lun Wu, *Existence of global solutions and invariant measures for stochastic differential equations driven by Poisson type noise with non-Lipschitz coefficients*, J. Math. Anal. Appl. **371** (2010), no. 1, 309–322, DOI 10.1016/j.jmaa.2010.05.039. MR2661009 (2011j:60180)

[3] Sergio Albeverio and Christof Cebulla, *Synchronizability of stochastic network ensembles in a model of interacting dynamical units*, Phys. A **386** (2007), no. 1, 503–512, DOI 10.1016/j.physa.2007.07.036. MR2585516 (2010j:92003)

[4] Sergio Albeverio, Claudio Cacciapuoti, and Domenico Finco, *Coupling in the singular limit of thin quantum waveguides*, J. Math. Phys. **48** (2007), no. 3, 032103, 21, DOI 10.1063/1.2710197. MR2314483 (2008a:81061)

[5] Albeverio, S. and Di Persio, L., Some stochastic dynamical models in neurobiology: recent developments, *European Communications in Mathematical and Theoretical Biology*, **14**, (2011), pp. 44–53.

[6] Sergio Albeverio, Luca Di Persio, and Elisa Mastrogiacomo, *Small noise asymptotic expansions for stochastic PDE's, I. The case of a dissipative polynomially bounded nonlinearity*, Tohoku Math. J. (2) **63** (2011), no. 4, 877–898, DOI 10.2748/tmj/1325886292. MR2872967

[7] Sergio Albeverio, Vadim Fatalov, and Vladimir I. Piterbarg, *Asymptotic behavior of the sample mean of a function of the Wiener process and the Macdonald function*, J. Math. Sci. Univ. Tokyo **16** (2009), no. 1, 55–93. MR2548933 (2011b:60319)

[8] Sergio Albeverio, Franco Flandoli, and Yakov G. Sinai, *SPDE in hydrodynamic: recent progress and prospects*, Lecture Notes in Mathematics, vol. 1942, Springer-Verlag, Berlin, 2008. Lectures given at the C.I.M.E. Summer School held in Cetraro, August 29–September 3, 2005; Edited by Giuseppe Da Prato and Michael Röckner. MR2459087 (2009g:76120)

[9] Sergio Albeverio and Song Liang, *Asymptotic expansions for the Laplace approximations of sums of Banach space-valued random variables*, Ann. Probab. **33** (2005), no. 1, 300–336, DOI 10.1214/009117904000001017. MR2118867 (2005m:62044)

[10] Sergio Albeverio, *Wiener and Feynman—path integrals and their applications*, Proceedings of the Norbert Wiener Centenary Congress, 1994 (East Lansing, MI, 1994), Proc. Sympos. Appl. Math., vol. 52, Amer. Math. Soc., Providence, RI, 1997, pp. 163–194. MR1440913 (98h:58033)

[11] Sergio Albeverio, Eugene Lytvynov, and Andrea Mahnig, *A model of the term structure of interest rates based on Lévy fields*, Stochastic Process. Appl. **114** (2004), no. 2, 251–263, DOI 10.1016/j.spa.2004.06.006. MR2101243 (2005h:60178)

[12] S. Albeverio, V. Mandrekar, and B. Rüdigér, *Existence of mild solutions for stochastic differential equations and semilinear equations with non-Gaussian Lévy noise*, Stochastic Process. Appl. **119** (2009), no. 3, 835–863, DOI 10.1016/j.spa.2008.03.006. MR2499860 (2010e:60113)

[13] Sergio Albeverio and Sonia Mazzucchi, *The trace formula for the heat semigroup with polynomial potential*, Seminar on Stochastic Analysis, Random Fields and Applications VI, Progr. Probab., vol. 63, Birkhäuser/Springer Basel AG, Basel, 2011, pp. 3–21, Edited by R. Dalang, M. Dozzi, F. Russo, DOI 10.1007/978-3-0348-0021-1_1. MR2857015 (2012j:60177)

[14] S. Albeverio and M. Röckner, *Stochastic differential equations in infinite dimensions: solutions via Dirichlet forms*, Probab. Theory Related Fields **89** (1991), no. 3, 347–386, DOI 10.1007/BF01198791. MR1113223 (92k:60123)

[15] Sergio Albeverio, Haio Röckle, and Victoria Steblovskaya, *Asymptotic expansions for Ornstein-Uhlenbeck semigroups perturbed by potentials over Banach spaces*, Stochastics Stochastics Rep. **69** (2000), no. 3-4, 195–238. MR1760977 (2001a:60064)

[16] S. Albeverio and V. Steblovskaya, *Asymptotics of infinite-dimensional integrals with respect to smooth measures. I*, Infin. Dimens. Anal. Quantum Probab. Relat. Top. **2** (1999), no. 4, 529–556, DOI 10.1142/S021902579900031X. MR1810812 (2002b:46075)

[17] Viorel Barbu and Giuseppe Da Prato, *Invariant measures and the Kolmogorov equation for the stochastic fast diffusion equation*, Stochastic Process. Appl. **120** (2010), no. 7, 1247–1266, DOI 10.1016/j.spa.2010.03.007. MR2639746 (2011g:60105)

[18] Stefano Bonaccorsi and Elisa Mastrogiacomo, *Analysis of the stochastic FitzHugh-Nagumo system*, Infin. Dimens. Anal. Quantum Probab. Relat. Top. **11** (2008), no. 3, 427–446, DOI 10.1142/S0219025708003191. MR2446518 (2009f:35174)

[19] Stefano Bonaccorsi, Carlo Marinelli, and Giacomo Ziglio, *Stochastic FitzHugh-Nagumo equations on networks with impulsive noise*, Electron. J. Probab. **13** (2008), no. 49, 1362–1379, DOI 10.1214/EJP.v13-532. MR2438810 (2010d:60143)

[20] Stefano Bonaccorsi and Delio Mugnolo, *Existence of strong solutions for neuronal network dynamics driven by fractional Brownian motions*, Stoch. Dyn. **10** (2010), no. 3, 441–464, DOI 10.1142/S0219493710003030. MR2671386 (2011d:92022)

[21] Zdzisław Brzeźniak and Dariusz Gątarek, *Martingale solutions and invariant measures for stochastic evolution equations in Banach spaces*, Stochastic Process. Appl. **84** (1999), no. 2, 187–225, DOI 10.1016/S0304-4149(99)00034-4. MR1719282 (2000m:60076)

[22] Claudio Cacciapuoti and Pavel Exner, *Nontrivial edge coupling from a Dirichlet network squeezing: the case of a bent waveguide*, J. Phys. A **40** (2007), no. 26, F511–F523, DOI 10.1088/1751-8113/40/26/F02. MR2344437 (2008f:81101)

[23] Sandra Cerrai, *Differentiability of Markov semigroups for stochastic reaction-diffusion equations and applications to control*, Stochastic Process. Appl. **83** (1999), no. 1, 15–37, DOI 10.1016/S0304-4149(99)00014-9. MR1705598 (2000g:60105)

[24] Sandra Cerrai and Mark Freidlin, *Smoluchowski-Kramers approximation for a general class of SPDEs*, J. Evol. Equ. **6** (2006), no. 4, 657–689, DOI 10.1007/s00028-006-0281-8. MR2267703 (2007k:60216)

[25] Stefano Cardanobile and Delio Mugnolo, *Analysis of a FitzHugh-Nagumo-Rall model of a neuronal network*, Math. Methods Appl. Sci. **30** (2007), no. 18, 2281–2308, DOI 10.1002/mma.880. MR2362954 (2008k:92009)

[26] Giuseppe Da Prato and Arnaud Debussche, *Asymptotic behavior of stochastic PDEs with random coefficients*, Discrete Contin. Dyn. Syst. **27** (2010), no. 4, 1553–1570, DOI 10.3934/dcds.2010.27.1553. MR2629537 (2011d:35531)

[27] Giuseppe Da Prato and Luciano Tubaro, *Self-adjointness of some infinite-dimensional elliptic operators and application to stochastic quantization*, Probab. Theory Related Fields **118** (2000), no. 1, 131–145, DOI 10.1007/PL00008739. MR1785456 (2002a:47032)

[28] Giuseppe Da Prato and Jerzy Zabczyk, *Stochastic equations in infinite dimensions*, Encyclopedia of Mathematics and its Applications, vol. 44, Cambridge University Press, Cambridge, 1992. MR1207136 (95g:60073)

[29] G. Da Prato and J. Zabczyk, *Ergodicity for infinite-dimensional systems*, London Mathematical Society Lecture Note Series, vol. 229, Cambridge University Press, Cambridge, 1996. MR1417491 (97k:60165)

[30] Jean-Dominique Deuschel and Daniel W. Stroock, *Large deviations*, Pure and Applied Mathematics, vol. 137, Academic Press Inc., Boston, MA, 1989. MR997938 (90h:60026)

[31] Amir Dembo and Ofer Zeitouni, *Large deviations techniques and applications*, 2nd ed., Applications of Mathematics (New York), vol. 38, Springer-Verlag, New York, 1998. MR1619036 (99d:60030)

[32] Jianfeng Feng, Mariya Shcherbina, and Brunello Tirozzi, *Stability of the dynamics of an asymmetric neural network*, Commun. Pure Appl. Anal. **8** (2009), no. 2, 655–671, DOI 10.3934/cpaa.2009.8.655. MR2461567 (2009m:62292)

[33] Benedetta Ferrario, *Invariant measures for a stochastic Kuramoto-Sivashinsky equation*, Stoch. Anal. Appl. **26** (2008), no. 2, 379–407, DOI 10.1080/07362990701857335. MR2399743 (2009j:60138)

[34] C. Rocşoreanu, A. Georgescu, and N. Giurgiţeanu, *The FitzHugh-Nagumo model*, Mathematical Modelling: Theory and Applications, vol. 10, Kluwer Academic Publishers, Dordrecht, 2000. Bifurcation and dynamics. MR1779040 (2002a:34059)

[35] Barbara Forster, Eva Lütkebohmert, and Josef Teichmann, *Absolutely continuous laws of jump-diffusions in finite and infinite dimensions with applications to mathematical finance*, SIAM J. Math. Anal. **40** (2008/09), no. 5, 2132–2153, DOI 10.1137/070708822. MR2471915 (2010g:60132)

[36] Masatoshi Fukushima, Yoichi Oshima, and Masayoshi Takeda, *Dirichlet forms and symmetric Markov processes*, Second revised and extended edition, de Gruyter Studies in Mathematics, vol. 19, Walter de Gruyter & Co., Berlin, 2011. MR2778606 (2011k:60249)

[37] Leszek Gawarecki and Vidyadhar Mandrekar, *Stochastic differential equations in infinite dimensions with applications to stochastic partial differential equations*, Probability and its Applications (New York), Springer, Heidelberg, 2011. MR2560625 (2012a:60179)

[38] Helge Holden, Bernt Øksendal, Jan Ubøe, and Tusheng Zhang, *Stochastic partial differential equations*, 2nd ed., Universitext, Springer, New York, 2010. A modeling, white noise functional approach. MR2571742 (2011c:60201)

[39] Hodgkin, A. and Huxley, A. (1952) A quantitative description of membrane current and its application to conduction and excitation in nerve. *J. Physiol.* 117:500–544.

[40] G. Kallianpur and R. Wolpert, *Infinite-dimensional stochastic differential equation models for spatially distributed neurons*, Appl. Math. Optim. **12** (1984), no. 2, 125–172, DOI 10.1007/BF01449039. MR764813 (86f:60074)

[41] Keener, J. and Sneyd, J., Mathematical physiology. Second edition. Interdisciplinary Applied Mathematics, 8/I. Springer, New York, 2009.

[42] Yuzuru Inahama and Hiroshi Kawabi, *Asymptotic expansions for the Laplace approximations for Itô functionals of Brownian rough paths*, J. Funct. Anal. **243** (2007), no. 1, 270–322, DOI 10.1016/j.jfa.2006.09.016. MR2291439 (2008b:60130)

[43] Yuzuru Inahama and Hiroshi Kawabi, *On the Laplace-type asymptotics and the stochastic Taylor expansion for Itô functionals of Brownian rough paths*, Proceedings of RIMS Workshop on Stochastic Analysis and Applications, RIMS Kôkyûroku Bessatsu, B6, Res. Inst. Math. Sci. (RIMS), Kyoto, 2008, pp. 139–152. MR2407560 (2009j:60123)

[44] Nobuyuki Ikeda and Shinzo Watanabe, *Stochastic differential equations and diffusion processes*, 2nd ed., North-Holland Mathematical Library, vol. 24, North-Holland Publishing Co., Amsterdam, 1989. MR1011252 (90m:60069)

[45] G. Jona-Lasinio and P. K. Mitter, *Large deviation estimates in the stochastic quantization of ϕ^4_2*, Comm. Math. Phys. **130** (1990), no. 1, 111–121. MR1055688 (91i:81094)

[46] G. Jona-Lasinio and P. K. Mitter, *On the stochastic quantization of field theory*, Comm. Math. Phys. **101** (1985), no. 3, 409–436. MR815192 (87j:81246)

[47] A. N. Kolmogorov and S. V. Fomin, *Elements of the theory of functions and functional analysis. Vol. 2: Measure. The Lebesgue integral. Hilbert space*, Translated from the first (1960) Russian ed. by Hyman Kamel and Horace Komm, Graylock Press, Albany, N.Y., 1961. MR0118796 (22 #9566a)

[48] Lehnertz, K. , Arnhold, J. , Grassberger, P. and Elger, C.E., Chaos in Brain ?, World Scientific, Singapore, (2000).

[49] G. E. Ladas and V. Lakshmikantham, *Differential equations in abstract spaces*, Academic Press, New York, 1972. Mathematics in Science and Engineering, Vol. 85. MR0460832 (57 #824)

[50] Zhi Ming Ma and Michael Röckner, *Introduction to the theory of (nonsymmetric) Dirichlet forms*, Universitext, Springer-Verlag, Berlin, 1992. MR1214375 (94d:60119)

[51] Robert Marcus, *Parabolic Itô equations*, Trans. Amer. Math. Soc. **198** (1974), 177–190. MR0346909 (49 #11630)

[52] Robert Marcus, *Parabolic Itô equations with monotone nonlinearities*, J. Funct. Anal. **29** (1978), no. 3, 275–286, DOI 10.1016/0022-1236(78)90031-9. MR512245 (81h:60085)

[53] Paul Malliavin and Setsuo Taniguchi, *Analytic functions, Cauchy formula, and stationary phase on a real abstract Wiener space*, J. Funct. Anal. **143** (1997), no. 2, 470–528, DOI 10.1006/jfan.1996.2989. MR1428825 (98e:60088)

[54] Carlo Marinelli and Michael Röckner, *On uniqueness of mild solutions for dissipative stochastic evolution equations*, Infin. Dimens. Anal. Quantum Probab. Relat. Top. **13** (2010), no. 3, 363–376, DOI 10.1142/S0219025710004152. MR2729590 (2011k:60220)

[55] Delio Mugnolo, *Gaussian estimates for a heat equation on a network*, Netw. Heterog. Media **2** (2007), no. 1, 55–79, DOI 10.3934/nhm.2007.2.55. MR2291812 (2008b:35113)

[56] Delio Mugnolo, *A variational approach to strongly damped wave equations*, Functional analysis and evolution equations, Birkhäuser, Basel, 2008, pp. 503–514, DOI 10.1007/978-3-7643-7794-6_30. MR2402747 (2009c:34123)

[57] Delio Mugnolo and Silvia Romanelli, *Dynamic and generalized Wentzell node conditions for network equations*, Math. Methods Appl. Sci. **30** (2007), no. 6, 681–706, DOI 10.1002/mma.805. MR2301840 (2007k:34197)

[58] El Maati Ouhabaz, *Analysis of heat equations on domains*, London Mathematical Society Monographs Series, vol. 31, Princeton University Press, Princeton, NJ, 2005. MR2124040 (2005m:35001)

[59] A. Pazy, *Semigroups of linear operators and applications to partial differential equations*, Applied Mathematical Sciences, vol. 44, Springer-Verlag, New York, 1983. MR710486 (85g:47061)

[60] Giorgio Parisi, *Statistical field theory*, Frontiers in Physics, vol. 66, Benjamin/Cummings Publishing Co. Inc. Advanced Book Program, Reading, MA, 1988. With a foreword by David Pines. MR1007980 (91f:81116)

[61] S. Peszat and J. Zabczyk, *Stochastic partial differential equations with Lévy noise*, Encyclopedia of Mathematics and its Applications, vol. 113, Cambridge University Press, Cambridge, 2007. An evolution equation approach. MR2356959 (2009b:60200)

[62] Claudia Prévôt and Michael Röckner, *A concise course on stochastic partial differential equations*, Lecture Notes in Mathematics, vol. 1905, Springer, Berlin, 2007. MR2329435 (2009a:60069)

[63] RALL, W., Branching dendritic trees and motoneurone membrane resistivity, *Exp. Neurol.* **1**, 491–527, 1959.

[64] RALL, W., Membrane potential transients and membrane time constant of motoneurons, *Exp. Neurol.*, **2**, 503–532, 1960.

[65] Luigi M. Ricciardi, *Diffusion processes and related topics in biology*, Springer-Verlag, Berlin, 1977. Notes taken by Charles E. Smith; Lecture Notes in Biomathematics, Vol. 14. MR0526557 (58 #26143)

[66] Carles Rovira and Samy Tindel, *Sharp Laplace asymptotics for a parabolic SPDE*, Stochastics Stochastics Rep. **69** (2000), no. 1-2, 11–30. MR1751716 (2000m:60028)

[67] M. Schilder, *Some asymptotic formulas for Wiener integrals*, Trans. Amer. Math. Soc. **125** (1966), 63–85. MR0201892 (34 #1770)

[68] Ichirō Shigekawa, *Existence of invariant measures of diffusions on an abstract Wiener space*, Osaka J. Math. **24** (1987), no. 1, 37–59. MR881745 (88f:28012)

[69] Barry Simon, *Functional integration and quantum physics*, 2nd ed., AMS Chelsea Publishing, Providence, RI, 2005. MR2105995 (2005f:81003)

[70] Luc Tartar, *An introduction to Sobolev spaces and interpolation spaces*, Lecture Notes of the Unione Matematica Italiana, vol. 3, Springer, Berlin, 2007. MR2328004 (2008g:46055)

[71] Henry C. Tuckwell, *Analytical and simulation results for the stochastic spatial FitzHugh-Nagumo model neuron*, Neural Comput. **20** (2008), no. 12, 3003–3033, DOI 10.1162/neco.2008.08-07-585. MR2467646 (2009m:92026)

[72] Henry C. Tuckwell, *Introduction to theoretical neurobiology. Vol. 1*, Cambridge Studies in Mathematical Biology, vol. 8, Cambridge University Press, Cambridge, 1988. Linear cable theory and dendritic structure. MR947344 (90a:92003a)

[73] Henry C. Tuckwell, *Introduction to theoretical neurobiology. Vol. 2*, Cambridge Studies in Mathematical Biology, vol. 8, Cambridge University Press, Cambridge, 1988. Nonlinear and stochastic theories. MR947345 (90a:92003b)

[74] Henry C. Tuckwell, *Random perturbations of the reduced FitzHugh-Nagumo equation*, Phys. Scripta **46** (1992), no. 6, 481–484, DOI 10.1088/0031-8949/46/6/001. MR1192067 (94d:92006)

[75] R. von Vintschger, *The existence of invariant measures for $C[0,1]$-valued diffusions*, Probab. Theory Related Fields **82** (1989), no. 2, 307–313, DOI 10.1007/BF00354766. MR998937 (90j:60077)

[76] Shinzo Watanabe, *Analysis of Wiener functionals (Malliavin calculus) and its applications to heat kernels*, Ann. Probab. **15** (1987), no. 1, 1–39. MR877589 (88h:60111)

[77] J. Zabczyk, *Symmetric solutions of semilinear stochastic equations*, Stochastic partial differential equations and applications, II (Trento, 1988), Lecture Notes in Math., vol. 1390, Springer, Berlin, 1989, pp. 237–256, DOI 10.1007/BFb0083952. MR1019609 (91a:60161)

DEPARTMENT OF APPLIED MATHEMATICS, UNIVERSITY OF BONN, GERMANY
E-mail address: albeverio@uni-bonn.de

DEPARTMENT OF MATHEMATICS, UNIVERSITY OF TRENTO, VIA SOMMARIVE 14, 38123 POVO, TRENTO, ITALY
E-mail address: dipersio@science.unitn.it

DEPARTMENT OF MATHEMATICS, UNIVERSITY OF MILANO BICOCCA, PIAZZA DELL' ATENEO NUOVO, 20133 MILANO, ITALY
E-mail address: elisa.mastrogiacomo@unimib.it

Proceedings of Symposia in Pure Mathematics
Volume **87**, 2013
http://dx.doi.org/10.1090/pspum/087/01440

On the spectra of large sparse graphs with cycles

D. Bollé, F. L. Metz, and I. Neri

This paper is dedicated to Fritz Gesztesy.

ABSTRACT. We present a general method for obtaining the spectra of large graphs with short cycles using ideas from statistical mechanics of disordered systems. This approach leads to an algorithm that determines the spectra of graphs up to a high accuracy. In particular, for (un)directed regular graphs with cycles of arbitrary length we derive exact and simple equations for the resolvent of the associated adjacency matrix. Solving these equations we obtain analytical formulas for the spectra and the boundaries of their support.

1. Introduction

In spectral graph theory one uses spectral analysis to study the interplay between the topology of a graph and the dynamical processes modelled through the graph [**1**–**4**]. In fact, various dynamical processes in disciplines ranging from physics, biology, information theory, and chemistry to technological and social sciences are modelled with graph theory. Therefore, graph theory forms a unified framework for their study [**5**, **6**]. In particular, one associates a certain matrix to a graph (e.g. the adjacency matrix, the Laplacian, the google matrix) and studies the connection between the spectral properties of this matrix and the properties of the dynamical processes governed through them. We mention some studies in this context: the stability of synchronization processes [**7**, **8**], the robustness and the effective resistance of networks [**9**], error-correcting codes [**10**], etc.

These examples illustrate how spectral graph theory relies to a large extent on the capability of determining spectra of sparse graphs. It is thus important to develop mathematical methods which allow to derive in a systematic way exact analytical as well as numerical results on spectra of large graphs. For an overview of analytical results on the spectra of infinite graphs we refer to the paper of Mohar and Woess [**4**].

Recently, the development of exact results for large sparse graphs has been reconsidered using ideas from statistical physics of disordered systems. In this approach one formulates the spectral analysis of graphs in a statistical-mechanics language using a complex valued measure [**11**]. The spectrum is given as the free energy density of this measure, which can be calculated using methods from disordered systems such as the replica method [**12**], the cavity method [**13**, **14**] or the

2010 *Mathematics Subject Classification.* Primary 05C63, 60B20, 82B44.
Key words and phrases. Infinite graphs, random matrices, disordered systems.

super-symmetric method [**15**]. This approach is exact for infinitely large graphs that do not contain cycles of finite length. In recent works the cavity method has been generalized to the study of spectra of graphs with a community structure [**16**] and spectra of small-world graphs [**17**]. The replica and cavity methods have also been used to derive the largest eigenvalue of sparse random matrix ensembles [**18**]. Although the cavity method is heuristic, it has been considered in a rigorous setting for undirected sparse graphs using the theory of local weak convergence [**19**]. However, for directed sparse graphs the asymptotic convergence of the spectrum to the resultant cavity expressions has not been shown [**20**]. Nevertheless, recent studies have shown the asymptotic convergence of the spectrum of highly connected sparse matrices to the circular law [**21–24**] and have proven the asymptotic convergence of the spectrum of Markov generators on higly connected random graphs [**25**]. These studies, however, do not concern finitely connected graphs.

In this work we extend the statistical-mechanics formulation, in particular the cavity method for the spectra of large sparse graphs which are locally tree-like [**13, 14, 19**], to large sparse graphs with many short cycles. Such an extension is relevant because cycles do appear in many real-world systems such as the internet [**26–28**]. We derive a set of resolvent equations for graph ensembles that contain many cycles of finite length. These equations are exact for infinitely large (un)directed Husimi graphs [**29, 30**] and solving them constitutes an algorithm for determining the spectral density of these graphs.

First, we show how this algorithm determines the spectra of irregular Husimi graphs up to a high accuracy, well corroborated by numerical simulations. Then we derive novel analytical results not only for the spectra of undirected regular Husimi graphs, but also for the spectra of directed regular Husimi graphs. In particular, we show that the boundary of the spectrum of directed Husimi graphs composed of cycles of length ℓ, is determined by a hypotrochoid with radii $R/r = \ell$ (R being the radius of the fixed circle and r of the moving circle). Not many analytical expressions for the spectra of directed sparse random graphs and non-Hermitian random matrices are known (besides some exceptions, see [**31**]). A short account of some of our results on regular graphs has appeared in [**32**].

2. Ensembles of graphs

Networks appearing in nature are usually modelled with theoretical ensembles of graphs. These ensembles consist of randomly constructed graphs with certain topological constraints on their connectivity. Model systems allow for a better understanding of the properties of the more complex real-world systems [**5, 6**].

In this work we consider simple graphs $\mathcal{G} = (V, E)$ of size $N = |V|$ consisting of a discrete set of vertices V and a set of edges $E \subset V \times V$. Simple graphs are uniquely defined by their adjacency matrix $\boldsymbol{A}$, with elements $[\boldsymbol{A}]_{ij} = A_{ij} \in \{0, 1\}$ for $i, j \in V$ and $[\boldsymbol{A}]_{ii} = 0$ for $\forall\, i$. We have $A_{ij} = 1$ when $(i, j) \in E$ and zero otherwise. An undirected edge is present between i and j when $A_{ij} = A_{ji} = 1$, a directed edge is present from i to j when $A_{ij} = 1$ and $A_{ji} = 0$, while $A_{ij} = A_{ji} = 0$ indicates that there is no edge present between i and j.

We define graph ensembles through a normalized distribution $P(\boldsymbol{A})$ for the adjacency matrices $\boldsymbol{A}$. Selecting a graph from the ensemble corresponds with drawing randomly an adjacency matrix from the distribution $P(\boldsymbol{A})$. We always consider ensembles of infinite graphs, for which $N \to \infty$. This is implicitly assumed throughout the whole paper. Below, we first define random graphs which are locally tree-like. Their spectral properties have been considered in several studies [**4, 13, 14, 19, 33, 34**]. Second, we define ensembles of graphs with many cycles: the cacti or Husimi graphs [**29, 30**].

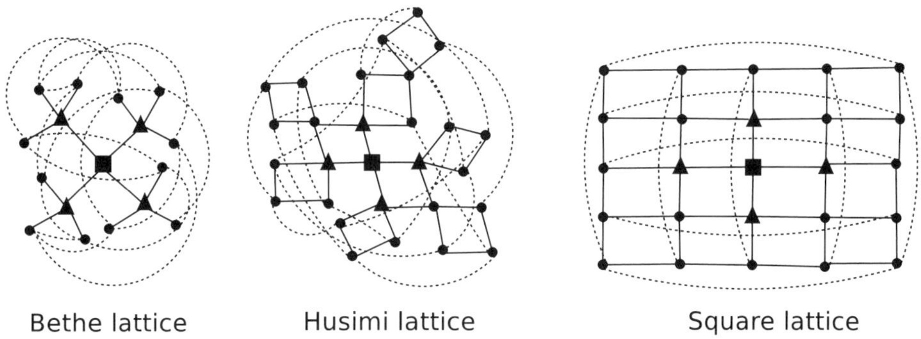

FIGURE 1. Three different regular graphs rooted around the central vertex. The central vertex is denoted by a square while its neighbours are denoted by triangles. The dashed lines indicate the edges between the leaves of the rooted graph. Left: a 4-regular graph. The rooted graph is locally tree-like. Center: a $(4, 2)$-regular Husimi graph. The rooted graph is not tree-like but does maintain an infinite-dimensional nature. Right: a non-random graph (the square lattice).

2.1. Graphs with a local tree structure. Ensembles of random graphs with only certain constraints on the vertex degrees are locally tree-like. Some well-studied examples with these characteristics are the ensemble of c-regular graphs (also called Bethe lattices) and the ensemble of irregular Erdös-Rényi graphs [**35**]. For undirected graphs, these ensembles can be formally defined as follows:

- c-regular graphs with fixed connectivity c:

$$P_{\rm reg}(\boldsymbol{A}; c) \sim \prod_{i=1}^{N} \delta(c_i; c) \prod_{i<j} \delta(A_{ij}; A_{ji}), \tag{2.1}$$

  with $c_i = \sum_j A_{ji}$ the degree of the i-th vertex. Figure 1 presents a sketch of a Bethe lattice.

- Erdös-Rényi graphs with mean connectivity c:

$$P_{\rm irreg}(\boldsymbol{A}; c) \sim \prod_{i<j} \left(\frac{c}{N} \delta\left(A_{ij}; 1\right) + \left(1 - \frac{c}{N}\right) \delta\left(A_{ij}; 0\right)\right) \prod_{i<j} \delta\left(A_{ij}; A_{ji}\right). \tag{2.2}$$

  In this ensemble the degrees fluctuate from vertex to vertex within the graph. The distribution of vertex degrees converges to a Poissonian distribution with mean c in its asymptotic limit $N \to \infty$.

Ensembles of directed graphs can be defined similarly by leaving out the symmetry constraints in the definitions eqs (2.1) and (2.2) and by taking into consideration the difference between A_{ij} and A_{ji}. For instance, we define a c-regular ensemble of directed graphs by the constraint $c_i^{\text{out}} = c_i^{\text{in}} = c$ for any i, with the indegree defined as $c_i^{\text{in}} \equiv \sum_j A_{ji}$ and the outdegree as $c_i^{\text{out}} \equiv \sum_j A_{ij}$.

The Erdös-Rényi and the c-regular ensemble have the local tree property, i.e. the probability to encounter a cycle of finite length in the local neighbourhood of a vertex becomes arbitrary small for $N \to \infty$ [**35**]. The local tree property of random graphs is an important characteristic which has allowed to determine the spectra of sparse graphs in previous works [**13, 14, 19**]. This property is illustrated in figure 1 for the Bethe lattice: cycles only appear at the leaves of the tree, where the vertex constraints have to be satisfied such that their typical length is of the order $\log(N)$.

2.2. Ensembles with cycles. Next, we define graph ensembles which contain many short cycles and are, therefore, not locally tree-like. A cycle of length ℓ is defined by a set of nodes $(i_1, i_2, \cdots i_\ell)$ with an edge connecting each pair $(i_n, i_{(n+1)(\text{mod } \ell)})$, for $n = 1, \ldots, \ell$. The set of all ℓ-tuples is denoted by V^ℓ. We define the following ensembles which are very similar to the Husimi graphs in [**29, 30**]:

- (ℓ, c)-regular Husimi graphs with fixed connectivity c and fixed cycle length ℓ:

$$P_{\text{reg}}(\boldsymbol{A}; c, \ell) \sim \prod_{i=1}^{N} \delta(c_i; c) \prod_{i<j} \delta(A_{ij}; A_{ji}), \tag{2.3}$$

  where c_i denotes now the number of ℓ-cycles of $\boldsymbol{A}$ incident to the i-th vertex. In figure 2 we show the typical local neighbourhood of an undirected (4,2)-regular Husimi graph.

- irregular Husimi graphs with fixed cycle length ℓ and mean loop connectivity c:

$$\begin{aligned} P_{\text{irreg}}(\boldsymbol{A}; c, \ell) \sim & \prod_{(i_1, i_2, \ldots, i_\ell) \in V^\ell} \left[\frac{c}{N^{\ell-1}} \delta\left(A_{i_1 i_\ell} \prod_{k=1}^{\ell-1} A_{i_k i_{k+1}}; 1 \right) \right. \\ & \left. + \left(1 - \frac{c}{N^{\ell-1}}\right) \delta\left(A_{i_1 i_\ell} \prod_{k=1}^{\ell-1} A_{i_k i_{k+1}}; 0 \right) \right] \times \prod_{i<j} \delta(A_{ij}; A_{ji}) \\ & \times \prod_{(i,j)} \delta\left(A_{ij} \sum_{(i,k_1,k_2,\cdots,k_{\ell-2},j) \in V_{ij}^\ell} A_{ik_1} \left(\prod_{n=1}^{\ell-3} A_{k_n k_{n+1}} \right) A_{k_{\ell-2} j}; A_{ij} \right). \end{aligned} \tag{2.4}$$

  The set $V_{ij}^\ell \subset V^\ell$ consists of all ℓ-tuples $(i, k_1, \cdots, k_{\ell-2}, j) \in V^\ell$. The first factor gives a weight $c/N^{\ell-1}$ to each randomly drawn cycle of length ℓ, the second factor is the constraint which makes the graph undirected and the last factor is a constraint which takes into consideration that each edge must belong to exactly one cycle of length ℓ.

We remark that in our notation of Husimi graphs the mean cycle connectivity c is the average number of cycles of length ℓ connected to a certain vertex, while $2c$ denotes the mean degree of each vertex. Note that for $\ell = 2$ the ensembles of Husimi graphs introduced here reduce to the corresponding ensembles of random graphs which are locally tree-like (discussed in the previous subsection).

As before, we can define as well directed ensembles of Husimi graphs by removing the symmetry constraint $\prod_{i<j} \delta(A_{ij}; A_{ji})$ and taking into consideration the

possible difference beween A_{ij} and A_{ji}. For instance, for a regular Husimi graph we set the vertex indegree and the vertex outdegree equal to c. In figure 2 we present the neighbourhood of a vertex in a directed (3,2)-regular Husimi graph.

Husimi graphs are not locally tree-like since they are composed of cycles, but they do have an infinite-dimensional character. Indeed, the local neighbourhood of a vertex contains different branches which are only connected by cycles of order $\log(N)$, see figures 1 and 2. This property allows us to present an exact spectral analysis.

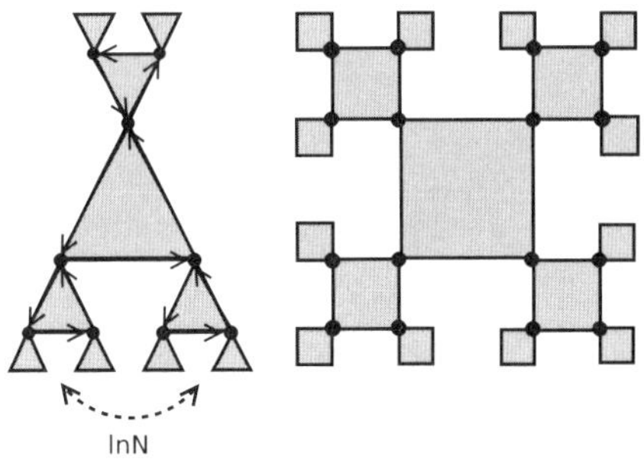

FIGURE 2. [**32**] Local structure of some (ℓ, c)-regular Husimi graphs, with ℓ the cycle length and c the cycle connectivity. Left: directed $(3, 2)$-regular Husimi graph. Right: undirected $(4, 2)$-regular Husimi graph. The Husimi graphs are infinite dimensional as different branches of a central node are separated by a distance of the order $\mathcal{O}(\ln N)$.

3. Spectral analysis of undirected graphs with cycles

In this section we show how to determine the spectral properties of undirected graphs with cycles. For regular graphs, our approach, briefly presented in [**32**], consists of four steps. First, we write the spectrum in statistical mechanics terms using a complex measure. Then we use the infinite-dimensional nature of the graphs to derive a closed expression in the submatrices of the resolvent. Next, we simplify the resolvent equations to the specific case of regular undirected Husimi graphs. Finally, we solve the resultant algebraic expressions to find analytical expressions for the spectra of regular Husimi graphs.

3.1. Statistical mechanics formulation of the spectrum. We define the resolvent of a real symmetric matrix $\boldsymbol{A}$ as

$$G(z) \equiv (\boldsymbol{A} - z\boldsymbol{I}_N)^{-1}, \tag{3.1}$$

where $\boldsymbol{I}_N$ is the $N \times N$ identity matrix and $z \in \mathbb{C}$, with $\mathrm{Re(z)} = \lambda$ and $\mathrm{Im(z)} > 0$. From the resolvent we can determine the asymptotic density $\rho(\lambda)$ of the real eigenvalues λ of an ensemble of matrices $\boldsymbol{A}$ [**4**]:

$$\rho(\lambda) = \lim_{\mathrm{Im(z)}\to 0^+} \lim_{N\to\infty} \mathrm{Im}\left[\frac{\mathrm{Tr}\boldsymbol{G}}{\pi N}\right]. \tag{3.2}$$

We have left out ensemble averages since we assume that the spectrum self-averages in the limit $N \to \infty$ to a deterministic value.

A common method in statistical physics is to write the resolvent elements and the spectrum as averages over a certain complex-valued measure $P(\boldsymbol{x};z)d\boldsymbol{x}$ [**11**]. Indeed, when defining the normalized Gaussian function

$$P(\boldsymbol{x};z) \sim \exp\left(\frac{i}{2}\boldsymbol{x}^T\boldsymbol{G}^{-1}(z)\boldsymbol{x}\right), \tag{3.3}$$

the diagonal elements $[\boldsymbol{G}]_{ii} = G_i$ are given by [**36**]

$$G_i = i\langle x_i^2\rangle, \tag{3.4}$$

with $\langle\cdot\rangle$ the average with respect to $P(\boldsymbol{x};z)$. In this formalism the spectrum (3.2) is written as [**12**]

$$\rho(\lambda) = -\lim_{\mathrm{Im(z)}\to 0^+}\lim_{N\to\infty}\frac{2}{\pi N}\mathrm{Im}\left[\frac{\partial}{\partial\lambda}\ln\mathcal{Z}\right], \tag{3.5}$$

with

$$\mathcal{Z} = \int d\boldsymbol{x}\exp\left(\frac{i}{2}\boldsymbol{x}^T\boldsymbol{G}^{-1}(z)\boldsymbol{x}\right). \tag{3.6}$$

such that

$$\rho(\lambda) \;=\; \lim_{\mathrm{Im[z]}\to 0^+}\frac{1}{\pi N}\sum_{i=1}^{N}\mathrm{Im}G_i(z), \quad \lambda = \mathrm{Re[z]}. \tag{3.7}$$

Hence, in the language of statistical mechanics the spectrum is the imaginary part of the derivative of the free energy density with respect to λ, as is seen explicitly in eq.(3.5).

3.2. The resolvent equations. We employ the cavity method to determine the spectral properties of locally tree-like graphs [**13, 19**]. We construct a matrix $\boldsymbol{A}^{(i)}$ (and its corresponding resolvent $\boldsymbol{G}^{(i)}$) through deletion of the i-th column and the i-th row in $\boldsymbol{A}$. The graph $\mathcal{G}^{(i)}$ induced by the matrix $\boldsymbol{A}^{(i)}$ is usually referred to as the cavity graph [**37**]. On this graph we use the Bethe-Peierls tree approximation [**37, 38**] to arrive at the following set of resolvent equations:

$$G_i^{(j)} = \frac{1}{z - \sum_{k\in\partial_i\setminus j} A_{ik}^2 G_k^{(i)}}, \quad G_i = \frac{1}{z - \sum_{k\in\partial_i} A_{ik}^2 G_k^{(i)}}, \tag{3.8}$$

where ∂_i is the set of vertices neighbouring i, and $\partial_i \setminus j$ denotes the set of all neighbours of i except for node j. In appendix A.1 we report the essential steps in deriving these resolvent equations.

The resolvent equations (3.8) have appeared in the literature for the first time, as far as we know, in the study of electron localization [**39**]. Similar equations have been derived in the context of Lévy matrices [**40**] and in the more general context of sparse random graphs [**13**]. The resolvent equations (3.8) form a set of $2|E|$ equations in $2|E|$-unknowns.

We interpret the set of equations for $\{G_i^{(j)}\}$ as a message-passing algorithm along the induced graph $\mathcal{G}$, similar to belief propagation which is widely used in statistical inference and information theory [**41**]. Here the messages are the diagonal elements of the resolvent of the cavity graph [**13**].

Next, we extend the resolvent equations (3.8) to Husimi graphs with many short cycles. Although the Bethe-Peierls approximation is no longer valid on these graphs, they still contain an infinite-dimensional character allowing for an analogous approximation. In statistical mechanics this forms the basis for the Kikuchi approximation [**42**, **43**]. For Husimi graphs the Kikuchi approximation gives us the resolvent equations:

$$G_{\alpha}^{(i)} = \left(z\boldsymbol{I}_{\ell-1} - \boldsymbol{D}_{\alpha}^{(i)} - \boldsymbol{L}_{\alpha} - \boldsymbol{L}_{\alpha}^{T}\right)^{-1}, \tag{3.9}$$

with $\alpha \in \partial_i^{(\ell-1)}$ and with $\partial_i^{(\ell-1)}$ the set of all $(\ell-1)$-tuples $(j_1, j_2, \cdots, j_{\ell-1})$ which form a cycle of length ℓ with the node i. The elements of the $(\ell-1)$-dimensional matrix $\boldsymbol{L}_{\alpha}$ are $[\boldsymbol{L}_{\alpha}]_{nm} = \left[\boldsymbol{L}_{(j_1,j_2,\cdots,j_{\ell-1})}\right]_{nm} = \delta_{n(m+1)}A_{j_n j_m}$, with $n, m = 1, ..., (\ell-1)$. The quantity $\boldsymbol{D}_{\alpha}^{(i)}$ is a diagonal matrix with elements

$$\left[\boldsymbol{D}_{\alpha}^{(i)}\right]_{kk} = \sum_{\beta\in\partial_{j_k}^{(\ell-1)}\setminus(i,j_1,\ldots,j_{k-1},j_{k+1},\ldots,j_{\ell-1})} \boldsymbol{A}_{j_k,\beta}^{T} G_{\beta}^{(j_k)} \boldsymbol{A}_{j_k,\beta},$$

with $\alpha = (j_1, j_2, \cdots, j_{\ell-1})$. We have also defined the $(\ell-1)$-dimensional vector $\boldsymbol{A}_{j_k,\beta}^{T} = \boldsymbol{A}_{j_k,(k_1,k_2,\cdots,k_{\ell-1})} = (A_{j_k k_1}\ 0\ \ldots\ 0\ A_{j_k k_{\ell-1}})$. The diagonal elements of the resolvent are given by

$$G_i = \frac{1}{z - \sum_{\alpha\in\partial_i^{(\ell-1)}} \boldsymbol{A}_{i,\alpha}^{T} G_{\alpha}^{(i)} \boldsymbol{A}_{i,\alpha}}. \tag{3.10}$$

In appendix A.2 we elaborate on the precise derivation of these equations.

The resolvent equations (3.9) determine the spectra of graphs with many short cycles of fixed length ℓ. Note that they can be straightforwardly generalized to graphs with variable cycle lengths and, even more general, to graphs composed from arbitrary figures, by replacing $\boldsymbol{L}_{\alpha}$ by the corresponding adjacency matrix of the figure α in the absence of a node i, and $\boldsymbol{A}_{j_k,\beta}$ by the submatrix of $\boldsymbol{A}$ connecting the figure β and the node j_k. In our case, the figures of the graph are the $(\ell-1)$-tuples α. The set of resolvent equations (3.9)-(3.10) can also be seen as a message-passing algorithm between regions of the graph, which is similar to the generalized belief propagation algorithm in information theory [**43**]. Here the messages of the algorithm are the $(\ell-1)\times(\ell-1)$ matrices $\boldsymbol{G}_{\alpha}^{(i)}$ sent from the $(\ell-1)$-tuples α to a vertex i, with whom the tuple α forms a cycle of length ℓ in the graph.

We have verified the exactness of the resolvent equations (3.9)-(3.10) on irregular Husimi graphs with $\ell = 3$ through a comparison with direct diagonalization results. This is presented in figure 3, where we also compare the spectrum of irregular Husimi graphs with mean cycle connectivity $c = 3$ with the spectrum of Erdös-Rényi random graphs with the same mean vertex degree $c = 6$. From these results it appears that the spectrum of Erdös-Rényi graphs converges faster to the Wigner semicircle for $c \to \infty$.

Finally, we point out that the ensemble definitions in section 2 are global by specifying $P(\boldsymbol{A})$. In the derivation of the resolvent equations (3.9)-(3.10) these global definitions are not explicitely used. Instead, our analysis uses the typical local neighbourhoods of the vertices in the graph. Therefore, our results are valid for all graphs which have a local neighbourhood similar to the one given by the resolvent equations. The connection between the global definitions and the resolvent equations can be made explicit through the distribution of local vertex neighbourhoods, see for instance [**19**]. This is how we have derived the results in figure 3.

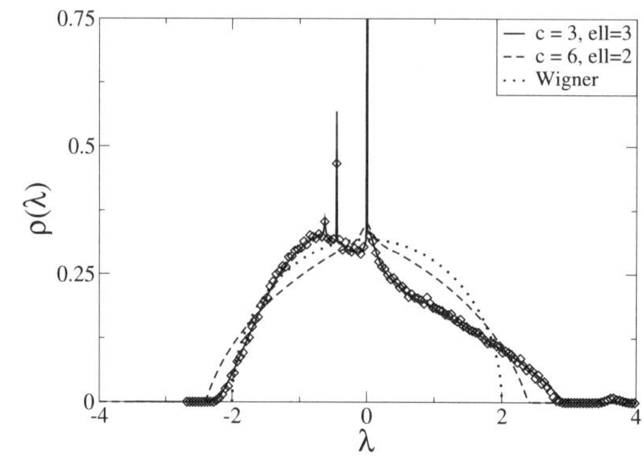

FIGURE 3. The density of eigenvalues λ for an irregular triangular Husimi graph ($\ell = 3$) of mean connectivity $c = 3$ with scaling $A_{ij} \to A_{ij}/\sqrt{2c-1}$ (solid lines) is compared with the eigenvalue density of an irregular Erdös-Rényi graph ($\ell = 2$) of mean connectivity $c = 6$ with scaling $A_{ij} \to A_{ij}/\sqrt{c-1}$ (dashed lines). Both ensembles have the same mean vertex degree. Results from solving the equations (3.9-3.10) for $\ell = 3$ with a regulator Im(z) $= 10^{-4}$ are compared with direct diagonalization results (markers). The dotted line is the Wigner semi-circle law.

In the next two sections we solve the resolvent equations for the case of undirected regular Husimi graphs for which they simplify considerably.

3.3. The resolvent equations for regular undirected graphs. In this section we consider graphs with a simple topology that allows to extract exact analytical solutions from the resolvent equations (3.9)-(3.10): regular Husimi graphs with cycles. For this ensemble of random graphs $\left|\partial_i^{(\ell-1)}\right| = c$ and $A_{ij} = 1$ for all connected pairs (i,j). In figure 2 the square undirected Husimi graph is illustrated.

Since there is no local disorder every node has the same local neighbourhood. For $N \to \infty$ the graph becomes transitive and we can set

$$G_\alpha^{(i)}(z) = \boldsymbol{C}(z), \qquad G_i(z) = G(z)\,, \tag{3.11}$$

in the resolvent equations (3.9). We get the following closed equations in the $(\ell-1)$-dimensional matrices $\boldsymbol{C}$:

$$\boldsymbol{C}(z) = \begin{pmatrix} z-(c-1)\mathcal{A}^T\boldsymbol{C}(z)\mathcal{A} & -1 & 0 & \cdots \\ -1 & z-(c-1)\mathcal{A}^T\boldsymbol{C}(z)\mathcal{A} & -1 & \cdots \\ 0 & -1 & \ddots & \\ \vdots & \vdots & & \ddots \end{pmatrix}^{-1}, \tag{3.12}$$

and the resolvent follows from

$$G(z) = \frac{1}{z - c\,\mathcal{A}^T\boldsymbol{C}(z)\mathcal{A}}, \tag{3.13}$$

where $\mathcal{A}^T = (1\ 0\ \ldots\ 0\ 1)$. It is convenient to introduce the scalar variable $G_s(z) = \mathcal{A}^T \boldsymbol{C}(z)\mathcal{A}$ and rewrite (3.12) as

$$G_s(z) = \mathcal{A}^T \begin{pmatrix} z-(c-1)G_s(z) & -1 & 0 & \cdots \\ -1 & z-(c-1)G_s(z) & -1 & \cdots \\ 0 & -1 & \ddots & \\ \vdots & \vdots & & \ddots \end{pmatrix}^{-1} \mathcal{A}. \tag{3.14}$$

The matrix in (3.14) has a simple structure. We have calculated the inverse analytically using specific methods for tridiagonal matrices [**44**]. This leads to our final formula for G_s when $\ell > 2$

$$G_s = \frac{2n_{\ell-2}+2}{n_{\ell-1}}. \tag{3.15}$$

The coefficients $n_2, \ldots, n_{\ell-1}$ are complex numbers given by the recurrence relation

$$n_i = n_1 n_{i-1} - n_{i-2}, \quad i \geq 2 \tag{3.16}$$

with initial values $n_0 = 1$ and $n_1 = z-(c-1)G_s$. Equations (3.15) and (3.16) are a set of very simple equations that determine the resolvent elements of undirected regular Husimi graphs. It is one of the main results of our previous work [**32**].

From the expression of G_s we find straightforwardly the Green function

$$G(z) = \frac{1}{z - c\,G_s(z)}, \tag{3.17}$$

and the spectrum, according to eq. (3.7) and (3.17),

$$\rho(\lambda) = \frac{c}{\pi}\frac{\mathrm{Im}G_s}{(\lambda - c\,\mathrm{Re}G_s)^2 + c^2\,(\mathrm{Im}G_s)^2}. \tag{3.18}$$

3.4. Spectra of regular undirected graphs. We solve the algebraic equations (3.15)-(3.16) to find the spectrum of (ℓ, c)-regular undirected Husimi graphs. Due to their simple linear structure, these recurrence relations can be solved exactly and allow to obtain the analytical form of the polynomials in the variable G_s for any value of $\ell > 2$:

- $\ell = 3$ corresponds with a quadratic equation:
$$G_s\,[z-(c-1)G_s] - G_s = 2.$$
- $\ell = 4$ gives a cubic equation:
$$G_s = \frac{2\,[z-(c-1)G_s]}{[z-(c-1)G_s]^2 - 2}.$$
- $\ell = 5$ is determined by a quintic equation:
$$G_s = \frac{2\,[z-(c-1)G_s]^3 - 4\,[z-(c-1)G_s] + 2}{[z-(c-1)G_s]^4 - 3\,[z-(c-1)G_s]^2 + 1}.$$
- $\ell = 6$ follows from a quartic equation:
$$G_s = \frac{2\,[z-(c-1)G_s]^2 - 4}{[z-(c-1)G_s]^3 - 3\,[z-(c-1)G_s]}.$$
- For $\ell > 6$, the polynomials have a degree larger than four.

The root of these polynomial equations which gives the expression for the spectrum is a stable fixed point of eq. (3.14). In general, one finds the roots of polynomials in terms of radicals up to degree four. For larger degrees, algebraic solutions in terms of radicals no longer exist, apart from some particular situations [**45**]. The roots of general polynomials with degree larger than four are given in terms of elliptic functions [**45**].

We have solved the above equations for $\ell = 3, 4, 6$ and have found the analytical expression for $\rho_\ell(\lambda)$, generalizing the Kesten-McKay law [**33, 46**] to regular graphs with cycles. For other values of ℓ, eqs. (3.15)-(3.16) are solved numerically in a straightforward way, allowing to study the spectrum as a function of ℓ with high accuracy.

From the solution of the resolvent equations, we have found the expressions for the spectra $\rho_\ell(\lambda)$ of (ℓ, c)-regular undirected Husimi graphs. We point out that for $\ell = 2$ we recover a sparse regular graph without short cycles. In this case we have the Kesten-Mckay eigenvalue-density distribution [**33, 46**]

$$\rho_2(\lambda) = \frac{c}{2\pi}\frac{\sqrt{4(c-1)-\lambda^2}}{c^2-\lambda^2}, \tag{3.19}$$

for $\lambda \in [-2\sqrt{c-1}, 2\sqrt{c-1}]$, and $\rho_2(\lambda) = 0$ otherwise. The expression (3.19) follows from solving eq. (3.14).

For $\ell = 3$ we recover the spectra $\rho_3(\lambda)$ of triangular Husimi graphs [**47**]

$$\rho_3(\lambda) = \frac{2(c-1)}{\pi c}\frac{\sqrt{D_3(\lambda)}}{D_3(\lambda) + \left[\lambda\frac{2(c-1)}{c} + (1-\lambda)\right]^2} + \frac{1}{3}\delta_{c,2}\,\delta(\lambda;-2),$$

for $D_3(\lambda) = 8(c-1) - (\lambda-1)^2 > 0$, and $\rho_3(\lambda) = 0$ otherwise. Therefore, the support of $\rho_3(\lambda)$ is $\lambda \in [1 - 2\sqrt{2(c-1)}, 1 + 2\sqrt{2(c-1)}]$.

For $\ell = 4$ we find the spectrum of a square Husimi graph [**32**]

$$\rho_4(\lambda) = \frac{6\sqrt{3}\,c\,(c-1)\,q_-(\lambda)}{\pi\left[2(c-3)\lambda + c\,q_+(\lambda)\right]^2 + 3\,\pi\,c^2\,q_-^2(\lambda),}$$

for $D_4(\lambda) > 0$ and $\rho_4(\lambda) = 0$ otherwise, where

$$q_\pm(\lambda) = s_+^{1/3} \pm s_-^{1/3}, \quad s_\pm(\lambda) = 9\lambda(c+1) - \lambda^3 \pm 9\sqrt{D_4(\lambda)},$$

and

$$D_4(\lambda) = -\frac{2}{3}\lambda^4 - \frac{\lambda^2}{3}\left(c^2 - 22c + 13\right) + \frac{8}{3}(c-2)^3.$$

The edges of $\rho_4(\lambda)$ are determined by finding the roots of $D_4(\lambda) = 0$. For $c \to \infty$, $\rho_4(\lambda)$ converges to the Wigner semicircle law. For $c = 2$, the spectrum contains a power-law divergence $\rho_4(\lambda) \sim |\lambda|^{-1/3}$ as $|\lambda| \to 0$.

We have also found the analytical expression for $\ell = 6$:

$$\rho_6(\lambda) = \frac{4c\sqrt{R(\lambda)F(\lambda)}}{\pi R(\lambda)\left[\frac{(c-4)}{(c-1)}|\lambda| + 2cR(\lambda)\right]^2 + \pi c^2 F(\lambda)},$$

for $D_6(\lambda) > 0$ and $\rho_6(\lambda) = 0$ otherwise, with

$$\begin{aligned} u_R(\lambda) &= \left(r(\lambda) + \sqrt{D_6(\lambda)}\right)^{\frac{1}{3}} + \left(r(\lambda) - \sqrt{D_6(\lambda)}\right)^{\frac{1}{3}} + \frac{[2c - 5 + 3\lambda^2]}{3(c-1)^2}, \\ R(\lambda) &= \sqrt{u_R(\lambda) - \frac{3\lambda^2}{4(c-1)^2} - \frac{(2c-5)}{(c-1)^2}}, \\ F(\lambda) &= -\frac{|\lambda|^3}{(c-1)^3} + \frac{4(2c+1)|\lambda|}{(c-1)^3} + 4R(\lambda)\left[\frac{(2c-5)}{(c-1)^2} - \frac{3\lambda^2}{2(c-1)^2} + u_R(\lambda)\right], \end{aligned}$$

and

$$\begin{aligned} D_6(\lambda) &= q^3(\lambda) + r^2(\lambda), \\ q(\lambda)\,(c-1)^4 &= \frac{-9\lambda^2 + 48(c-1) - (2c-5)^2}{9}, \\ r(\lambda)\,(c-1)^5 &= \left[\frac{(4c-7)^2 - (4c-7)(2c-5)}{2(c-1)} + \frac{(2c-5)^2}{3(c-1)} + \frac{(34-16c)}{3}\right]\lambda^2 \\ &\quad + \frac{16(2c-5)}{3} + \frac{(2c-5)^3}{27(c-1)}. \end{aligned}$$

The function $D_6(\lambda)$ is the discriminant of the cubic polynomial associated to the

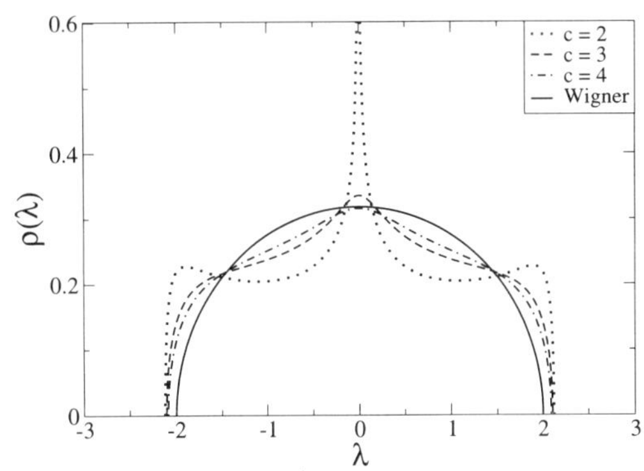

FIGURE 4. Spectra of $(4, c)$-regular undirected Husimi graphs of connectivities $c = 2, 3, 4$ and with a scaling $A_{ij} \to A_{ij}/\sqrt{2c-1}$. The spectrum converges to the Wigner semi-circle law for $c \to \infty$.

quartic polynomial for G_s. The edges of $\rho_6(\lambda)$ are obtained from the roots of $D_6(\lambda) = 0$. The spectrum converges to the Wigner semicircle law when $c \to \infty$.

For $\ell = 5$ and $\ell \geq 7$ we obtain the spectrum by solving numerically the set of eqs. (3.15)-(3.16). One can show that $\rho_\ell(\lambda)$ converges to the Wigner law in the limit $c \to \infty$, and converges to the Kesten-Mckay law in the limit $\ell \to \infty$ (this limit corresponds to a graph without short cycles).

In figure 4 we present the evolution of the spectrum as a function of the connectivity c for the case $\ell = 4$. We notice indeed the fast convergence to the Wigner semi-circle law for $c \to \infty$. In figure 5 we present the evolution of the spectrum as a

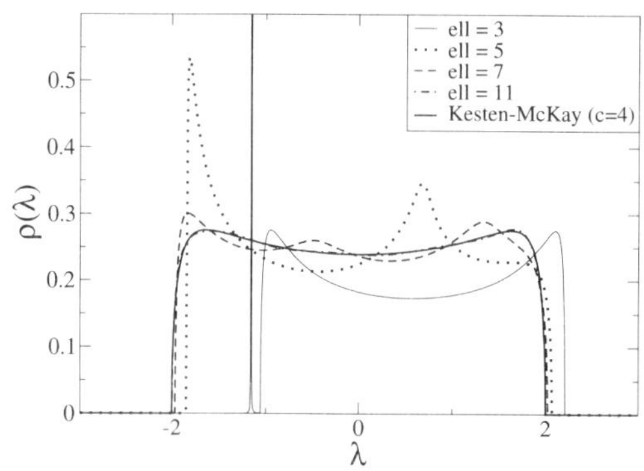

FIGURE 5. Spectra of $(\ell, 2)$ regular undirected Husimi graphs with a scaling $A_{ij} \to A_{ij}/\sqrt{2c-1}$ and with cycle lengths $\ell = 3, 5, 7, 11$. The spectrum converges to the Kesten-McKay law for $\ell \to \infty$.

function of the cycle length ℓ, for fixed $c = 2$. We see how the spectrum converges rapidly to the Kesten-McKay law for $\ell \to \infty$.

4. Spectral analysis of directed graphs with cycles

In our study of spectral properties of regular directed graphs with cycles we follow again four steps. First, we write the spectrum in statistical mechanics terms [**14**]. This is done by mapping the resolvent calculation of a directed graph on a resolvent calculation of a related undirected graph, using the Hermitization procedure [**48**]. The resolvent of this undirected graph is then analyzed using the methodology presented in section 3. Thereby, the infinite-dimensional nature of the resultant graphs again allows to derive a closed expression in submatrices of the resolvent matrix. Thirdly, we show how this closed set of equations determines the spectrum of directed regular Husimi graphs and we derive their explicit form in this case. Finally, the resultant algebraic expressions are solved to arrive at explicit analytical expressions in the support of the spectra.

4.1. Statistical mechanics formulation of the spectrum. The eigenvalues $\lambda_1, \ldots, \lambda_N$ of the $N \times N$ adjacency matrix $\boldsymbol{A}$ of directed Husimi graphs are distributed in the complex plane, contrary to the real eigenvalues for undirected Husimi graphs. By introducing $\partial^* = \frac{1}{2}\left(\frac{\partial}{\partial x} + i\frac{\partial}{\partial y}\right)$ and using the relation $\partial^*(x+iy)^{-1} = \pi\delta(x)\delta(y)$, the density of states $\rho(\lambda) = N^{-1}\sum_{\mu=1}^{N}\delta(x - \mathrm{Re}\lambda_\mu)\delta(y - \mathrm{Im}\lambda_\mu)$ at a certain point $\lambda = x + iy$ can be written formally as $\rho(\lambda) = \lim_{N\to\infty}(N\pi)^{-1}\partial^*\mathrm{Tr}\boldsymbol{G}(\lambda)$, where $\boldsymbol{G}(\lambda)$ is the resolvent $\boldsymbol{G}(\lambda) = (\lambda - \boldsymbol{A})^{-1}$. The operation $(\cdot)^*$ denotes complex conjugation. The $N \times N$ resolvent $\boldsymbol{G}(\lambda)$ is the central object of interest and its non-analytic behavior at the eigenvalues of $\boldsymbol{A}$ poses difficulties in applying various techniques well-developed for Hermitian matrices [**48**]. An elegant way to avoid this problem is the Hermitization method. We define a $2N \times 2N$ block matrix

$$\boldsymbol{H}_\epsilon(\lambda) = \begin{pmatrix} \epsilon \boldsymbol{I}_N & -i(\lambda - \boldsymbol{A}) \\ -i(\lambda^* - \boldsymbol{A}^T) & \epsilon \boldsymbol{I}_N \end{pmatrix}, \tag{4.1}$$

where $\epsilon > 0$ is a regularizer and $\boldsymbol{I}_N$ is the N-dimensional identity matrix. The lower-left block of $\lim_{\epsilon\to 0^+} \boldsymbol{H}_\epsilon^{-1}(\lambda)$ is precisely the matrix $\boldsymbol{G}(\lambda)$. Thus, the problem reduces to calculating the diagonal matrix elements $\mathcal{G}_j(\lambda,\epsilon) = \left[\boldsymbol{H}_\epsilon^{-1}(\lambda)\right]_{j+N,j}$ ($j = 1,\ldots,N$), from which the spectrum is determined according to

$$\rho(\lambda) = -\frac{i}{N\pi} \lim_{\epsilon\to 0^+, N\to\infty} \sum_{j=1}^{N} \partial^* \mathcal{G}_j(\lambda,\epsilon)\,. \tag{4.2}$$

The form of the enlarged matrix $\boldsymbol{H}$ depends on the problem at hand and different proposals have appeared in the literature [**48–51**]. The form (4.1) is particularly convenient here, since its Hermitian part is a positive-definite matrix and one can represent $\boldsymbol{H}^{-1}$ as a Gaussian integral. For this purpose we introduce a set of two-dimensional column vectors $\{\boldsymbol{\Phi}_i\}_{i=1,\ldots,N}$ with complex elements and the "Hamiltonian" function

$$\mathcal{H}(\{\boldsymbol{\Phi}_j, \boldsymbol{\Phi}_j^\dagger\};\lambda) = \sum_{i=1}^{N} \boldsymbol{\Phi}_i^\dagger.\boldsymbol{S}_\epsilon(\lambda)\boldsymbol{\Phi}_i + i \sum_{i,j=1}^{N} \boldsymbol{\Phi}_i^\dagger.\mathcal{J}_{ij}\boldsymbol{\Phi}_j\,, \tag{4.3}$$

where

$$\mathcal{J}_{ij} = \begin{pmatrix} 0 & A_{ij} \\ A_{ji} & 0 \end{pmatrix}, \tag{4.4}$$

and $\boldsymbol{S}_\epsilon(\lambda) = [\epsilon \boldsymbol{I}_2 - i\,(x\sigma_x - y\sigma_y)]$, with (σ_x, σ_y) denoting Pauli matrices.

A graphical representation using an induced graph is again useful in these calculations. For a non-Hermitian matrix with real entries the induced graph is directed, contrary to an undirected graph for real symmetric matrices. Graphically the matrix elements A_{ij} correspond then with a directed edge from node i to node j.

Combining the representation of $\left[\boldsymbol{H}_\epsilon^{-1}(\lambda)\right]_{ij}$ as a Gaussian integral with eq. (4.3), $\mathcal{G}_k(\lambda,\epsilon)$ reads

$$\mathcal{G}_k(\lambda,\epsilon) \;=\; \frac{1}{2}\int \left[\prod_{j=1}^{N} d\boldsymbol{\Phi}_j d\boldsymbol{\Phi}_j^\dagger\right] \left[\boldsymbol{\Phi}_k^\dagger.(\sigma_x + i\sigma_y)\boldsymbol{\Phi}_k\right] \mathcal{P}(\{\boldsymbol{\Phi}_j, \boldsymbol{\Phi}_j^\dagger;\lambda\})\,, \tag{4.5}$$

where we have introduced the complex-valued measure $\prod_{j=1}^{N} \mathcal{P}(\{\boldsymbol{\Phi}_j, \boldsymbol{\Phi}_j^\dagger\};\lambda) d\boldsymbol{\Phi}_j d\boldsymbol{\Phi}_j^\dagger$

$$\mathcal{P}(\{\boldsymbol{\Phi}_i, \boldsymbol{\Phi}_i^\dagger\};\lambda) = \frac{\exp\left[-\mathcal{H}(\{\boldsymbol{\Phi}_i, \boldsymbol{\Phi}_i^\dagger\};\lambda)\right]}{\int \left[\prod_{j=1}^{N} d\boldsymbol{\Phi}_j d\boldsymbol{\Phi}_j^\dagger\right] \exp\left[-\mathcal{H}(\{\boldsymbol{\Phi}_i, \boldsymbol{\Phi}_i^\dagger\};\lambda)\right]}\,. \tag{4.6}$$

Equation (4.5) shows that the local marginals $\{\mathcal{P}(\boldsymbol{\Phi}_k, \boldsymbol{\Phi}_k^\dagger;\lambda)\}_{k=1,\ldots,N}$ determine the functions $\{\mathcal{G}_k\}_{k=1,\ldots,N}$, from which the spectrum follows through eq. (4.2).

4.2. The resolvent equations. We use again the infinite dimensional nature of Husimi graphs to derive an exact equation in the resolvent elements. Due to the sparse structure of $\boldsymbol{A}$, the average number of nodes in a path connecting two randomly chosen cycles scales as $\ln N$ (see figure 8). This fundamental property allows to compute $\{\mathcal{P}(\boldsymbol{\Phi}_k, \boldsymbol{\Phi}_k^\dagger)\}_{k=1,\ldots,N}$ using the cavity method, as demonstrated in appendix A.2. The main difference with the undirected case resides in the "state variables" describing the nodes of the graph and the corresponding "Hamiltonian". While in the undirected case they are scalar real variables, here there is a two-dimensional complex-vector $\boldsymbol{\Phi}_i$ associated to each vertex and they mutually interact through $\mathcal{J}_{ij}$, see eq. (4.3). Hence the resolvent equations involve matrices composed of two-dimensional dimensional blocks.

Extending the cavity formulation [**14**] to directed graphs with cycles, see appendix A.2, we have derived the following equation

$$\mathcal{G}_i(\lambda, \epsilon) = \left(\boldsymbol{S}_\epsilon(\lambda) + \sum_{\alpha \in \partial_i^{(\ell-1)}} \mathbb{A}_{i\alpha}^T \mathcal{D}_\alpha^{(i)} \mathbb{A}_{i\alpha} \right)_{21}^{-1}, \tag{4.7}$$

where $\partial_i^{(\ell-1)}$ is the set of all $(\ell-1)$ tuples $(j_1, j_2, \cdots, j_{\ell-1})$ which form a cycle of length ℓ with node i. The $2 \times 2(\ell-1)$ block matrix $\mathbb{A}_{i\alpha}^T = (\mathcal{J}_{ij_1}\ \mathbf{0}\ \ldots\ \mathbf{0}\ \mathcal{J}_{ij_{\ell-1}})$ encodes the interaction between i and a given tuplet $\alpha = (j_1, j_2, \cdots, j_{\ell-1})$. The matrix $\mathbf{0}$ is a two-dimensional matrix filled with zeros. The $2(\ell-1) \times 2(\ell-1)$ block matrices $\mathcal{D}_\alpha^{(i)} = \mathcal{D}_{(j_1, j_2, \cdots, j_{\ell-1})}^{(i)}$ fulfill the cavity equations

$$\mathcal{D}_\alpha^{(i)} = \left(\boldsymbol{S}_\epsilon(\lambda) \otimes \boldsymbol{I}_{\ell-1} + \mathcal{B}_\alpha^{(i)} + i\mathbb{L}_\alpha + i\mathbb{L}_\alpha^T \right)^{-1}, \tag{4.8}$$

where $i = 1, \ldots, N$ and $\alpha \in \partial_i^{(\ell-1)}$. The matrix $\mathbb{L}_\alpha$ is composed of 2×2 block elements defined by $[\mathbb{L}_\alpha]_{nm} = [\mathbb{L}_{(j_1, j_2, \cdots, j_{\ell-1})}]_{nm} = \delta_{n,m+1} \mathcal{J}_{j_n, j_m}$, where $n = 2, \ldots, \ell-1$. The matrix $\mathcal{B}_\alpha^{(i)}$ is a diagonal matrix formed by the following 2×2 block elements

$$[\mathcal{B}_\alpha^{(i)}]_{kk} = \sum_{\beta \in \partial_{j_k}^{(\ell-1)} \setminus (i, j_1, \ldots, j_{k-1}, j_{k+1}, \ldots, j_{\ell-1})} \mathbb{A}_{j_k\beta}^T \mathcal{D}_\beta^{(j_k)} \mathbb{A}_{j_k\beta}, \tag{4.9}$$

with $k = 1, \ldots, \ell-1$ and $\alpha = (j_1, j_2, \cdots, j_{\ell-1})$.

Once eqs. (4.8) have been solved, the spectrum follows from eqs. (4.7) and (4.2). The cavity equations have an interpretation in terms of a message-passing algorithm: the matrix $\mathcal{D}_\alpha^{(i)}$ is seen as the message sent by the $(\ell-1)$ nodes of cycle α to node i of the same cycle [**43**]. This completes the general solution of the problem.

4.3. The resolvent equations for regular directed graphs. We determine now the resolvent equations for the spectrum ρ_ℓ of infinitely large (ℓ, c)-regular directed Husimi graphs. These graphs have $|\partial_i^{(\ell-1)}| = c$ for $i = 1, \ldots, N$, i.e., each vertex is incident to c cycles of length ℓ. We set $A_{ij} = 1$ and $A_{ji} = 0$ when there is a directed edge from node i to j, such that the corresponding matrix $\mathcal{J}_{ij}$ assumes the form $\mathcal{J} = \frac{1}{2}(\sigma_x + i\sigma_y)$. As a consequence, the matrices $\{\mathcal{D}_\beta^{(i)}, \mathbb{A}_{i\beta}, \mathbb{L}_\beta\}$ become independent of the indices (i, β). It is convenient to define the two-dimensional matrix $\boldsymbol{G}_A = \mathbb{A}^T \mathcal{D} \mathbb{A}$, where $\mathbb{A}^T = (\mathcal{J}\ \mathbf{0}\ \ldots\ \mathbf{0}\ \mathcal{J}^T)$. We write $\rho(\lambda)$ in terms of $\boldsymbol{G}_A$ as follows

$$\rho_\ell(\lambda) = \frac{1}{i\pi} \lim_{\epsilon \to 0} \partial^* \left[\boldsymbol{S}_\epsilon(\lambda) + c\,\boldsymbol{G}_A \right]_{21}^{-1}. \tag{4.10}$$

From eqs. (4.8) and (4.9) one obtains that, for $\ell > 2$, the two-dimensional matrix $\boldsymbol{G}_A$ solves the equation

$$\boldsymbol{G}_A = \mathbb{A}^T \Big[\big(\boldsymbol{S}_\epsilon(\lambda) + (c-1)\boldsymbol{G}_A \big) \otimes \boldsymbol{I}_{l-1} + i\mathcal{J} \otimes \boldsymbol{L}_{\ell-1} + i\mathcal{J}^T \otimes \boldsymbol{L}_{\ell-1}^T \Big]^{-1} \mathbb{A}, \tag{4.11}$$

where $\boldsymbol{L}_{\ell-1}$ is a $(\ell-1)$-dimensional matrix with elements $[\boldsymbol{L}_{\ell-1}]_{ij} = \delta_{i,j+1}$. The derivative of eq. (4.11) yields an equation in $\partial^* \boldsymbol{G}_A$, which has to be solved together with (4.11) to find $\rho_\ell(\lambda)$ through eq. (4.10).

Equation (4.11) allows to derive accurate numerical results for the spectrum of directed Husimi graphs as a function of ℓ. As an illustration, we present in figure 6 some cuts of $\rho_3(\lambda)$ and $\rho_4(\lambda)$ along the real direction for fixed values of y. These results correspond very well with direct diagonalization, confirming the exactness of (4.11). For a three-dimensional graph of $\rho_3(\lambda)$ we refer the reader to [**32**]. Analogously to undirected Husimi graphs, $\rho_\ell(\lambda)$ converges to the spectrum of

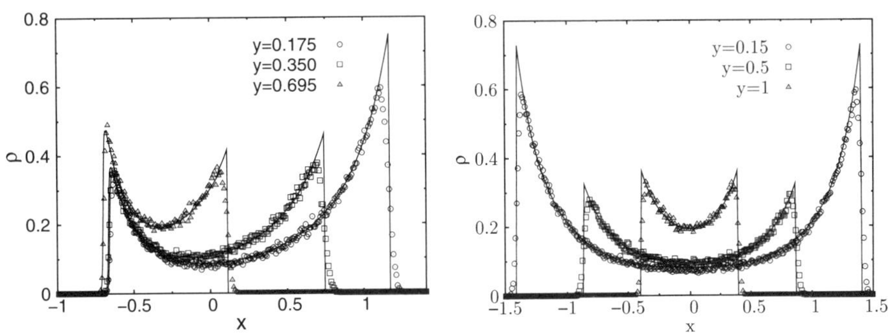

(a) $\ell = 3$

(b) $\ell = 4$

FIGURE 6. Three cuts of the spectrum $\rho_\ell(\lambda)$ of (ℓ, c)-regular directed Husimi graphs along the real direction for $c = 2$, $\ell = 3$ [**32**] and $c = 2$, $\ell = 4$. These results (solid lines) are obtained from the numerical solution of eq. (4.11) and they are compared with direct diagonalization results (markers) for an ensemble of 3×10^4 matrices of dimension $N = 10^3$.

a directed regular graph without short cycles for $\ell \to \infty$ [**20, 31, 52**]:

$$\rho_\infty(\lambda) = \frac{c-1}{\pi} \left(\frac{c}{c^2 - |\lambda|^2} \right)^2, \tag{4.12}$$

for $|\lambda|^2 < c$, and $\rho_\infty(\lambda) = 0$ otherwise. This convergence is shown numerically in [**32**]. By rescaling $A_{ij} \to A_{ij}/\sqrt{c-1}$, the solution of (4.11) leads to Girko's law in the limit $c \to \infty$.

4.4. The spectral boundaries for regular directed graphs. In order to derive analytical equations for the support of $\rho_\ell(\lambda)$, we determine the inverse of the $2(\ell-1) \times 2(\ell-1)$ matrix present in eq. (4.11). Since this matrix has a tridiagonal block structure its inverse can be computed analytically using the method in [**44**]. Applying this scheme to the matrix in eq. (4.11), we have simplified (4.11) into an equation involving sums and products of only two-dimensional matrices. This equation forms the equivalent for directed regular graphs of the equation (3.15) for undirected regular graphs. The resultant equations in $\boldsymbol{G}_A$ can be solved using the following *ansatz* [**14**]

$$\boldsymbol{G}_A = \begin{pmatrix} a & ib \\ ib^* & d \end{pmatrix}, \tag{4.13}$$

where b is complex and a and d are both real variables. The Hermitian part of this matrix is positive-definite provided that a and d are positive. This condition ensures that the Gaussian integrals in the cavity method are convergent [**14**]. Solving numerically eq. (4.11) for a finite regularizer $\epsilon > 0$, we find that $a > 0$ and $d > 0$. In the limit $\epsilon \to 0^+$, a and d vanish at the boundary of $\rho(\lambda)$. Therefore, setting $\epsilon \to 0^+$ and $a = d = 0$ in eq. (4.11), and solving the resulting equations for b leads to an analytical expression for the support of $\rho_\ell(\lambda)$. In this way we obtain the following equations for b and λ at the boundaries of the support:

- $\ell = 3$:

$$b = \frac{[\lambda^* - (c-1)b^*]^2}{|\lambda - (c-1)b|^4}, \tag{4.14}$$

$$|\lambda - (c-1)b|^4 = (c-1)\left[1 + |\lambda - (c-1)b|^2\right]. \tag{4.15}$$

- $\ell = 4$:

$$b = \frac{[\lambda^* - (c-1)b^*]^2}{|\lambda - (c-1)b|^6}, \tag{4.16}$$

$$|\lambda - (c-1)b|^6 = (c-1)\left[1 + |\lambda - (c-1)b|^2 + |\lambda - (c-1)b|^4\right]. \tag{4.17}$$

The solutions of the polynomials (4.15) and (4.17) in the variable $s \equiv |\lambda - (c-1)b|$ are given by

- $\ell = 3$:

$$s = \left[\frac{(c-1)}{2}\left(1 + \sqrt{1 + \frac{4}{(c-1)}}\right)\right]^{\frac{1}{2}}, \tag{4.18}$$

- $\ell = 4$:

$$s = \left[[R_+(c)]^{\frac{1}{3}} + [R_-(c)]^{\frac{1}{3}} + \frac{1}{3}(c-1)\right]^{\frac{1}{2}}, \tag{4.19}$$

where we have defined

$$R_\pm(c) = \frac{1}{27}(c-1)^3 + \frac{1}{3}(c-1)^2\left(\frac{1}{2} \pm F(c)\right) + \frac{1}{2}(c-1),$$

$$F(c) = \left[\frac{1}{81}(c-1)^2 + \frac{1}{9}(c-1)\kappa + \frac{1}{4}\kappa^2 - \frac{1}{81}(c-1)^2\kappa^3\right]^{\frac{1}{2}},$$

with $\kappa = 1 + \frac{3}{(c-1)}$. By parametrizing $\lambda - (c-1)b = s\exp(it)$, with $t \in [0, 2\pi]$, and substituting this form in eqs. (4.14) and (4.16), we find the following expressions for the boundary of the support $\lambda_\ell(t)$ of triangular ($\ell = 3$) and square ($\ell = 4$) regular directed Husimi graphs:

- $\ell = 3$:

$$\lambda_3(t) = s\exp(it) + \frac{(c-1)}{s^2}\exp(-2it), \tag{4.20}$$

- $\ell = 4$:

$$\lambda_4(t) = s\exp(it) + \frac{(c-1)}{s^3}\exp(-3it). \tag{4.21}$$

These are the parametric equations which describe, for each corresponding ℓ, an hypotrochoid in the complex plane. The parameter s as a function of the cycle degree c for $\ell = 3$ and $\ell = 4$ is given by, respectively, eqs. (4.18) and (4.19).

A hypotrochoid is a cyclic function in the complex plane which is drawn by rotating a small circle of radius r in a larger circle of radius R [**53**]. The support of triangular and square Husimi graphs is therefore given by hypotrochoids with, respectively, $R/r = 3$ and $R/r = 4$. These analytical results for the support of the spectra of directed Husimi graphs for $\ell = 3$ and $\ell = 4$ are shown in the lower graphs of figure 7. The agreement with direct diagonalization results for $c = 2$ is excellent, confirming the exactness of our analytical results. Based on the form of

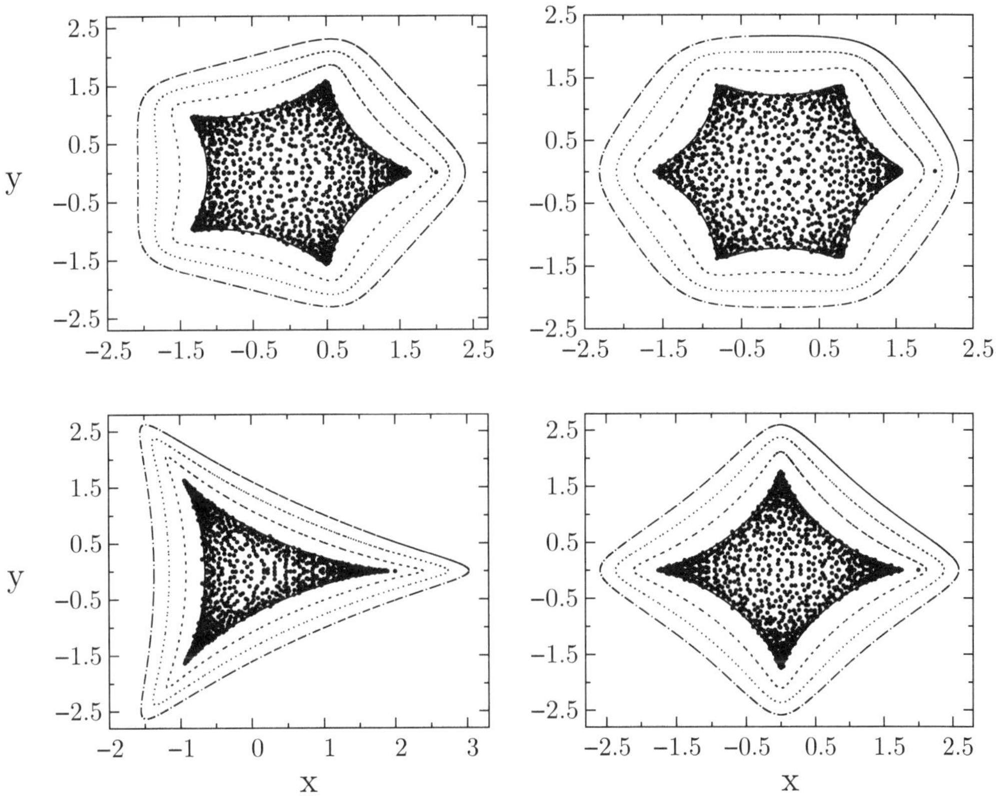

FIGURE 7. Analytical solutions for the boundary of the support of directed (ℓ, c)-regular Husimi graphs for several values of the cycle length ℓ and the following values c of the number of cycles incident to each vertex: $c = 2$ (solid line), $c = 3$ (dashed line), $c = 4$ (dotted line) and $c = 5$ (dot-dashed line). The hypotrochoids have a rotational symmetry by the angle $2\pi/\ell$, from which one obtains the value of the cycle length ℓ. Direct diagonalization results (dots) for $N \times N$ matrices with $N = 1000$ and $c = 2$ are shown.

eqs. (4.14-4.17), we conjecture that, for a given ℓ and c, the following equations are fulfilled at the boundary of the support of $\rho_\ell(\lambda)$

$$b = \frac{\left[\lambda^* - (c-1)b^*\right]^{\ell-1}}{s^{2(\ell-1)}}, \tag{4.22}$$

$$s^{2(\ell-1)} = (c-1)\sum_{n=0}^{\ell-2} s^{2n}. \tag{4.23}$$

Substituting $\lambda - (c-1)b = s\exp(it)$ ($t \in [0, 2\pi]$) in eq. (4.22) reads

$$\lambda_\ell(t) = s\exp(it) + \frac{(c-1)}{s^{\ell-1}}\exp\left[-i(\ell-1)t\right] \tag{4.24}$$

for the boundary of the support of a (ℓ, c) directed regular Husimi graph. Remarkably, eq. (4.24) is a hypotrochoid with a fraction $R/r = \ell$. The parameter s is determined from the roots of a polynomial of degree ℓ in the variable s^2, see eq. (4.23). Equation (4.23) can also be written as

$$s^2 = c - \frac{(c-1)}{s^{2(\ell-1)}}. \tag{4.25}$$

We have found an analytical expression for the roots for $\ell = 3$ and $\ell = 4$. For larger values of ℓ, we have solved eq. (4.25) numerically and, by choosing the stable solution, we have derived accurate values for the parameters of the hypotrochoids. We present explicit results for $\ell = 5$ and $\ell = 6$ in the upper graphs of figure 7. Direct diagonalization results exhibit once more an excellent agreement with the theoretical results, strongly supporting our conjecture that the support of the spectrum of directed regular Husimi graphs for general ℓ is given by eqs. (4.24-4.25). The support of regular Husimi graphs converges to the circle $|\lambda|^2 = c$ in the limit $l \to \infty$, corresponding with the expression (4.12) for a graph without short cycles.

5. Conclusion

In this work we have obtained the spectra of (un)directed Husimi graphs. The main result is a set of exact equations which determines a belief-propagation like algorithm in the resolvent elements of the matrix. For irregular graphs we have shown a very good correspondence between direct diagonalization results and our approach. For regular graphs we have derived several novel analytical expressions for the spectrum of undirected Husimi graphs and the boundaries of the spectrum of directed Husimi graphs. Remarkably, the boundaries of directed regular Husimi graphs consist of hypotrochoid functions in the complex plane.

Our results indicate that, at high connectivities, the spectrum of undirected random graphs converges to the Wigner semicircle law, while the spectrum of directed random graphs converges to Girko's circular law. This convergence seems to be rather universal and independent of the specific graph topology. It would be interesting to better understand the conditions under which finitely connected graphs converge to these limiting laws [**21**, **24**, **54**, **55**].

Finally, we point out that the eigenvalues of random unistochastic matrices are distributed over hypocycloids in the complex plane [**56**]. This close similarity with our results suggests an interesting connection between the spectra of $\ell \times \ell$ unistochastic matrices and regular directed Husimi graphs with cycles of length ℓ.

Acknowledgments

This paper is dedicated to Fritz Gesztesy, on the occasion of his 60th birthday. The first author wants to thank Fritz not only for many years of stimulating and fruitful collaborations but especially for a lifetime friendship! The second author is indebted to Karol Życzkowski for illuminating discussions.

Appendix A. Cavity method applied to random matrices

We present the essential steps to determine the resolvent equations using the cavity method [**37, 57**]. This method is based on the introduction of cavities in a graph $\mathcal{G}$ forming subgraphs $\mathcal{G}^{(i)}$, where the node i and all its incident edges have been removed, see figure 8. In analogy one can also remove the i-th column and the i-th row from a matrix $\boldsymbol{A}$ to obtain the submatrix $\boldsymbol{A}^{(i)}$.

A.1. Resolvent equations for locally tree-like graphs. The cavity method is based on the consideration that a probability distribution $P(\boldsymbol{x}; z)$ defined on a locally tree-like graph has the factorization property:

$$P^{(j)}_{\partial_j}(x_{\partial_j}; z) = \prod_{k\in\partial_j} P^{(j)}_k(x_k; z). \tag{A.1}$$

The quantity $P^{(j)}_i$ is the i-th marginal of $P^{(j)}(\boldsymbol{x}; z)$ on the cavity subgraph $\mathcal{G}^{(j)}$ of $\mathcal{G}$, and $P^{(j)}_{\partial_j}$ is the marginal of $P^{(j)}(\boldsymbol{x}; z)$ with respect to the set of variables ∂_j. In the language of spin models, the factorization property follows from the locally tree-like structure of a typical neighbourhood in the graph, see figure 8.

Following the derivation as presented in [**13**], we find a set of closed equations in the marginals $P^{(j)}_i$

$$\begin{aligned} P^{(j)}_i(x_i; z) &\sim \exp\left(-i\frac{zx_i^2}{2}\right) \\ &\times \int \left(\prod_{k\in\partial i\setminus j} dx_k P^{(i)}_k(x_k; z)\right) \exp\left(ix_i \sum_{k\in\partial_i\setminus j} A_{ik}x_k\right), \end{aligned} \tag{A.2}$$

from which the marginals P_i follow:

$$\begin{aligned} P_i(x_i; z) &\sim \exp\left(-i\frac{zx_i^2}{2}\right) \\ &\times \int \left(\prod_{k\in\partial i} dx_k P^{(i)}_k(x_k; z)\right) \exp\left(ix_i \sum_{k\in\partial_i} A_{ik}x_k\right). \end{aligned} \tag{A.3}$$

Finally, we use the fact that the $P^{(j)}_i$ are Gaussian functions

$$P^{(j)}_i(x_i; z) = \sqrt{\frac{i}{2\pi G^{(j)}_i(z)}} \exp\left(-i\frac{x_i^2}{2G^{(j)}_i(z)}\right), \tag{A.4}$$

to recover the resolvent equations (3.8), after substitution of (A.4) in (A.2) and (A.3).

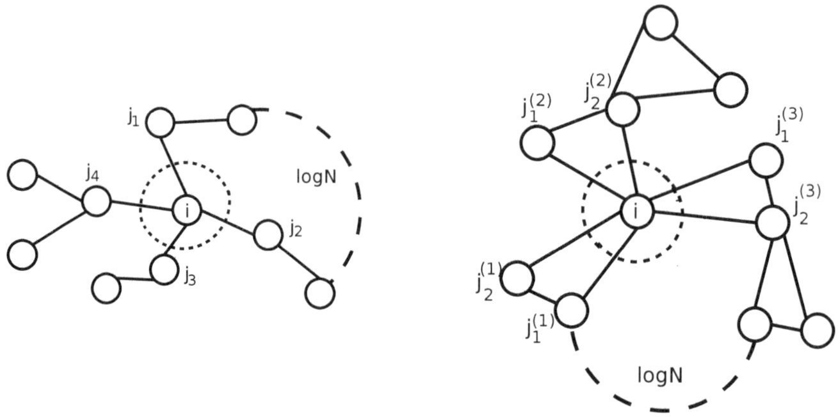

FIGURE 8. Left: a graph with a locally tree-like structure. Introducing a cavity at the vertex i decouples neighbouring vertices (in the sense that their mutual distance scales with $\log(N)$). Right: a Husimi graph. After introducing a cavity at the vertex i neighbouring pair of nodes become decoupled.

A.2. Resolvent equations for Husimi graphs. We apply now a similar logic to graphs with many cycles which have an infinite-dimensional structure. In this case we use the factorization property

$$P^{(i)}_{\partial_i^{(\ell-1)}}\left(x_{\partial_i^{(\ell-1)}};z\right) = \prod_{\alpha\in\partial_i^{(\ell-1)}} P^{(i)}_{\alpha}\left(x_{\alpha};z\right). \tag{A.5}$$

on the marginals of the distribution $P(\boldsymbol{x};z)$. The factorization property follows from the fact that the average distance between different branches connected to i is of the order $\ln N$ after removal of i, see figure 8. Note that $P^{(i)}_{\alpha}$ is the marginal of $P(\boldsymbol{x};z)$ for a $(\ell-1)$ tuple $\alpha=(j_1,\ldots,j_{\ell-1})$ forming a cycle with the i-th vertex.

We have generalized the derivation for regular Husimi graphs presented in [**32**] to the case of arbitrary Husimi graphs. We find a set of closed equations in the marginals $P^{(i)}_{\alpha}$:

(A.6)

$$\begin{aligned} P^{(i)}_{\alpha}\left(x_{\alpha};z\right) \sim{}& \exp\left(-\frac{iz}{2}\sum_{k=1}^{\ell-1} x^2_{j_k^{(\alpha)}} + i\sum_{k=1}^{\ell-2} A_{j_k^{(\alpha)}\, j_{k+1}^{(\alpha)}}\, x_{j_k^{(\alpha)}} x_{j_{k+1}^{(\alpha)}}\right) \\ &\times \int \prod_{j\in\alpha}\left[\prod_{\beta\in\partial_j^{(\ell-1)}\setminus\alpha} dx_\beta P^{(j)}_{\beta}\left(x_\beta;z\right)\right] \\ &\times \prod_{j\in\alpha}\left[\prod_{\beta\in\partial_j^{(\ell-1)}\setminus\alpha} \exp\left(i\sum_{j\in\alpha}\sum_{\beta\in\partial_j^{(\ell)}\setminus\alpha}\left(A_{jj_1^{(\beta)}}\, x_{j_1^{(\beta)}} x_j + A_{j_{\ell-1}^{(\beta)} j}\, x_{j_{\ell-1}^{(\beta)}} x_j\right)\right)\right], \end{aligned}$$

with the $(\ell-1)$-tuple $\alpha = \left(j_1^{(\alpha)}, j_2^{(\alpha)}, \cdots, j_{\ell-1}^{(\alpha)}\right)$. The marginals P_i are given as a function of the marginals $P_\alpha^{(i)}$:

(A.7)
$$P_i(x_i; z) \sim \exp\left(-i\frac{z x_i^2}{2}\right) \times \int \prod_{\alpha\in\partial_i^{(\ell-1)}} dx_\alpha P_\alpha^{(i)}(x_\alpha; z) \prod_\alpha \exp\left(i\, A_{i j_1^{(\alpha)}}\, x_{j_1^{(\alpha)}}\, x_i + i\, A_{j_{\ell-1}^{(\alpha)} i}\, x_{j_{\ell-1}^{(\alpha)}}\, x_i\right).$$

We now use the Gaussian ansatz for the marginals $P_\alpha^{(i)}(x_\alpha; z)$:

$$P_\alpha^{(i)}(x_\alpha; z) \sim \exp\left(-\frac{i}{2}\boldsymbol{x}^T \left(\boldsymbol{G}_\alpha^{(i)}\right)^{-1} \boldsymbol{x}\right), \tag{A.8}$$

with $\boldsymbol{G}_\alpha^{(i)}$ the $(\ell-1)\times(\ell-1)$ submatrix of the resolvent $\boldsymbol{G}^{(i)}$ of the cavity matrix $\boldsymbol{A}^{(i)}$:

$$\boldsymbol{G}_\alpha^{(i)} = \begin{pmatrix} G^{(i)}_{j_1^{(\alpha)} j_1^{(\alpha)}} & G^{(i)}_{j_1^{(\alpha)} j_2^{(\alpha)}} & \cdots & G^{(i)}_{j_1^{(\alpha)} j_{\ell-1}^{(\alpha)}} \\ G^{(i)}_{j_2^{(\alpha)} j_1^{(\alpha)}} & G^{(i)}_{j_2^{(\alpha)} j_2^{(\alpha)}} & \cdots & G^{(i)}_{j_2^{(\alpha)} j_{\ell-1}^{(\alpha)}} \\ \vdots & \vdots & \cdots & \vdots \\ G^{(i)}_{j_{\ell-1}^{(\alpha)} j_1^{(\alpha)}} & G^{(i)}_{j_{\ell-1}^{(\alpha)} j_2^{(\alpha)}} & \cdots & G^{(i)}_{j_{\ell-1}^{(\alpha)} j_{\ell-1}^{(\alpha)}} \end{pmatrix}. \tag{A.9}$$

We also use a Gaussian ansatz of the type (A.4) for the marginal P_i. Substitution of these ansätze in (A.6) and (A.7) gives the resolvent equations (3.9) and (3.10).

We remark that the resolvent equations can also be derived using methods from random matrix theory , i.e. by recursively applying the Schur-complement formula [**19**, **31**].

References

[1] Bojan Mohar, *Some applications of Laplace eigenvalues of graphs*, Graph symmetry (Montreal, PQ, 1996), NATO Adv. Sci. Inst. Ser. C Math. Phys. Sci., vol. 497, Kluwer Acad. Publ., Dordrecht, 1997, pp. 225–275. MR1468791 (98g:05096)

[2] Fan R. K. Chung, *Spectral graph theory*, CBMS Regional Conference Series in Mathematics, vol. 92, Published for the Conference Board of the Mathematical Sciences, Washington, DC, 1997. MR1421568 (97k:58183)

[3] Andries E. Brouwer and Willem H. Haemers, *Spectra of graphs*, Universitext, Springer, New York, 2012. MR2882891

[4] Bojan Mohar and Wolfgang Woess, *A survey on spectra of infinite graphs*, Bull. London Math. Soc. **21** (1989), no. 3, 209–234, DOI 10.1112/blms/21.3.209. MR986363 (90d:05162)

[5] Alain Barrat, Marc Barthélemy, and Alessandro Vespignani, *Dynamical processes on complex networks*, Cambridge University Press, Cambridge, 2008. MR2797803 (2012e:05001)

[6] M. E. J. Newman, *Networks*, Oxford University Press, Oxford, 2010. An introduction. MR2676073 (2011h:05002)

[7] M. Baharona, L. M. Pecora, *Synchronization in small-world systems*, Phys. Rev. Lett. **89** (2002), 054101.

[8] S. F. Edwards, D. R. Wilkinson, *The surface statistics of granular aggregate*, Proc. R. Soc. London, Ser. A **381**, 17 (1982); B. Kozma, M. B. Hastings, G. Korniss, *Roughness scaling for Edwards-Anderson relaxation in small-world networks*, Phys. Rev. Lett. **92**, 108701 (2004).

[9] D. J. Klein and M. Randić, *Resistance distance*, J. Math. Chem. **12** (1993), no. 1-4, 81–95, DOI 10.1007/BF01164627. Applied graph theory and discrete mathematics in chemistry (Saskatoon, SK, 1991). MR1219566 (94d:94041)

[10] Shlomo Hoory, Nathan Linial, and Avi Wigderson, *Expander graphs and their applications*, Bull. Amer. Math. Soc. (N.S.) **43** (2006), no. 4, 439–561 (electronic), DOI 10.1090/S0273-0979-06-01126-8. MR2247919 (2007h:68055)

[11] S. F. Edwards, R. C. Jones, *The Eigenvalue Spectrum of a Large Symmetric Random Matrix*, J. Phys. A **9** (1976), 1595-1603.

[12] Reimer Kühn, *Spectra of sparse random matrices*, J. Phys. A **41** (2008), no. 29, 295002, 21, DOI 10.1088/1751-8113/41/29/295002. MR2455271 (2010d:82052)

[13] Tim Rogers, Isaac Pérez Castillo, Reimer Kühn, and Koujin Takeda, *Cavity approach to the spectral density of sparse symmetric random matrices*, Phys. Rev. E (3) **78** (2008), no. 3, 031116, 6, DOI 10.1103/PhysRevE.78.031116. MR2496088 (2010c:82005)

[14] T. Rogers, I. P. Castillo, *Cavity approach to the spectral density of non-hermitian sparse matrices*, Phys. Rev. E **79** (2009), 012101.

[15] Yan V. Fyodorov and A. D. Mirlin, *On the density of states of sparse random matrices*, J. Phys. A **24** (1991), no. 9, 2219–2223. MR1117870 (92f:82022)

[16] Tim Rogers, Conrad Pérez Vicente, Koujin Takeda, and Isaac Pérez Castillo, *Spectral density of random graphs with topological constraints*, J. Phys. A **43** (2010), no. 19, 195002, 20, DOI 10.1088/1751-8113/43/19/195002. MR2639125 (2011h:82032)

[17] Reimer Kühn and Jort van Mourik, *Spectra of modular and small-world matrices*, J. Phys. A **44** (2011), no. 16, 165205, 18, DOI 10.1088/1751-8113/44/16/165205. MR2787089 (2012a:05192)

[18] Yoshiyuki Kabashima and Hisanao Takahashi, *First eigenvalue/eigenvector in sparse random symmetric matrices: influences of degree fluctuation*, J. Phys. A **45** (2012), no. 32, 325001, 19, DOI 10.1088/1751-8113/45/32/325001. MR2959198

[19] Charles Bordenave and Marc Lelarge, *Resolvent of large random graphs*, Random Structures Algorithms **37** (2010), no. 3, 332–352, DOI 10.1002/rsa.20313. MR2724665 (2011g:60063)

[20] Charles Bordenave and Djalil Chafaï, *Around the circular law*, Probab. Surv. **9** (2012), 1–89, DOI 10.1214/11-PS183. MR2908617

[21] Terence Tao and Van Vu, *Random matrices: the circular law*, Commun. Contemp. Math. **10** (2008), no. 2, 261–307, DOI 10.1142/S0219199708002788. MR2409368 (2009d:60091)

[22] Terence Tao and Van Vu, *Random matrices: universality of ESDs and the circular law*, Ann. Probab. **38** (2010), no. 5, 2023–2065, DOI 10.1214/10-AOP534. With an appendix by Manjunath Krishnapur. MR2722794 (2011e:60017)

[23] Friedrich Götze and Alexander Tikhomirov, *The circular law for random matrices*, Ann. Probab. **38** (2010), no. 4, 1444–1491, DOI 10.1214/09-AOP522. MR2663633 (2012a:60011)

[24] Philip Matchett Wood, *Universality and the circular law for sparse random matrices*, Ann. Appl. Probab. **22** (2012), no. 3, 1266–1300. MR2977992

[25] Charles Bordenave, Pietro Caputo, and Djalil Chafaï, *Circular law theorem for random Markov matrices*, Probab. Theory Related Fields **152** (2012), no. 3-4, 751–779, DOI 10.1007/s00440-010-0336-1. MR2892961

[26] P. M. Gleiss, P. F. Stadler, A. Wagner, D. A. Fell, *Small cycles in small worlds*, arxiv:cond-mat/0009124 (2000).

[27] G. Bianconi, G. Caldarelli, A. Capocci, *Loops structure of the internet at the autonomous system level*, Phys. Rev. E **71** (2005), 066116.

[28] G. Bianconi, N. Gulbahce, A. E. Motter, *Local structure of directed networks*, Phys. Rev. Lett. **100** (2008), 118701.

[29] Kodi Husimi, *Note on Mayers' theory of cluster integrals*, J. Chem. Phys. **18** (1950), 682–684. MR0038903 (12,467i)

[30] Frank Harary and George E. Uhlenbeck, *On the number of Husimi trees. I*, Proc. Nat. Acad. Sci. U. S. A. **39** (1953), 315–322. MR0053893 (14,836a)

[31] I. Neri, F. L. Metz, *Spectra of sparse non-Hermitian random matrices: an analytical solution*, Phys. Rev. Lett. **109** (2012), 030602

[32] F. L. Metz, I. Neri, D. Bollé, *Spectra of sparse regular graphs with loops*, Phys. Rev. E **84** (2011), 055101(R).

[33] Brendan D. McKay, *The expected eigenvalue distribution of a large regular graph*, Linear Algebra Appl. **40** (1981), 203–216, DOI 10.1016/0024-3795(81)90150-6. MR629617 (84h:05089)

[34] O. Khorunzhy, M. Shcherbina, and V. Vengerovsky, *Eigenvalue distribution of large weighted random graphs*, J. Math. Phys. **45** (2004), no. 4, 1648–1672, DOI 10.1063/1.1667610. MR2043849 (2005h:05186)

[35] Béla Bollobás, *Random graphs*, 2nd ed., Cambridge Studies in Advanced Mathematics, vol. 73, Cambridge University Press, Cambridge, 2001. MR1864966 (2002j:05132)
[36] F. L. Metz, I. Neri, and D. Bollé, *Localization transition in symmetric random matrices*, Phys. Rev. E (3) **82** (2010), no. 3, 031135, 10, DOI 10.1103/PhysRevE.82.031135. MR2787990 (2012e:82030)
[37] Marc Mézard and Andrea Montanari, *Information, physics, and computation*, Oxford Graduate Texts, Oxford University Press, Oxford, 2009. MR2518205 (2010k:94019)
[38] H. A. Bethe, *Statistical theory of superlattices*, Proc. Roy. Soc. London Ser A **150** (1935), 552-575.
[39] R. Abou-Chacra, D. J. Thouless, P. W. Anderson, *A selfconsistent theory of localization*, J. Phys. C: Solid State Phys. **6** (1973), 1734
[40] P. Cizeau, J. P. Bouchaud,*Theory of Lévy matrices*, Phys. Rev. E **50** (1994), 1810-1822.
[41] Judea Pearl, *Probabilistic reasoning in intelligent systems: networks of plausible inference*, The Morgan Kaufmann Series in Representation and Reasoning, Morgan Kaufmann, San Mateo, CA, 1988. MR965765 (90g:68003)
[42] Ryoichi Kikuchi, *A theory of cooperative phenomena*, Physical Rev. (2) **81** (1951), 988–1003. MR0041069 (12,788f)
[43] W. T. Yedidia, J. S. Freeman, Y. Weiss, *Bethe free energy, kikuchi approximations, and belief propagation algorithms*, Technical Report TR-2001-16, Mitsubishi Electric Reseach, 2001.
[44] Y. Huang and W. F. McColl, *Analytical inversion of general tridiagonal matrices*, J. Phys. A **30** (1997), no. 22, 7919–7933, DOI 10.1088/0305-4470/30/22/026. MR1616951 (99h:15007)
[45] R. Bruce King, *Beyond the quartic equation*, Birkhäuser Boston Inc., Boston, MA, 1996. MR1401346 (97g:12002)
[46] Harry Kesten, *Symmetric random walks on groups*, Trans. Amer. Math. Soc. **92** (1959), 336–354. MR0109367 (22 #253)
[47] M. Eckstein, M. Kollar, K. Byczuk, D. Vollhardt, *Hopping on the Bethe lattice: Exact results for density of states and dynamical mean-field theory*, Phys. Rev. B **71**, 235119 (2005); M. Galiceanu, A. Blumen, *Spectra of Husimi cacti: exact results and applications*, J. Chem. Phys.**127**, 134904 (2007).
[48] Joshua Feinberg and A. Zee, *Non-Hermitian random matrix theory: method of Hermitian reduction*, Nuclear Phys. B **504** (1997), no. 3, 579–608, DOI 10.1016/S0550-3213(97)00502-6. MR1488584 (98g:82019)
[49] J. T. Chalker, Z. J. Wang, *Diffusion in a random velocity field: Spectral properties of a non-hermitian Fokker-Planck operator*, Phys. Rev. Lett. **79**, 1797 (1997).
[50] J. T. Chalker, B. Mehlig, *Eigenvector statistics in non-hermitian random matrix ensembles*, Phys. Rev. Lett. **81**, 3367 (1998).
[51] R. A. Janik, W. Nörenberg, M. A. Nowak, G.Papp, I. Zahed, *Correlation of eigenvectors for non-hermitian random matrix models*, Phys. Rev. E **60**, 2699 (1999).
[52] T.Rogers, *New results on the spectral density of random matrices*, thesis (2010).
[53] J. D. Lawrence, *A Catalog of Special Plane Curves*, New York: Dover, pp. 165-168, (1972).
[54] I. Dumitriu, S. Pal, *Sparse regular random graphs: spectral density and eigenvectors* , arXiv:0910.5306.
[55] L. Tran, V. Vu, K. Wang, *Sparse random graphs: Eigenvalues and Eigenvectors*, arXiv:1011.6646
[56] Karol Życzkowski, Marek Kuś, Wojciech Słomczyński, and Hans-Jürgen Sommers, *Random unistochastic matrices*, J. Phys. A **36** (2003), no. 12, 3425–3450, DOI 10.1088/0305-4470/36/12/333. Random matrix theory. MR1986428 (2004f:15044)
[57] Marc Mézard, Giorgio Parisi, and Miguel Angel Virasoro, *Spin glass theory and beyond*, World Scientific Lecture Notes in Physics, vol. 9, World Scientific Publishing Co. Inc., Teaneck, NJ, 1987. MR1026102 (91k:82066)

Instituut voor Theoretische Fysica, KULeuven, Celestijnenlaan 200D, B-3001 Leuven, Belgium
E-mail address: desire.bolle@fys.kuleuven.be

Dipartimento di Fisica, Sapienza Università di Roma, Piazzale A. Moro 2, 00185 Roma, Italy
E-mail address: fmetzfmetz@gmail.com

Université Montpellier 2, Laboratoire Charles Coulomb UMR 5221, F-34095, Montpellier, France
E-mail address: izaakneri@gmail.com

Proceedings of Symposia in Pure Mathematics
Volume **87**, 2013
http://dx.doi.org/10.1090/pspum/087/01439

Jones' extension operator on Sobolev spaces with partially vanishing traces

Kevin Brewster, Dorina Mitrea, Irina Mitrea, and Marius Mitrea

Dedicated to Fritz Gesztesy's 60-th birthday

ABSTRACT. We prove that for each (ε,δ)-domain $\Omega\subseteq\mathbb{R}^n$, any closed subset D of $\overline{\Omega}$, any $k\in\mathbb{N}$ and any $p\in[1,\infty]$, P.W. Jones' extension operator Λ_k maps the space $W^{k,p}_D(\Omega)$ boundedly into $W^{k,p}_D(\mathbb{R}^n)$, where with $\mathcal{O}$ denoting either Ω or $\mathbb{R}^n$, the space $W^{k,p}_D(\mathcal{O})$ is defined as the completion in the classical Sobolev space $W^{k,p}(\mathcal{O})$ of (restrictions to $\mathcal{O}$ of) functions from $\mathscr{C}^\infty_c(\mathbb{R}^n)$ whose supports are disjoint from D. In turn, this result is used to provide a structural theorem pertaining to the nature of the space $W^{k,p}_D(\Omega)$ in an (ε,δ)-domain $\Omega\subseteq\mathbb{R}^n$.

1. Motivation

We debut by discussing some background definitions and results. Fix a space dimension $n\in\mathbb{N}$, $n\geq 2$, and denote by $\mathscr{L}^n$ the n-dimensional Lebesgue measure in $\mathbb{R}^n$. Given a Lebesgue measurable set $\mathcal{O}$ in $\mathbb{R}^n$, we let $L^p(\mathcal{O},\mathscr{L}^n)$, $0<p\leq\infty$, stand for the scale of (equivalent classes of) Lebesgue-measurable functions which are p-th power $\mathscr{L}^n$-integrable in $\mathcal{O}$. Also, given an open set $\mathcal{O}\subseteq\mathbb{R}^n$, for each $p\in(0,\infty]$ denote by $L^p_{loc}(\mathcal{O},\mathscr{L}^n)$ the space of Lebesgue-measurable functions u in $\mathcal{O}$ with the property that $u\big|_K\in L^p(K,\mathscr{L}^n)$ for every compact subset K of $\mathcal{O}$.

With $\mathbb{N}$ denoting the collection of all (strictly) positive integers, we shall abbreviate $\mathbb{N}_0:=\mathbb{N}\cup\{0\}$. In particular, $\mathbb{N}_0^n$ may be regarded as the set of all multi-indices $\{\alpha=(\alpha_1,...,\alpha_n):\alpha_i\in\mathbb{N}_0,\ 1\leq i\leq n\}$. As usual, for each multi-index $\alpha=(\alpha_1,...,\alpha_n)\in\mathbb{N}_0^n$ we denote by $|\alpha|:=\alpha_1+\cdots+\alpha_n$ its length. Also, write $\partial^\alpha:=\partial^{\alpha_1}_{x_1}\cdots\partial^{\alpha_n}_{x_n}$. For an arbitrary set $E\subseteq\mathbb{R}^n$ we shall denote by E°, $\overline{E}$, $\operatorname{diam}E$, $\operatorname{dist}(\cdot,E)$, and E^c, respectively the interior, closure, diameter, distance to, and complement of E in $\mathbb{R}^n$. In addition, $\operatorname{dist}(F,E)$, denotes the distance from F to E. As usual, $B(x,r):=\{y\in\mathbb{R}^n:|x-y|<r\}$ for each $x\in\mathbb{R}$ and $r\in(0,\infty)$, where $|\cdot|$ denotes the standard Euclidean norm. Finally, given an open subset $\mathcal{O}$ of $\mathbb{R}^n$, we shall denote by $\mathscr{C}^\infty_c(\mathcal{O})$ the collection of all indefinitely differentiable functions in $\mathcal{O}$, which vanish outside of a compact subset of $\mathcal{O}$.

2010 *Mathematics Subject Classification.* Primary: 46E35; Secondary 42B35.

Key words and phrases. Higher-order Sobolev space, linear extension operator, (ε,δ)-domain, uniform domain.

This author has been supported in part by Simons Foundation grant #200750.

This author has been supported in part by NSF grant DMS-1201736.

Assume now that $\Omega \subseteq \mathbb{R}^n$ is an arbitrary, nonempty, open set. Then for each positive integer $k \in \mathbb{N}$ and integrability exponent $p \in [1, \infty]$, the L^p-based Sobolev space of order k in Ω is defined intrinsically by

$$
\begin{aligned}
W^{k,p}(\Omega) := \big\{ u \in L^1_{loc}(\Omega, \mathscr{L}^n) : \partial^\alpha u \in L^p(\Omega, \mathscr{L}^n) \\
\text{for each } \alpha \in \mathbb{N}_0^n \ \text{ with } |\alpha| \leq k \big\},
\end{aligned}
\tag{1.1}
$$

where the derivatives are taken in the sense of distributions. As is well-known (cf., e.g., [**AF**, Theorem 3.3, p. 60]), $W^{k,p}(\Omega)$ becomes a Banach space when equipped with the natural norm

$$
\|u\|_{W^{k,p}(\Omega)} := \sum_{|\alpha| \leq k} \|\partial^\alpha u\|_{L^p(\Omega, \mathscr{L}^n)} \quad \text{for each } \ u \in W^{k,p}(\Omega). \tag{1.2}
$$

For future purposes let us also define

$$
\mathring{W}^{k,p}(\Omega) := \text{the closure of } \mathscr{C}_c^\infty(\Omega) \text{ in } \Big(W^{k,p}(\Omega) \, , \, \| \cdot \|_{W^{k,p}(\Omega)} \Big). \tag{1.3}
$$

For a wealth of information regarding the general topic of Sobolev spaces, the reader is referred to the informative discussion in the monograph [**M**].

Extension results for Sobolev spaces defined on open subsets of the Euclidean space are important tools in analysis. For example, a classical result in harmonic analysis asserts that any Lipschitz domain $\Omega \subseteq \mathbb{R}^n$ is a $W^{k,p}$-extension domain for all $k \in \mathbb{N}$ and $p \in [1, \infty]$. This means that there exists a bounded linear operator

$$
E : W^{k,p}(\Omega) \longrightarrow W^{k,p}(\mathbb{R}^n) \tag{1.4}
$$

with the property that $(Eu)\big|_\Omega = u$ for each $u \in W^{k,p}(\Omega)$. This theorem was proved by A.P. Calderón in [**C**] in the case $p \in (1, \infty)$. Subsequently, E.M. Stein extended it to the case $p \in [1, \infty]$ via a different proof (cf. [**S**, Theorem 5, p. 181]). However, the original proofs of Calderón's and Stein's theorems make essential use of the fact that Lipschitz domains satisfy a uniform cone property. The latter property actually characterizes Lipschitzianity, so new ideas must be involved if the goal is to establish extension results beyond this class of domains.

Via a conceptually novel approach, P.W. Jones succeeded (cf. [**J**, Theorem 1, p. 73]) in generalizing the results of Calderón and Stein to a much larger class of sets, which he called (ε, δ)-domains. Jones also proved that a finitely connected open set $\Omega \subseteq \mathbb{R}^2$ is a $W^{k,p}$-extension domain for all $k \in \mathbb{N}$ and $p \in [1, \infty]$ if and only if Ω is an (ε, δ)-domain for some values $\varepsilon, \delta > 0$ (cf. [**J**, Theorem 3, p. 74]). Since Jones' class of domains is going to be of basic importance for the goals we have in mind, below we record its actual definition.

Definition 1.1. Given $\varepsilon \in (0, \infty)$ and $\delta \in (0, \infty]$, call an nonempty, open, proper subset Ω of $\mathbb{R}^n$ an (ε, δ)-**domain** if for any $x, y \in \Omega$ with $|x - y| < \delta$ there exists a rectifiable curve $\gamma : [0, 1] \to \Omega$ such that

$$
\begin{aligned}
&\gamma(0) = x, \quad \gamma(1) = y, \quad \text{length}(\gamma) \leq \tfrac{1}{\varepsilon}|x - y|, \quad \text{and} \\
&\frac{\varepsilon |z - x| \, |z - y|}{|x - y|} \leq \text{dist}\,(z, \partial\Omega), \quad \forall \, z \in \gamma([0, 1]).
\end{aligned}
\tag{1.5}
$$

Informally, the third condition in (1.5) says that Ω is locally connected in some quantitative sense, while the fourth condition in (1.5) says that there exists some type of "tube" T, with $\gamma([0, 1]) \subset T \subset \Omega$ and the width of T at a point z on the curve is of the order $\min\{|z - x|, |z - y|\}$.

Examples of (ε,δ)-domains include bi-Lipschitz images of Lipschitz domains, open sets whose boundaries are given locally as graphs of functions in the Zygmund class Λ_1, or of functions with gradients in the John-Nirenberg space BMO, as well as the classical van Koch snowflake domain of conformal mapping theory. The boundary of an (ε,δ)-domain can be highly nonrectifiable and, in general, no regularity condition on $\partial\Omega$ can be inferred from the (ε,δ) property described in Definition 1.1. The fact that, in general, (ε,δ)-domains are not sets of finite perimeter can be seen from the fact that the classical van Koch snowflake domain does not have finite perimeter. Also, as pointed out in [**J**, p. 73], for each $\alpha\in[n-1,n)$ there exists an open set $\Omega\subseteq\mathbb{R}^n$ such that Ω is an (ε,∞)-domain for some $\varepsilon=\varepsilon(\alpha)\in(0,\infty)$ and $\mathcal{H}^\alpha(\mathcal{O}\cap\partial\Omega)>0$ for all open subsets $\mathcal{O}\subseteq\mathbb{R}^n$ satisfying $\mathcal{O}\cap\partial\Omega\neq\emptyset$, where $\mathcal{H}^\alpha$ denotes the α-dimensional Hausdorff measure in $\mathbb{R}^n$. This being said, it has been shown in [**J**, Lemma 2.3, p. 77] that

$$\text{any } (\varepsilon,\delta)\text{-domain } \Omega\subseteq\mathbb{R}^n \text{ satisfies } \mathscr{L}^n(\partial\Omega)=0. \tag{1.6}$$

Jones' (ε,δ)-domains interface tightly with the category of uniform domains considered a little earlier by O. Martio and J. Sarvas in [**MS**]. Recall that a nonempty, proper, open subset Ω of $\mathbb{R}^n$ is said to be a `uniform domain` provided there exists a constant $c=c(\Omega)\in[1,\infty)$ with the property that each pair of points $x_1,x_2\in\Omega$ can be joined by a rectifiable curve γ in Ω for which

$$\begin{aligned} &\text{length}\,(\gamma)\leq c\,|x_1-x_2| \quad\text{and}\\ &\min_{j=1,2}|x_j-x|\leq c\,\text{dist}\,(x,\partial\Omega) \quad\text{for each}\quad x\in\gamma. \end{aligned}\tag{1.7}$$

Then a nonempty, proper, open subset of the Euclidean space is an (ε,δ)-domain for some $\varepsilon,\delta>0$ if and only if it is a locally uniform domain.

The idea underpinning the construction of Jones' extension operator from [**J**, Theorem 1, p. 73] is to glue together, via a scale-sensitive partition of unity associated with a Whitney decomposition of the interior of the complement of the domain, certain polynomials which best fit the given function on the corresponding reflected cube across the boundary. While all this is made precise in Theorem 1.2 below, for the time being we briefly digress in order to clarify some terminology.

According to Whitney's decomposition lemma (cf. [**S**, Theorem 1, p. 167]), it is possible to associate to any open, nonempty, proper subset $\mathcal{O}$ of $\mathbb{R}^n$ a family $\mathcal{W}(\mathcal{O})=\{Q_j\}_{j\in\mathbb{N}}$ of countably many closed dyadic cubes from $\mathbb{R}^n$ such that

$$\mathcal{O}=\bigcup_{j\in\mathbb{N}}Q_j, \tag{1.8}$$

$$\sqrt{n}\,\ell(Q_j)\leq\text{dist}\,(Q_j,\partial\mathcal{O})\leq4\sqrt{n}\,\ell(Q_j),\quad\text{for all}\quad j\in\mathbb{N}, \tag{1.9}$$

$$Q_j^\circ\cap Q_k^\circ=\emptyset,\quad\text{for all}\quad j,k\in\mathbb{N}\quad\text{with}\quad j\neq k, \tag{1.10}$$

$$\tfrac14\,\ell(Q_j)\leq\ell(Q_k)\leq4\,\ell(Q_j),\quad\text{for all}\quad j,k\in\mathbb{N}\quad\text{with}\quad Q_j\cap Q_k\neq\emptyset. \tag{1.11}$$

Above, $\ell(Q)$ denotes the side-length of the cube Q, and Q° stands for the interior of Q. Also, given a positive number λ and a cube Q, we denote by λQ the cube with the same center x_Q as Q, and side-length $\lambda\ell(Q)$. With this convention, it is then straightforward to check that (1.9) implies

$$\text{if } \lambda\in(0,3) \text{ then } \lambda Q_j\subseteq\mathcal{O}\quad\text{for all}\quad j\in\mathbb{N}. \tag{1.12}$$

In fact, for each $\lambda \in (0,3)$ there exists $c_\lambda \in (0,1)$ such that

$$c_\lambda \leq \frac{\operatorname{dist}\big(\lambda Q_j, \partial\mathcal{O}\big)}{\ell(Q_j)} \leq c_\lambda^{-1}, \quad \text{for all } \ j \in \mathbb{N}. \tag{1.13}$$

Here is the result advertised earlier. It is a detailed statement of [**J**, Theorem 1, p. 73].

THEOREM 1.2 (P.W. Jones's Extension Theorem). *Let Ω be a finitely connected (ε,δ)-domain in $\mathbb{R}^n$ and fix $k \in \mathbb{N}$. Also, pick a Whitney decomposition $\mathcal{W}(\Omega)$ of Ω along with a Whitney decomposition $\mathcal{W}\big((\Omega^c)^\circ\big)$ of $(\Omega^c)^\circ$, and consider the collection of all small cubes in the latter, i.e., define*

$$\mathcal{W}_s\big((\Omega^c)^\circ\big) := \big\{Q \in \mathcal{W}\big((\Omega^c)^\circ\big) : \ell(Q) \leq \varepsilon\delta/(16n)\big\}. \tag{1.14}$$

For any function $u \in L^1_{loc}(\Omega, \mathscr{L}^n)$ and any dyadic cube $Q \in \mathcal{W}(\Omega)$ let $P_Q(u)$ denote the unique polynomial of degree $k-1$ which best fits u on Q in the sense that

$$\int_Q \partial^\alpha\big(u - P_Q(u)\big)\, d\mathcal{L}^n = 0 \ \text{ for each } \ \alpha \in \mathbb{N}_0^n \ \text{ with } \ |\alpha| \leq k-1. \tag{1.15}$$

To each $Q \in \mathcal{W}_s\big((\Omega^c)^\circ\big)$ assign a cube $Q^* \in \mathcal{W}(\Omega)$ satisfying (cf. [**J**, Lemma 2.4, p. 77]*)*

$$\ell(Q) \leq \ell(Q^*) \leq 4\,\ell(Q) \ \text{ and } \ \operatorname{dist}(Q, Q^*) \leq C_{n,\varepsilon}\,\ell(Q). \tag{1.16}$$

Finally, to each $u \in L^1_{loc}(\Omega, \mathscr{L}^n)$ associate the function $\Lambda_k u$ defined $\mathscr{L}^n$-a.e. in $\mathbb{R}^n$ by

$$\Lambda_k u := \begin{cases} u & \text{in } \ \Omega, \\[6pt] \displaystyle\sum_{Q \in \mathcal{W}_s((\Omega^c)^\circ)} P_{Q^*}(u)\,\varphi_Q & \text{in } \ \big(\Omega^c\big)^\circ, \end{cases} \tag{1.17}$$

where the family $\{\varphi_Q\}_{Q \in \mathcal{W}_s((\Omega^c)^\circ)}$ consists of functions satisfying

$$\begin{gathered} \varphi_Q \in \mathscr{C}_c^\infty(\mathbb{R}^n), \quad \operatorname{supp}\varphi_Q \subseteq \tfrac{17}{16}Q, \quad 0 \leq \varphi_Q \leq 1, \\ \big|\partial^\alpha \varphi_Q\big| \leq C_\alpha \ell(Q)^{-|\alpha|}, \quad \forall\, \alpha \in \mathbb{N}_0^n, \end{gathered} \tag{1.18}$$

for every $Q \in \mathcal{W}_s\big((\Omega^c)^\circ\big)$, as well as

$$\sum_{Q \in \mathcal{W}_s((\Omega^c)^\circ)} \varphi_Q \equiv 1 \qquad \text{on} \qquad \bigcup_{Q \in \mathcal{W}_s((\Omega^c)^\circ)} Q. \tag{1.19}$$

Then for every $p \in [1,\infty]$ the operator Λ_k satisfies

$$\Lambda_k : W^{k,p}(\Omega) \longrightarrow W^{k,p}(\mathbb{R}^n) \quad \text{linearly and boundedly}, \tag{1.20}$$

with operator norm controlled in terms of $\varepsilon, \delta, n, p, k$, and

$$\Lambda_k u\big|_\Omega = u, \quad \mathscr{L}^n\text{-a.e. on } \Omega \text{ for every } u \in W^{k,p}(\Omega). \tag{1.21}$$

As remarked by P.W. Jones, his extension operator Λ_k lacks certain distinguished features which Stein's and Calderón's extension operators enjoy. For example, while Stein's extension operator is universal (i.e., it does not depend on k), Jones's extension operator is not[1]. Moreover, as opposed to Calderón's extension

[1]This aspect has been addressed by L.G. Rogers in [**R**], where he has succeeded in suitably modifying Jones' construction of Λ_k presented in Theorem 1.2 as to make it independent of k.

operator, Jones' extension operator does not necessarily map functions from $\mathscr{C}_c^\infty(\Omega)$ into functions supported in $\overline{\Omega}$.

Here we wish to take a closer look at the latter phenomenon and show that, despite failing to satisfy the aforementioned support containment property, Jones' extension operator continues to exhibit a useful related feature. Specifically, in the next section we shall show that if D is a given closed subset of the closure of a finitely connected (ε, δ)-domain $\Omega \subseteq \mathbb{R}^n$ then for any function $u \in W^{k,p}(\Omega)$ which vanishes on D, its Jones extension $\Lambda_k u \in W^{k,p}(\mathbb{R}^n)$ also vanishes on D. In turn, this property is used to prove that Jones' extension operator Λ_k continues to be a well-defined linear and bounded mapping from $W_D^{k,p}(\Omega)$ into $W_D^{k,p}(\mathbb{R}^n)$. These spaces are introduced in Definition 2.2 via a completion procedure which makes it natural to interpret them as Sobolev spaces with partially vanishing traces on the set D. The main result of the paper is formally stated in Theorem 2.4 and its main consequence, a structural theorem for the space $W_D^{k,p}(\Omega)$, is recorded in Corollary 2.5.

2. Main results

Our first result in this paper brings to the forefront a salient feature of Jones' extension operator Λ_k from Theorem 1.2, namely the property that for any function $u \in W^{k,p}(\Omega)$ the support of its extension $\Lambda_k u \in W^{k,p}(\mathbb{R}^n)$ does not touch $\partial\Omega$ outside the region where the support of u itself makes contact with $\partial\Omega$. In order to make this precise, we need to introduce some notation.

Generally speaking, given an open set $\mathcal{O} \subseteq \mathbb{R}^n$ and an $\mathscr{L}^n$-measurable function v on $\mathcal{O}$, define

$$\operatorname{supp} v := \big\{x \in \overline{\mathcal{O}} : \text{there is no } r > 0 \text{ such that}$$

$$v \equiv 0 \ \ \mathscr{L}^n\text{-a.e. in } B(x,r) \cap \mathcal{O}\big\}. \tag{2.1}$$

Note that while the function v is known to be defined only in $\mathcal{O}$, the set $\operatorname{supp} v$ (itself a closed subset of $\mathbb{R}^n$) is contained in $\overline{\mathcal{O}}$. It is also clear from the above definition that if $\mathcal{O} \subseteq \mathbb{R}^n$ is an open set and v is an $\mathscr{L}^n$-measurable function defined on $\mathcal{O}$, then

$$\operatorname{supp}\big(v\big|_U\big) \subseteq \overline{U} \cap \operatorname{supp} v, \qquad \text{for any open subset } U \text{ of } \mathcal{O}. \tag{2.2}$$

Moreover, since every open cover of $\mathcal{O} \setminus \operatorname{supp} v$ has a countable subcover (given that the open set in question is σ-compact), it follows that

$$v \text{ vanishes } \mathscr{L}^n\text{-a.e. on } \mathcal{O} \setminus \operatorname{supp} v. \tag{2.3}$$

PROPOSITION 2.1. Let Ω be an (ε, δ)-domain in $\mathbb{R}^n$ and fix some $k \in \mathbb{N}$. Then the Jones' extension operator Λ_k from Theorem 1.2 has the property that, given any $p \in [1, \infty]$, one has

$$\overline{\Omega} \cap \operatorname{supp}(\Lambda_k u) = \operatorname{supp} u, \qquad \forall\, u \in W^{k,p}(\Omega). \tag{2.4}$$

Moreover, for every $u \in W^{k,p}(\Omega)$, one has

$$\operatorname{supp} u \ \text{ compact} \Longrightarrow \operatorname{supp}(\Lambda_k u) \ \text{ compact}. \tag{2.5}$$

PROOF. Fix $k \in \mathbb{N}$, $p \in [1, \infty]$, and select an arbitrary $u \in W^{k,p}(\Omega)$. Then combining (1.21) with (2.2) (used here with $v := \Lambda_k u$, $\mathcal{O} := \mathbb{R}^n$, and $U := \Omega$) yields

$$\operatorname{supp} u = \operatorname{supp}\big(\Lambda_k u\big|_\Omega\big) \subseteq \overline{\Omega} \cap \operatorname{supp} \Lambda_k u, \tag{2.6}$$

proving the right-to-left inclusion in (2.4). Proceeding in the opposite direction, we first claim that

$$\operatorname{supp}\Lambda_k u \subseteq \operatorname{supp} u \cup \overline{F_u}, \tag{2.7}$$

where

$$F_u := \bigcup_{\substack{Q\in\mathcal{W}_s\left((\Omega^c)^\circ\right),\\ Q^*\cap\operatorname{supp} u\neq\emptyset}} \tfrac{17}{16}Q. \tag{2.8}$$

To justify this claim, assume that $x_o \in \mathbb{R}^n \setminus (\operatorname{supp} u \cup \overline{F_u})$. Then there exists $r > 0$ with the property that

$$u \text{ vanishes } \mathscr{L}^n\text{-a.e. in } B(x_o,r)\cap\Omega, \tag{2.9}$$

and such that $B(x_o,r) \cap F_u = \emptyset$. The latter condition further entails (in concert with (1.18)) that, on the one hand,

$$\sum_{\substack{Q\in\mathcal{W}_s\left((\Omega^c)^\circ\right),\\ Q^*\cap\operatorname{supp} u\neq\emptyset}} P_{Q^*}(u)\,\varphi_Q \equiv 0 \quad \text{in} \quad B(x_o,r)\cap\left(\Omega^c\right)^\circ. \tag{2.10}$$

On the other hand, we claim that we also have

$$\sum_{\substack{Q\in\mathcal{W}_s\left((\Omega^c)^\circ\right),\\ Q^*\cap\operatorname{supp} u=\emptyset}} P_{Q^*}(u)\,\varphi_Q \equiv 0 \quad \text{in} \quad B(x_o,r)\cap\left(\Omega^c\right)^\circ. \tag{2.11}$$

Indeed, since for every $Q \in \mathcal{W}_s\big((\Omega^c)^\circ\big)$ the best fit polynomial $P_{Q^*}(u)$ has degree $k-1$, condition (1.15) entails

$$P_{Q^*}(u)(x) = \sum_{|\alpha|\leq k-1} \frac{x^\alpha}{\alpha!} ⨍_{Q^*} \partial^\alpha u\, d\mathscr{L}^n, \qquad \forall\, x\in\mathbb{R}^n. \tag{2.12}$$

Here and elsewhere, $⨍$ stands for integral average. In particular, this and (2.3) show that

$$P_{Q^*}(u) \equiv 0 \quad \text{for every} \;\; Q\in\mathcal{W}_s\big((\Omega^c)^\circ\big) \;\; \text{with} \;\; Q^*\cap\operatorname{supp} u=\emptyset, \tag{2.13}$$

and (2.11) readily follows from (2.13). Together, (2.10) and (2.11) imply that

$$\sum_{Q\in\mathcal{W}_s\left((\Omega^c)^\circ\right)} P_{Q^*}(u)\,\varphi_Q \equiv 0 \quad \text{in} \quad B(x_o,r)\cap\left(\Omega^c\right)^\circ. \tag{2.14}$$

From (1.17), (2.9), (2.14), and (1.6), we may then deduce that

$$\Lambda_k u \text{ vanishes } \mathscr{L}^n\text{-a.e. in } B(x_o,r). \tag{2.15}$$

Hence, $x_o \notin \operatorname{supp}\Lambda_k u$ which finishes the proof of (2.7). Having established (2.7), we next claim that

$$\overline{\Omega}\cap\overline{F_u} \subseteq \operatorname{supp} u. \tag{2.16}$$

To justify this, reason by contradiction and assume that there exists a point $x_o \in \overline{\Omega}\cap\overline{F_u}$ such that $x_o \notin \operatorname{supp} u$. In particular, the latter condition entails the existence of some $r > 0$ for which

$$u \text{ vanishes } \mathscr{L}^n\text{-a.e. in } B(x_o,r)\cap\Omega. \tag{2.17}$$

Let us take a closer look at the fact that $x_o \in \overline{\Omega} \cap \overline{F_u}$. For starters, the fact that $F_u \subseteq (\Omega^c)^\circ$ (as seen from (1.12) and (2.8)) forces

$$\overline{\Omega} \cap \overline{F_u} \subseteq \overline{\Omega} \cap \overline{(\Omega^c)^\circ} = \overline{\Omega} \setminus (\overline{\Omega})^\circ \subseteq \overline{\Omega} \setminus \Omega = \partial\Omega, \tag{2.18}$$

hence

$$x_o \in \partial\Omega. \tag{2.19}$$

Next, the membership of x_o to the closure of the set F_u defined in (2.8) entails the existence of a sequence of dyadic cubes $\{Q_j\}_j \subseteq \mathcal{W}_s\big((\Omega^c)^\circ\big)$ and a sequence $\{x_j\}_j$ of points in $\mathbb{R}^n$ satisfying

$$x_j \in \tfrac{17}{16}Q_j \quad \text{for every} \quad j, \tag{2.20}$$

$$Q_j^* \cap \operatorname{supp} u \neq \emptyset \quad \text{for every} \quad j, \tag{2.21}$$

$$\lim_j x_j = x_o. \tag{2.22}$$

Now, (2.20) forces

$$|x_j - x_{Q_j}| \leq \sqrt{n}\,\tfrac{17}{16}\,\ell(Q_j) \quad \text{for every} \quad j, \tag{2.23}$$

while from (1.13) we conclude that there exists $c \in (0,1)$ such that

$$c\,\ell(Q_j) \leq \operatorname{dist}\big(\tfrac{17}{16}Q_j, \partial\Omega\big) \leq \operatorname{dist}\big(x_j, \partial\Omega\big) \leq |x_j - x_o|, \quad \text{for all} \quad j, \tag{2.24}$$

where the last equality uses (2.19). From (1.16) we also deduce that

$$\ell(Q_j) \leq \ell(Q_j^*) \leq 4\,\ell(Q_j) \quad \text{and} \quad |x_{Q_j} - x_{Q_j^*}| \leq C_{n,\varepsilon}\,\ell(Q_j), \quad \text{for all} \quad j. \tag{2.25}$$

Combining now (2.22)-(2.25) yields

$$\lim_j x_{Q_j^*} = x_o \quad \text{and} \quad \lim_j \ell(Q_j^*) = 0. \tag{2.26}$$

In turn, from (2.26) we deduce that

$$\text{there exists } j \text{ such that } Q_j^* \subseteq B(x_o, r) \cap \Omega \tag{2.27}$$

which, in light of (2.17), implies that

$$\text{there exists } j \text{ such that } u \text{ vanishes } \mathscr{L}^n\text{-a.e. in } Q_j^*. \tag{2.28}$$

This, however, contradicts (2.21). The proof of (2.16) is therefore complete.

With (2.16) in hand, and availing ourselves of (2.7) we may write

$$\overline{\Omega} \cap \operatorname{supp} \Lambda_k u \subseteq \big(\overline{\Omega} \cap \operatorname{supp} u\big) \cup \big(\overline{\Omega} \cap \overline{F_u}\,\big) \subseteq \operatorname{supp} u, \tag{2.29}$$

which proves the left-to-right inclusion in (2.4).

At this stage, in order to complete the proof of the proposition, there remains to establish (2.5). To this end, assume that $u \in W^{k,p}(\Omega)$ is such that $\operatorname{supp} u$ is a compact subset of $\overline{\Omega}$, and consider an arbitrary point $x_1 \in F_u$. Then there exists some cube $Q \in \mathcal{W}_s\big((\Omega^c)^\circ\big)$ such that $x_1 \in \frac{17}{16}Q$ and $Q^* \cap \operatorname{supp} u \neq \emptyset$. Pick some $x_2 \in Q^* \cap \operatorname{supp} u$ and note that, thanks to (1.14)-(1.16), we may estimate

$$|x_1 - x_Q| \leq 2\sqrt{n}\,\tfrac{17}{16}\,\ell(Q) \leq C(n,\varepsilon,\delta), \tag{2.30}$$

$$|x_Q - x_{Q^*}| \leq \operatorname{dist}(Q, Q^*) \leq C_{n,\varepsilon}\ell(Q) \leq C(n,\varepsilon,\delta), \tag{2.31}$$

$$|x_{Q^*} - x_2| \leq 2\sqrt{n}\,\ell(Q^*) \leq 8\sqrt{n}\,\ell(Q) \leq C(n,\varepsilon,\delta), \tag{2.32}$$

for some finite constant $C(n,\varepsilon,\delta)>0$. Collectively, (2.30)-(2.32) imply that

$$\begin{aligned} \mathrm{dist}\,(x_1,\mathrm{supp}\,u) &\leq |x_1-x_2| \leq |x_1-x_Q|+|x_Q-x_{Q^*}|+|x_{Q^*}-x_2| \\ &\leq R:=3\,C(n,\varepsilon,\delta). \end{aligned} \tag{2.33}$$

Since x_1 has been arbitrarily chosen in F_u, this proves that

$$F_u\subseteq\big\{x\in\mathbb{R}^n:\,\mathrm{dist}\,(x,\mathrm{supp}\,u)\leq R\big\}. \tag{2.34}$$

Going further, from (2.34) and (2.7) we deduce that

$$\mathrm{supp}\,\Lambda_k u\subseteq\big\{x\in\mathbb{R}^n:\,\mathrm{dist}\,(x,\mathrm{supp}\,u)\leq R\big\}. \tag{2.35}$$

Given that $\mathrm{supp}\,u$ is compact, this shows that $\mathrm{supp}\,\Lambda_k u$ is a bounded set, hence also compact, as desired. □

Moving on, we make the following definition.

DEFINITION 2.2. Given a nonempty open set Ω in $\mathbb{R}^n$ and a closed subset D of $\overline{\Omega}$, consider

$$\mathscr{C}^\infty_D(\Omega):=\big\{\widetilde{u}\big|_\Omega:\,u\in\mathscr{C}^\infty_c(\mathbb{R}^n\setminus D)\big\}, \tag{2.36}$$

where tilde denotes the extension by zero outside the support to $\mathbb{R}^n$, and for each $k\in\mathbb{N}$, $p\in[1,\infty]$, define

$$W^{k,p}_D(\Omega):=\text{the closure of } \mathscr{C}^\infty_D(\Omega) \text{ in } \Big(W^{k,p}(\Omega)\,,\,\|\cdot\|_{W^{k,p}(\Omega)}\Big). \tag{2.37}$$

Hence, by design,

$$\mathscr{C}^\infty_D(\Omega) \text{ is a dense linear subspace of } W^{k,p}_D(\Omega) \tag{2.38}$$

and, given the completeness of $\Big(W^{k,p}(\Omega)\,,\,\|\cdot\|_{W^{k,p}(\Omega)}\Big)$,

$$W^{k,p}_D(\Omega) \text{ is a closed linear subspace of } W^{k,p}(\Omega). \tag{2.39}$$

As a consequence,

$$\Big(W^{k,p}_D(\Omega)\,,\,\|\cdot\|_{W^{k,p}(\Omega)}\Big) \text{ is a Banach space.} \tag{2.40}$$

It is also clear from definitions that

$$\Big\{u\big|_\Omega:\,u\in W^{k,p}_D(\mathbb{R}^n)\Big\}\subseteq W^{k,p}_D(\Omega), \tag{2.41}$$

and that, corresponding to the case when $D=\partial\Omega$, we have

$$W^{k,p}_{\partial\Omega}(\Omega)=\mathring{W}^{k,p}(\Omega), \tag{2.42}$$

since

$$\mathscr{C}^\infty_{\partial\Omega}(\Omega)=\mathscr{C}^\infty_c(\Omega). \tag{2.43}$$

The special case $D=\emptyset$ is considered below.

LEMMA 2.3. *Let Ω be an (ε,δ)-domain in $\mathbb{R}^n$ and fix some $k\in\mathbb{N}$ along with some $p\in[1,\infty)$. Then, corresponding to the case when $D=\emptyset$, one has*

$$W^{k,p}_\emptyset(\Omega)=W^{k,p}(\Omega). \tag{2.44}$$

PROOF. This follows from the approximation result proved in [**J**, § 4, pp. 83-85] and a straightforward truncation argument. □

We are now in a position to state the principal result of this paper, which may be regarded as a suitable analogue of the property of Calderón's extension operator of not increasing the support of functions $u \in W^{k,p}(\Omega)$ which vanish near $\partial\Omega$, in the case of Jones' extension operator.

THEOREM 2.4 (Extension Theorem for Sobolev Spaces with Partially Vanishing Traces). *Let Ω be a finitely connected (ε, δ)-domain in $\mathbb{R}^n$ and fix an arbitrary number $k \in \mathbb{N}$. Then Jones' extension operator Λ_k (from Theorem 1.2) has the property that, for each closed subset D of $\overline{\Omega}$ and each $p \in [1, \infty]$,*

$$\Lambda_k : W_D^{k,p}(\Omega) \longrightarrow W_D^{k,p}(\mathbb{R}^n) \quad \textit{linearly and boundedly,} \tag{2.45}$$

with operator norm controlled solely in terms of $n, \varepsilon, \delta, k, p$, and the size of the connected components of Ω.

It is instructive to note that if $1 \leq p < \infty$ then Theorem 2.4 reduces to Jones' extension theorem recorded in Theorem 1.2 precisely when $D = \emptyset$ (here Lemma 2.3 is used). Before presenting the proof of this result we record the following immediate consequence, of independent interest.

COROLLARY 2.5. Let $k \in \mathbb{N}$ and $p \in [1, \infty]$. Also, assume that Ω is a finitely connected (ε, δ)-domain in $\mathbb{R}^n$ and that D is a closed subset of $\overline{\Omega}$. Then

$$W_D^{k,p}(\Omega) = \Big\{u\big|_\Omega : u \in W_D^{k,p}(\mathbb{R}^n)\Big\}. \tag{2.46}$$

PROOF. The right-to-left inclusion is contained in (2.41), while the left-to-right inclusion is a direct consequence of Theorem 2.4 and (1.21). □

We now turn to the task of providing the

PROOF OF THEOREM 2.4. Assume that $\Omega \subseteq \mathbb{R}^n$, $D \subseteq \overline{\Omega}$, $k \in \mathbb{N}$, and $p \in [1, \infty]$ are as in the statement of the theorem. Then from (2.39) and (1.20) we deduce that

$$\Lambda_k : W_D^{k,p}(\Omega) \longrightarrow W^{k,p}(\mathbb{R}^n) \quad \text{linearly and boundedly.} \tag{2.47}$$

On account of this and (2.39) it therefore suffices to show that (recall from (2.40) that $W_D^{k,p}(\mathbb{R}^n)$ is complete)

$$\Lambda_k \mathscr{C}_D^\infty(\Omega) \subseteq W_D^{k,p}(\mathbb{R}^n). \tag{2.48}$$

With this in mind, pick an arbitrary $u \in \mathscr{C}_D^\infty(\Omega)$. In particular, $u \in W^{k,p}(\Omega)$ and there exists a compact subset K of $\overline{\Omega}$ such that

$$K \cap D = \emptyset \quad \text{and} \quad \operatorname{supp} u \subseteq K. \tag{2.49}$$

Thanks to this and Proposition 2.1 we therefore have

$$\operatorname{supp} \Lambda_k u \quad \text{is a compact set in } \mathbb{R}^n, \tag{2.50}$$

and

$$\begin{aligned} D \cap \operatorname{supp} \Lambda_k u &= \big(D \cap \overline{\Omega}\big) \cap \operatorname{supp} \Lambda_k u \\ &= D \cap \Big(\overline{\Omega} \cap \operatorname{supp} \Lambda_k u\Big) = D \cap \operatorname{supp} u = \emptyset. \end{aligned} \tag{2.51}$$

Since D is closed and $\operatorname{supp} \Lambda_k u$ is compact, (2.51) forces

$$\operatorname{dist}\big(D\,, \operatorname{supp} \Lambda_k u\big) > 0. \tag{2.52}$$

At this stage, choose a function $\theta \in \mathscr{C}_c^\infty(\mathbb{R}^n)$ such that $\int_{\mathbb{R}^n} \theta \, d\mathscr{L}^n = 1$ and, for each $j \in \mathbb{N}$, define $\theta_j(x) := j^n \theta(jx)$ for every $x \in \mathbb{R}^n$. Finally, set

$$\varphi_j := \theta_j * \Lambda_k u \quad \text{in} \quad \mathbb{R}^n, \quad \text{for each} \quad j \in \mathbb{N}. \tag{2.53}$$

Then

$$\begin{aligned} &\varphi_j \in \mathscr{C}_c^\infty(\mathbb{R}^n \setminus D) \quad \text{if } j \text{ is large enough, and} \\ &\varphi_j \longrightarrow \Lambda_k u \quad \text{in} \quad W^{k,p}(\mathbb{R}^n) \quad \text{as} \quad j \to \infty, \end{aligned} \tag{2.54}$$

by virtue of (2.50), (2.52), and the fact that $\Lambda_k u \in W^{k,p}(\mathbb{R}^n)$. In light of the definition of $W_D^{k,p}(\mathbb{R}^n)$ (cf. (2.37)), the approximation result in (2.54) implies that actually $\Lambda_k u \in W_D^{k,p}(\mathbb{R}^n)$, as desired. □

References

[AF] Robert A. Adams and John J. F. Fournier, *Sobolev spaces*, 2nd ed., Pure and Applied Mathematics (Amsterdam), vol. 140, Elsevier/Academic Press, Amsterdam, 2003. MR2424078 (2009e:46025)

[C] A.-P. Calderón, *Lebesgue spaces of differentiable functions and distributions*, Proc. Sympos. Pure Math., Vol. IV, American Mathematical Society, Providence, R.I., 1961, pp. 33–49. MR0143037 (26 #603)

[J] Peter W. Jones, *Quasiconformal mappings and extendability of functions in Sobolev spaces*, Acta Math. **147** (1981), no. 1-2, 71–88, DOI 10.1007/BF02392869. MR631089 (83i:30014)

[MS] O. Martio and J. Sarvas, *Injectivity theorems in plane and space*, Ann. Acad. Sci. Fenn. Ser. A I Math. **4** (1979), no. 2, 383–401. MR565886 (81i:30039)

[M] Vladimir G. Maz'ja, *Sobolev spaces*, Springer Series in Soviet Mathematics, Springer-Verlag, Berlin, 1985. Translated from the Russian by T. O. Shaposhnikova. MR817985 (87g:46056)

[R] Luke G. Rogers, *Degree-independent Sobolev extension on locally uniform domains*, J. Funct. Anal. **235** (2006), no. 2, 619–665, DOI 10.1016/j.jfa.2005.11.013. MR2225465 (2008j:46027)

[S] Elias M. Stein, *Singular integrals and differentiability properties of functions*, Princeton Mathematical Series, No. 30, Princeton University Press, Princeton, N.J., 1970. MR0290095 (44 #7280)

Department of Mathematics, University of Missouri, Columbia, Missouri 65211
E-mail address: kjb886@mail.mizzou.edu

Department of Mathematics, University of Missouri, Columbia, Missouri 65211
E-mail address: mitread@missouri.edu

Department of Mathematics, Temple University, Philadelphia, Pennsylvania 19122
E-mail address: imitrea@temple.edu

Department of Mathematics, University of Missouri, Columbia, Missouri 65211
E-mail address: mitream@missouri.edu

Proceedings of Symposia in Pure Mathematics
Volume **87**, 2013
http://dx.doi.org/10.1090/pspum/087/01438

Some spectral properties of Rooms and Passages domains and their skeletons

B. M. Brown, W. D. Evans, and I. G. Wood

Dedicated to Fritz Gesztesy on the occasion of his sixtieth birthday.

Abstract. In this paper we investigate spectral properties of Laplacians on Rooms and Passages domains. In the first part, we use Dirichlet-Neumann bracketing techniques to show that for the Neumann Laplacian in certain Rooms and Passages domains the second term of the asymptotic expansion of the counting function is of order $\sqrt{\lambda}$. For the Dirichlet Laplacian our methods only give an upper estimate of the form $\sqrt{\lambda}$. In the second part of the paper, we consider the relationship between Neumann Laplacians on Rooms and Passages domains and Sturm-Liouville operators on the skeleton.

1. Introduction

Let $-\Delta_{N,\Omega}$ denote the Neumann Laplacian on a bounded open subset Ω of $\mathbb{R}^n, n \geq 2$, and let $E : H^1(\Omega) \to L^2(\Omega)$ be the canonical embedding, where $H^1(\Omega)$ is the standard Sobolov space. Then $(-\Delta_{N,\Omega} + 1)^{-1} = EE^*$, and $-\Delta_{N,\Omega}$ has a compact resolvent (and thus a discrete spectrum) if and only if E is compact. The compactness, or otherwise, of E is determined by the nature of the boundary $\partial\Omega$ of Ω. In [**1**] it was shown that E being compact is equivalent to a *generalised extension property*, which is that *there exists a function space* $\Lambda(\mathbb{R}^n)$ *on* $\mathbb{R}^n$ *which is compactly embedded in* $L^2(B)$ *for any ball* $B \subset \mathbb{R}^n$, *and is such that there is a continuous extension* $\mathcal{E} : H^1(\Omega) \to \Lambda(\mathbb{R}^n)$. In general $\Lambda(\mathbb{R}^n)$ is a space of arbitrary smoothness. If $\partial\Omega \in Lip_\alpha, 0 < \alpha < 1$, i.e., $\partial\Omega$ coincides with the subgraph of a Lip_α function in a neighbourhood of each point, then $\Lambda(\mathbb{R}^n) = H^\alpha(\mathbb{R}^n)$, the Sobolev space of order α, so that, in this case, there is a reduction of smoothness in going from Ω to $\mathbb{R}^n$. For domains with singular boundaries, like the "Rooms and Passages" domain defined in Section 2, the compactness of the embedding E can be taken as a measure of the smoothness of the boundary, which is otherwise difficult to describe. When E is not compact, [**7**] shows that for any closed subset S of the non-negative real numbers, there exists a modified "Rooms and Passages" or a "Comb" domain such that S equals the essential spectrum of $-\Delta_{N,\Omega}$.

Suppose that $-\Delta_{N,\Omega}$ has a discrete spectrum and denote the number of its eigenvalues less than λ by $N_N(\lambda)$. The problem of determining the asymptotic

2010 *Mathematics Subject Classification.* Primary.

The authors would like to thank the referee for carefully reading an earlier version of the manuscript and Dr. Rob Davies for producing the figures.

behaviour of $N_N(\lambda)$ as $\lambda \to \infty$ has a long history. If the boundary $\partial\Omega$ of Ω is sufficiently smooth, it has long been known that one has the asymptotic formula

$$N_N(\lambda) = \omega_n(2\pi)^{-n}|\Omega|\lambda^{n/2} + o(\lambda^{n/2}), \tag{1.1}$$

where ω_n is the volume of the unit ball in $\mathbb{R}^n$ and $|\Omega|$ is the volume of the domain Ω. For general Ω the Weyl term $\omega_n(2\pi)^{-n}|\Omega|\lambda^{n/2}$ dominates, and in the error $R(\lambda) := N_N(\lambda) - \omega_n(2\pi)^{-n}|\Omega|\lambda^{n/2}$, it is the boundary $\partial\Omega$, rather than any other topological or geometrical feature of Ω which is dominant. For instance, in [**9**], it is shown that if $\partial\Omega \in Lip_\alpha, 0 < \alpha < 1$, the remainder $R(\lambda) = O(\lambda^{(n-1)/2\alpha})$ and this is order sharp. Here and in the following, the $O(f(\lambda))$- and $o(f(\lambda))$-notation is to be understood in the limit $\lambda \to \infty$. Moreover, in [**4**], a domain of von-Koch snowflake type Ω is exhibited for which $R(\lambda) \asymp \lambda^{d_o}$, where d_o denotes the outer Minkowski dimension of the boundary and $A(\lambda) \asymp B(\lambda)$ means that $|A(\lambda)/B(\lambda)|$ is bounded above and below by positive constants as $\lambda \to \infty$. Specific information about the error is hard to come by. For general domains one of the few available techniques is the Courant-Weyl variational method involving "Dirichlet-Neumann bracketing". This effectively reduces the problem to estimating the corresponding counting functions $N_D(\lambda)$ and $N_N(\lambda)$ for Dirichlet and Neumann Laplacians on a set of cubes which cover Ω. A variant of this technique is used in [**9**] in which cubes are replaced by other relatively simple sets for which the Dirichlet and Neumann Laplacian counting functions can be estimated from above and below. The first part of this paper is a contribution to the study of how the error term $R(\lambda)$ depends on the boundary $\partial\Omega$. We look in detail at the much studied "Rooms and Passages" domain, in which the Weyl formula (1.1) holds, being particularly concerned with upper and lower bounds for $R(\lambda)\lambda^{-1/2}$.

In [**5**] it was shown that for a rather restricted class of domains Ω, (which does not include Rooms and Passages) $-\Delta_{N,\Omega}$ has a compact resolvent if and only if a Sturm-Liouville operator defined on the *skeleton* of Ω has a compact resolvent. Recall that the *skeleton* of an open set Ω is the complement of the set of points $\mathbf{x}$ in Ω for which there exists a unique point $\mathbf{y}$ on $\partial\Omega$ such that $|\mathbf{y} - \mathbf{x}|$ is equal to the distance of $\mathbf{x}$ from $\mathbb{R}^n \setminus \Omega$. The result in [**5**] was motivated by Theorem 3.3 in [**2**] in which Ω is a horn, whose skeleton is a half-line. In the second part of the paper (Sections 4 and 5) we investigate this problem for general Rooms and Passages domains.

2. Rooms & Passages domains

We consider a Rooms and Passages (R&P for short) domain Ω defined as the union of square rooms R_i (i odd) of size $h_i \times h_i$ joined by rectangular passages P_i (i even) of size $h_i \times \delta_i$ with $\delta_i < \min\{h_{i-1}, h_{i+1}\}$. We assume $h_i \to 0$ as $i \to \infty$. See Figure 1.

In this section and the next, we further restrict the R&P domains under consideration by assuming that $h_i = C^i$ and $\delta_i = kC^{i\alpha}$, where $C < 1$, $\alpha > 1$ and k are constants, with $k < C^{3-2\alpha}$. This guarantees that each passage is narrower than both adjacent rooms. It follows from the general result in [**3**], Example 6.1.1, that the embedding $E : H^1(\Omega) \to L^2(\Omega)$ is compact if and only if $\alpha < 3$.

PROPOSITION 2.1. *For $\alpha > 3$ we have $0 \in \sigma_{ess}(-\Delta_{N,\Omega})$ where σ_{ess} denotes the essential spectrum of the operator.*

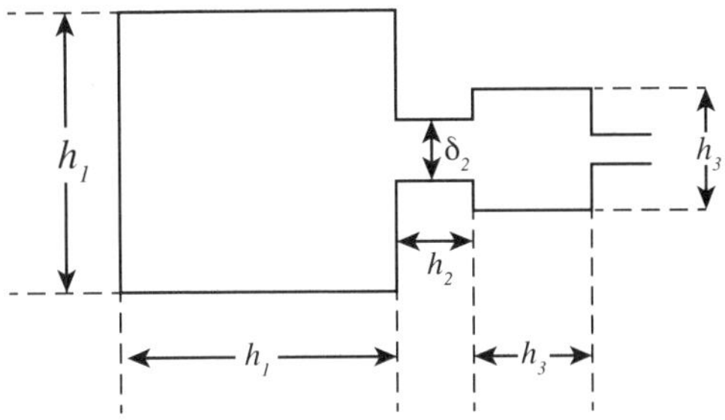

FIGURE 1. The start of a general Rooms and Passages domain.

PROOF. We give a simple singular sequence proof. By Ω_j we denote the domain consisting of the first j rooms and passages while $T_j = \Omega \setminus \Omega_j$ denotes the "tail". Choose a sequence of C^∞ functions φ_j which depend only on the x-variable such that supp $\varphi_j \subseteq \Omega_{4j} \setminus \Omega_{2j-1}$,

$$(2.1)\quad \varphi_j = \begin{cases} 1 & \text{on } \Omega_{4j-1} \setminus \Omega_{2j}, \\ 0 & \text{on } \Omega_{4j}^c \cup \Omega_{2j-1}, \end{cases} \quad \text{and } |\varphi_j'| = \begin{cases} O(C^{-2j}) & \text{on } \Omega_{2j} \setminus \Omega_{2j-1}, \\ O(C^{-4j}) & \text{on } \Omega_{4j} \setminus \Omega_{4j-1}. \end{cases}$$

In other words, the function φ_j is initially zero, increases to 1 in the j-th passage and falls back to zero in the $2j$-th passage. The support of the sequence of the φ_j 'disappears' into the tail and each φ_j satisfies the Neumann boundary condition on $\partial\Omega$. The singular sequence we wish to consider is given by

$$(2.2)\qquad f_j(x, y) = \frac{1}{\|\varphi_j\|_{L^2(\Omega)}} \varphi_j(x) \quad \text{for } (x, y) \in \Omega.$$

Obviously, $\|f_j\|_{L^2(\Omega)} = 1$, while f_j weakly converges to 0 in $L^2(\Omega)$ as $j \to \infty$.

We have the following:

$$(2.3)\qquad \|\varphi_j\|_{L^2(\Omega)} \asymp (|\Omega_{4j}| - |\Omega_{2j}|)^{1/2} \quad \text{as } j \to \infty$$

and

$$\begin{aligned} |\Omega_{4j}| &= \sum_{i=1,\ i\ odd}^{4j-1} C^{2i} + \sum_{i=1,\ i\ even}^{4j} kC^{i(1+\alpha)} \\ (2.4)\qquad &= C^2 \frac{1 - C^{8j}}{1 - C^4} + kC^{2(1+\alpha)} \frac{1 - C^{4j(1+\alpha)}}{1 - C^{2(1+\alpha)}}. \end{aligned}$$

Therefore,

$$|\Omega_{4j}| - |\Omega_{2j}| = C^2 \frac{C^{4j} - C^{8j}}{1 - C^4} + kC^{2(1+\alpha)} \frac{C^{2j(1+\alpha)} - C^{4j(1+\alpha)}}{1 - C^{2(1+\alpha)}} \asymp C^{4j} \text{ as } j \to \infty,$$

and for large j, $\|\varphi_j\|_{L^2(\Omega)}$ behaves like C^{2j}.

As φ_j' is supported only on two passages we get from (2.1),

$$(2.5)\qquad \|\varphi_j'\|^2_{L^2(\Omega)} = O(C^{-8j} kC^{4j(1+\alpha)} + C^{-4j} kC^{2j(1+\alpha)}) = O(C^{-2j(1-\alpha)}) \text{ as } j \to \infty.$$

Hence, as $j \to \infty$,
(2.6)

$$\|f_j'\|_{L^2(\Omega)} = \frac{1}{\|\varphi_j\|_{L^2(\Omega)}}\|\varphi_j'\|_{L^2(\Omega)} = O(C^{-j(1-\alpha)}C^{-2j}) = O(C^{-j(3-\alpha)}) \to 0$$

if $\alpha > 3$. This implies that $(-\Delta)^{1/2}f_j \to 0$ as $j \to \infty$ although $\|f_j\|_{L^2(\Omega)} = 1$. Therefore, we have $0 \in \sigma_{ess}((-\Delta_{N,\Omega})^{1/2})$, which implies $0 \in \sigma_{ess}(-\Delta_{N,\Omega})$ when $\alpha > 3$. □

3. Eigenvalue asymptotics of the Dirichlet and Neumann Laplacians

The special class of R&P domains introduced in the previous section will be considered. Let $\alpha < 3$ and denote by $N_D(\lambda), N_N(\lambda)$, respectively, the counting functions of the Dirichlet and Neumann Laplacians on the R&P domain Ω. To determine their asymptotic behaviour, we shall apply the Dirichlet-Neumann bracketing technique.

We begin this analysis of the spectrum of the Laplacian on an R&P domain Ω by discussing the contribution from the tail. As before, let $\Omega = \Omega_{2M} \cup T_{2M}$ where Ω_{2M} consists of the first $2M$ rooms and passages and T_{2M} denotes the tail. On applying Theorems 4.6 and 5.1 to Example 6.1.1 in [**3**], it follows[1] that the optimal constant $K(T_{2M})$ in the Poincaré inequality

$$\|f - f_{T_{2M}}\|_{L^2(T_{2M})} \le K(T_{2M})\,\|\nabla f\|_{L^2(T_{2M})}, \quad (f \in H^1(T_{2M})),$$

satisfies

$$K(T_{2M}) \le cC^{(3-\alpha)M}$$

where c is a positive constant.

It follows that the Neumann Laplacian on T_{2M} will only contribute the trivial eigenvalue 0 if

$$\lambda < (1/c)^2 C^{2(\alpha-3)M}, \quad \text{i.e.} \quad M > \frac{\log(c^2\lambda)}{2(3-\alpha)\log C^{-1}}. \tag{3.1}$$

Since $\alpha > 1$, the tail after M passages has area

$$|\,T_{2M}\,| = \frac{C^{2+4M}}{1-C^4} + \frac{kC^{2(1+\alpha)}C^{2M(1+\alpha)}}{1-C^{2(1+\alpha)}} \asymp C^{4M} \text{ as } M \to \infty.$$

It follows that for M as in (3.1)

$$|\,T_{2M}\,| = O(\lambda^{-2/(3-\alpha)})$$

and so

$$|\Omega| = |\Omega_{2M}| + o(\sqrt{\lambda}).$$

Now, for M satisfying (3.1), the counting function for the Neumann-Laplacian on Ω differs by at most 1 from that on Ω_{2M}. It follows that

$$R(\lambda) - R_{2M}(\lambda) = o(\sqrt{\lambda}), \tag{3.2}$$

where $R(\lambda)$ and $R_{2M}(\lambda)$ denote the error term for the problem on Ω and Ω_{2M}, respectively. As we have that the $(n+1)$-th Neumann eigenvalue is a lower bound for the n-th Dirichlet eigenvalue (see [**6**]), for λ satisfying (3.1), the tail cannot

[1] By Theorem 4.6, the bound on $K(T_{2M})$ involves a quantity $c(J_1)$, which according to Theorem 5.1 is bounded by $\sqrt{N_a}$. Noting that their constant C corresponds to our C^{-1}, Theorem 6.1 and Example 6.1.1 with $k = 2M$ then give $N_a \asymp C^{2M(3-\alpha)}$.

contribute any Dirichlet eigenvalue, so the same reasoning as for the Neumann case implies that (3.2) also holds in the Dirichlet case.

3.1. Asymptotics for $N_N(\lambda)$. Our strategy here is to partition the domain and use the Dirichlet Neumann bracketing technique to obtain the required estimates. In order to obtain these estimates we first obtain a lower bound for the number of eigenvalues of the Neumann-Laplacian, we partition the rooms into five sections imposing Neumann boundary conditions on the boundary of Ω and Dirichlet boundary conditions on all artificially introduced internal boundaries (see Figure 2). An upper bound is obtained by only introducing an artificial boundary to separate the room from the neighbouring passages and imposing Neumann conditions on all the boundaries (see Figure 3). This is a simple consequence of the variational principle.

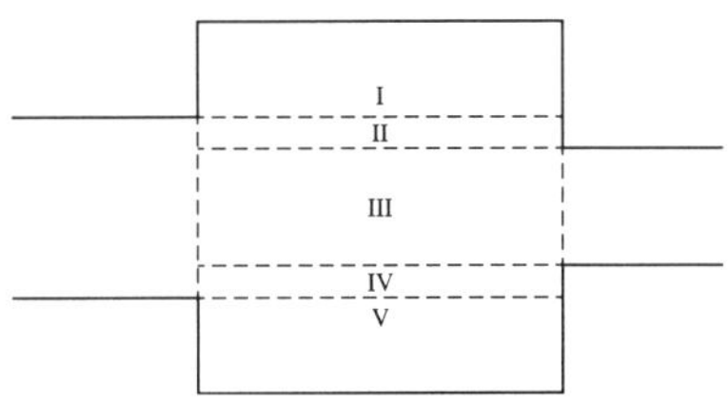

FIGURE 2. Artificial boundaries and sub-regions of a room for the lower bound: Neumann conditions on the dotted lines.

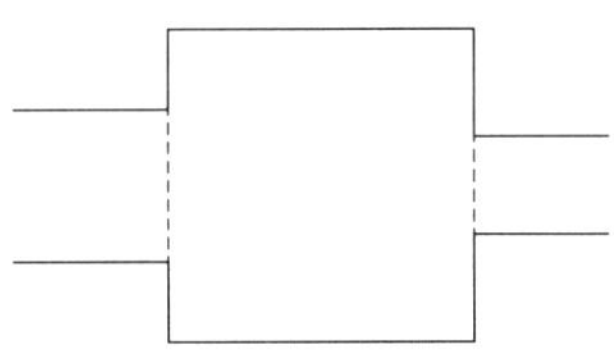

FIGURE 3. Artificial boundaries of a room for the upper bound: Dirichlet conditions on the dotted lines.

We first consider the one-dimensional problem on $[-a, a]$ and have the following eigenfunctions and eigenvalues:

(a) Dirichlet conditions at both end points:

$$\psi_m(x) = c \sin \frac{m\pi(x+a)}{2a}, \quad \lambda_m = \frac{m^2\pi^2}{4a^2}, \quad m \geq 1.$$

(b) Dirichlet conditions at $-a$ and Neumann at a:

$$\psi_m(x) = c \sin \frac{(2m+1)\pi(x+a)}{4a}, \quad \lambda_m = \frac{(2m+1)^2\pi^2}{16a^2}, \quad m \geq 0.$$

(c) Neumann conditions at both end points:

$$\psi_m(x) = c \cos \frac{m\pi(x+a)}{2a}, \quad \lambda_m = \frac{m^2\pi^2}{4a^2}, \quad m \geq 0.$$

For the lower estimate for a room, this leads to the following.

(1) In regions I and V, we have the set of eigenvalues

$$\left\{\lambda_{m,n} = \frac{m^2\pi^2}{4a^2} + \frac{(2n+1)^2\pi^2}{16b^2} \,:\, m,n \geq 0\right\} \quad \text{with} \quad a = \frac{C^j}{2},\ b = \frac{C^j - kC^{\alpha(j-1)}}{4}.$$

(2) In regions II and IV, we have the set of eigenvalues

$$\left\{\lambda_{m,n} = \tfrac{(2m+1)^2\pi^2}{16a^2} + \tfrac{n^2\pi^2}{4b^2} \,:\, m \geq 0,\ n \geq 1\right\}$$
$$\text{with} \quad a = \tfrac{C^j}{2},\ b = \tfrac{k}{4}\left(C^{\alpha(j-1)} - C^{\alpha(j+1)}\right).$$

(3) In region III, we have the set of eigenvalues

$$\left\{\lambda_{m,n} = \frac{m^2\pi^2}{4a^2} + \frac{n^2\pi^2}{4b^2} \,:\, m,n \geq 1\right\} \quad \text{with} \quad a = \frac{C^j}{2},\ b = k\frac{C^{\alpha(j+1)}}{2}.$$

We now need to count the integer lattice points in the first quadrant satisfying $\lambda_{m,n} \leq \lambda$. By [**8**], the number of integer lattice points in a plane region $X \cdot R$, where X is a real scaling parameter, is given by

$$\mathcal{N}(X) = AX^2 + o(X^{\frac{131}{208}+\varepsilon}) \text{ as } X \to \infty$$

for any $\varepsilon > 0$, where A is the area of the region R. This gives us the following:

(1) In regions I and V, we have an ellipse with half axes $\frac{2a\sqrt{\lambda}}{\pi}$ and $\frac{2b\sqrt{\lambda}}{\pi}$ which is shifted by $-1/2$ in the y-direction. As an upper estimate for the area below the x-axis we use the area of the rectangle which is subtracted from the area of the quarter ellipse (the error from this can easily be seen to be $o(\sqrt{\lambda})$). We then add the contributions along the two axes to get

$$\begin{aligned}
&\operatorname{card}\left\{\lambda_{m,n} = \frac{m^2\pi^2}{4a^2} + \frac{(2n+1)^2\pi^2}{16b^2} \,:\, m,n \geq 0,\ \lambda_{m,n} \leq \lambda\right\} \\
&\geq \frac{\pi}{4}\frac{2a\sqrt{\lambda}}{\pi}\frac{2b\sqrt{\lambda}}{\pi} - \frac{1}{2}\frac{2a\sqrt{\lambda}}{\pi} - o(\sqrt{\lambda}) + \frac{2a\sqrt{\lambda}}{\pi} + \frac{2b\sqrt{\lambda}}{\pi} \\
&= \frac{ab}{\pi}\lambda + \frac{(2b+a)\sqrt{\lambda}}{\pi} - o(\sqrt{\lambda}).
\end{aligned}$$

Here, card A denotes the cardinality of the finite set A.

(2) In regions II and IV, we again have an ellipse with half axes $\frac{2a\sqrt{\lambda}}{\pi}$ and $\frac{2b\sqrt{\lambda}}{\pi}$ which this time is shifted by $-1/2$ in the x-direction. As an upper estimate for the area left of the y-axis we use the area of the rectangle which is subtracted from the total area. We then add the contributions along the y-axis to get

$$\begin{aligned}
&\operatorname{card}\left\{\lambda_{m,n} = \frac{(2m+1)^2\pi^2}{16a^2} + \frac{n^2\pi^2}{4b^2} \,:\, m \geq 0,\ n \geq 1,\ \lambda_{m,n} \leq \lambda\right\} \\
&\geq \frac{\pi}{4}\frac{2a\sqrt{\lambda}}{\pi}\frac{2b\sqrt{\lambda}}{\pi} - \frac{1}{2}\frac{2b\sqrt{\lambda}}{\pi} - o(\sqrt{\lambda}) + \frac{2b\sqrt{\lambda}}{\pi} \\
&= \frac{ab}{\pi}\lambda + \frac{b\sqrt{\lambda}}{\pi} - o(\sqrt{\lambda}).
\end{aligned}$$

(3) In region III, we count the lattice points in an ellipse with half axes $\frac{2a\sqrt{\lambda}}{\pi}$ and $\frac{2b\sqrt{\lambda}}{\pi}$ to give

$$\begin{aligned}&\operatorname{card}\left\{\lambda_{m,n} = \frac{m^2\pi^2}{4a^2} + \frac{n^2\pi^2}{4b^2} : m,n \geq 1,\ \lambda_{m,n} \leq \lambda\right\}\\ &\geq \frac{\pi}{4}\frac{2a\sqrt{\lambda}}{\pi}\frac{2b\sqrt{\lambda}}{\pi} - o(\sqrt{\lambda}) = \frac{ab}{\pi}\lambda - o(\sqrt{\lambda}).\end{aligned}$$

Collecting these results, we obtain a lower estimate for the contribution of the j-th room to the counting function of the form

$$N_j \geq \frac{C^{2j}}{4\pi}\lambda + \left[2C^j - \frac{k}{2}\left(C^{\alpha(j+1)} + C^{\alpha(j-1)}\right)\right]\frac{\sqrt{\lambda}}{\pi} - o(\sqrt{\lambda}).$$

For the upper estimate of the counting function, we simply need to consider the eigenvalues

$$\left\{\lambda_{m,n} = \frac{m^2\pi^2}{4a^2} + \frac{n^2\pi^2}{4b^2} : m,n \geq 0\right\}$$

of the Neumann-Laplacian on the square, where $a = b = \frac{C^j}{2}$. In order to count the integer lattice points, we take the area of the ellipse with half axes $\frac{2a\sqrt{\lambda}}{\pi}$ and $\frac{2b\sqrt{\lambda}}{\pi}$ and add the additional points along the x- and y-axes:

$$\begin{aligned}&\operatorname{card}\left\{\lambda_{m,n} = \frac{m^2\pi^2}{4a^2} + \frac{n^2\pi^2}{4b^2} : m,n \geq 0,\ \lambda_{m,n} \leq \lambda\right\}\\ &\leq \frac{\pi}{4}\frac{2a\sqrt{\lambda}}{\pi}\frac{2b\sqrt{\lambda}}{\pi} + \frac{2a\sqrt{\lambda}}{\pi} + \frac{2b\sqrt{\lambda}}{\pi} + o(\sqrt{\lambda})\\ &= \frac{ab}{\pi}\lambda + \frac{2a\sqrt{\lambda}}{\pi} + \frac{2b\sqrt{\lambda}}{\pi} + o(\sqrt{\lambda}).\end{aligned}$$

As an upper estimate for the contribution of the j-th room, we therefore get

$$N_j \leq \frac{C^{2j}}{4\pi}\lambda + \frac{2C^j}{\pi}\sqrt{\lambda} + o(\sqrt{\lambda}).$$

The calculations involving the first room are a little different and we only use three partitions (see Figure 4).

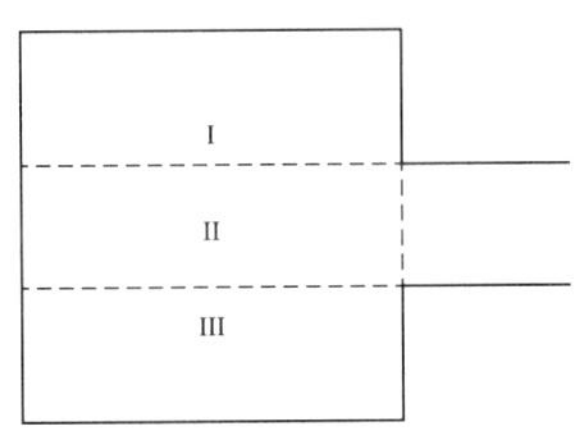

FIGURE 4. Subdivisions of the first room for the lower bound: Neumann conditions on the dotted lines.

(1) In regions I and III, we have the set of eigenvalues

$$\left\{\frac{m^2\pi^2}{4a^2} + \frac{(2n+1)^2\pi^2}{16b^2} : m,n \geq 0\right\} \quad \text{with } a = \frac{C}{2},\ b = \frac{C - kC^{2\alpha}}{4}.$$

(2) In region II, we have the set of eigenvalues

$$\left\{\frac{(2m+1)^2\pi^2}{16a^2}+\frac{n^2\pi^2}{4b^2}\;:\; m\geq 0,\; n\geq 1\right\} \quad \text{with } a=\frac{C}{2},\; b=\frac{kC^{2\alpha}}{2}.$$

Proceeding as for the other rooms we again need to count integer lattice points in the first quadrant.

- In regions I and III, we have

$$\operatorname{card}\left\{\lambda_{m,n}=\frac{m^2\pi^2}{4a^2}+\frac{(2n+1)^2\pi^2}{16b^2}\;:\; m,n\geq 0,\; \lambda_{m,n}\leq\lambda\right\}$$
$$\geq \frac{ab}{\pi}\lambda+\frac{(2b+a)\sqrt{\lambda}}{\pi}-o(\sqrt{\lambda}).$$

- In region II, we have

$$\operatorname{card}\left\{\lambda_{m,n}=\frac{(2m+1)^2\pi^2}{16a^2}+\frac{n^2\pi^2}{4b^2}\;:\; m\geq 0,\; n\geq 1,\; \lambda_{m,n}\leq\lambda\right\}$$
$$\geq \frac{ab}{\pi}\lambda+\frac{b\sqrt{\lambda}}{\pi}-o(\sqrt{\lambda}).$$

Combining these we get a lower estimate for the contribution of the first room

$$N_1 \;\geq\; \frac{\lambda C^2}{4\pi}+\left(2C-\frac{k}{2}C^{2\alpha}\right)\frac{\sqrt{\lambda}}{\pi}-o(\sqrt{\lambda}).$$

We next sum over the rooms (omitting the $o(\sqrt{\lambda})$-term for convenience). Denoting the volume of the first M rooms by $V(R_M)$, we see

$$\begin{aligned}\sum_{j=1,j\ odd}^{2M} N_j \;&\geq\; \sum_{j=1,j\ odd}^{2M}\frac{\lambda C^{2j}}{4\pi}\\ &+\left(2\sum_{j=1,j\ odd}^{2M}C^j-\frac{k}{2}\left(C^{2\alpha}+\sum_{j=3,j\ odd}^{2M}C^{\alpha(j-1)}\left(1+C^{2\alpha}\right)\right)\right)\frac{\sqrt{\lambda}}{\pi}\\ &=\;\frac{V(R_M)}{4\pi}\lambda+\left(2C\frac{1-C^{2M}}{1-C^2}-\frac{k}{2}\cdot\frac{2C^{2\alpha}-C^{2\alpha M}-C^{2\alpha(M+1)}}{1-C^{2\alpha}}\right)\frac{\sqrt{\lambda}}{\pi}.\end{aligned}$$

Also,

$$\sum_{j=1,j\ odd}^{2M} N_j\leq\sum_{j=1,j\ odd}^{2M}\left(\frac{\lambda C^{2j}}{4\pi}+\frac{2\sqrt{\lambda}C^j}{\pi}\right)=\frac{V(R_M)}{4\pi}\lambda+\frac{2\sqrt{\lambda}}{\pi C}\left(\frac{C^2-C^{2(M+1)}}{1-C^2}\right).$$

We now proceed to estimate the counting function in a passage. Here,

$$a=\frac{C^j}{2},\; b=\frac{kC^{\alpha j}}{2} \quad\text{and}\quad \lambda_{m,n}=\frac{\pi^2m^2}{C^{2j}}+\frac{\pi^2n^2}{C^{2\alpha j}k^2},\; m\geq 1, n\geq 0.$$

Then, since the lattice point counting estimate is on the ellipse with semi axes $\frac{2a\sqrt{\lambda}}{\pi}$ and $\frac{2b\sqrt{\lambda}}{\pi}$, we have

$$N_j\geq\frac{\lambda}{\pi}ab-o\left(\sqrt{\lambda}\right)+\frac{2a\sqrt{\lambda}}{\pi}=\frac{kC^{(1+\alpha)j}}{4\pi}\lambda+\frac{C^j}{\pi}\sqrt{\lambda}-o\left(\sqrt{\lambda}\right).$$

For the upper bound we count the eigenvalues

$$\lambda_{m,n}=\frac{\pi^2m^2}{C^{2j}}+\frac{\pi^2n^2}{C^{2\alpha j}k^2},\; m\geq 0, n\geq 0.$$

Thus

$$N_j \leq \frac{\lambda}{\pi}ab + o\left(\sqrt{\lambda}\right) + \frac{2a\sqrt{\lambda}}{\pi} + \frac{2b\sqrt{\lambda}}{\pi} = \frac{kC^{(1+\alpha)j}}{4\pi}\lambda + \left(\frac{kC^{\alpha j}}{\pi} + \frac{C^j}{\pi}\right)\sqrt{\lambda} + o\left(\sqrt{\lambda}\right).$$

Summing, we obtain bounds for the counting function in the passages

$$\begin{aligned}\sum_{j=1,\ even}^{2M} N_j &\geq \sum_{j=1,\ even}^{2M} \frac{kC^{(1+\alpha)j}}{4\pi}\lambda + \frac{C^j}{\pi}\sqrt{\lambda} = \frac{V(P_M)}{4\pi}\lambda + \sum_{i=1}^{M}\frac{C^{2i}}{\pi}\sqrt{\lambda} \\ &= \frac{V(P_M)}{4\pi}\lambda + \frac{\sqrt{\lambda}}{\pi}\frac{C^2 - C^{2(M+1)}}{1 - C^2},\end{aligned}$$

where $V(P_M)$ denotes the volume of the first M passages. We have the upper bound

$$\begin{aligned}\sum_{j=1,\ even}^{2M} N_j &\leq \frac{V(P_M)}{4\pi}\lambda + \sum_{j=1,\ j\ even}^{2M}\left(\frac{kC^{\alpha j}}{\pi} + \frac{C^j}{\pi}\right)\sqrt{\lambda} \\ &= \frac{V(P_M)}{4\pi}\lambda + \frac{\sqrt{\lambda}}{\pi}\left(k\sum_{i=1}^{M}C^{2\alpha i} + \sum_{i=1}^{M}C^{2i}\right) \\ &= \frac{V(P_M)}{4\pi}\lambda + \left(k\frac{C^{2\alpha} - C^{2\alpha(1+M)}}{1 - C^{2\alpha}} + \frac{C^2 - C^{2(1+M)}}{1 - C^2}\right)\frac{\sqrt{\lambda}}{\pi}.\end{aligned}$$

We now are in a position to estimate the bounds of the counting function for the domain Ω_{2M} Summing the contributions from the rooms and the passages, we get the lower estimate

$$\frac{\mid \Omega_{2M} \mid}{4\pi}\lambda + \left(\frac{2C + C^2 - 2C^{2M+1} - C^{2M+2}}{1 - C^2} - \frac{k}{2}\frac{2C^{2\alpha} - C^{2\alpha M} - C^{2\alpha(M+1)}}{1 - C^{2\alpha}}\right)\frac{\sqrt{\lambda}}{\pi}$$

with the upper estimate given by

$$\frac{\mid \Omega_{2M} \mid}{4\pi}\lambda + \left(\frac{2C + C^2 - 2C^{2M+1} - C^{2M+2}}{1 - C^2} + k\frac{C^{2\alpha} - C^{2\alpha(1+M)}}{1 - C^{2\alpha}}\right)\frac{\sqrt{\lambda}}{\pi}.$$

As $M \to \infty$, since there is no contribution from the tail and $C < 1$ we get

$$\frac{\sqrt{\lambda}}{\pi}\left(\frac{2C + C^2}{1 - C^2} - k\frac{C^{2\alpha}}{1 - C^{2\alpha}}\right) \leq N_N(\lambda) - \frac{\mid \Omega \mid}{4\pi}\lambda \leq \frac{\sqrt{\lambda}}{\pi}\left(\frac{2C + C^2}{1 - C^2} + \frac{kC^{2\alpha}}{1 - C^{2\alpha}}\right).$$

These yield the following result.

THEOREM 3.1. *As* $\lambda \to \infty$,

$$C_1\frac{\sqrt{\lambda}}{\pi} \leq N_N(\lambda) - \frac{\mid \Omega \mid}{4\pi}\lambda \leq C_2\frac{\sqrt{\lambda}}{\pi}$$

where

$$C_1 = \frac{2C + C^2}{1 - C^2} - k\frac{C^{2\alpha}}{1 - C^{2\alpha}} > 0, \quad C_2 = \frac{2C + C^2}{1 - C^2} + \frac{kC^{2\alpha}}{1 - C^{2\alpha}}$$

and

$$C_2 - C_1 = 2k\frac{C^{2\alpha}}{1 - C^{2\alpha}} \to 0 \quad as \quad k \to 0.$$

Therefore, the error is precisely of order $\sqrt{\lambda}$ *and in the limit* $k \to 0$*, we obtain the precise constant.*

3.2. Asymptotics for $N_D(\lambda)$. We now estimate bounds for the counting function of the Dirichlet-Laplacian. In order to do this we again look separately at the rooms and the passages. The lower bound is obtained by choosing Dirichlet conditions on all the boundaries. The set of eigenvalues is given by

$$\left\{\frac{m^2\pi^2}{4a^2}+\frac{n^2\pi^2}{4b^2}\ :\ m,n\geq 1\right\} \quad \text{with } a=\frac{C^j}{2},\ b=\frac{1}{2}\begin{cases} C^j & \text{for rooms,}\\ kC^{\alpha j} & \text{for passages.}\end{cases}$$

Therefore, the contribution to the lower bound from each room or passage is $\frac{ab}{\pi}\lambda - o(\sqrt{\lambda})$. Adding all contributions, we get the lower bound

$$N_D(\lambda)\geq \frac{|\,\Omega\,|}{4\pi}\lambda+o(\sqrt{\lambda}).$$

To get an upper bound in the first room, we choose Dirichlet conditions on three sides of the boundary and Neumann conditions on the right side. This gives us the set of eigenvalues

$$\left\{\frac{(2m+1)^2\pi^2}{16a^2}+\frac{n^2\pi^2}{4b^2}\ :\ m\geq 0,\ n\geq 1\right\} \quad \text{with } a=b=\frac{C}{2}.$$

Counting lattice points gives $\frac{ab}{\pi}\lambda+\frac{b\sqrt{\lambda}}{\pi}+o(\sqrt{\lambda})$, so as an upper estimate for the contribution of the first room, we get

$$\frac{\lambda C^2}{4\pi}+\frac{C}{2}\frac{\sqrt{\lambda}}{\pi}+o(\sqrt{\lambda}).$$

For the remaining rooms and passages we choose Dirichlet conditions on the horizontal boundaries and Neumann conditions on the vertical ones. This gives us the set of eigenvalues

$$\left\{\frac{m^2\pi^2}{4a^2}+\frac{n^2\pi^2}{4b^2}\ :\ m\geq 0,\ n\geq 1\right\}.$$

Again counting lattice points gives $\frac{ab}{\pi}\lambda+\frac{2b\sqrt{\lambda}}{\pi}+o(\sqrt{\lambda})$. Summing all contributions yields as the upper estimate (omitting the $o(\sqrt{\lambda})$-term).

$$\begin{aligned}&\frac{|\Omega|}{4\pi}\lambda+\left(\frac{C}{2}+\sum_{j=3,j\ odd}^{2M}C^j+k\sum_{j=1,j\ even}^{2M}C^{j\alpha}\right)\frac{\sqrt{\lambda}}{\pi}\\ &=\frac{|\Omega|}{4\pi}\lambda+\left(C\frac{1-C^{2M}}{1-C^2}-\frac{C}{2}+kC^{2\alpha}\frac{1-C^{2\alpha M}}{1-C^{2\alpha}}\right)\frac{\sqrt{\lambda}}{\pi}.\end{aligned}$$

Therefore, letting $M\to\infty$, we find the following result.

THEOREM 3.2. *As* $\lambda\to\infty$

$$o(\sqrt{\lambda})\leq N_D(\lambda)-\frac{|\,\Omega\,|}{4\pi}\lambda\leq\left(\frac{C(C^2+1)}{2(1-C^2)}+\frac{k}{C^{-2\alpha}-1}\right)\frac{\sqrt{\lambda}}{\pi}+o(\sqrt{\lambda}).$$

We remark that the lower bound is given by having Dirichlet boundary conditions everywhere which does not give a $\sqrt{\lambda}$-term. Therefore, we can only get an $o(\sqrt{\lambda})$ error estimate and not determine the sign of the $\sqrt{\lambda}$-term.

4. A related problem on the skeleton

We define $\mathbf{y} \in \partial\Omega$ to be a near point of $\mathbf{x} \in \Omega$ if $|\mathbf{x} - \mathbf{y}| = \operatorname{dist}(\mathbf{x}, \partial\Omega)$, the distance of $\mathbf{x}$ to the boundary of Ω. Therefore, denoting by $\mathcal{N}(\mathbf{x})$ the set of near points of $\mathbf{x}$, the skeleton Γ of Ω is the set

$$\mathcal{S}(\Omega) := \{\mathbf{x} \in \Omega : \operatorname{card} \mathcal{N}(\mathbf{x}) > 1\}.$$

It follows that the skeleton of the R&P domain Ω is the union of a sequence of line segments and parabolic arcs, $\Gamma = \{e_j\}_{j \in \mathbb{N}}$ say, connecting points in Ω; see Figure 5. For any $\mathbf{x} \in \Omega$, there exists $\mathbf{t} \in \Gamma$ such that $\mathbf{x}$ lies on one of two line segments $C_+(\mathbf{t}), C_-(\mathbf{t})$ connecting $\mathbf{t}$ to its 2 near points $\mathbf{y}_+, \mathbf{y}_-$: set $\tau : \Omega \to \Gamma, \mathbf{x} \to \mathbf{t}$. If $\tau(\mathbf{x}) = \mathbf{t} \in e \in \Gamma$, we may therefore define the following co-ordinate system on $\tau^{-1}(e)$:

$$\mathbf{x} = \mathbf{x}(\sigma, s), \quad \tau(\mathbf{x}) = \mathbf{t} = \mathbf{t}(\sigma), \quad s \in (-l(\sigma), l(\sigma)), \tag{4.1}$$

where σ denotes arc length along e, s is the distance from $\mathbf{t}$ to $\mathbf{x}$ along $C_+(\mathbf{t}(\sigma))$ or $C_-(\mathbf{t}(\sigma))$, $l(\sigma)$ is the length of $C_+(\mathbf{t}(\sigma))$ and $C_-(\mathbf{t}(\sigma))$ and we set $0 \le s \le l(\mathbf{t}(\sigma))$ along $C_+(\mathbf{t}(\sigma))$ and $-l(\mathbf{t}(\sigma)) \le s \le 0$ along $C_-(\mathbf{t}(\sigma))$. With $\mathbf{x} = (x, y)$ and $\tau(\mathbf{x}) = (\tau_1(\mathbf{x}), \tau_2(\mathbf{x}))$ it is shown in [**5**, (2.4)], that the determinant of the Jacobian

$$J := \left| \frac{\partial(x, y)}{\partial(\sigma, s)} \right| = \frac{1}{|\nabla\tau(\sigma, s)|} := \frac{1}{[|\nabla\tau_1(\mathbf{x})|^2 + |\nabla\tau_2(\mathbf{x})|^2]^{1/2}}. \tag{4.2}$$

For a measurable subset Γ_0 of Γ and $\Omega_0 := \tau^{-1}(\Gamma_0)$, it then follows that, for any $f \in L^2(\Omega_0)$ with $f = 0$ outside Ω_0,

$$\int_{\Omega_0} f(\mathbf{x}) d\mathbf{x} = \sum_{j \in \mathbb{N}} \int_{e_j} d\sigma \int_{-l(\sigma)}^{l(\sigma)} f(\sigma, s) \frac{1}{|\nabla\tau(\sigma, s)|} ds. \tag{4.3}$$

This implies, in particular, that for $f = F \circ \tau$ with $F \in L^2(\Gamma_0)$,

$$\int_{\Omega_0} F \circ \tau(\mathbf{x}) d\mathbf{x} = \sum_{j \in \mathbb{N}} \int_{e_j} F(\sigma) d\sigma \int_{-l(\sigma)}^{l(\sigma)} \frac{1}{|\nabla\tau(\sigma, s)|} ds = \int_{\Gamma_0} F(\sigma)\alpha(\sigma) d\sigma,$$

where

$$\alpha(\sigma) := \int_{-l(\sigma)}^{l(\sigma)} \frac{1}{|\nabla\tau(\sigma, s)|} ds. \tag{4.4}$$

The integral

$$\beta(\sigma) := \int_{-l(\sigma)}^{l(\sigma)} |\nabla\tau(\sigma, s)| ds \tag{4.5}$$

will also feature in certain specific regions of Ω.

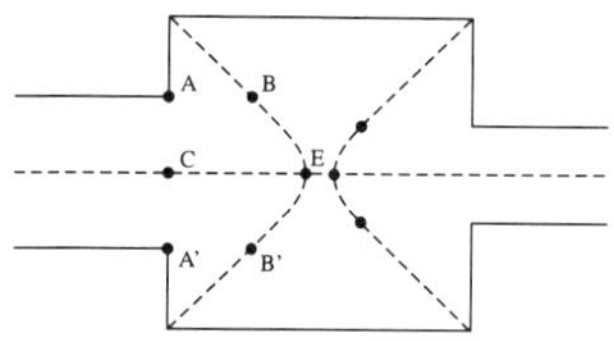

FIGURE 5. The skeleton.

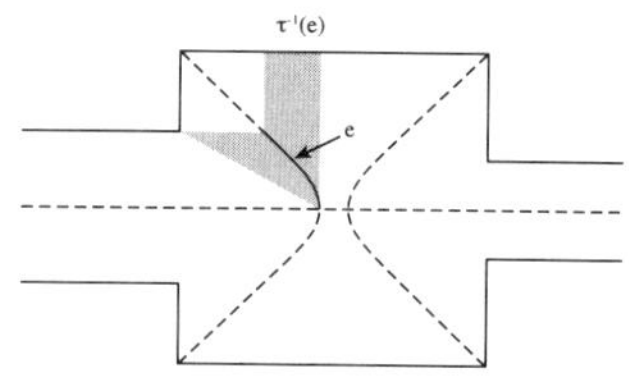

FIGURE 6. $\tau^{-1}(e)$ for a parabolic edge.

We shall be considering a general R&P domain hereafter, and not the special case of Sections 2 and 3. Therefore we allow for the possibility that the Neumann Laplacian does not have a discrete spectrum. Our first task is to make explicit the change of co-ordinates (4.1) in each region $\tau^{-1}(e), e \in \Gamma$ and then determine the map τ. The edges fall into 3 groups which have to be handled separately. In what follows below, we denote the height of a room by h and of a passage by δ.

Group 1 This consists of edges which are either in a passage or lie in the centre of a room with adjacent parabolic edges. Here, $\sigma = x$, $s = \pm y$, so the determinant of the Jacobian in (4.2) equals 1 and

$$l(\sigma) = \begin{cases} h/2 & \text{in rooms,} \\ \delta/2 & \text{in passages.} \end{cases}$$

Hence,

$$\alpha(\sigma) = \beta(\sigma) = \begin{cases} h & \text{in rooms,} \\ \delta & \text{in passages.} \end{cases} \tag{4.6}$$

Group 2 These are the edges in a room which are straight line segments along the diagonals. In Figure 5 with the origin at O, the edge on the diagonal of the square given by $0 < x < \frac{h-\delta}{2}$ and $y > \delta/2$ lies in this group and the analysis that follows is typical for all edges in this group. First, consider the triangle below the bisecting line, i.e. $\delta/2 < y < h/2 - x$. Here, we re-parameterize points $(x, y) \in \Omega$ by (σ, s), where σ is the arc length along the skeleton measured from the corner and s is the negative horizontal distance of the point from the skeleton. Thus,

$$(x, y) = \left(\frac{\sigma}{\sqrt{2}} + s, -\frac{\sigma}{\sqrt{2}} + \frac{h}{2}\right), \quad (\sigma, s) = \left(\sqrt{2}(\frac{h}{2} - y), -\frac{h}{2} + x + y\right).$$

In the triangle above the diagonal, where $\frac{h}{2} - x < y < \frac{h}{2}$, we choose s to be the vertical distance to the skeleton. Here,

$$(x, y) = \left(\frac{\sigma}{\sqrt{2}}, -\frac{\sigma}{\sqrt{2}} + \frac{h}{2} + s\right), \quad (\sigma, s) = \left(\sqrt{2}x, -\frac{h}{2} + x + y\right).$$

We note that in this whole square, we have that

$$\tau(x, y) = \left(\frac{\sigma}{\sqrt{2}}, \frac{h}{2} - \frac{\sigma}{\sqrt{2}}\right),$$

with $0 < \sigma < \frac{h-\delta}{\sqrt{2}}$. Moreover, the determinant of the Jacobian $J = 1/\sqrt{2}$ and $l(\sigma) = \sigma/\sqrt{2}$. Therefore,

$$\alpha(\sigma) = \sigma, \quad \beta(\sigma) = 2\sigma. \tag{4.7}$$

Group 3 These are edges which are such that every point on the edge has a re-entrant corner as one of its two near points. Thus in Figure 5, the parabolic edges BE and the line segment OE belong to this group associated with the re-entrant corner A. We consider the parabolic edge BE. For $y > 0$ this is determined by $|AQ| = |QQ'|$, where $Q = (x_0, y_0)$ is a point on the parabola and $Q' = (x_0, h/2)$. This gives

$$x_0^2 + \left(y_0 - \frac{\delta}{2}\right)^2 = \left(y_0 - \frac{h}{2}\right)^2.$$

After a little algebra this leads to

$$y_0 = -\frac{x_0^2}{h-\delta} + \frac{1}{4}(h+\delta). \tag{4.8}$$

In particular, the parabola intersects the x-axis at $E = \left(\frac{\sqrt{h^2-\delta^2}}{2}, 0\right)$.

Consider the part of the domain emanating from the re-entrant corner to the parabolic part of the skeleton. Let $\tau(x,y) = (x_0, y_0)$. Then in addition to lying on the parabola, (x_0, y_0) satisfies

$$y_0 = \frac{y - \delta/2}{x} x_0 + \frac{\delta}{2}. \tag{4.9}$$

The arc length along the parabola

$$\sigma = \int_{\frac{1}{2}(h-\delta)}^{x_0} \sqrt{1 + \left(\frac{-2x}{h-\delta}\right)^2}\, dx = \frac{h-\delta}{2} \int_1^{t_0} \sqrt{1+t^2}\, dt, \tag{4.10}$$

with $t_0 = \frac{2x_0}{h-\delta}$. A straightforward calculation gives that

$$\sigma = \frac{1}{4}(h-\delta)\left(t_0\sqrt{t_0^2+1} + \sinh^{-1}(t_0) - \sqrt{2} - \sinh^{-1}(1)\right). \tag{4.11}$$

In particular, the length of the parabolic edge is

$$|CE| = \frac{\sqrt{2}}{4}\sqrt{h(h+\delta)} - \frac{\sqrt{2}}{4}(h-\delta) + \frac{h-\delta}{4}\left(\sinh^{-1}\left(\sqrt{\frac{h+\delta}{h-\delta}}\right) - \sinh^{-1}(1)\right).$$

We now use (4.8) and (4.9) to express t_0 in terms of x and y. Eliminating y_0 in (4.8), we get

$$(h-\delta)\left(\frac{y-\delta/2}{x}x_0 + \frac{\delta}{2} - \frac{1}{4}(h+\delta)\right) + x_0^2 = 0.$$

This yields

$$t_0^2 + \frac{2y-\delta}{x}t_0 - 1 = 0, \tag{4.12}$$

so that

$$t_0 = \frac{\frac{\delta}{2} - y}{x} + \sqrt{\left(\frac{\frac{\delta}{2} - y}{x}\right)^2 + 1}. \tag{4.13}$$

For the distance s from (x, y) to (x_0, y_0) this then gives

$$s^2 = \left(x - \frac{h-\delta}{2}t_0\right)^2 + \left(y - \frac{y-\delta/2}{x}\frac{h-\delta}{2}t_0 - \frac{\delta}{2}\right)^2 = \frac{\left(t_0^2+1\right)^2\left(x - \frac{h-\delta}{2}t_0\right)^2}{4t_0^2}$$

and hence

$$s = -\frac{\left(t_0^2+1\right)\left(x-\frac{h-\delta}{2}t_0\right)}{2t_0}.$$

We need to calculate the determinant of the Jacobian $\frac{\partial(\sigma,s)}{\partial(x,y)}$. Several terms in this Jacobian vanish and we have

$$\left|\frac{\partial(\sigma,s)}{\partial(x,y)}\right| = \left|\frac{\partial\sigma}{\partial t_0}\frac{\partial t_0}{\partial y}\frac{\partial s}{\partial x}\right|.$$

A calculation gives

$$\frac{\partial\sigma}{\partial t_0} = \frac{1}{2}\sqrt{t_0^2+1}(h-\delta), \quad \frac{\partial s}{\partial x} = -\frac{t_0^2+1}{2t_0}, \quad \frac{\partial t_0}{\partial y} = -\frac{1}{x} - \frac{\delta-2y}{x\sqrt{(\delta-2y)^2+4x^2}}.$$

Therefore,

$$\left|\frac{\partial(\sigma,s)}{\partial(x,y)}\right| = \frac{\left(t_0^2+1\right)^{3/2}(h-\delta)\left(\frac{(\delta-2y)}{\sqrt{(\delta-2y)^2+4x^2}}+1\right)}{4t_0x}.$$

Substitution of t_0 in terms of x and y gives

$$\left|\frac{\partial(\sigma,s)}{\partial(x,y)}\right| = \sqrt{2}(h-\delta)\frac{\left((\delta-2y)^2+4x^2\right)^{\frac{1}{4}}}{\left(\sqrt{(\delta-2y)^2+4x^2}-\delta+2y\right)^{3/2}}. \tag{4.14}$$

To analyse the behaviour of $J^{-1} = \left|\frac{\partial(\sigma,s)}{\partial(x,y)}\right|$ near the re-entrant corner, let

$$x = r\cos\theta, \quad y = \frac{\delta}{2} - r\sin\theta \quad (i.e.\ \delta - 2y = 2r\sin\theta).$$

Then

$$\left|\frac{\partial(\sigma,s)}{\partial(x,y)}\right| = \sqrt{2}(h-\delta)\frac{\left(4r^2\right)^{\frac{1}{4}}}{(2r-2r\sin\theta)^{3/2}} = \frac{h-\delta}{\sqrt{2}r\,(1-\sin\theta)^{3/2}},$$

so J^{-1} behaves like $1/r$ where r is the distance from the corner. This behaviour of J^{-1} implies that on the parabolic edge

$$\alpha(\sigma) < \infty, \text{ but } \beta(\sigma) = \infty. \tag{4.15}$$

For this reason, we will need to make sure that the weight β does not appear in the analysis on those edges (like the parabolic edges) whose points have re-entrant near points. How we do this will be made apparent in the next section.

5. A Sturm-Liouville operator

We denote the set of edges of Γ which belong to groups 1 and 2 by Γ_{reg} and those in Group 3 by Γ_{sing}. Note that Γ_{sing} consists of the parabolic edges and ones like the edge OE in Figure 5 which connect an end of a passage and the parabolic edges. The map τ in (4.2) maps a re-entrant corner onto every point on a singular edge, which motivates us to define any function f on $e \in \Gamma_{\text{sing}}$ to be constant.

The underlying Hilbert spaces on Γ are as follows:

$$\widetilde{L}^2(\Gamma) = \bigoplus_{e\in\Gamma_{\text{reg}}} L^2(e;\alpha(\sigma)d\sigma) \bigoplus_{e\in\Gamma_{\text{sing}}} \mathbb{C}\mathbf{1}_e, \tag{5.1}$$

where $\mathbf{1}_e$ is the characteristic function of the edge e, $L^2(e;\alpha(\sigma)d\sigma)$ is the weighted Lebesgue space with inner-product

$$\int_e f(\sigma)\overline{g(\sigma)}\alpha(\sigma)d\sigma, \tag{5.2}$$

and with $f = (f_e) \in \widetilde{L}^2(\Gamma)$, we have $f_e =$ constant for $e \in \Gamma_{\text{sing}}$;

$$\widetilde{H}^1(\Gamma) := \bigoplus_{e\in\Gamma_{\text{reg}}} H^1(e;\alpha,\beta) \bigoplus_{e\in\Gamma_{\text{sing}}} \mathbb{C}\mathbf{1}_e, \tag{5.3}$$

where for $e \in \Gamma_{\text{reg}}$, $H^1(e;\alpha,\beta)$ is the weighted Sobolev space with inner product

$$(f_e, g_e)_{H^1(e;\alpha,\beta)} = \int_e \left\{\nabla f_e(\sigma)\overline{\nabla g_e(\sigma)}\beta(\sigma) + f_e(\sigma)\overline{g_e(\sigma)}\alpha(\sigma)\right\} d\sigma. \tag{5.4}$$

The inner-products on $\widetilde{L}^2(\Gamma)$ and $\widetilde{H}^1(\Gamma)$ are respectively,

$$(f, g) = \sum_{e\in\Gamma} \int_e f(\sigma)\overline{g(\sigma)}\alpha(\sigma)d\sigma$$

and

$$\begin{aligned} (f, g)_{\widetilde{H}^1(\Gamma)} &= \sum_{e\in\Gamma_{\text{reg}}} \int_e \left([\nabla f(\sigma)\cdot\overline{\nabla g(\sigma)}]\beta(\sigma) + [f(\sigma)\overline{g(\sigma)}]\alpha(\sigma)\right) d\sigma \\ &+ \sum_{e\in\Gamma_{\text{sing}}} \int_e [f(\sigma)\overline{g(\sigma)}]\alpha(\sigma)d\sigma. \end{aligned} \tag{5.5}$$

We define $H_\Gamma + I$ to be the self-adjoint operator in $\widetilde{L}^2(\Gamma)$ associated with the $\widetilde{H}^1(\Gamma)$ inner-product, where I is the identity operator on $\widetilde{L}^2(\Gamma)$. The following theorem is readily proved by a standard argument; cf., [**5**], Theorem 3.3.

THEOREM 5.1. *The domain $D(H_\Gamma)$ of H_Γ consists of sequences $u = (u_e) \in \widetilde{H}^1(\Gamma)$ which satisfy the following :*

(1) *$u_e =$ constant for $e \in \Gamma_{sing}$;*
(2) *for each $e \in \Gamma_{reg}$, $\beta u_e'$ is locally absolutely continuous on the interior of e and*

$$\lim\{\beta(\sigma)\frac{du_e}{d\sigma}\} = 0 \tag{5.6}$$

as σ tends to the end points of e;

(3) *for $e \in \Gamma_{sing}$, $(H_\Gamma u)_e = 0$;*
(4) *for $e \in \Gamma_{reg}$,*

$$(H_\Gamma u)_e(\sigma) = -\frac{1}{\alpha(\sigma)}\frac{d}{d\sigma}\left[\beta(\sigma)\frac{du_e}{d\sigma}\right].$$

An important part in the analysis will be played by the following operator which maps functions on the skeleton to functions on the R&P domain. Define

$$T_0 : \widetilde{L}^2(\Gamma) \to L^2(\Omega) \text{ by } T_0 f = f\circ\tau \text{ for } f \in \widetilde{L}^2(\Gamma). \tag{5.7}$$

LEMMA 5.2. *Let $e \in \Gamma$. Then for $g \in L^2(\tau^{-1}(e))$ and $\mathbf{t}(\sigma) \in e$ we have*

$$(T_0^* g)(\sigma) = \frac{1}{\alpha(\sigma)}\int_{-l(\sigma)}^{l(\sigma)} g(\sigma, s)\frac{1}{|\nabla\tau(\sigma, s)|}\, ds. \tag{5.8}$$

PROOF. For $F \in L^2(e)$,

$$\int_e (T_0^* g)(\sigma)\overline{F(\sigma)}d\sigma = \int_{\tau^{-1}(e)} g(\sigma)\overline{(T_0 F)(\sigma)}d\sigma$$
$$= \int_e \overline{F(\sigma)}\left(\frac{1}{\alpha(\sigma)}\int_{-l(\sigma)}^{l(\sigma)} \frac{g(\sigma,s)}{|\nabla\tau(\sigma,s)|}\, ds\right)\alpha(\sigma)\, d\sigma,$$

proving the result. □

REMARK 5.3. Note that in particular, for $\mathbf{t}(\sigma) \in e \in \Gamma_{\text{sing}}$, the function $(T_0^* g)(\sigma)$ is constant and takes the value

$$(T_0^* g)(\sigma) = \frac{1}{|\tau^{-1}(e)|}\int_{\tau^{-1}(e)} g(x)\, dx. \tag{5.9}$$

LEMMA 5.4. *T_0 is an isometry and so $T_0^* T_0 = I$, the identity on $\widetilde{L}^2(\Gamma)$.*

PROOF. Let $F \in \widetilde{L}^2(\Gamma)$. Then for any edge e, we have

$$\int_{\tau^{-1}(e)} |(F\circ\tau)(x)|^2\, dx = \int_e |F(\sigma)|^2 \int_{-l(\sigma)}^{l(\sigma)} \frac{1}{|\nabla\tau(\sigma,s)|}\, ds\, d\sigma$$
$$= \int_e |F(\sigma)|^2\alpha(\sigma)\, d\sigma. \tag{5.10}$$

Adding the contributions from all edges shows that $\|T_0 F\|_{L^2(\Omega)} = \|F\|_{\widetilde{L}^2(\Gamma)}$ which completes the proof. □

The proof of the next lemma on how T_0 interacts with derivatives is the same as that of [**5**, Lemma 3.2].

LEMMA 5.5. *For $F \in \widetilde{H}^1(\Gamma)$,*

$$\sum_{e\in\Gamma}\int_{\tau^{-1}(e)} |\nabla(F\circ\tau)(x)|^2\, dx = \sum_{e\in\Gamma_{\text{reg}}}\int_e |F'(\sigma)|^2\beta(\sigma)\, d\sigma. \tag{5.11}$$

Let $\hat{H}^1(\Omega) := \bigoplus_{e\in\Gamma} H^1(\tau^{-1}(e))$.

COROLLARY 5.6. *The map $T_1 : \widetilde{H}^1(\Gamma) \to \hat{H}^1(\Omega)$ given by $T_1 f = f\circ\tau$ for $f \in \widetilde{H}^1(\Gamma)$ is an isometry.*

In the final theorem, E_1 denotes the natural embedding of $\hat{H}^1(\Omega)$ into the space $\bigoplus_{e\in\Gamma} L^2(\tau^{-1}(e))$, and H_Ω denotes the selfadjoint operator associated with the $\hat{H}^1(\Omega)$ inner-product; thus H_Ω is the orthogonal sum

$$H_\Omega = \bigoplus_{e\in\Gamma}(-\Delta_{N,\tau^{-1}(e)}).$$

THEOREM 5.7. (1) *The operator $(H_\Gamma + I)^{-1}$ is not compact on $\widetilde{L}^2(\Gamma)$ and $0 \in \sigma_{ess}(H_\Gamma)$.*
(2) *E_1 is not compact and $0 \in \sigma_{ess}(H_\Omega)$.*
(3) *If $E_1(T_1 T_1^* - I) : \hat{H}^1(\Omega) \to L^2(\Omega)$ is compact, then $\sigma_{ess}(H_\Omega) \subseteq \sigma_{ess}(H_\Gamma)$.*

PROOF. (1) On the singular edges $(H_\Gamma + I)^{-1}$ is just I. As there are infinitely many singular edges, $(H_\Gamma + I)^{-1}$ is not compact. Moreover, by considering functions supported on one singular edge, we see that 0 is an eigenvalue of H_Γ of infinite multiplicity.

(2) Take an infinite sequence of edges $(e_n)_{n\in\mathbb{N}}$ and consider the sequence of normalised characteristic functions
$$\varphi_n(\mathbf{x}) = \frac{1}{\sqrt{|\tau^{-1}(e_n)|}}\chi_{\tau^{-1}(e_n)}(\mathbf{x}) \text{ for } \mathbf{x} \in \Omega.$$
This bounded sequence in $\hat{H}^1(\Omega)$ has no convergent subsequence in $L^2(\Omega)$. Moreover, every φ_n is an eigenfunction of H_Ω with eigenvalue 0.
(3) The proof of this is similar to that of Theorem 4.4 in [**5**].

□

It is interesting to compare the results we get in Theorem 5.7 with those obtained in [**2**] for horn shaped domains where the authors are able to relate the essential spectrum of the Neumann Laplacian on the horn to that of a Schrödinger operator on the skeleton. In our case the re-entrant corners of the Rooms and Passages domain force us to introduce singular edges which cause the operator on the skeleton to decouple and prevent such a detailed result from being obtained.

References

[1] V. I. Burenkov, *Extension of functions with preservation of certain smoothness, and the compactness of embeddings for spaces of differentiable functions*, Tr. Mat. Inst. Steklova **248** (2005), no. Issled. po Teor. Funkts. i Differ. Uravn., 74–85 (Russian, with Russian summary); English transl., Proc. Steklov Inst. Math. **1 (248)** (2005), 69–80. MR2165917 (2006d:46043)

[2] E. B. Davies and B. Simon, *Spectral properties of Neumann Laplacian of horns*, Geom. Funct. Anal. **2** (1992), no. 1, 105–117, DOI 10.1007/BF01895707. MR1143665 (93g:35099)

[3] W. D. Evans and D. J. Harris, *Sobolev embeddings for generalized ridged domains*, Proc. London Math. Soc. (3) **54** (1987), no. 1, 141–175, DOI 10.1112/plms/s3-54.1.141. MR872254 (88b:46056)

[4] W. D. Evans and D. J. Harris, *Fractals, trees and the Neumann Laplacian*, Math. Ann. **296** (1993), no. 3, 493–527, DOI 10.1007/BF01445117. MR1225988 (94k:35218)

[5] W. D. Evans and Yoshimi Saitō, *Neumann Laplacians on domains and operators on associated trees*, Q. J. Math. **51** (2000), no. 3, 313–342, DOI 10.1093/qjmath/51.3.313. MR1782097 (2001j:47052)

[6] N. Filonov, *On an inequality for the eigenvalues of the Dirichlet and Neumann problems for the Laplace operator*, Algebra i Analiz **16** (2004), no. 2, 172–176, DOI 10.1090/S1061-0022-05-00857-5 (Russian); English transl., St. Petersburg Math. J. **16** (2005), no. 2, 413–416. MR2068346 (2005f:35228)

[7] Rainer Hempel, Luis A. Seco, and Barry Simon, *The essential spectrum of Neumann Laplacians on some bounded singular domains*, J. Funct. Anal. **102** (1991), no. 2, 448–483, DOI 10.1016/0022-1236(91)90130-W. MR1140635 (93h:35144)

[8] M. N. Huxley, *Exponential sums and lattice points. III*, Proc. London Math. Soc. (3) **87** (2003), no. 3, 591–609, DOI 10.1112/S0024611503014485. MR2005876 (2004m:11127)

[9] Yu. Netrusov and Yu. Safarov, *Weyl asymptotic formula for the Laplacian on domains with rough boundaries*, Comm. Math. Phys. **253** (2005), no. 2, 481–509, DOI 10.1007/s00220-004-1158-8. MR2140257 (2006a:35222)

Cardiff School of Computer Science and Informatics, Cardiff University, Queen's Buildings, 5 The Parade, Roath, Cardiff CF24 3AA

E-mail address: malcolm@cs.cf.ac.uk

School of Mathematics Cardiff University, Senghennydd Road, Cardiff, Wales, United Kingdom CF24 4AG

E-mail address: EvansWD@cardiff.ac.uk

School of Mathematics, Statistics and Actuarial Science, Cornwallis Building, University of Kent, Canterbury, Kent, CT2 7NF

E-mail address: i.wood@kent.ac.uk

Proceedings of Symposia in Pure Mathematics
Volume **87**, 2013
http://dx.doi.org/10.1090/pspum/087/01429

Finite gap Jacobi matrices: A review

Jacob S. Christiansen, Barry Simon, and Maxim Zinchenko

1. Introduction

Perhaps the most common theme in Fritz Gesztesy's broad opus is the study of problems with periodic or almost periodic finite gap differential and difference equations, especially those connected to integrable systems. The present paper reviews recent progress in the understanding of finite gap Jacobi matrices and their perturbations. We'd like to acknowledge our debt to Fritz as a collaborator and friend. We hope Fritz enjoys this birthday bouquet!

We consider Jacobi matrices, J, on $\ell^2(\{1,2,\dots,\})$ indexed by $\{a_n,b_n\}_{n=1}^\infty$, $a_n>0$, $b_n\in\mathbb{R}$, where ($u_0\equiv 0$)

$$(Ju)_n = a_n u_{n+1} + b_n u_n + a_{n-1}u_{n-1} \tag{1.1}$$

or its two-sided analog on $\ell^2(\mathbb{Z})$ where a_n,b_n,u_n are indexed by $n\in\mathbb{Z}$ and J is still given by (1.1) (we refer to "Jacobi matrix" for the one-sided objects and "two-sided Jacobi matrix" for the $\mathbb{Z}$ analog). Here the a's and b's parametrize the operator J and $\{u_n\}\in\ell^2$.

We recall that associated to each bounded Jacobi matrix, J, there is a unique probability measure, μ, of compact support in $\mathbb{R}$ characterized by either of the equivalent

(a) J is unitarily equivalent to multiplication by x on $L^2(\mathbb{R},d\mu)$ by a unitary with $(U\delta_1)(x)\equiv 1$.

(b) $\{a_n,b_n\}_{n=1}^\infty$ are the recursion parameters for the orthogonal polynomials for μ.

We'll call μ the spectral measure for J.

By a finite gap Jacobi matrix, we mean one whose essential spectrum is a finite union

$$\sigma_{\text{ess}}(J) = \mathfrak{e} \equiv [\alpha_1,\beta_1]\cup\cdots\cup[\alpha_{\ell+1},\beta_{\ell+1}] \tag{1.2}$$

where

$$\alpha_1<\beta_1<\cdots<\alpha_{\ell+1}<\beta_{\ell+1} \tag{1.3}$$

2010 *Mathematics Subject Classification.* Primary 47B36, 42C05, 58J53, 34L15.

Key words and phrases. Isospectral torus, Orthogonal polynomials, Szegő's theorem, Szegő asymptotics, Lieb–Thirring bounds.

The first author was supported in part by a Steno Research Grant (09-064947) from the Danish Research Council for Nature and Universe.

The second author was supported in part by NSF grant DMS-0968856.

The third author was supported in part by NSF grant DMS-0965411.

ℓ counts the number of gaps.

We will see that for each such $\mathfrak{e}$, there is an ℓ-dimensional torus of two-sided J's with $\sigma(J) = \mathfrak{e}$ and J almost periodic and regular in the sense of Stahl–Totik [**56**]. We'll present the theory of perturbations of such J that decay but not too slowly. Our interest will be in spectral types, Lieb–Thirring bounds on the discrete eigenvalues and on orthogonal polynomial asymptotics. We begin in Section 2 with a discussion of the case $\ell = 0$ where we may as well take $\mathfrak{e} = [-2, 2]$, in which the (0-dimensional) torus is the single point with $a_n \equiv 1$, $b_n \equiv 0$. We'll discuss the theory in that case as background.

Section 3 describes the isospectral torus. Section 4 discusses the results for general finite gap sets with a mention of the special results that occur if each $[\alpha_j, \beta_j]$ has rational harmonic measure, in which case the isospectral torus contains only periodic J's. Section 5 discusses a method for the general finite gap case which relies on the realization of $\mathbb{C} \cup \{\infty\} \setminus \mathfrak{e}$ as the quotient of the unit disk in $\mathbb{C}$ by a Fuchsian group—a method pioneered by Peherstorfer–Sodin–Yuditskii [**42, 55**], who were motivated by earlier work of Widom [**64**] and Aptekarev [**4**].

While we focus on the finite gap case, we note there are some results on general compact $\mathfrak{e}$'s in $\mathbb{R}$ with various restrictive conditions on $\mathfrak{e}$ (e.g., Parreau–Widom). Peherstorfer–Yuditskii [**42**] discuss homogeneous sets and Christiansen [**8, 9**] proves versions of Theorems 4.3 and 4.5 below for suitable infinite gap $\mathfrak{e}$'s. See [**16, 65**] for discussion of properties of some $\mathfrak{e}$'s and examples relevant to this area.

These works suggest forms of two conditions in the finite gap case suitable for generalization. Let $\rho_{\mathfrak{e}}$ be the equilibrium measure for $\mathfrak{e}$ and $G_{\mathfrak{e}}(z)$ its Green's function $(-\mathcal{E}(\rho_{\mathfrak{e}}) - \Phi_{\rho_{\mathfrak{e}}}(z)$ in terms of (3.1)/(3.2)). Then (4.5) should read

$$\sum_{n=1}^{N} G_{\mathfrak{e}}(x_n) < \infty \tag{1.4}$$

(which for finite gap $\mathfrak{e}$ is equivalent to (4.5)). Similarly, (4.6) should read

$$\int \log[f(x)]\, d\rho_{\mathfrak{e}}(x) > -\infty \tag{1.5}$$

(again, for finite gap $\mathfrak{e}$ equivalent to (4.6)).

2. The Zero Gap Case

The Jacobi matrix, J_0, with $a_n \equiv 1$, $b_n \equiv 0$ is called the free Jacobi matrix. It is easy to see that the solutions of $J_0 u = \lambda u$ are given by solving

$$\alpha + \alpha^{-1} = \lambda \tag{2.1}$$

for $\lambda \in \mathbb{C}$ and setting

$$u_n = \frac{1}{2i}\,(\alpha^n - \alpha^{-n}) \tag{2.2}$$

This is polynomially bounded in n if and only if $|\alpha| = 1$. If $\alpha = e^{ik}$, then

$$\lambda = 2\cos k, \qquad u_n = \sin(kn) \tag{2.3}$$

Thus,

$$\sigma(J_0) = [-2, 2], \quad \lambda \in (-2, 2) \Rightarrow \text{all eigenfunctions bounded} \tag{2.4}$$

(by all eigenfunctions here, we mean without the boundary condition $u_0 = 0$).

In identifying the spectral type, the following is useful:

THEOREM 2.1. *Let J be a Jacobi matrix with $a_n + a_n^{-1} + |b_n|$ bounded. Suppose all solutions of $(Ju)_n = \lambda u_n$ (where u_0, u_1 are arbitrary) are bounded for $\lambda \in S \subset \mathbb{R}$. Then the spectrum of J on S is purely a.c. in the sense that if μ is the spectral measure of J and $|\cdot|$ is Lebesgue measure, then*

$$\mu_s(S) = 0, \qquad T \subset S \text{ and } |T| > 0 \Rightarrow \mu_{ac}(T) > 0 \tag{2.5}$$

REMARK. The modern approach to this theorem would use the inequalities of Jitomirskaya–Last [**28**, **29**] or Gilbert–Pearson subordinacy theory [**23**, **24**, **30**, **40**] to handle μ_s and the results of Last–Simon [**36**] for the a.c. spectrum. The simplest proof for this special case (where the above ideas are overkill) is perhaps Simon [**49**].

A simple variation of parameters in the difference equation implies that under ℓ^1 perturbations, eigenfunctions remain bounded when $\lambda \in (-2, 2)$, that is,

THEOREM 2.2. *Let J be a Jacobi matrix with*

$$\sum_{n=1}^{\infty} |a_n - 1| + |b_n| < \infty \tag{2.6}$$

Then $\sigma_{ess}(J) = [-2, 2]$ and the spectrum on $(-2, 2)$ is purely a.c.

REMARK. The continuum analog of Theorem 2.2 goes back to Titchmarsh [**60**].

Thus, the spectrum outside $[-2, 2]$ is a set of eigenvalues $\{x_n\}_{n=1}^N$ where $N \in \mathbb{N} \cup \{\infty\}$. (2.6) has implications for these eigenvalues.

THEOREM 2.3. *Let $\{x_n\}_{n=1}^N$ be the eigenvalues of a Jacobi matrix. Then*

$$\sum_{n=1}^{N} (x_n^2 - 4)^{1/2} \le \sum_{n=1}^{\infty} |b_n| + 4 \sum_{n=1}^{\infty} |a_n - 1| \tag{2.7}$$

REMARKS. 1. This implies

$$\sum_{n=1}^{N} \operatorname{dist}(x_n, [-2, 2])^{1/2} \le \frac{1}{2} \Big(\sum_{n=1}^{\infty} |b_n| + 4 \sum_{n=1}^{\infty} |a_n - 1| \Big) \tag{2.8}$$

2. The analog of (2.8) in the continuum case is due to Lieb–Thirring [**37**] who proved it when the power $1/2$ is replaced by $p > 1/2$ and the right side is replaced by $|b_n|^{p+1/2}$, $|a_n - 1|^{p+1/2}$ and $1/2$ by a suitable constant. They proved the analog is false if $p < 1/2$ and conjectured the result if $p = 1/2$. This conjecture was proven by Weidl [**63**] with an alternate proof and optimal constant by Hundermark–Lieb–Thomas [**25**]. (2.8) and its $p > 1/2$ analogs are called Lieb–Thirring inequalities after [**37**].

3. This theorem is a result of Hundertmark–Simon [**26**] who used a method inspired by [**25**].

4. (2.7) is optimal in the sense that its $p < 1/2$ analog is false and one cannot put a constant $\gamma < 1$ in front of neither the b sum nor the $a - 1$ sum. The same also applies to (2.8).

5. (2.7) implies $p > 1/2$ analogs by an argument of Aizenman–Lieb [**3**].

6. The one-half power in (2.7)/(2.8) is especially significant for the following reason:

$$x(z) = z + z^{-1} \tag{2.9}$$

maps $\mathbb{D}$ to $\mathbb{C} \cup \{\infty\} \setminus [-2,2]$. Its inverse

$$z(x) = \tfrac{1}{2}\left(x - \sqrt{x^2-4}\right) \tag{2.10}$$

has a square root singularity at $x = \pm 2$. Thus, the finiteness of the left side of (2.7)/(2.8) is equivalent to a Blaschke condition

$$\sum_{n=1}^{N} (1 - |z(x_n)|) < \infty \tag{2.11}$$

THEOREM 2.4. *Let J be a Jacobi matrix with $\sigma_{ess}(J) = [-2,2]$ and Jacobi parameters $\{a_n, b_n\}_{n=1}^{\infty}$. Suppose its spectral measure has the form*

$$d\mu = f(x)\,dx + d\mu_s \tag{2.12}$$

where $d\mu_s$ is singular with respect to dx. Suppose that $\{x_n\}_{n=1}^{N}$ are its pure points outside $[-2,2]$. Consider the three conditions:

$$(a) \qquad \sum_{n=1}^{N} \operatorname{dist}(x_n, [-2,2])^{1/2} < \infty \tag{2.13}$$

$$(b) \qquad \int_{-2}^{2} (4-x^2)^{-1/2} \log[f(x)]\,dx > -\infty \tag{2.14}$$

$$(c) \qquad \lim_{n\to\infty} a_1 \dots a_n \text{ exists in } (0,\infty) \tag{2.15}$$

Then any two conditions imply the third. Moreover, in that case,

$$(d) \qquad \sum_{n=1}^{\infty} (a_n - 1)^2 + b_n^2 < \infty \tag{2.16}$$

$$(e) \qquad \lim_{K\to\infty} \sum_{n=1}^{K} (a_n - 1) \text{ and } \lim_{K\to\infty} \sum_{n=1}^{K} b_n \text{ exist} \tag{2.17}$$

REMARKS. 1. (2.13) is called a critical Lieb–Thirring inequality. (2.14) is the Szegő condition.

2. Since $f \in L^1$, the integral in (2.14) can only diverge to $-\infty$. That is, the integral over $\log_+$ is always finite and (2.14) is equivalent to the integral converging absolutely.

3. By a result of Ullman [**62**], $\sigma_{\text{ess}}(J) = [-2,2]$ and $f(x) > 0$ for a.e. x in $[-2,2]$ implies $\lim_{n\to\infty}(a_1 \dots a_n)^{1/n} = 1$, so (2.15) can be thought of as a second term in the asymptotics of $\frac{1}{n}\log(a_1 \dots a_n)$.

4. Condition (c) can be thought of as three statements: $\limsup < \infty$, $\liminf > 0$, and $\limsup = \liminf$. The full strength of (c) is not always needed. For example, (a) plus $\limsup > 0$ implies (b) and the rest of (c).

5. This result can be thought of as an analog of a theorem of Szegő for OPUC [**57**] (see also [**50**, Ch. 2]). That (b) $\Rightarrow$ (c), if there are no eigenvalues, is due to Shohat [**47**] and that (b) $\Leftrightarrow$ (c), if there are finitely many x's, is due to Nevai [**38**]. The general (a) + (b) $\Rightarrow$ (c) is due to Peherstorfer–Yuditskii [**41**] and the essence of this theorem is from Killip–Simon [**32**], although the precise theorem is from Simon–Zlatoš [**54**].

COROLLARY 2.5. *If (2.6) holds, then so does (2.14).*

PROOF. (2.6) implies $\prod_{n=1}^{\infty} a_n$ converges absolutely and, by Theorem 2.3, it implies (2.13). Thus, (2.14) holds by Theorem 2.4. □

REMARKS. 1. This result was a conjecture of Nevai [**39**].

2. It was proven by Killip–Simon [**32**]. It was the need to complete the proof of this that motivated Hundertmark–Simon [**26**].

There is a close connection between these conditions and asymptotics of the OPRL:

THEOREM 2.6. *Let* $\{p_n(x)\}_{n=0}^{\infty}$ *be the orthonormal polynomials for a Jacobi matrix, J, obeying the conditions* (a)–(c) *of Theorem* 2.4. *Then uniformly for x in compact subsets of* $\mathbb{C} \cup \{\infty\} \setminus [-2,2]$,

$$\lim_{n\to\infty} \frac{p_n(x)}{\left[\frac{1}{2}(x+\sqrt{x^2-4}\,)\right]^n} \tag{2.18}$$

exists and is analytic with zeros only at the x_n's.

REMARKS. 1. When there are no x_n's, this is essentially a result of Szegő [**57**, **58**]. For the general case, see Peherstorfer–Yuditskii [**41**].

2. This is called Szegő asymptotics.

3. The reason for the different sign in (2.10) and (2.18) is that, as $n \to \infty$, $p_n(x) \to \infty$, $|z(x)| < 1$ so $z(x)^n p_n(x)$ is bounded. The other solution of (2.9) is $z(x)^{-1}$ and it is that solution that appears in the denominator of (2.18).

While conditions (a)–(c) of Theorem 2.4 are sufficient for Szegő asymptotics, they are not necessary:

THEOREM 2.7. *Let J be a Jacobi matrix whose parameters obey* (2.16) *and* (2.17). *Then* (2.18) *holds on compact subsets of* $\mathbb{C} \cup \{\infty\} \setminus [-2,2]$. *Conversely, if* (2.18) *holds uniformly on the circle $|x| = R$ for some $R > 2$, then* (2.16) *and* (2.17) *hold.*

REMARKS. 1. This is a result of Damanik–Simon [**14**].

2. There exist examples where (2.16) and (2.17) hold but both (2.13) and (2.14) fail.

THEOREM 2.8. *For a Jacobi matrix, J, with parameters $\{a_n, b_n\}_{n=1}^{\infty}$, spectral measure obeying* (2.12), *and discrete eigenvalues $\{x_n\}_{n=1}^{N}$, one has*

$$\sum_{n=1}^{\infty} (a_n - 1)^2 + b_n^2 < \infty \tag{2.19}$$

if and only if

$$(a) \qquad \sigma_{ess}(J) = [-2,2] \tag{2.20}$$

$$(b) \qquad \sum_{n=1}^{N} \operatorname{dist}(x_n, [-2,2])^{3/2} < \infty \tag{2.21}$$

$$(c) \qquad \int_{-2}^{2} (4-x^2)^{+1/2} \log[f(x)]\, dx > -\infty \tag{2.22}$$

REMARKS. 1. This theorem is due to Killip–Simon [**32**]. They call (a) Blumenthal–Weyl, (b) Lieb–Thirring, and (c) quasi-Szegő.

2. The continuous analog of (2.19) $\Rightarrow$ (2.21) is due to Lieb–Thirring [**37**].

THEOREM 2.9. *Let J be a Jacobi matrix with $\sigma_{ess}(J) = [-2, 2]$ and spectral measure, $d\mu$, given by* (2.12)*. Suppose $f(x) > 0$ for a.e. x in $[-2, 2]$. Then*

$$\lim_{n\to\infty} |a_n - 1| + |b_n| = 0 \tag{2.23}$$

REMARK. This is often called the Denisov–Rakhmanov theorem after [**15, 44, 45**]. The result is due to Denisov. Rakhmanov had the analog for OPUC which implies the weak version of Theorem 2.9, where $\sigma_{\rm ess}(J) = [-2, 2]$ is replaced by $\sigma(J) = [-2, 2]$. That the result as stated was true was a long-standing conjecture settled by Denisov.

Conditions on the spectrum combined with weak conditions on the Jacobi parameters have strong consequences. For example, the existence of $\lim_{n\to\infty} a_1 \dots a_n$ clearly has no implication for the b's, but if combined with $\sigma(J) = [-2, 2]$ implies, by Theorems 2.4 and 2.8, that $\sum_{n=1}^\infty b_n^2 < \infty$. Similarly, one has

THEOREM 2.10. *Suppose $\sigma_{ess}(J) = [-2, 2]$ and*

$$\lim_{n\to\infty} (a_1 \dots a_n)^{1/n} = 1 \tag{2.24}$$

Then

$$\lim_{N\to\infty} \frac{1}{N} \sum_{n=1}^{N} (a_n - 1)^2 + b_n^2 = 0 \tag{2.25}$$

REMARKS. 1. (2.24) says that the underlying measure is regular in the sense of Ullman–Stahl–Totik; see the discussion in Section 3.

2. This theorem is a result of Simon [**52**].

3. The Isospectral Torus

Let $\mathfrak{e}$ be a finite gap set with ℓ gaps and $\ell + 1$ components, $\mathfrak{e}_j = [\alpha_j, \beta_j]$, $j = 1, \dots, \ell + 1$. There is associated to $\mathfrak{e}$ a natural ℓ-dimensional torus, $\mathcal{T}_\mathfrak{e}$, of almost periodic Jacobi matrices. If $\{a_n, b_n\}_{n=-\infty}^\infty$ are almost periodic sequences, they are determined by their values for $n \geq 1$ so we can view the elements of $\mathcal{T}_\mathfrak{e}$ as either one- or two-sided Jacobi matrices. There are at least three different ways to think of $\mathcal{T}_\mathfrak{e}$:

(a) As reflectionless two-sided Jacobi matrices, J, with $\sigma(J) = \mathfrak{e}$. This is the approach of [**5, 7, 21, 22, 42, 53, 55, 59**].
(b) As one-sided Jacobi matrices whose m-functions are minimal Herglotz functions on the Riemann surface of $\big[\prod_{j=1}^{\ell+1}(z-\alpha_j)(z-\beta_j)\big]^{1/2}$. This is the approach of [**10**].
(c) As two-sided almost periodic J which are regular in the sense of Stahl–Totik [**56**] with $\sigma(J) = \mathfrak{e}$. This is the approach of [**35**].

In understanding these notions, some elementary aspects of potential theory are relevant, so we begin by discussing them. For discussion of potential theory ideas in spectral theory, see Stahl–Totik [**56**] or Simon [**51**].

On our finite gap set, $\mathfrak{e}$, there is a unique probability measure, $\rho_\mathfrak{e}$, called the equilibrium measure which minimizes

$$\mathcal{E}(\rho) = \int \log|x-y|^{-1}\, d\rho(x)d\rho(y) \tag{3.1}$$

among all probability measures supported on $\mathfrak{e}$. The corresponding equilibrium potential is

$$\Phi_{\rho_\mathfrak{e}}(x) = \int \log|x-y|^{-1}\, d\rho_\mathfrak{e}(x) \tag{3.2}$$

The capacity, $C(\mathfrak{e})$, is defined by

$$C(\mathfrak{e}) = \exp(-\mathcal{E}(\rho_\mathfrak{e})) \tag{3.3}$$

A Jacobi matrix with $\sigma_{\text{ess}}(J) = \mathfrak{e}$ has

$$\limsup (a_1 \dots a_n)^{1/n} \le C(\mathfrak{e}) \tag{3.4}$$

J is called regular if one has equality in (3.4). We call a two-sided Jacobi matrix regular if each of the (one-sided) Jacobi matrices

$$J_+ \text{ (resp. } J_-) \text{ with parameters } \{a_n, b_n\}_{n=1}^\infty \text{ (resp. } \{a_{-n}, b_{-n+1}\}_{n=1}^\infty) \tag{3.5}$$

is regular. $\rho_\mathfrak{e}$ is the density of zeros for any regular J with $\sigma_{\text{ess}}(J) = \mathfrak{e}$.

The $\ell+1$ numbers $\rho_\mathfrak{e}([\alpha_j, \beta_j])$, $j = 1, \dots, \ell+1$, which sum to 1 are called the harmonic measures of the bands. We also recall that a bounded function, ψ, on $\mathbb{Z}$ is called almost periodic if $\{S^k\psi\}_{k\in\mathbb{Z}}$, where $(S^k\psi)_n = \psi_{n-k}$, has compact closure in ℓ^∞ (see the appendix to Section 5.13 in [**53**] for more on this class). Such ψ's are associated to a continuous function, Ψ, on a torus of finite or countably infinite dimension so that

$$\psi_n = \Psi(e^{2\pi i n\omega_1}, e^{2\pi i n\omega_2}, \dots) \tag{3.6}$$

The set of $\{n_0 + \sum_{k=1}^K n_k\omega_k \colon n_0, n_k \in \mathbb{Z},\ \sum_{k=1}^K |n_k| < \infty\}$ is called the frequency module of ψ when there is no proper submodule (over $\mathbb{Z}$) that includes all the nonvanishing Bohr–Fourier coefficients. This set for arbitrary $\{\omega_k\}_{k=1}^K$ is called the frequency module generated by $\{\omega_k\}_{k=1}^K$.

With $J_\pm$ given by (3.5), we define $m_\pm(z)$ for $z \in \mathbb{C}\setminus\mathbb{R}$ by

$$m_\pm(z) = \langle \delta_1, (J_\pm - z)^{-1}\delta_1\rangle \tag{3.7}$$

One has for a two-sided Jacobi matrix that

$$\langle \delta_0, (J-z)^{-1}\delta_0\rangle = -(a_0^2 m_+(z) - m_-(z)^{-1})^{-1} \tag{3.8}$$

An important fact is that $J_\pm$ are determined by $m_\pm$, essentially because $m_\pm$ determine the spectral measures $\mu_\pm$ via their Herglotz representations,

$$m_\pm(z) = \int \frac{d\mu_\pm(x)}{x-z} \tag{3.9}$$

and $\mu_\pm$ determine the a's and b's via recursion coefficients for OPRL. Alternatively, the Jacobi parameters can be read off a continued fraction expansion of $m_\pm(z)$ at $z = \infty$.

It is sometimes useful to let $\widetilde{J}_-$ have parameters $\{a_{-n-1}, b_{-n}\}_{n=1}^\infty$, in which case

$$\langle \delta_0, (J-z)^{-1}\delta_0\rangle = -(z - b_0 + a_0^2 m_+(z) + a_{-1}^2 \tilde{m}_-(z))^{-1} \tag{3.10}$$

We can now turn to the descriptions of the isospectral torus. A two-sided Jacobi matrix, J, is called reflectionless on $\mathfrak{e}$ if for a.e. $\lambda \in \mathfrak{e}$ and all n,

$$\operatorname{Re}\langle \delta_n, (J - (\lambda + i0))^{-1} \delta_n \rangle = 0 \tag{3.11}$$

($g(\lambda + i0)$ means $\lim_{\varepsilon \downarrow 0} g(\lambda + i\varepsilon)$). It is known that this is equivalent to

$$a_0^2 m_+(\lambda + i0)\, \overline{m_-(\lambda + i0)} = 1 \text{ for a.e. } \lambda \in \mathfrak{e} \tag{3.12}$$

First Definition of the Isospectral Torus. A two-sided Jacobi matrix, J, is said to lie in the isospectral torus, $\mathcal{T}_{\mathfrak{e}}$, for $\mathfrak{e}$ if $\sigma(J) = \mathfrak{e}$ and J is reflectionless on $\mathfrak{e}$.

$G_{00}(z) = \langle \delta_0, (J - z)^{-1} \delta_0 \rangle$ is determined by $\operatorname{Im} \log(G_{00}(x + i0))$ via an exponential Herglotz representation. This argument is $\pi/2$ on $\mathfrak{e}$, 0 on $(-\infty, \alpha_1)$, and π on $(\beta_{\ell+1}, \infty)$. G_{00} is real in each gap and monotone, so G_{00} has at most one zero and that zero determines $\operatorname{Im} \log(G_{00}(x + i0))$ on that gap. If $G_{00} > 0$ on (β_j, α_{j+1}) we'll say the zero is at β_j and if $G_{00} < 0$ on (β_j, α_{j+1}) the zero is at α_{j+1}. Thus, the zeros of G_{00} determine G_{00} and so $\operatorname{Im} G_{00}(\lambda + i0)$ on $\mathfrak{e}$.

By (3.10), G_{00} has a zero at λ_0 if and only if m_+ or $\tilde{m}_-$ has a pole at λ_0, and one can show that m_+ and $\tilde{m}_-$ have no common poles. The residue of the pole is determined by the derivative of G_{00} at $\lambda = \lambda_0$. The reflectionless condition determines $\operatorname{Im} m_+$ and $\operatorname{Im} \tilde{m}_-$ on $\mathfrak{e}$, so $a_0, a_{-1}, b_0, m_+, \tilde{m}_-$, and thus J, are uniquely determined by knowing the position of the zero and if they are in the gaps (as opposed to the edges) whether the poles are in m_+ or $\tilde{m}_-$. Hence, for each gap, we have the two copies of (β_j, α_{j+1}) glued at the ends, that is, a circle. Thus, given that one can show each possibility occurs, $\mathcal{T}_{\mathfrak{e}}$ is a product of ℓ circles, that is, a torus. It is not hard to show that the Jacobi parameters depend continuously on the positions of the zeros of G_{00} and $m_+/\tilde{m}_-$ data.

We turn to the second approach. Any G_{00} as above is purely imaginary on the bands which implies, by the reflection principle, that it can be meromorphically continued to a matching copy of $\mathcal{S}_+ \equiv \mathbb{C} \cup \{\infty\} \setminus \mathfrak{e}$. This suggests meromorphic functions on $\mathcal{S}$, two copies of $\mathcal{S}_+$ glued together along $\mathfrak{e}$, will be important. $\mathcal{S}$ is precisely the compactified Riemann surface of $\sqrt{R(z)}$, where

$$R(z) = \prod_{j=1}^{\ell+1} (z - \alpha_j)(z - \beta_j) \tag{3.13}$$

$\mathcal{S}$ is a Riemann surface of genus ℓ. Meromorphic functions on the surface that are not functions symmetric under interchange of the sheets (i.e., meromorphic on $\mathbb{C}$) have degree at least $\ell + 1$.

By a minimal meromorphic Herglotz function, we mean a meromorphic function of degree $\ell + 1$ on $\mathcal{S}$ that obeys

(i) $\operatorname{Im} f > 0$ on $\mathcal{S}_+ \cap \mathbb{C}_+$ ($\mathbb{C}_+ = \{z\colon \operatorname{Im} z > 0\}$)
(ii) f has a zero at ∞ on $\mathcal{S}_+$ and a pole at ∞ on $\mathcal{S}_-$.

Such functions must have their ℓ other poles on $\mathbb{R}$ in the gaps on one sheet or the other and are uniquely, up to a constant, determined by these ℓ poles, one per gap. Each "gap," when you include the two sheets and branch points at the gap edges, is a circle. So if we normalize by $m(z) = -z^{-1} + O(z^{-2})$ near ∞ on $\mathcal{S}_+$, the set of such minimal normalized Herglotz functions is an ℓ-dimensional torus. Each such Herglotz function can be written on $\mathcal{S}_+ \cap \mathbb{C}_+$ as

$$m(z) = \int \frac{d\mu(x)}{x - z} \tag{3.14}$$

where μ is supported on $\mathfrak{e}$ plus the poles of m in the gaps on $\mathcal{S}_+$. μ then determines a Jacobi matrix.

Second Definition of the Isospectral Torus. The isospectral torus, $\mathcal{T}_\mathfrak{e}$, is the set of one-sided J's whose m-functions are minimal Herglotz functions on the compact Riemann surface $\mathcal{S}$ of $\sqrt{R}$ given by (3.13).

The relation between the two definitions is that the restrictions of the two-sided J's to the one-sided are these J given by minimal Herglotz functions. In the other direction, each J is almost periodic and so has a unique almost periodic two-sided extension.

Third Definition of the Isospectral Torus. The isospectral torus is the almost periodic two-sided J's with $\sigma(J) = \mathfrak{e}$ and which are regular.

This is equivalent to the reflectionless definition since regularity implies the Lyapunov exponent is zero and then Kotani theory [**33**, **48**] implies J is reflectionless.

As noted, the J's in the isospectral torus are all almost periodic. Their frequency module is generated by the harmonic measures of the bands. In particular, the elements of the isospectral torus are periodic if and only if all harmonic measures are rational. Their spectra are purely a.c. and all solutions of $Ju = \lambda u$ are bounded for any $\lambda \in \mathfrak{e}^{\text{int}}$.

Szegő asymptotics is more complicated than in the $\ell = 0$ case. One has for the OPRL associated to a point in the isospectral torus (thought of as a one-sided Jacobi matrix) that for all $z \in \mathbb{C} \setminus \sigma(J)$,

$$p_n(z) \exp(-n\Phi_{\rho_\mathfrak{e}}(z)) \tag{3.15}$$

is asymptotically almost periodic as a function of n with magnitude bounded away from 0 for all n. The frequency module is z-dependent (as written, this is even true if $\ell = 0$ as can bee seen from the free case): the frequencies come from the harmonic measures of the bands plus one that comes from the conjugate harmonic function of $\Phi_{\rho_\mathfrak{e}}(z)$ in $\mathbb{C}_+$ (which gives the z-dependence of the frequency module). The limit of (3.15) on $\mathfrak{e}$, where $\Phi_{\rho_\mathfrak{e}}(x) = 0$, yields the boundedness of solutions of $(J - \lambda)u = 0$. There is also a limit at $z = \infty$: $a_1 \dots a_n / C(\mathfrak{e})^n$ which is almost periodic.

4. Results in the Finite Gap Case

As we've seen, if $\tilde{J}$ is in the isospectral torus for $\mathfrak{e}$ and $\lambda \in \mathfrak{e}^{\text{int}}$, then all solutions of $\tilde{J}u = \lambda u$ are bounded. This remains true under ℓ^1 perturbations by a variation of parameters, so Theorem 2.1 is applicable and we have

THEOREM 4.1. *Let $\mathfrak{e}$ be a finite gap set and $\tilde{J}$, with parameters $\{\tilde{a}_n, \tilde{b}_n\}_{n=1}^\infty$, an element of $\mathcal{T}_\mathfrak{e}$, the isospectral torus for $\mathfrak{e}$. Let J be a Jacobi matrix with*

$$\sum_{n=1}^\infty |a_n - \tilde{a}_n| + |b_n - \tilde{b}_n| < \infty \tag{4.1}$$

Then $\sigma_{ess}(J) = \mathfrak{e}$ and the spectrum on $\mathfrak{e}^{int}$ is purely a.c.

REMARK. We are not aware of this appearing explicitly in the literature, although it follows easily from results in [**10**, **42**].

As for eigenvalues in $\mathbb{R} \setminus \mathfrak{e}$:

THEOREM 4.2. *There is a constant C depending only on $\mathfrak{e}$ so that for any Jacobi matrix, J, obeying* (4.1) *for some $\tilde{J} \in \mathcal{T}_{\mathfrak{e}}$, we have, with $\{x_n\}_{n=1}^N$ the eigenvalues of J,*

$$\sum_{n=1}^{N} \operatorname{dist}(x_n, \mathfrak{e})^{1/2} \leq C_0 + C\biggl(\sum_{n=1}^{\infty} |a_n - \tilde{a}_n| + |b_n - \tilde{b}_n|\biggr) \tag{4.2}$$

where

$$C_0 = \sum_{j=1}^{\ell} \left| \frac{\alpha_{j+1} - \beta_j}{2} \right|^{1/2} \tag{4.3}$$

REMARKS. 1. This result is essentially in Frank–Simon [**18**]. They are only explicit about perturbations of two-sided Jacobi matrices where $\tilde{J}$ has no eigenvalues. They mention that one can use interlacing to then get results for the one-sided case—this makes that idea explicit.

2. Prior to [**18**], Frank–Simon–Weidl [**19**] proved such a bound on the x_n in $\mathbb{R} \setminus [\alpha_1, \beta_{\ell+1}]$ and Hundertmark–Simon [**27**] if $1/2$ in the power of $\operatorname{dist}(\dots)^{1/2}$ is replaced by $p > 1/2$ and 1 in the power of $|a_n - \tilde{a}_n|$ and $|b_n - \tilde{b}_n|$ by $p + 1/2$, that is, noncritical Lieb–Thirring bounds.

THEOREM 4.3. *Let J be a Jacobi matrix with $\sigma_{ess}(J) = \mathfrak{e}$ and Jacobi parameters $\{a_n, b_n\}_{n=1}^\infty$. Suppose its spectral measure has the form*

$$d\mu = f(x)\, dx + d\mu_s \tag{4.4}$$

where $d\mu_s$ is singular with respect to dx. Suppose $\{x_n\}_{n=1}^N$ are the pure points of $d\mu$ outside $\mathfrak{e}$. Consider the three conditions:

$$\text{(a)} \qquad \sum_{n=1}^{N} \operatorname{dist}(x_n, \mathfrak{e})^{1/2} < \infty \tag{4.5}$$

$$\text{(b)} \qquad \int_{\mathfrak{e}} \operatorname{dist}(x, \mathbb{R} \setminus \mathfrak{e})^{-1/2} \log[f(x)]\, dx > -\infty \tag{4.6}$$

$$\text{(c)} \qquad \text{For some constant } R > 1,\ R^{-1} \leq \frac{a_1 \dots a_n}{C(\mathfrak{e})^n} \leq R \tag{4.7}$$

Then any two imply the third, and if they hold, there exists $\tilde{J} \in \mathcal{T}_{\mathfrak{e}}$ so that

$$\lim_{n \to \infty} |a_n - \tilde{a}_n| + |b_n - \tilde{b}_n| = 0 \tag{4.8}$$

Moreover,

$$\text{(d)} \qquad \lim_{n \to \infty} \frac{a_1 \dots a_n}{\tilde{a}_1 \dots \tilde{a}_n} \text{ exists in } (0, \infty) \tag{4.9}$$

$$\text{(e)} \qquad \lim_{K \to \infty} \sum_{n=1}^{K} (b_n - \tilde{b}_n) \text{ exists in } \mathbb{R} \tag{4.10}$$

REMARKS. 1. Depending on which implications one looks at, only part of (c) is needed. For example, if (a) holds,

$$\text{(b)} \Leftrightarrow \limsup_{n\to\infty} \frac{a_1 \dots a_n}{C(\mathfrak{e})^n} > 0 \tag{4.11}$$

(that is, indeed, lim sup and not lim inf).

2. As stated, this theorem (except for (e); see below) is due to Christiansen–Simon–Zinchenko [**11**], but parts of it were known. While [**11**] focus on Szegő asymptotics (see below), the work of Widom [**64**] and Aptekarev [**4**] implied if there are no or finitely many x_n's, then (b) $\Rightarrow$ (c), and Peherstorfer–Yuditskii [**42**] proved (a) + (b) $\Rightarrow$ (c) (and as noted to us privately by Peherstorfer, combining their results and an idea of Garnett [**20**] yields (4.11)).

3. That (e) holds does not seem to have been noted before, although it follows easily from the results in [**11**]. For $g_n(z) \equiv p_n(z)/\tilde{p}_n(z)$ has a limit as $n \to \infty$ on $\mathbb{C} \setminus [\alpha_1, \beta_{\ell+1}]$ and that limit also exists and is analytic and nonzero at infinity (see Theorem 4.5 below). Since

$$z^{-n} p_n(z) = (a_1 \dots a_n)^{-1} \biggl(1 - \Bigl(\sum_{j=1}^{n} b_j\Bigr) z^{-1} + O(z^{-2})\biggr) \tag{4.12}$$

near $z = \infty$,

$$\log(g_n(z)) = -\log\biggl(\frac{a_1 \dots a_n}{\tilde{a}_1 \dots \tilde{a}_n}\biggr) - \biggl[\sum_{j=1}^{n} (b_j - \tilde{b}_j)\biggr] z^{-1} + O(z^{-2}) \tag{4.13}$$

so convergence of the analytic functions uniformly near ∞ implies convergence of the $O(z^{-1})$ term.

Theorems 4.2 and 4.3 immediately imply:

COROLLARY 4.4. *If* (4.1) *holds, so does* (4.6).

PROOF. Since $\tilde{a}_1 \dots \tilde{a}_n / C(\mathfrak{e})^n$ is almost periodic bounded away from 0 and ∞, and $\sum_{n=1}^{\infty} |a_n - \tilde{a}_n| < \infty$ and $\tilde{a}_n$, $\tilde{a}_n^{-1}$ bounded imply $\sum_{n=1}^{\infty} |1 - a_n/\tilde{a}_n| < \infty$, we have (4.9), which implies (4.7). By Theorem 4.2, (4.1) $\Rightarrow$ (4.5), so Theorem 4.3 implies (4.6). $\square$

REMARK. This is a result of [**18**], although [**11**] conjectured Theorem 4.2 and noted it would imply this corollary.

THEOREM 4.5. *If the conditions* (a)–(c) *of Theorem* 4.3 *hold, then for all* $z \in \mathbb{C} \cup \{\infty\} \setminus [\alpha_1, \beta_{\ell+1}]$, $\lim_{n\to\infty} p_n(z)/\tilde{p}_n(z)$ *exists and the limit is analytic with zeros only at the* x_n *in* $\mathbb{R} \setminus [\alpha_1, \beta_{\ell+1}]$.

REMARKS. 1. In this form, this result is from [**11**], although earlier it appeared implicitly in Peherstorfer–Yuditskii [**42, 43**], and special cases (with stronger assumptions on the x_n's) are in [**4, 64**]. See also [**53**].

2. There is also an asymptotic result on $\mathfrak{e}$ not pointwise but in $L^2(d\mu)$ sense; see [**11**].

3. Asymptotics results for orthogonal polynomials on finite gap sets have been pioneered by Akhiezer and Tomčuk [**1, 2**].

We do not know an analog of the "if and only if" statement of Theorem 2.7, but there is one direction:

THEOREM 4.6. *Let $\{\tilde{a}_n, \tilde{b}_n\}_{n=1}^\infty$ be an element of the isospectral torus, $\mathcal{T}_{\mathfrak{e}}$, of a finite gap set, $\mathfrak{e}$. Let $\{a_n, b_n\}_{n=1}^\infty$ be another set of Jacobi parameters and $\delta a_n, \delta b_n$ given by*

$$\delta a_n = a_n - \tilde{a}_n, \qquad \delta b_n = b_n - \tilde{b}_n$$

Suppose that
(a)

$$\sum_{n=1}^\infty |\delta a_n|^2 + |\delta b_n|^2 < \infty \tag{4.14}$$

(b) *For any $\mathbf{k} \in \mathbb{Z}^\ell$,*

$$\sum_{n=1}^N e^{2\pi i(\mathbf{k}\cdot\boldsymbol{\omega})n}\delta a_n \quad \textit{and} \quad \sum_{n=1}^N e^{2\pi i(\mathbf{k}\cdot\boldsymbol{\omega})n}\delta b_n \tag{4.15}$$

have (finite) limits as $N \to \infty$.
(c) *For every $\varepsilon > 0$,*

$$\sup_N \left\{ \left| \sum_{n=1}^N e^{2\pi i(\mathbf{k}\cdot\boldsymbol{\omega})n}\delta a_n \right| + \left| \sum_{n=1}^N e^{2\pi i(\mathbf{k}\cdot\boldsymbol{\omega})n}\delta b_n \right| \right\} \le C_\varepsilon \exp(\varepsilon|\mathbf{k}|) \tag{4.16}$$

Let $p_n(z)$ (resp. $\tilde{p}_n(z)$) be the orthonormal polynomials for $\{a_n, b_n\}_{n=1}^\infty$ (resp. $\{\tilde{a}_n, \tilde{b}_n\}_{n=1}^\infty$). Then for any $z \in \mathbb{C} \setminus \mathbb{R}$,

$$\lim_{n\to\infty} \frac{p_n(z)}{\tilde{p}_n(z)} \tag{4.17}$$

exists and is finite and nonzero.

REMARKS. 1. Here $\boldsymbol{\omega} = (\omega_1, \dots, \omega_\ell)$ is the ℓ-tuple of harmonic measures (i.e., $\omega_j = \rho_{\mathfrak{e}}([\alpha_j, \beta_j])$) and $\mathbf{k}\cdot\boldsymbol{\omega} = \sum_{j=1}^\ell k_j\omega_j$. We thus require infinitely many conditions.

2. This result is from [**12**].

3. If the torus consists of period p elements (i.e., each $\rho_{\mathfrak{e}}([\alpha_j, \beta_j])$ is k_j/p, where there is no common factor for $p, k_1, \dots, k_\ell$), then the infinity of conditions (4.15) reduces to the finitely many conditions that for $j = 1, 2, \dots, p$, $\sum_{n=0}^N \delta a_{np+j}$ and $\sum_{n=0}^N \delta b_{np+j}$ have finite limits and (4.16) becomes automatic.

4. [**12**] uses this theorem to construct examples where Szegő asymptotics holds, but both (4.5) and (4.6) fail to hold.

An analog of Theorem 2.8 is not known for general $\mathfrak{e}$ but is known in one special case. We say $\mathfrak{e}$ is p-periodic with all gaps open if $\ell = p - 1$, and for $j = 1, \dots, p$, $\rho_{\mathfrak{e}}([\alpha_j, \beta_j]) = 1/p$.

We also need a notion of approach to the isospectral torus rather than a single element. Given two Jacobi matrices, we define

$$d_m(J, J') = \sum_{k=0}^\infty e^{-|k|}(|a_{m+k} - a'_{m+k}| + |b_{m+k} - b'_{m+k}|) \tag{4.18}$$

and

$$d_m(J, \mathcal{T}_{\mathfrak{e}}) = \inf_{J' \in \mathcal{T}_{\mathfrak{e}}} d_m(J, J') \tag{4.19}$$

THEOREM 4.7. *Let $\mathfrak{e}$ be p-periodic with all gaps open. Let J be a Jacobi matrix with spectral measure obeying* (4.4) *and eigenvalues $\{x_n\}_{n=1}^N$ outside $\mathfrak{e}$. Then*

$$\sum_{m=1}^{\infty} d_m(J, \mathcal{T}_\mathfrak{e})^2 < \infty \tag{4.20}$$

if and only if

$$(a) \qquad \sigma_{ess}(J) = \mathfrak{e} \tag{4.21}$$

$$(b) \qquad \sum_{n=1}^{N} \operatorname{dist}(x_n, \mathfrak{e})^{3/2} < \infty \tag{4.22}$$

$$(c) \qquad \int_\mathfrak{e} \operatorname{dist}(x, \mathbb{R} \setminus \mathfrak{e})^{+1/2} \log[f(x)]\, dx > -\infty \tag{4.23}$$

REMARK. This theorem is due to Damanik–Killip–Simon [**13**]. Their method is specialized to the periodic case, and in that case, proves some of the earlier results of this section, such as Theorem 4.2.

THEOREM 4.8. *Suppose J is a Jacobi matrix with $\sigma_{ess}(J) = \mathfrak{e}$ and so that the f of* (4.4) *is a.e. strictly positive on $\mathfrak{e}$. Then*

$$\lim_{m\to\infty} d_m(J, \mathcal{T}_\mathfrak{e}) = 0 \tag{4.24}$$

REMARKS. 1. This is a result of Remling [**46**]. For the periodic case, it was proven earlier by [**13**], who conjecture the result for general $\mathfrak{e}$.

2. Remling replaces (4.24) by the assertion that every right limit of J (i.e., limit point of $\{a_{n+r}, b_{n+r}\}_{n=1}^{\infty}$ as $r \to \infty$) is in $\mathcal{T}_\mathfrak{e}$. By a compactness argument, it is easy to see that this is equivalent to (4.24).

THEOREM 4.9. *Let $\mathfrak{e}$ be a finite gap set and J a Jacobi matrix so that*

$$(a) \qquad \sigma_{ess}(J) = \mathfrak{e} \tag{4.25}$$

$$(b) \qquad J \text{ is regular, i.e., } \lim_{n\to\infty} (a_1 \dots a_n)^{1/n} = C(\mathfrak{e}) \tag{4.26}$$

Then

$$\lim_{M\to\infty} \frac{1}{M} \sum_{m=1}^{M} d_m(J, \mathcal{T}_\mathfrak{e})^2 = 0 \tag{4.27}$$

REMARKS. 1. This result was proven in case all harmonic measures are rational by Simon [**52**], who conjectured the result in general. It was proven by Krüger [**34**].

2. By the Schwarz inequality, (4.27) is equivalent to

$$\lim_{M\to\infty} \frac{1}{M} \sum_{m=1}^{M} d_m(J, \mathcal{T}_\mathfrak{e}) = 0 \tag{4.28}$$

We close this section on results with a list of some open questions:

(1) Do (a)–(c) of Theorem 4.3 imply that

$$\sum_{n=1}^{\infty} (a_n - \tilde{a}_n)^2 + (b_n - \tilde{b}_n)^2 < \infty \tag{4.29}$$

as is true in the case $\mathfrak{e} = [-2, 2]$?

(2) Is there an extension of Theorem 4.7 to the general $\mathfrak{e}$ case?
(3) Is there a converse to Theorem 4.6? This would be interesting even in the periodic case.

5. Methods

The theory of regular Jacobi matrices says one expects the leading growth of $P_n(z)$ as $n \to \infty$ to be $\exp(n\Phi_{\rho_\mathfrak{e}}(z))$. $\Phi_{\rho_\mathfrak{e}}$ is harmonic on $\mathbb{C} \cup \{\infty\} \setminus \mathfrak{e}$ so we can locally define a harmonic conjugate and so $\widetilde{\Phi}_{\rho_\mathfrak{e}}(z)$ analytic with $\operatorname{Re} \widetilde{\Phi}_{\rho_\mathfrak{e}} = \Phi_{\rho_\mathfrak{e}}$. If you circle around x, $\log(z-x)$ changes by $2\pi i$, so circling around the band $[\alpha_j, \beta_j]$, we expect $\int \log(z-x)\, d\rho_\mathfrak{e}(x)$ to change by $2\pi i \rho_\mathfrak{e}([\alpha_j, \beta_j])$ and $\exp(-\widetilde{\Phi}_{\rho_\mathfrak{e}}(z))$ to have a change of phase by $\exp(-2\pi i \rho_\mathfrak{e}([\alpha_j, \beta_j]))$. Thus, we are led to consider analytic functions on $\mathbb{C}_+$ which we can continue along any curve in $\mathbb{C} \cup \{\infty\} \setminus \mathfrak{e}$.

To get a single-valued function, we need to lift to the universal covering space of $\mathbb{C} \cup \{\infty\} \setminus \mathfrak{e}$ and $\exp(-\widetilde{\Phi}_{\rho_\mathfrak{e}}(z))$ will transform under the homotopy group via a character of this group.

So long as $\ell \neq 0$, this cover, as a Riemann surface, is the disk, $\mathbb{D}$, and the deck transformations act as a family of fractional linear transformations on the disk, that is, a Fuchsian group. The use of these Fuchsian groups is thus critical to the theory and used to prove several of the theorems of Section 4 (Theorems 4.7, 4.8, and 4.9 are exceptions).

For more on Fuchsian groups, see Beardon [**6**], Ford [**17**], Katok [**31**], Simon [**53**], and Tsuji [**61**]. The pioneers in this approach were Sodin–Yuditskii [**55**]. See [**10**–**12**, **42**, **53**] for applications of these techniques.

Acknowledgements

The first and third authors would like to thank Caltech for its hospitality where this manuscript was written.

References

[1] N. I. Ahiezer, *Orthogonal polynomials on several intervals*, Dokl. Akad. Nauk SSSR **134** (1960), 9–12. (Russian); translated as Soviet Math. Dokl. **1** (1960), 989–992.
[2] N. I. Ahiezer and Ju. Ja. Tomčuk, *On the theory of orthogonal polynomials over several intervals*, Dokl. Akad. Nauk SSSR **138** (1961), 743–746. (Russian); translated as Soviet Math. Dokl. **2** (1961), 687–690.
[3] M. Aizenman and E.H. Lieb, *On semi-classical bounds for eigenvalues of Schrödinger operators*, Phys. Lett. **66A** (1978), 427–429.
[4] A.I. Aptekarev, *Asymptotic properties of polynomials orthogonal on a system of contours, and periodic motions of Toda chains*, Math. USSR Sb. **53** (1986), 233–260; Russian original in Mat. Sb. (N.S.) **125(167)** (1984), 231–258. Math. USSR-Sb. 53 (1986), 233–260.
[5] V. Batchenko and F. Gesztesy, *On the spectrum of Jacobi operators with quasi-periodic algebro-geometric coefficients*, Int. Math. Res. Papers **No. 10** (2005), 511–563.
[6] A.F. Beardon, *The Geometry of Discrete Groups*, corrected reprint of the 1983 original, Graduate Texts in Mathematics, **91**, Springer-Verlag, New York, 1995.
[7] W. Bulla, F. Gesztesy, H. Holden, and G. Teschl, *Algebro-geometric quasi-periodic finite-gap solutions of the Toda and Kac–van Moerbeke hierarchies*, Memoirs Amer. Math. Soc. **135**, No. 641 (1998).
[8] J.S. Christiansen, *Szegő's theorem on Parreau–Widom sets*, Adv. Math. **229** (2012), 1180–1204.
[9] J.S. Christiansen, *Szegő asymptotics on Parreau–Widom sets*, in preparation.
[10] J.S. Christiansen, B. Simon, and M. Zinchenko, *Finite gap Jacobi matrices, I. The isospectral torus*, Constr. Approx. **32** (2010), 1–65.

[11] J.S. Christiansen, B. Simon, and M. Zinchenko, *Finite gap Jacobi matrices, II. The Szegő class*, Constr. Approx. **33** (2011), 365–403.
[12] J.S. Christiansen, B. Simon, and M. Zinchenko, *Finite gap Jacobi matrices, III. Beyond the Szegő class*, Constr. Approx. **35** (2012), 259–272.
[13] D. Damanik, R. Killip, and B. Simon, *Perturbations of orthogonal polynomials with periodic recursion coefficients*, Annals of Math. (2) **171** (2010), 1931–2010.
[14] D. Damanik and B. Simon, *Jost functions and Jost solutions for Jacobi matrices, I. A necessary and sufficient condition for Szegő asymptotics*, Invent. Math. **165** (2006), 1–50.
[15] S.A. Denisov, *On Rakhmanov's theorem for Jacobi matrices*, Proc. Amer. Math. Soc. **132** (2004), 847–852.
[16] A. Eremenko and P. Yuditskii, *Comb functions*, Contemp. Math. **578** (2012), 99–118.
[17] L.R. Ford, *Automorphic Functions*, 2nd ed., Chelsea, New York, 1951.
[18] R. Frank and B. Simon, *Critical Lieb–Thirring bounds in gaps and the generalized Nevai conjecture for finite gap Jacobi matrices*, Duke Math. J. **157** (2011), 461–493.
[19] R. Frank, B. Simon, and T. Weidl, *Eigenvalue bounds for perturbations of Schrödinger operators and Jacobi matrices with regular ground states*, Comm. Math. Phys. **282** (2008), 199–208.
[20] J.B. Garnett, *Bounded Analytic Functions*, Pure and Applied Math., **96**, Academic Press, New York, 1981.
[21] F. Gesztesy and H. Holden, *Soliton Equations and Their Algebro-Geometric Solutions. Vol. I: $(1+1)$-Dimensional Continuous Models*, Cambridge Studies in Advanced Mathematics, **79**, Cambridge University Press, Cambridge, 2003.
[22] F. Gesztesy, H. Holden, J. Michor, and G. Teschl, *Soliton Equations and Their Algebro-Geometric Solutions. Vol. II: $(1+1)$-Dimensional Discrete Models*, Cambridge Studies in Advanced Mathematics, **114**, Cambridge University Press, Cambridge, 2008.
[23] D.J. Gilbert, *On subordinacy and analysis of the spectrum of Schrödinger operators with two singular endpoints*, Proc. Roy. Soc. Edinburgh Sect. A **112** (1989), 213–229.
[24] D.J. Gilbert and D.B. Pearson, *On subordinacy and analysis of the spectrum of one-dimensional Schrödinger operators*, J. Math. Anal. Appl. **128** (1987), 30–56.
[25] D. Hundertmark, E.H. Lieb, and L.E. Thomas, *A sharp bound for an eigenvalue moment of the one-dimensional Schrödinger operator*, Adv. Theor. Math. Phys. **2** (1998), 719–731.
[26] D. Hundertmark and B. Simon, *Lieb–Thirring inequalities for Jacobi matrices*, J. Approx. Theory **118** (2002), 106–130.
[27] D. Hundertmark and B. Simon, *Eigenvalue bounds in the gaps of Schrödinger operators and Jacobi matrices*, J. Math. Anal. Appl. **340** (2008), 892–900.
[28] S. Jitomirskaya and Y. Last, *Power-law subordinacy and singular spectra, I. Half-line operators*, Acta Math. **183** (1999), 171–189.
[29] S. Jitomirskaya and Y. Last, *Power law subordinacy and singular spectra, II. Line operators*, Comm. Math. Phys. **211** (2000), 643–658.
[30] S. Kahn and D.B. Pearson, *Subordinacy and spectral theory for infinite matrices*, Helv. Phys. Acta **65** (1992), 505–527.
[31] S. Katok, *Fuchsian Groups*, University of Chicago Press, Chicago, 1992.
[32] R. Killip and B. Simon, *Sum rules for Jacobi matrices and their applications to spectral theory*, Annals of Math. (2) **158** (2003), 253–321.
[33] S. Kotani, *Ljapunov indices determine absolutely continuous spectra of stationary random one-dimensional Schrödinger operators*, Stochastic Analysis (Katata/Kyoto, 1982), pp. 225–247, North–Holland Math. Library, **32**, North–Holland, Amsterdam, 1984.
[34] H. Krüger, *Probabilistic averages of Jacobi operators*, Comm. Math. Phys. **295** (2010), 853–875.
[35] H. Krüger and B. Simon, *Cantor polynomials and some related classes of OPRL*, in preparation.
[36] Y. Last and B. Simon, *Eigenfunctions, transfer matrices, and absolutely continuous spectrum of one-dimensional Schrödinger operators*, Invent. Math. **135** (1999), 329–367.
[37] E.H. Lieb and W. Thirring, *Inequalities for the moments of the eigenvalues of the Schrödinger Hamiltonian and their relation to Sobolev inequalities*, in "Studies in Mathematical Physics. Essays in Honor of Valentine Bargmann," pp. 269–303, Princeton University Press, Princeton, NJ, 1976.
[38] P. Nevai, *Orthogonal polynomials*, Mem. Amer. Math. Soc. **18** (1979), no. 213, 185 pp.

[39] P. Nevai, *Orthogonal polynomials, recurrences, Jacobi matrices, and measures*, in "Progress in Approximation Theory" (Tampa, FL, 1990), pp. 79–104, Springer Ser. Comput. Math., **19**, Springer, New York, 1992.

[40] D.B. Pearson, *Quantum Scattering and Spectral Theory*, Academic Press, London, 1988.

[41] F. Peherstorfer and P. Yuditskii, *Asymptotics of orthonormal polynomials in the presence of a denumerable set of mass points*, Proc. Amer. Math. Soc. **129** (2001), 3213–3220.

[42] F. Peherstorfer and P. Yuditskii, *Asymptotic behavior of polynomials orthonormal on a homogeneous set*, J. Anal. Math. **89** (2003), 113–154.

[43] F. Peherstorfer and P. Yuditskii, *Remark on the paper "Asymptotic behavior of polynomials orthonormal on a homogeneous set"*, arXiv math.SP/0611856.

[44] E.A. Rakhmanov, *On the asymptotics of the ratio of orthogonal polynomials*, Math. USSR Sb. **32** (1977), 199–213.

[45] E.A. Rakhmanov, *On the asymptotics of the ratio of orthogonal polynomials, II*, Math. USSR Sb. **46** (1983), 105–117.

[46] C. Remling, *The absolutely continuous spectrum of Jacobi matrices*, Annals of Math. (2) **174** (2011), 125–171.

[47] J.A. Shohat, *Théorie Générale des Polinomes Orthogonaux de Tchebichef*, Mémorial des Sciences Mathématiques, **66**, pp. 1–69, Paris, 1934.

[48] B. Simon, *Kotani theory for one dimensional stochastic Jacobi matrices*, Comm. Math. Phys. **89** (1983), 227–234.

[49] B. Simon, *Bounded eigenfunctions and absolutely continuous spectra for one-dimensional Schrödinger operators*, Proc. Amer. Math. Soc. **124** (1996), 3361–3369.

[50] B. Simon, *Orthogonal Polynomials on the Unit Circle, Part 1: Classical Theory*, AMS Colloquium Series, **54.1**, American Mathematical Society, Providence, RI, 2005.

[51] B. Simon, *Equilibrium measures and capacities in spectral theory*, Inverse Problems and Imaging **1** (2007), 713–772.

[52] B. Simon, *Regularity and the Cesàro–Nevai class*, J. Approx. Theory **156** (2009), 142–153.

[53] B. Simon, *Szegő's Theorem and Its Descendants: Spectral Theory for L^2 Perturbations of Orthogonal Polynomials*, Princeton University Press, Princeton, NJ, 2011.

[54] B. Simon and A. Zlatoš, *Sum rules and the Szegő condition for orthogonal polynomials on the real line*, Comm. Math. Phys. **242** (2003), 393–423.

[55] M. Sodin and P. Yuditskii, *Almost periodic Jacobi matrices with homogeneous spectrum, infinite-dimensional Jacobi inversion, and Hardy spaces of character-automorphic functions*, J. Geom. Anal. **7** (1997), 387–435.

[56] H. Stahl and V. Totik, *General Orthogonal Polynomials*, in "Encyclopedia of Mathematics and its Applications," **43**, Cambridge University Press, Cambridge, 1992.

[57] G. Szegő, *Beiträge zur Theorie der Toeplitzschen Formen I, II*, Math. Z. **6** (1920), 167–202; **9** (1921), 167–190.

[58] G. Szegő, *Über den asymptotischen Ausdruck von Polynomen, die durch eine Orthogonalitätseigenschaft definiert sind*, Math. Ann. **86** (1922), 114–139.

[59] G. Teschl, *Jacobi Operators and Completely Integrable Nonlinear Lattices*, Mathematical Surveys and Monographs, **72**, American Mathematical Society, Providence, RI, 2000.

[60] E.C. Titchmarsh, *On expansions in eigenfunctions, IV.*, Quart. J. Math., Oxford Ser. **12** (1941), 33–50.

[61] M. Tsuji, *Potential Theory in Modern Function Theory*, reprint of the 1959 original, Chelsea, New York, 1975.

[62] J.L. Ullman, *On the regular behaviour of orthogonal polynomials*, Proc. London Math. Soc. (3) **24** (1972), 119–148.

[63] T. Weidl, *On the Lieb–Thirring constants $L_{\gamma,1}$ for $\gamma \geq 1/2$*, Comm. Math. Phys. **178** (1996), 135–146.

[64] H. Widom, *Extremal polynomials associated with a system of curves in the complex plane*, Adv. in Math. **3** (1969), 127–232.

[65] P. Yuditskii, *On the direct Cauchy theorem in Widom domains: Positive and negative examples*, Comput. Methods Funct. Theory **11** (2011), 395–414.

Department of Mathematical Sciences, University of Copenhagen, Universitetsparken 5, DK-2100 Copenhagen, Denmark
E-mail address: stordal@math.ku.dk

Mathematics 253-37, California Institute of Technology, Pasadena, CA 91125, USA
E-mail address: bsimon@caltech.edu

Department of Mathematics and Statistics, University of New Mexico, Albuquerque, NM 87131, USA
E-mail address: maxim@math.unm.edu

Proceedings of Symposia in Pure Mathematics
Volume **87**, 2013
http://dx.doi.org/10.1090/pspum/087/01427

Momentum operators on graphs

Pavel Exner

With congratulations to my friend Fritz Gesztesy on the occasion of his 60th birthday.

ABSTRACT. We discuss ways in which momentum operators can be introduced on an oriented metric graph. A necessary condition appears to the balanced property, or a matching between the numbers of incoming and outgoing edges; we show that a graph without an orientation, locally finite and at most countably infinite, can made balanced oriented *iff* the degree of each vertex is even. On such graphs we construct families of momentum operators; we analyze their spectra and associated unitary groups. We also show that the unique continuation principle does not hold here.

1. Introduction

Our writings have their own fates once they left our hands and one can only guess how successful they will eventually be. Fritz bibliography is extensive, some two hundred items if not more, and covers many areas. All his texts are pleasure to read, deep in contents and perfectly organized. Nevertheless, one of them made much larger impact than any others, namely the monograph [**AGHH**] first published in 1988. It is a collective work but Fritz hand in unmistakably present in the exposition, and I add that it makes me proud to be a part of the second edition.

Thinking about how the book resonated in the community, one has to come to the conclusion that that it did not happen by a chance, rather the subject of solvability struck some important needs. Such models are complex and versatile enough to be applicable to numerous physical situations and at the same time, they are mathematically accessible and allow derive conclusions without involving a heavy machinery. This concerns both the proper subject of the book [**AGHH**] and its various extensions, among them the theory of quantum graphs which is developing rapidly; a broad overview of recent developments can be found in [**EKKST**].

This is the frame into which the present little *étude* belongs. While from the quantum mechanical point of view the operators most frequently studied are Hamiltonians which are, mathematically speaking, typically Laplacians, their singular perturbations and modifications, there are other observables of interest too. In this

---

2010 *Mathematics Subject Classification.* Primary 54C40, 14E20; Secondary 46E25, 20C20.
Key words and phrases. Differential geometry, algebraic geometry.
The research was supported by the Czech Science Foundation within the project P203/11/0701. I am indebted to Peter Kuchment for a discussion on his book in preparation and to the referee for reading the manuscript carefully.

paper we deal with momentum operators acting locally as imaginary multiples of the first derivative in the appropriate functional spaces.

This is not to say that first-order operators on graphs have not been studied before. On the one hand, momentum-type operators on graphs have been discussed recently in [**FKW07**] from the viewpoint of appropriate index theorems, see also [**BK**] where their use in quantum-chaos studies based on bond scattering matrices is described. Momentum operators are considered also in [**Ca99, Eg11**], see [**ES10**] for related analysis of Berry-Keating operator. On the other hand, there is a recent series of voluminous papers treating such operators on two or a larger number of intervals [**JPT11, JPT12a, JPT12b**]; they come from a different literature background, only weakly connected to the applications of self-adjoint extensions in modeling quantum dynamics such as those mentioned above or collected in [**AEG**].

Our discussion in this paper necessarily overlaps in part with the indicated studies looking at the problem from a bit different perspective. In contrast to graph Laplacians definition of momentum operators require the graph to be oriented, not just arbitrarily but in a *balanced* way, with matching numbers of incoming and outgoing edges in each vertex. We show that an undirected graph, locally finite and at most countably infinite, can be given such orientations, in general different ones, if and only if the degrees of its vertices are even, and describe possible momentum operators. We derive some properties of their spectra, describe unitary groups associated with them, and present an example showing that the unique continuation principle is not valid here.

2. Preliminaries

Let us first introduce some notions we will need. We consider a graph Γ consisting of a family of *vertices*, $\mathcal{V} = \{v_j : j \in I^v\}$ indexed by a set $I \equiv I^v$, and a family $\mathcal{E}$ of edges[1]. The latter includes *finite* (or *internal*) edges, $\mathcal{E}_{\mathrm{fin}} = \{e_k : k \in I^e_{\mathrm{fin}}\}$ and *semi-infinite* edges (alternatively *external* ones or *leads*), $\mathcal{E}_\infty = \{h_k : k \in I^e_\infty\}$. If each pair of vertices is connected by at most one edge, we can identify I^e_{fin} with a subset of $I \times I$ specified the adjacency matrix, and if no more than one lead is attached to each vertex, I^e_∞ can be identified with a subset of I^v. Note that a graph can be always modified to satisfy these requirements by inserting dummy vertices to the "superfluous" edges, however, we will not need these assumptions.

The graphs we are going to consider are *metric* and *oriented*. The first notion means that each finite edge e_k can be identified with a line segment $[0, l_k]$ and a semi-infinite one with a halfline. The second one says that the orientation of the edge parametrization is not arbitrary; each finite edge e_k has its *starting point* associated with $x = 0$ and *endpoint* to which $x = l_k$ corresponds. In the standard terminology of graph theory oriented graph form a subclass among the *directed* ones in which each edge has a single orientation. In contrast to the usual graph theory, our graphs can also have semi-infinite edges and we have to take account of them. We divide $\mathcal{E}_\infty$ into the family $\mathcal{E}^+_\infty = \{h_k : k \in I^e_{\infty,+}\}$ of *outgoing* edges parametrized by $[0, \infty)$, with zero referring to the vertex, and $\mathcal{E}^-_\infty$ consisting of *incoming* ones, parametrized by $(-\infty, 0]$.

We can thus count edges meeting at a given vertex v_j: there are $d_j^{\mathrm{fin},\pm}$ finite edges starting and ending there, respectively, together $d_j^{\infty,+}$ outgoing and $d_j^{\infty,-}$

[1]The graph constituent symbols can be labelled by Γ but we mostly refrain from doing that.

incoming leads. The vertex degree d_j is the sum of those four numbers which can be split into the incoming and outgoing part, $d_j^{\pm} := d_j^{\mathrm{fin},\pm} + d_j^{\infty,\pm}$. We also introduce $d_j^{\mathrm{fin}} := d_j^{\mathrm{fin},+} + d_j^{\mathrm{fin},-}$ and $d_j^{\infty} := d_j^{\infty,+} + d_j^{\infty,-}$ and set

$$(2.1) \qquad N_{\mathrm{fin}} := \sum_{j\in I} d_j^{\mathrm{fin},+} = \sum_{j\in I} d_j^{\mathrm{fin},-}, \quad N_\infty^{\pm} := \sum_{j\in I} d_j^{\infty,\pm}, \quad N_\infty = N_\infty^+ + N_\infty^- .$$

In all the paper we suppose that Γ is either *locally finite* and *at most countably infinite* meaning that I is at most countable and

$$(2.2) \qquad \exists\, c_d > 0 \; : \; d_j \le c_d \quad \text{for } \forall j \in I .$$

This includes several categories, in particular, *finite* graphs having $\sharp I < \infty$ and $N_\infty = 0$, further *finite-core* graphs with $\sharp I < \infty$ and $N_\infty > 0$, and finally *properly countably infinite* meaning that I is countable. In the latter case the values of N_{fin} and N_∞ can be zero, a finite number, or infinite, in all possible combinations. In case of infinite graphs we shall also suppose that.

$$(2.3) \qquad \inf_{j\in I} l_j > 0 .$$

The graph will be called *balanced*, or alternatively *balanced oriented*, provided

$$(2.4) \qquad d_j^{\mathrm{fin},+} = d_j^{\mathrm{fin},-} \quad \text{and} \quad d_j^{\infty,+} = d_j^{\infty,-} \quad \text{for } \forall j \in I ;$$

the said notion is again common in oriented graph theory and we remark that our definition is consistent with the standard terminology since one can amend our graphs having $N_\infty^{\pm} \neq 0$ with a vertex at infinity where the outgoing semi-infinite edges "end" and the incoming ones "start".

3. Balanced orientability

In physical models where graphs are employed to describe spatial structures and motion of particles or fields on them usually no orientation is prescribed. Before proceeding further let us thus ask whether and how an undirected graph can be given orientation. The question has to be made more precise because one can always put arrows to such a graph edges; in graph theory one usually asks about existence of oriented paths connecting any pair of graph vertices. In contrast to that for us the local balance will matter; we call the a metric graph *balanced orientable* if one can parametrize its edges in a way which makes it an oriented and balanced graph.

THEOREM 3.1. *An undirected graph Γ satisfying the above countability assumption and (2.2) is balanced orientable* iff *the degree of any vertex $v \in \mathcal{V}_\Gamma$ is even.*

PROOF. The necessary condition is obvious. To check the sufficient one we introduce some notions. A *path* in an oriented graph is a family of subsequently adjacent edges such that at each vertex on it an incoming edge meets an outgoing one; the edges constituting a path can be parametrized by adjacent intervals of the real axis. A *free path* is a path which can be followed in both directions without termination; it can be either *infinite* parametrizable by $\mathbb{R}$ or a *loop* consisting of a finite number of finite edges which brings one to the initial point.

By assumption the vertices $v \in \mathcal{V}_\Gamma$ can be numbered. Take $v_1 \in \mathcal{V}_\Gamma$, pick two edges emanating from it and give them orientations making one of them incoming and the other outgoing. Follow the latter to the other end, choose one of the (even number of the) edges emanating from that vertex and make it outgoing, and do the same "backwards" with the chosen edge incoming at v_1. Proceeding in the same way we get either an infinite free path or a loop in case the forward and backward branch meet at a vertex; it is clear from the construction that the loop will have a definite orientation. We delete the constructed free path from Γ obtaining a graph Γ' which satisfies the assumptions of the theorem having $d_1' \le d_1 - 2$ and $d_j' \le d_j$ for $j \ge 2$. If $d_1' \ne 0$ we construct in Γ' another free path through v_1, delete it from the graph, and continue until this vertex is eliminated entirely.

Keeping the original vertex numbering we proceed to the vertex with lowest index value in the "reduced" graph obtained in this way, construct a free path going through it, eliminate it from the graph, and go on with such path eliminations until this vertex is fully removed from the graph; continuing the procedure we exhaust after an at most countable number of steps all the vertices which reduces the undirected graph remainder to an empty set proving thus the result. □

REMARKS 3.2. (a) If $N^{\text{fin}} < \infty$ the graph balanced orientability requires the number of semi-infinite edges to be even. This is seen well if we use the "flower" model [**Ku08**, **EL10**] of Γ with all the vertices put together. Since the degree of this "grand vertex" is the sum of degrees of the original vertices and every finite edge contributes to this quantity by an even number, and furthermore, the number of external edges has to be finite in view of (2.2), it must be even. Another simple consequence of the above result is that a balanced orientable graph cannot have "loose ends", i.e. finite edges with a vertex of degree one.

(b) The simplest situation from the orientability point view occurs if the graph edges can be paired; this is the case, in particular, if one thinks of an undirected graph edge as of a pair of *bonds* having opposite orientations. This is a useful trick, employed for instance in studies of quantum chaos on graphs [**KS03**], allowing one to express properties of the original undirected graph in terms of "bond scattering matrices" – cf. [**BK**, Sec. 2.2].

(c) The construction employed in the proof shows that a balanced oriented graph can be regarded as *a union of free paths*, disjoint except for the vertices in which they intersect. In general, an undirected graph satisfying the assumptions of the theorem can be oriented in different ways. In case of a *free chain graph* with all the vertices of degree two the ambiguity is trivial consisting of a choice of one the two possible orientations, while in case of nontrivial *branchings* meaning existence of a $j \in I$ with $d_j > 2$, the number of ways to orient Γ is larger.

(d) If $N^{\text{fin}} < \infty$ any infinite free paths begin and end with an external edge, provided these are present, which confirms the claim made above. On the other hand, graphs with $N^{\text{fin}} = \infty$ can be balanced orientable even with an odd number of leads. A trivial example illustrating this claim is a graph Γ isometric to the line with the vertices at the points $x = 0, 1, 2, \ldots$. A less trivial example is obtained if we replace the positive halfline by a tree graph with all the vertices of degree four. At each branching we have one edge to the left of the vertex and three to the right, of which we choose two of the same orientation as their left neighbor and one opposite. It is easy to see that such a graph with one semi-infinite edge is balanced orientable and it can be identified with a family of infinite free paths.

(e) Admissible orientations can be very different, in particular, if $\sharp I = \infty$. As an example, consider a square-lattice graph corresponding naturally to the subset $\{(x,y) \,:\, x \in \mathbb{R},\, y = j\ell,\, j \in \mathbb{Z}\} \cup \{(x,y) \,:\, y \in \mathbb{R},\, x = i\ell,\, i \in \mathbb{Z}\}$ of $\mathbb{R}^2$ for a fixed $\ell > 0$. One can orient it by making all the horizontal and vertical lines infinite free paths. An alternative is to regard the lattice as a checkerboard pattern and make free paths of the perimeters of all the "black" squares; it that case the oriented graph can be identified with the corresponding infinite family of loops. Furthermore, it is not difficult to find ways of orientation in which the two types of free paths, the infinite ones and the loops, are combined.

4. Momentum operators

After these preliminaries let us pose our main question. We will suppose that Γ is the configuration space of a quantum system, namely that a spinless quantum particle lives on the graph, and ask whether one can define for it a momentum-type observable, that is, a self-adjoint operator which acts as $\psi_k \mapsto -i\psi_k'$ on the kth edge. The state Hilbert space of such a system will naturally be

$$L^2(\Gamma) := \bigoplus_{k\in I^e_{\mathrm{fin}}} L^2(0,l_k) \oplus \bigoplus_{k\in I^e_\infty} L^2(0,\infty) \tag{4.1}$$

the elements of which we write as columns $\psi = \big(\{\psi^{\mathrm{fin}}_{j,i} : j\in I,\, i=1,\dots,d^{\mathrm{fin}}_j\},\, \psi^\infty_{j,i} : j\in I,\, i=1,\dots,d^\infty_j\}\big)^{\mathrm{T}}$, or $\psi = \big(\{\psi^{\mathrm{fin}}_k : k\in I^e_{\mathrm{fin}}\}, \{\psi^\infty_k : k\in I^e_\infty\}\big)^{\mathrm{T}}$ if we number the edges as we go, with the scalar product

$$(\phi,\psi) := \sum_{k\in I^e_{\mathrm{fin}}} \int_0^{l_k} |\psi^{\mathrm{fin}}_k(x)|^2 \mathrm{d}x + \sum_{k\in I^e_\infty} \int_0^\infty |\psi^\infty_k(x)|^2 \mathrm{d}x\,. \tag{4.2}$$

As a starting point of the construction we take the operator $P_0 : P_0\psi = -i\psi'$ with the domain $D(P_0)$ consisting of $W^{1,2}$ functions vanishing at the vertices, or more explicitly being equal to

$$\{\psi^{\mathrm{fin}}_k \in W^{1,2}(e_k) : \psi^{\mathrm{fin}}_k(0) = \psi^{\mathrm{fin}}_k(l_k) = 0\} \cup \{\psi^\infty_k \in W^{1,2}(\mathbb{R}^+) : \psi^\infty_k(0) = 0\}\,. \tag{4.3}$$

It is straightforward to check that P_0 defined in this way is symmetric and it adjoint acts as $P_0^*\psi = -i\psi'$ on $D(P_0^*) = W^{1,2}(\Gamma\setminus\mathcal{V})$.

It is natural that candidates for the role of momentum operator are to be looked for among the self-adjoint extensions of P_0. To construct them we need the corresponding boundary form; an easy argument using integration by parts gives

$$\begin{aligned}(P^*\phi,\psi) - (\phi, P^*\psi) &= i(\phi^{\mathrm{out}},\psi^{\mathrm{out}}) - i(\phi^{\mathrm{in}},\psi^{\mathrm{in}}) \\ &= i\sum_{k\in I^e_{\mathrm{fin}}} \Big(\overline{\phi}^{\mathrm{fin}}_k(l_k-)\psi^{\mathrm{fin}}_k(l_k-) - \overline{\phi}^{\mathrm{fin}}_k(0+)\psi^{\mathrm{fin}}_k(0+)\Big) \\ &\quad -i\sum_{k\in I^e_{\infty,+}} \overline{\phi}^\infty_k(0+)\psi^\infty_k(0+) + i\sum_{k\in I^e_{\infty,-}} \overline{\phi}^\infty_k(0-)\psi^\infty_k(0-)\end{aligned} \tag{4.4}$$

for any $\phi,\psi\in D(P_0^*)$, where we have employed the following shorthands,

$$\psi^{\mathrm{out}} := \Big(\psi^{\mathrm{fin}}_1(0+),\dots,\psi^{\mathrm{fin}}_{N_{\mathrm{fin}}}(0+), \psi^\infty_1(0+),\dots,\psi^\infty_{N^+_\infty}(0+)\Big)^{\mathrm{T}},$$

$$\psi^{\mathrm{in}} := \Big(\psi^{\mathrm{fin}}_1(l_1-),\dots,\psi^{\mathrm{fin}}_{N_{\mathrm{fin}}}(l_{N_{\mathrm{fin}}}-), \psi^\infty_1(0-),\dots,\psi^\infty_{N^-_\infty}(0-)\Big)^{\mathrm{T}}.$$

PROPOSITION 4.1. *Self-adjoint extensions of P_0 always exist provided $N_{\mathrm{fin}}=\infty$; in case $N_{\mathrm{fin}}<\infty$ they exist* iff $N^+_\infty = N^-_\infty$. *They are characterized by the condition $\psi^{\mathrm{out}} = U\psi^{\mathrm{in}}$ where U is a unitary operator of dimension $N_{\mathrm{fin}} + N^+_\infty$.*

PROOF. It is straightforward to check that the deficiency indices of P_0 are $(N_{\text{fin}}+N_\infty^+, N_{\text{fin}}+N_\infty^-)$ and that the boundary form (4.4) vanishes *iff* the boundary values satisfy $\psi^{\text{out}} = U\psi^{\text{in}}$ for any $\psi \in D(P_0^*)$. □

However, not every self-adjoint extension P_U of P_0 is a good candidate for the operators we are interested in. We have to add the requirement of *locality* which means that the boundary conditions couple the boundary values

$$\psi_j^{\text{out}} := \left(\psi_{j,1}^{\text{fin}}(0+), \dots, \psi_{j,d_j^{\text{fin},+}}^{\text{fin}}(0+), \psi_{j,1}^{\infty}(0+), \dots, \psi_{j,d_j^{\infty,+}}^{\infty}(0+)\right)^{\mathrm{T}},$$

$$\psi_j^{\text{in}} := \left(\psi_{j,1}^{\text{fin}}(l_{j,1}-), \dots, \psi_{j,d_j^{\text{fin},-}}^{\text{fin}}(l_{j,d_j^{\text{fin},-}}-), \psi_{j,1}^{\infty}(0-), \dots, \psi_{j,d_j^{\infty,-}}^{\infty}(0-)\right)^{\mathrm{T}},$$

where $l_{j,k}$, $k = 1, \dots, d_j^{\text{fin},-}$ are the lengths of the incoming finite edges at the vertex v_j. We will call an extension P_U a *momentum operator* on the graph Γ if the unitary operator U is block diagonal, $U = \operatorname{diag}\{U_j : j \in I\}$, with the unitary matrix blocks coupling the vertex boundary values

$$\psi_j^{\text{out}} = U_j \psi_j^{\text{in}}, \quad j \in I. \tag{4.5}$$

This definition leads to the following conclusion.

THEOREM 4.2. *An oriented graph Γ supports momentum operators* iff *it is balanced. In such a case they are characterized by the conditions (4.5).*

PROOF. One has to compare the dimensions of the boundary value spaces. □

Before proceeding further it is appropriate to say a few words about the physical meaning of the self-adjointness requirement (4.5), in particular, in comparison with the more common problem of constructing self-adjoint Laplacians on graphs [**EŠ89**, **Ku08**, **BK**]. In the latter case self-adjointness ensures conservation of probability current the components of which, $(\psi_k, -i\psi_k') = \frac{1}{2i}\operatorname{Im}(\psi_k, \psi_k')$ on the k-th edge, enter the appropriate boundary form. Here, in contrast, the right-hand side of eq. (4.4) contains *probabilities* of finding the particle at the incoming and outgoing edge points; the importance of this fact will become more obvious when we shall construct the unitary groups associated with the operators P_U in Sec. 7 below[2].

Another thing which deserves a comment is the locality requirement we have made. We have mentioned in Remark 3.2a that it is sometimes useful to replace a given graph Γ by another one in which the vertices are identified; such an identification naturally extends the class of admissible momentum operators. These considerations can have a practical meaning. An elementary example concerns a momentum operator on a finite interval which in itself is not a balanced orientable graph, however, it arises naturally when we factorize it from an infinite periodic system on line [**AGHH**, Sec. III.2]; turning it into a torus — i.e., a loop in this case — by identifying the endpoints, we get a family of momentum operators conventionally referred to as *quasimomentum* in this case. A less trivial example of this type is represented by two-interval momentum operators discussed in [**JPT11**] in connection with the Fuglede conjecture.

[2]Note also the difference in the size of the matrices which determine the coupling. For a graph Laplacian the extensions are described by $d_j \times d_j$ matrices in the vertex v_j, cf. [**Ku08**, Thm. 5], while here we deal with unitary matrices which are $d_j^\pm \times d_j^\pm$, i.e. half the size.

Note also that the matrix U defining a momentum operator can be subject to other restrictions, in addition to those imposed by the locality requirement. If there is a proper subgraph $\Gamma' \subset \Gamma$ which can be in the sense of Remark 3.2c identified with a family of paths and P_U is reduced by the subspace $L^2(\Gamma') \subset L^2(\Gamma)$ we say that P_U is *decomposable*, in the opposite case we call the momentum operator *indecomposable.* A decomposable P_U can be regarded as a collection of momentum operators on the appropriate subgraphs which can be analyzed independently.

A similar approach can also be applied locally: if a subspace of the boundary-value space at a given vertex reduces the corresponding coupling matrix U_j we may regard the vertex in question as a family of vertices, each of them connecting only the edges "talking to each other". i.e. referring to the same invariant subspace of U_j. It is useful to stress that here we mean not just any subspace but such that the vectors associated with the edges involved form its basis, in other words, the matrix U_j is block-diagonal after a suitable permutation of its rows and columns.

5. Momentum operator spectra: finite graphs

From now on we shall consider only balanced oriented graphs and investigate properties of momentum operators on them. The first question concerns their spectra; it is not surprising that the graph finiteness plays here a decisive role.

THEOREM 5.1. *If a graph Γ is finite, then any momentum operator P_U on it has a purely discrete spectrum. Moreover, $N_U(\lambda) := \sharp(\sigma(P_U) \cap (-\lambda, \lambda)) \le \frac{1}{\pi} L\lambda + \mathcal{O}(1)$ holds as the window half-width $\lambda \to \infty$, where L is the total length of Γ.*

PROOF. A finite balanced oriented graph can be in view of Remarks 3.2 identified with a finite family of loops. Choosing U_0 for which all of them are mutually disconnected and the functions on each of them are smoothly connected at the vertices, we get operator P_{U_0} with a purely discrete spectrum — in fact, one can write explicitly $\sigma(P_{U_0}) = \bigcup_j \{\frac{2\pi m}{\ell_j} : m \in \mathbb{Z}\}$ where ℓ_j are the lengths of the loops. Since by Krein's formula any other self-adjoint extension P_U differs from P_{U_0} by a finite-rank perturbation in the resolvent, the character of the spectrum is preserved.

Furthermore, using the explicit spectrum of P_{U_0} we find that the number of its eigenvalues in $(-\lambda, \lambda)$ can be estimated from above by $\sharp I + \sum_j \frac{\lambda \ell_j}{\pi}$ and since by [**We**, Sec. 8.3] any other P_U can have at most $\sharp I$ eigenvalues, counting multiplicity, in each gap of $\sigma(P_{U_0})$ we arrive at the estimate

$$N_U(\lambda) \le 3\sharp I + \frac{1}{\pi}\lambda \sum_j \ell_j \,,$$

which yields the sought 'semiclassical' result. □

EXAMPLE 5.2. Consider a *loop graph* of total length L with vertices at points $0 < x_1 < x_2 < \cdots < x_N = L$ with the points $x = 0$ and $x = L$ identified. Since the degree of each vertex is two, all the admissible momentum operators are characterized by N-tuples $\alpha = \{\alpha_1, \dots, \alpha_N\} \in (-\pi, \pi]^N \subset \mathbb{R}^N$ through the conditions $\psi(x_j+) = \mathrm{e}^{i\alpha_j}\psi(x_j-)\,,\ j = 1, \dots, N$. The spectrum is easily found to be

$$\sigma(P_U) = \left\{ \frac{1}{L}\Big(2\pi n - \sum_{j=1}^{N} \alpha_j\Big) : n \in \mathbb{Z} \right\}.$$

It is obvious that the family of all P_U decomposes into isospectral classes specified by the values $\det U$ of the diagonal matrices U.

The spectrum may become more complex once the graph Γ has a branching, i.e. there is at least one vertex of degree $d_j \geq 4$; recall that $d_j = 3$ is excluded by the orientability requirement as discussed in Sec. 3.

EXAMPLE 5.3. A *figure-eight graph* is equivalent in the sense of [**Ku08, EL10**] to the "two-interval" case discussed in detail in [**JPT11**]. The eigenvalues of any P_U here have maximum multiplicity two but their distribution is more complicated than in the previous example, in particular, it is purely aperiodic provided the length ratio of the two loops of the graph is irrational.

The asymptotic estimate given in Theorem 5.1 can be in fact made two-sided.

THEOREM 5.4. *For a finite Γ we have $N_U(\lambda) = \frac{1}{\pi}L\lambda + \mathcal{O}(1)$ as $\lambda \to \infty$.*

PROOF. We can identify the loops from the proof of Theorem 5.1 with the edges of Γ in the sense of Remark 3.2a. Substituting the boundary values $\psi^{\text{out}} = \{c_1, \dots, c_{N_{\text{fin}}}\}^{\text{T}}$ and $\psi^{\text{in}} = \{c_1 e^{ik\ell_1}, \dots, c_{N_{\text{fin}}} e^{ik\ell_{N_{\text{fin}}}}\}^{\text{T}}$ into the condition (4.5) we get N_{fin} conditions to satisfy, $\sum_{m=1}^{N_{\text{fin}}} \left(U_{jm} e^{ik\ell_m} - \delta_{jm}\right) c_m = 0\,,\; j = 1, \dots, N_{\text{fin}}\,$; note that the left-hand side can vanish at real values of k only since U is unitary. This system of linear equations can be solved *iff*

$$\det \left(U_{jm} e^{ik\ell_m} - \delta_{jm}\right)_{j,m=1}^{N_{\text{fin}}} = 0\,.$$

The left-hand side of the last relation is a trigonometric polynomial with the senior and junior terms $\prod_j U_{jj}\, e^{ikL}$ and $(-1)^{N_{\text{fin}}}$, respectively. Assume first that the senior coefficient $\prod_j U_{jj} \neq 0$. According to Langer's classical result [**La31**] the number of zeros in the interval $(-\lambda, \lambda)$ behaves then as $\frac{1}{\pi}L\lambda + \mathcal{O}(1)$ as $\lambda \to \infty$. If the assumption is not satisfied we use the fact that by the implicit-function theorem the zeros as functions of the entries U_{jm} of U are analytic, hence for such a P_U there is a family of zeros with the 'correct' asymptotics arbitrarily close to $\sigma(P_U)$.

Let us add that there is a subtle point in the last argument. In general it may happen that the vanishing senior coefficient in a spectral condition in the form of a trigonometric expression may change the asymptotic distribution of zeros if some of the latter escape to infinity in the complex plane as the parameters vary; an example of such a behavior can be found in [**DEL10**]. Here it cannot happen, however, because we know from the outset that all the zeros lie at the real axis. □

6. Momentum operator spectra: infinite graphs

The situation changes substantially if we abandon the finiteness assumption.

THEOREM 6.1. *If $N_\infty > 0$ we have $\sigma(P_U) = \sigma_{\text{ess}}(P_U) = \mathbb{R}$ for any momentum operator P_U. The same is true if $\sharp I = \infty$ and there is a unitary operator U' with $\dim \operatorname{Ran}(U - U') < \infty$ such that $P_{U'}$ is decomposable containing at least one infinite free path; in such a case we also have $\sigma_{\text{ac}}(P_U) = \mathbb{R}$.*

PROOF. Suppose that the graph has at least one external edge. Since the graph remains balanced oriented if we switch the orientation of each edge, we may assume without loss of generality that it is an outgoing one parametrized by $[0, \infty)$. For a fixed $k \in \mathbb{R}$ we take the following family of functions,

$$\psi_{y,\epsilon} : \psi_{y,\epsilon}(x) = \epsilon^{1/2} e^{ikx} \phi(\epsilon(x - y))\,, \quad \epsilon > 0\,, \tag{6.1}$$

with $\phi \in C_0^\infty(-1,1)$ such that $\|\phi\| = 1$. We obviously have $\|\psi_{y,\epsilon}\| = 1$ and the supports lie in the positive halfline provided $y > \epsilon^{-1}$. We use the same symbol for functions on Γ which are zero on all the other edges; them $P_U\psi_{y,\epsilon}$ is independent of U and an easy computation gives

$$\|(P_U - k)\psi_{2\epsilon^{-1},\epsilon}\|^2 = \epsilon^2 \int_{-1}^{1} |\phi'(u)|^2 \,\mathrm{d}u \longrightarrow 0$$

as $\epsilon \to 0$. Choosing $\epsilon_n = 2^{-2n}$, $n \in \mathbb{N}$, we achieve that the functions with different n have disjoint supports, hence $\psi_{2\epsilon_n^{-1},\epsilon_n} \to 0$ in the weak topology of $L^2(\Gamma)$, and by Weyl's criterion we infer that $k \in \sigma_{\mathrm{ess}}(P_U)$.

In the second case the motion on Γ corresponding to $P_{U'}$ can be decoupled and it contains a component which is unitarily equivalent to the momentum operator on $\mathbb{R}$, hence $\sigma_{\mathrm{ac}}(P_{U'}) = \mathbb{R}$; since by assumption the resolvents of $P_{U'}$ and the original P_U differ by a finite-rank operator, their absolutely continuous spectra coincide. □

The second claim can ensure that the absolutely continuous spectrum covers the real axis even in absence of external edges; examples are easily found. On the other hand, conclusions of the theorem naturally do not mean that momentum operators on finite-core graphs must have a purely absolutely continuous spectrum.

EXAMPLE 6.2. Consider a graph consisting of a line to which a loop of length $\ell > 0$ is attached at one point. Modulo mirror transformations, there is essentially one way to orient such a graph in a balanced way. Consider two different matrices coupling the boundary values $(\psi_1^\infty(0-), \psi^{\mathrm{fin}}(\ell-))^{\mathrm{T}}$ and $(\psi_2^\infty(0+), \psi^{\mathrm{fin}}(0+))^{\mathrm{T}}$. If we take $U = \left(\begin{smallmatrix} 0 & 1 \\ 1 & 0 \end{smallmatrix}\right)$ the corresponding P_U is unitarily equivalent to the momentum operator on the line and has a purely *ac* spectrum. On the other hand, $U' = I$ leads to full decoupling of the loop from the line giving rise to an infinite family of eigenvalues, $\{\frac{2\pi m}{\ell} : m \in \mathbb{Z}\}$ embedded in $\sigma_{\mathrm{ac}}(P_{U'}) = \mathbb{R}$.

Different ways of orienting a given undirected graph can give rise to momentum operators with different spectra. Nevertheless, some properties are invariant with respect to the choice of orientation, for instance, the conclusions related to the the presence or absence of external edges in the above two theorems. The differences can be more dramatic if $\sharp I = \infty$ as the following example shows.

EXAMPLE 6.3. Consider the square-lattice graph of Remark 3.2e. If the orientation follows the horizontal and vertical lines we can choose U which identifies at each vertex the limits in the two directions separately. The respective operator P_U is isomorphic to an infinite direct sum of identical copies of the operator $-i\frac{\mathrm{d}}{\mathrm{d}x}$ on $L^2(\mathbb{R})$, and consequently, $\sigma(P_U) = \mathbb{R}$ with infinite multiplicity. On the other hand, consider the "checkerboard" orientation which identifies the graph with an infinite family of loops of length 4ℓ. Choosing U' which gives rise to self-adjoint extension $P_{U'}$ the domain of which are functions which are $W^{1,2}$ locally and continuous on each square loop, we get $\sigma(P_{U'}) = \{\frac{\pi m}{2\ell} : m \in \mathbb{Z}\}$, again infinitely degenerate.

7. Groups associated with momentum operators

In order to describe the associated groups let us first introduce the notion of a *route* $r_{x',x}$ from a point $x' \in \Gamma$ to a point $x \in \Gamma$. By this we mean a finite curve $r_{x',x} : [0, \ell(r_{x',x})] \to \Gamma$ with $r_{x',x}(0) = x$ and $r_{x',x}(\ell(r_{x',x})) = x'$ respecting the orientation of Γ; the number $\ell(r_{x',x})$ is called the *length* of $r_{x',x}$. We stress the difference between a *path* in Γ considered in Section 3 above and a route: the former is a sequence of edges through which the particle can travel respecting the orientation, the latter is the actual journey it made. If Γ contains a loop, for instance, for a path it is not important whether one circles it repeatedly or not, while for a route it certainly makes a difference.

We fix a momentum operator P_U on Γ determined by a matrix U, and define the *route factor* for $r_{x',x}$ with $x', x \notin \mathcal{E}$. It is equal to one if there is no $x'' \in (0, \ell(r_{x',x}))$ such that $r_{x',x}(x'') \in \mathcal{E}$, i.e. if a traveler on this route meets no vertex. In the opposite case we take in each vertex on the route the factor coming from the coupling — if the traveler passes from m-th ingoing edge to the j-th outgoing the factor is the U_{jm} element of the matrix — and define the route factor as the product of the factors coming from all the vertices passed on the route; we denote it as $U(r_{x',x})$. It may be zero if some of the matrix elements on the way vanishes.

Given $x \in \Gamma$ which is not a vertex we denote by $\mathcal{R}_a(x)$ the set of routes of length a ending at x. It may consists of a single route which happens if going back we meet no vertex — it happens for sure if a is small enough — or of a larger number of routes, however, in view of the assumption (2.3) their number is finite. Equipped with these notions we can construct the group associated with P_U.

We will write elements of $L^2(\Gamma)$ as $\psi = \{\psi_k : k \in I^e\}^{\mathrm{T}}$ without distinguishing now internal and external edges. Choosing a point $x = \{x_k : x_k \in e_k,\, k \in I^e\}$ we write values of a function representing element $\psi \in L^2(\Gamma)$ at the point x as $\psi(x) = \{\psi_k(x_k) : k \in I^e\}^{\mathrm{T}}$ and define the operator $\mathcal{U}(a) : L^2(\Gamma) \to L^2(\Gamma)$ by

$$(7.1) \qquad \big(\mathcal{U}(a)\psi\big)(x) := \Bigg\{ \sum_{r_{x'_{k'},x_k} \in \mathcal{R}_a(x_k)} U(r_{x'_{k'},x_k})\psi_{k'}(x'_{k'}) \,:\, k \in I^e \Bigg\}^{\mathrm{T}}$$

for those $x \in \Gamma$ and a for which none of the points x_k and x'_k involved coincides with a vertex of Γ; note that for each $a \in \mathbb{R}$ this requirement excludes an at most countable subset of graph points. It is easy to check that these operators form a group, $\mathcal{U}(a)\mathcal{U}(a)' = \mathcal{U}(a + a')$ for all $a, a' \in \mathbb{R}$. It is less obvious that the operator defined by (7.1) is unitary. To check it let us first look how does the map

$$\psi_k \mapsto \sum_{r_{x'_{k'},x_k} \in \mathcal{R}_a(x_k)} U(r_{x'_{k'},x_k})\psi_{k'}$$

act in the space $\ell^2(I^e)$. The set $\mathcal{R}_a(x)$ can be regarded as a 'backward' tree with every route from 'tip' $x'_{k'}$ to the 'root' x_k having length a. Since by construction the incoming and outgoing edges at every vertex are related by a unitary matrix, the norm is preserved and images of orthogonal elements remain orthogonal going 'backward' through the tree. Consequently, we have

$$\Bigg| \sum_{r_{x'_{k'},x_k} \in \mathcal{R}_a(x_k)} U(r_{x'_{k'},x_k})\psi_{k'}(x'_{k'}) \Bigg|^2 = |\psi_k(x_k)|^2$$

and contributions to $\mathcal{U}(a)\psi$ coming from different edge components are mutually orthogonal. Since the integration over x_k and $x'_{k'}$ on each edge is taken with respect to the same Lebesgue measure, it follows that $\|\mathcal{U}(a)\psi\|^2 = \|\psi\|^2$ holds for any $\psi \in L^2(\Gamma)$ which we wanted to prove.

One has to keep in mind that the group depends on the choice of the operator P_U and if there is need to stress this fact we should write its elements as $\mathcal{U}_U(a)$.

In general the group actions are shifts along the graph edges, however, in contrast to the trivial situation when Γ is a line, vertex coupling plays a role, especially if there is a nontrivial branching, $d_j > 2$. This returns us to the question of physical meaning of different self-adjoint extension we have touched briefly in Sec. 4.

EXAMPLE 7.1. To illustrate the action the group consider a *star-shaped graph* consisting of an equal number n of incoming and outgoing semi-infinite leads, $N_\infty^\pm = n \geq 2$, connected in a single vertex, in which the coupling is described by an $n \times n$ unitary matrix U. Take a vector ψ on the first incoming lead identified with the interval $(-\infty, 0]$ assuming that it has a compact support and denote $b := -\inf \operatorname{supp} \psi$. If $a > b$ the function $\mathcal{U}(a)\psi$ is supported on the outgoing leads only, its component on the j-th one being $U_{j1}\psi(\cdot - a)$. In other words, the incoming wave packets splits into scaled copies of the original one with the weights which guarantee that the probability after the shift through the vertex[3] is preserved. If the initial wave packet is supported on more than a single edge, the resulting one is naturally a superposition of those coming from the involved incoming contributions.

More generally, it is easy to see that a state represented by a function of compact support will remain compactly supported if operators $\mathcal{U}(a)$ are applied to it. This does not mean, however, that the support will keep its properties, in particular, that its Lebesgue measure should have a bound independent of a.

EXAMPLE 7.2. Consider again the graph of Example 6.2. In the two situation mentioned there the group action is simple: for $U = \begin{pmatrix} 0 & 1 \\ 1 & 0 \end{pmatrix}$ the operator P_U is unitarily equivalent to the momentum operator on the line along which the corresponding $\mathcal{U}(a)$ shifts function, while for $U = I$ the group decomposes into shifts along the line and cyclic motion on the loop. The situation is different if all the elements of U are nonzero. Consider ψ with the support on the incoming lead, $\operatorname{supp} \psi \subset \left(-\frac{3}{4}\ell, -\frac{1}{4}\ell\right)$ and apply $\mathcal{U}(n\ell)$ to it. It is not difficult to see that the loop component keeps the shape changing just the "size" being $\left(\mathcal{U}(n\ell)\psi\right)_2(x) = (U_{22})^{n-1} U_{21}\psi(x - \ell)$ while

$$\left(\mathcal{U}(n\ell)\psi\right)_1(x) = U_{11}\psi(x - n\ell) + \sum_{k=1}^{n-1} U_{12}(U_{22})^{n-1-k} U_{21}\psi(x - k\ell)$$

holds for the line part. In other words, the function is a linear combination of shifted copies of the original function with the number of components increasing with n, in particular, we have $\operatorname{diam} \operatorname{supp} \left(\mathcal{U}(n\ell)\psi\right)_1 \geq (n-1)\ell$.

The action of $\mathcal{U}(a)$ on an indecomposable graph, even a finite one, can be more more complicated if such a Γ has more than one loop and their lengths are incommensurable. Note also that a related example can be found in **[JPT12a]** where, however, the model is interpreted in terms of Lax-Phillips scattering theory.

8. A two-loop example

To get another insight into spectral properties of the operators P_U, let us analyze one more simple example. The graph Γ in this case will consists of two leads, Γ_0 and Γ_4 identified with the halflines $(-\infty, 0]$ and $(]0, \infty)$, respectively, and three finite edges Γ_j of lengths ℓ_j, $j = 1, 2, 3$, connecting the endpoints of Γ_0 and Γ_4; the first two are oriented from Γ_0 to Γ_4 and the third one in the opposite direction. Such

[3]We avoid the term "passing through" to stress that no time evolution is involved here, at least as long as we think about our model in terms of standard quantum mechanics.

a graph is obviously balanced oriented so one can construct momentum operators on it; we choose the one for which the coupling at both vertices (coupling respectively the edges $(3,0) \to (1,2)$ and $(1,2) \to (3,4)$) is described by the same matrix,

$$U = U^* = \frac{1}{\sqrt{2}} \begin{pmatrix} 1 & 1 \\ 1 & -1 \end{pmatrix}. \tag{8.1}$$

Looking for eigenfunctions, proper or generalized, of the momentum operator — which we denote for the sake of simplicity P_U again — we use the Ansätze $\psi_j(x) = c_j\, \mathrm{e}^{-ikx}$ on the k-th edge, $j = 0, \dots, 4$.

Consider first the situation when the two "right-looking" edges are of the same length, $\ell_1 = \ell_2$. Specifying the conditions (4.5) to the present case and excluding the coefficients c_1, c_2 we arrive at the relation

$$\mathrm{e}^{-ik\ell_1} \begin{pmatrix} c_3 \\ c_4 \end{pmatrix} = \begin{pmatrix} c_3\, \mathrm{e}^{ik\ell_3} \\ c_0 \end{pmatrix}.$$

It has two possible independent solutions. If $c_0 = c_4$ we require $c_3 = c_3\, \mathrm{e}^{ik\ell_3}$ which gives rise to an infinite series of eigenvalues,

$$k = \frac{2\pi n}{\ell_1 + \ell_3}, \qquad n \in \mathbb{Z}; \tag{8.2}$$

the coefficients for the remaining edges are $c_1 = c_2 = \frac{1}{\sqrt{2}}\, c_3\, \mathrm{e}^{-ik\ell_1}$. On the other hand, if $c_3 = 0$ the solution exists for any $k \in \mathbb{R}$ with the coefficients $c_4 = c_0\, \mathrm{e}^{ik\ell_1}$ and $c_1 = -c_2 = \frac{1}{\sqrt{2}}\, c_0$. In other words, the spectrum of P_U consists of absolutely continuous part covering the real axis and the series of embedded eigenvalues (8.2).

This looks like the spectrum we have found in Example 6.2, however, there is an important difference. The operator $P_{U'}$ there was decomposable, hence its embedded point spectrum was in a sense trivial. Here it is not the case and still we have compactly supported eigenfunctions on a infinite graph. Moreover, the example can be modified by replacing the two semi-infinite leads by a single finite edge of length ℓ_4 connecting the two vertices again. The solutions symmetric with respect to permutation of ψ_j, $j = 1, 2$, are as before, with eigenfunctions vanishing on the new edge. The antisymmetric ones, on the other hand, require the relations $c_0 = c_4\, \mathrm{e}^{ik\ell_4} = c_0\, \mathrm{e}^{ik(\ell_1+\ell_4)}$ to be satisfied, giving rise to eigenvalues

$$k = \frac{2\pi n}{\ell_1 + \ell_4}, \qquad n \in \mathbb{Z},$$

replacing the absolutely continuous spectral component of the original example. What is important is that they correspond to $c_3 = 0$, hence all the eigenfunctions in this case vanish at some edge of the modified, now compact graph. These examples allow us to make the following conclusion.

PROPOSITION 8.1. *For momentum operators on a balanced graph the unique continuation principle is in general not valid.*

We note that a similar result holds for graph Laplacians. Here, however, the claim is less obvious; Laplacian eigenfunctions corresponding to positive eigenvalues are trigonometric functions and as such they can have zeros which for a proper choice of geometry may coincide with graph vertices [**EL10**], while momentum operator eigenfunctions on an edge cannot vanish being of the form $c\,\mathrm{e}^{ikx}$.

Returning to the original example we note that the embedded spectrum emerged as the result of the assumption $\ell_1 = \ell_2$. If it is not valid, the problem is more complicated. Writing down the coupling conditions and excluding c_1, c_2 we arrive at

$$c_3 = c_0\,\frac{\mathrm{e}^{ik\ell_2} - \mathrm{e}^{ik\ell_1}}{\mathrm{e}^{ik\ell_3}\big(\mathrm{e}^{ik\ell_2} + \mathrm{e}^{ik\ell_1}\big) - 2}$$

together with the condition $c_0\,F\big(\mathrm{e}^{ik\ell_1}, \mathrm{e}^{ik\ell_2}, \mathrm{e}^{ik\ell_3}\big) = c_4$ where F is a rational function with same denominator. Embedded eigenvalues correspond to the k's for which the latter vanishes, hence such a spectrum is present under suitable commensurability relations between the edges lengths.

One can naturally also ask what happens if we change those parameters. In case of graph Laplacians we know that violation of rationality turns in general embedded eigenvalues into resonances. Let us look what we have here; for simplicity we consider the situation when $\ell_1 = \ell_3 = \ell$ and $\ell_2 = \ell + \delta$. The mentioned denominator vanishes when $4i \sin k\ell + \mathrm{e}^{ik\ell}\big(\mathrm{e}^{ik\delta} - 1\big)$ does, hence for small δ it yields

$$k = \frac{\pi n}{\ell}\left(1 + \frac{(-1)^n}{4i}\,\delta + \mathcal{O}(\delta^2)\right) \tag{8.3}$$

in the vicinity of the eigenvalues (8.2), i.e. complex "resonance" solutions; in distinction to Laplacian resonances one cannot localize them in one complex halfplane.

References

[AGHH] S. Albeverio, F. Gesztesy, R. Høegh-Krohn, and H. Holden, *Solvable models in quantum mechanics*, 2nd ed., AMS Chelsea Publishing, Providence, RI, 2005. With an appendix by Pavel Exner. MR2105735 (2005g:81001)

[BK] G. Berkolaiko and P. Kuchment, *An Introduction to Quantum Graphs*, American Mathematical Society, Providence, RI, 2013.

[Ca99] Robert Carlson, *Inverse eigenvalue problems on directed graphs*, Trans. Amer. Math. Soc. **351** (1999), no. 10, 4069–4088, DOI 10.1090/S0002-9947-99-02175-3. MR1473434 (99m:34189)

[DEL10] E. Brian Davies, Pavel Exner, and Jiří Lipovský, *Non-Weyl asymptotics for quantum graphs with general coupling conditions*, J. Phys. A **43** (2010), no. 47, 474013, 16, DOI 10.1088/1751-8113/43/47/474013. MR2738108 (2011m:81112)

[AEG] G. Dell'Antonio, P. Exner, V. Geyler, eds., *Special Issue on Singular Interactions in Quantum Mechanics: Solvable Models*, J. Phys. A: Math. Gen. **38** (2005), No. 22.

[Eg11] S. Egger *né* Endres: The solution of the "constant term problem" and the ζ-regularized determinant for quantum graphs, *PhD thesis*, Universität Ulm 2011.

[ES10] Sebastian Endres and Frank Steiner, *The Berry-Keating operator on $L^2(\mathbb{R}_>, \mathrm{d}x)$ and on compact quantum graphs with general self-adjoint realizations*, J. Phys. A **43** (2010), no. 9, 095204, 33, DOI 10.1088/1751-8113/43/9/095204. MR2592329 (2011g:81081)

[EKKST] *Analysis on graphs and its applications*, Proceedings of Symposia in Pure Mathematics, vol. 77, American Mathematical Society, Providence, RI, 2008. Papers from the program held in Cambridge, January 8–June 29, 2007; Edited by Pavel Exner, Jonathan P. Keating, Peter Kuchment, Toshikazu Sunada and Alexander Teplyaev. MR2459860 (2009i:00008)

[EL10] Pavel Exner and Jiří Lipovský, *Resonances from perturbations of quantum graphs with rationally related edges*, J. Phys. A **43** (2010), no. 10, 105301, 21, DOI 10.1088/1751-8113/43/10/105301. MR2593999 (2011e:81102)

[EŠ89] P. Exner and P. Šeba, *Free quantum motion on a branching graph*, Rep. Math. Phys. **28** (1989), no. 1, 7–26, DOI 10.1016/0034-4877(89)90023-2. MR1109248 (92e:81019)

[FKW07] S. A. Fulling, P. Kuchment, and J. H. Wilson, *Index theorems for quantum graphs*, J. Phys. A **40** (2007), no. 47, 14165–14180, DOI 10.1088/1751-8113/40/47/009. MR2438118 (2009e:34075)

[JPT11] P.E.T. Jorgensen, S. Pedersen, Feng Tian, *Momentum operators in two intervals: spectra and phase transition*, `arXiv:1110.5948`.

[JPT12a] Palle Jorgensen, Steen Pedersen, and Feng Tian, *Translation representations and scattering by two intervals*, J. Math. Phys. **53** (2012), no. 5, 053505, 49. MR2964262

[JPT12b] P.E.T. Jorgensen, S. Pedersen, Feng Tian, *Spectral theory of multiple intervals*, `arXiv:1201.4120`

[KS03] Tsampikos Kottos and Uzy Smilansky, *Quantum graphs: a simple model for chaotic scattering*, J. Phys. A **36** (2003), no. 12, 3501–3524, DOI 10.1088/0305-4470/36/12/337. Random matrix theory. MR1986432 (2004g:81064)

[Ku08] Peter Kuchment, *Quantum graphs: an introduction and a brief survey*, Analysis on graphs and its applications, Proc. Sympos. Pure Math., vol. 77, Amer. Math. Soc., Providence, RI, 2008, pp. 291–312. MR2459876 (2010b:81058)

[La31] R. E. Langer, *On the zeros of exponential sums and integrals*, Bull. Amer. Math. Soc. **37** (1931), no. 4, 213–239, DOI 10.1090/S0002-9904-1931-05133-8. MR1562129

[We] Joachim Weidmann, *Linear operators in Hilbert spaces*, Graduate Texts in Mathematics, vol. 68, Springer-Verlag, New York, 1980. Translated from the German by Joseph Szücs. MR566954 (81e:47001)

Department of Theoretical Physics, Nuclear Physics Institute, Czech Academy of Sciences, 25068 Řež near Prague, and Doppler Institute for Mathematical Physics and Applied Mathematics, Czech Technical University, Břehová 7, 11519 Prague, Czechia

E-mail address: `exner@ujf.cas.cz`

Proceedings of Symposia in Pure Mathematics
Volume **87**, 2013
http://dx.doi.org/10.1090/pspum/087/01432

Asymptotic parabolicity for strongly damped wave equations

Genni Fragnelli, Gisèle Ruiz Goldstein, Jerome A. Goldstein,
and Silvia Romanelli

This paper is dedicated to Fritz Gesztesy on his 60th birthday .

ABSTRACT. For S a positive selfadjoint operator on a Hilbert space,

$$\frac{d^2u}{dt}(t) + 2F(S)\frac{du}{dt}(t) + S^2u(t) = 0$$

describes a class of wave equations with strong friction or damping if F is a positive Borel function. Under suitable hypotheses, it is shown that

$$u(t) = v(t) + w(t)$$

where v satisfies

$$2F(S)\frac{dv}{dt}(t) + S^2v(t) = 0$$

and

$$\frac{w(t)}{\|v(t)\|} \to 0, \text{ as } t \to +\infty.$$

The required initial condition $v(0)$ is given in a canonical way in terms of $u(0)$, $u'(0)$.

1. Introduction

Let S be an injective nonnegative selfadjoint operator on a complex Hilbert space $\mathcal{H}$. That is $S = S^* \geq 0$, $0 \notin \sigma_p(S)$. Consider the damped wave equation

$$u''(t) + 2Bu'(t) + S^2u(t) = 0, \quad t \geq 0, \tag{1.1}$$

with initial conditions

$$u(0) = f, \quad u'(0) = g; \tag{1.2}$$

here $' = d/dt$. When $B = 0$, (1.1) reduces to the wave equation, and the corresponding heat equation normally considered is

$$v'(t) + S^2v(t) = 0.$$

In this paper we take B to be a positive selfadjoint operator which commutes with S and is "smaller than S". More precisely, we assume that

$$0 = \inf \sigma(S), \tag{1.3}$$

2010 *Mathematics Subject Classification.* Primary 35L45, 35B40, 47B25, 34G10, 35C05, 35K25, 35K35, 35L05, 35L25, 47A60, 47D03.

Key words and phrases. Damped wave equation, asymptotic behavior, positive selfadjoint operators, (C_0) semigroups and groups of operators, strong friction, telegraph equation, strongly damped waves, asymptotic parabolicity.

i.e., 0 is the spectrum of $S = S^* \geq 0$, but is not an eigenvalue, F is a continuous function from $(0, +\infty)$ to $(0, +\infty)$, $B = F(S)$, and F satisfies: there exists $\gamma > 0$ such that

$$\begin{cases} F(x) > x & \text{for } 0 < x < \gamma, \\ F(\gamma) = \gamma, & \\ F(x) < x & \text{for } x > \gamma, \\ \limsup_{x \to 0^+} F(x) < +\infty, & \\ \liminf_{x \to +\infty} ((1-\delta)x - F(x)) \geq 0, & \text{for some } \delta > 0. \end{cases} \tag{1.4}$$

We make a further comment on (1.3). Think of S^2 as $-\Delta$ with suitable boundary conditions, acting on $H = L^2(\Omega)$ where Ω is a domain in $\mathbb{R}^N$. Then (1.3) implies that S cannot have compact resolvent, and we are thus led to work exclusively in unbounded domains Ω.

The operator B represents a general friction coefficient. The most common case is the telegraph equation in which case

$$B = aI$$

where a is a positive constant. In this case (1.4) holds with $\gamma = a$. Another simple case is

$$B = aS^\alpha$$

where the constants a, α satisfy

$$a > 0, \quad \alpha \in [0, 1).$$

In this case, $\gamma = a^{\frac{1}{1-\alpha}}$ in (1.4). The only interesting case is when S is unbounded. The *strongly damped wave equation* refers to the case when B is also unbounded.

Our main result, Theorem 3.1, can be stated informally as follows. Let $S, B = F(S)$ be as above and suppose f, g are such that (1.1), (1.2) has a unique solution u. Consider the corresponding first order (in t) equation, obtained by erasing $u''(t)$ in (1.1) and replacing u by v:

$$2Bv'(t) + S^2 v(t) = 0, \quad t \geq 0, \tag{1.5}$$

with initial condition

$$v(0) = h. \tag{1.6}$$

This vector h is given by

$$h = \frac{1}{2}\chi_{(0,\gamma)}(S)\{(F(S)^2 - S^2)^{\frac{1}{2}}(F(S)f + g) + f\},$$

a formula which will be derived and explained later; and our conclusion requires that f, g are such that $h \neq 0$. We will show that

$$l(t) := \frac{\|u(t) - v(t)\|}{\|v(t)\|} \to 0, \tag{1.7}$$

as $t \to 0$, and we will find closed subspaces $\mathbb{E}_n$ of $\mathcal{H}$ such that $\mathbb{E}_n \subset \mathbb{E}_{n+1}$,

$$\bigcup_{n=1}^{\infty} \mathbb{E}_n \text{ is dense in } \mathcal{H},$$

and

$$l(t) \leq C_n e^{-\epsilon_n t}$$

for $f, g \in \mathbb{E}_n$ where C_n, ϵ_n are positive constants. In general, there is no rate of convergence in (1.2) that works for all solutions.

The point of the theorem is that, for large times the solution of the "hyperbolic equation" (1.1) looks like the solution of the "parabolic equation" (1.5). In the telegraph equation case when $B = F(S) = aI$, (1.5) becomes

$$2av'(t) + S^2 v(t) = 0,$$

which, with (1.6), is solved by

$$v(t) = e^{-\frac{t}{2a}S^2} h.$$

In the case of strong damping, the solution of the limiting parabolic problem is

$$v(t) = e^{-\frac{t}{2}B^{-1}S^2} h.$$

Think of $S = (-\Delta)^{\frac{1}{2}}$ on $L^2(\mathbb{R}^N)$ and

$$B = aS^\alpha = a(-\Delta)^{\frac{\alpha}{2}},$$

$0 < \alpha < 1$. Then $B^{-1}S^2 = \dfrac{1}{a}(-\Delta)^{1-\frac{\alpha}{2}}$ with domain

$$D(B^{-1}S^2) = H^{2-\alpha}(\mathbb{R}^N),$$

while

$$D(S^2) = H^2(\mathbb{R}^N);$$

here we use the standard Sobolev space notation. Thus $B^{-1}S^2$ is a pseudodifferential operator of lower order $2 - \alpha$ than that of the Laplacian, unless $\alpha = 0$ in which case we have the telegraph equation.

It has long been "known" that the telegraph equation is an asymptotic approximant for the heat equation, especially in the case of $S^2 = -d^2/dx^2$ on $L^2(\mathbb{R})$. The pioneers in this area include G. I. Taylor [**12**] in 1922, S. Goldstein [**8**] in 1938 and M. Kac [**9**] in 1956.

Some of the associated random walk ideas are discussed in [**4**], which eventually led to [**1**], the main theorem in which is the special case of our main theorem here with $B = aI$. The importance of the interesting case of strong damping was recognized by Fritz Gesztesy and is discussed in detail in [**5**]. For additional results on strong damping, see [**7**].

Section 2 reviews some spectral theory. Section 3 contains the statement and the proof of our main result. The proof is patterned after that in [**1**], but it in fact is simplified and streamlined. Section 4 contains examples.

2. Selfadjoint and normal operators

Let S be a selfadjoint operator on $\mathcal{H}$ with spectrum $\sigma(S)$. By the spectral Theorem there exists an L^2 space $L^2(\Lambda, \Sigma, \nu)$ and a unitary operator

$$W : \mathcal{H} \to L^2(\Lambda, \Sigma, \nu)$$

such that S is unitarily equivalent, via W, to the maximally defined operator of multiplication by a $\Sigma-$measurable function

$$m : \Lambda \to \sigma(S) \subset \mathbb{R},$$

i.e.,

$$Sf = W^{-1} M_m W f,$$

for

$$f \in D(S) = \{W^{-1}g \in \mathcal{H} : \; mg \in L^2(\Lambda, \Sigma, \nu)\}$$

and

$$(M_m g)(x) = m(x)g(x), \quad \text{for } x \in \Lambda, g \in L^2(\Lambda, \Sigma, \nu).$$

Two selfadjoint operators S_1, S_2 *commute* if and only if the bounded operators

$$(\lambda_1 I - S_1)^{-1}, (\lambda_2 I - S_2)^{-1}$$

commute for all $\lambda_1, \lambda_2 \in \mathbb{C} \setminus \mathbb{R}$ if and only if

$$e^{itS_1} \quad \text{and} \quad e^{isS_2}$$

commute for all $t, s \in \mathbb{R}$. Similarly, two normal operators N_1, N_2 with

$$\sup \operatorname{Re} \sigma(N_j) < +\infty, \quad j = 1, 2$$

commute if and only if e^{tN_1} and e^{sN_2} commute for all $t, s \geq 0$; here N is normal means $N = S_1 + iS_2$ where S_1, S_2 are commuting selfadjoint operators.

The functional calculus for S selfadjoint says that for every Borel measurable function F from $\sigma(S) \subset \mathbb{R}$ to $\mathbb{C}$, $F(S)$ defined by

$$F(S) = W^{-1} M_{F(m)} W$$

is normal, and any two of these operators commute. Moreover,

$$F \to F(S)$$

is linear and is an algebra homomorphism, thus

$$F_1(S)F_2(S) = (F_1 F_2)(S),$$

etc. Also, $F(S)$ is bounded on $\mathcal{H}$ if and only if F is bounded on $\sigma(S)$, and $F(S)$ is selfadjoint if and only if F is real valued. And for $S = S^*$, $F(S)$ is semibounded (above or below) if and only if $F(\sigma(S))$ is, in $\mathbb{R}$.

In particular, for Γ a Borel set in $[0, +\infty)$, $P_\Gamma = \chi_\Gamma(S)$ is the orthogonal projection of $\mathcal{H}$ onto $\chi_\Gamma(S)(\mathcal{H})$; and

$$P_\Gamma F(S) = F(S) P_\Gamma = P_\Gamma F(S P_\Gamma)$$

is the part of $F(S)$ in Γ, and its spectrum is contained in Γ.

If F_1, F_2 are complex Borel functions on $\sigma(S)$ that are bounded above, it follows that $F_j(S)$ and $\Sigma_{k=1}^n F_k(S)$ generate (C_0) semigroups on $\mathcal{H}$ and

$$e^{t\Sigma_{k=1}^n F_k(S)} = \Pi_{k=1}^n e^{tF_k(S)}, \tag{2.1}$$

and the product can be taken in any order. Finally, if $L = F(S) = L^* \geq 0$, then $[L]^{\frac{1}{2}}$ denotes the unique nonnegative square root of L.

For more on the spectral theorem, see the books [**6**], [**10**], [**11**].

3. The main result

Consider the problem (1.1), (1.2), which we rewrite as

$$u'' + 2Bu' + S^2 u = 0, \quad t \geq 0, \tag{3.1}$$

$$u(0) = f, u'(0) = g, \tag{3.2}$$

where $S = S^* \geq 0$ on $\mathcal{H}$,

$$\inf \sigma(S) = 0 \notin \sigma_p(S), \tag{3.3}$$

$B = F(S)$ where F is a continuous function from $(0,+\infty)$ to $(0,+\infty)$ which is bounded near zero and strictly less than the identity function near infinity, in the sense that for some $\delta > 0$,

$$\liminf_{x\to+\infty}((1-\delta)x - F(x)) \geq 0. \tag{3.4}$$

We also assume there exists $\gamma > 0$ such that (1.4) holds, namely

$$\begin{cases} F(x) > x & \text{for } 0 < x < \gamma, \\ F(\gamma) = \gamma, \\ F(x) < x & \text{for } x > \gamma, \\ \limsup_{x\to 0^+} F(x) < +\infty. \end{cases} \tag{3.5}$$

Let Γ be the open interval $(0,\gamma)$ and let

$$P_\Gamma = \chi_\Gamma(S). \tag{3.6}$$

THEOREM 3.1. *Assume all the statements in the above paragraph. Let v be the solution of the corresponding "parabolic" equation*

$$2Bv' + S^2 v = 0, \tag{3.7}$$

obtained by deleting the second derivative term in (3.1). *Let*

$$v(0) = h := \frac{1}{2}P_\Gamma(f + [(B^2 - S^2)P_\Gamma]^{\frac{1}{2}}(Bf + g)). \tag{3.8}$$

Then, for u the solution of (3.1), (3.2),

$$u(t) = v(t)(1 + o(t)) \tag{3.9}$$

holds as $t \to +\infty$, provided $h \neq 0$. Moreover, if $\Gamma_n = \left[\frac{1}{n}, \delta - \frac{1}{n}\right]$ and if $0 \neq h \in P_{\Gamma_n}(\mathcal{H})$ for some $n \in \mathbb{N}$, then

$$u(t) = v(t)(1 + o(e^{-\epsilon_n t})) \tag{3.10}$$

for some $\epsilon_n > 0$.

REMARK 3.2. Note that

$$P_{\Gamma_n}(\mathcal{H}) \subset P_{\Gamma_{n+1}}(\mathcal{H}),$$

and $\bigcup_{n=1}^{\infty} P_{\Gamma_n}(\mathcal{H})$ is dense in $P_\Gamma(\mathcal{H})$.

PROOF. First recall that the square root of $(B^2 - S^2)P_\Gamma$ refers to the non-negative square root. We first show that the problem (3.1), (3.2) is wellposed, by showing that it is governed by a (C_0) contraction semigroup.

We first treat the case of $B = 0$ in (3.1).

Rewrite (3.1), (3.2) as, for $U = \begin{pmatrix} Su \\ u' \end{pmatrix}$,

$$\begin{gathered} U' = \begin{pmatrix} Su' \\ u'' \end{pmatrix} = \begin{pmatrix} 0 & S \\ -S & 0 \end{pmatrix}\begin{pmatrix} Su \\ u' \end{pmatrix} = GU, \\ U(0) = L = \begin{pmatrix} Sf \\ g \end{pmatrix}. \end{gathered} \tag{3.11}$$

Let $\mathcal{K}$ be the completion of $D(S) \oplus D(S)$ in the norm

$$\left\| \begin{pmatrix} m \\ n \end{pmatrix} \right\|_{\mathcal{K}} = (\|m\|^2 + \|n\|^2)^{\frac{1}{2}},$$

where $\|\cdot\|$ is the norm on $\mathcal{H}$. Then G defined by (3.11) is skewadjoint on $\mathcal{K}$ and thus generates a (C_0) unitary group by Stone's Theorem (cf. [**6**], [**10**], [**11**]). We consider this as a semigroup since we are only concerned with times $t \geq 0$. Next, (3.1) can be written in $\mathcal{K}$ as

$$U' = (G + P)U$$

where

$$P = \begin{pmatrix} 0 & 0 \\ 0 & -2B \end{pmatrix}.$$

Since $B = B^* \geq 0$,

$$\begin{aligned} \left\| P \begin{pmatrix} m \\ n \end{pmatrix} \right\|_{\mathcal{K}} &= \|2Bn\| \leq (1-\epsilon)\|Sn\| + C_\epsilon \|n\| \\ &\leq (1-\epsilon) \left\| P \begin{pmatrix} m \\ n \end{pmatrix} \right\|_{\mathcal{K}} + C_\epsilon \left\| \begin{pmatrix} m \\ n \end{pmatrix} \right\|_{\mathcal{K}} \end{aligned}$$

for some $\epsilon > 0$ and a corresponding $C_\epsilon > 0$, for all $n \in D(S) \subset D(B)$, thanks to (3.4) and the last line in (3.5) (or (1.4)). Namely, write $B = B_1 + B_2 := BP_{(0,M)} + BP_{[M,\infty)}$, where M is such that

$$x \geq F(x) + \delta x,$$

i.e.

$$F(x) \leq (1-\delta)x$$

for $x \geq M$ and $F(x)$ is bounded in $[0, M]$. Thus B_1 is bounded, $B_2 = B_2^* \geq 0$ and

$$\|B_2 n\| \leq (1-\delta)\|Sn\| + \|B_2\|\|n\|$$

for all $n \in D(S)$. Thus for $N = \begin{pmatrix} m \\ n \end{pmatrix}$,

$$\|PN\|_{\mathcal{K}} \leq (1-\delta)\|GN\|_{\mathcal{K}} + M\|N\|_{\mathcal{K}}$$

where $\delta > 0$ and $M = \|B_2\|$. It follows (cf. e.g. [**6**], [**10**], [**11**]) that $G + P$ is m−dissipative and generates a (C_0) contraction semigroup on $\mathcal{K}$, since P is obviously dissipative. Then (3.1), (3.2) has a unique strongly C^2 solution (resp. mild solution) if $f \in D(S^2), g \in D(S)$ (resp. $f \in D(S), g \in \mathcal{H}$): cf. [**6**, Chapter 2, Theorem 7.8].

We shall express the unique solution using d'Alembert's formula. We seek solution of the form

$$u(t) = e^{tC}h$$

where C is a Borel function of S. By (3.1), C must satisfy

$$C^2 + 2BC + S^2 = 0.$$

Formally,

$$C = C_\pm = -B \pm (B^2 - S^2)^{\frac{1}{2}}.$$

Selfadjoint operators have many square roots, but nonnegative selfadjoint operators have unique nonnegative square roots. Thus we uniquely define $C_\pm$ by

$$C_\pm = -B \pm (Q_0 + iQ) \tag{3.12}$$

where

$$Q_0 = [(B^2 - S^2)\chi_{(0,\gamma)}(S)]^{\frac{1}{2}}, Q = [(S^2 - B^2)|\chi_{[\gamma,+\infty)}(S)]^{\frac{1}{2}}. \tag{3.13}$$

Thus the solution u of (3.1), (3.2) can be written as

$$u(t) = e^{tC_+}h_+ + e^{tC_-}h_-,$$

where $C_\pm$ are defined by (3.12), (3.13). There are strong C^2 solutions (resp. mild solutions) if and only if $h_\pm \in D(S^2)$ (resp. $h_\pm \in D(S)$).

Given $f = u(0), g = u'(0)$, we obtain $h_\pm$ by inverting the 2×2 system

$$f = h_+ + h_-$$
$$g = C_+h_+ + C_-h_-.$$

An elementary calculation gives

$$h_- = \frac{1}{2}(f - (Q_0 + iQ)^{-1}(Bf + g)) \tag{3.14}$$

$$h_+ = \frac{1}{2}(f + (Q_0 + iQ)^{-1}(Bf + g)). \tag{3.15}$$

Write

$$u = u_1 + u_2 + u_3$$

where

$$u_1(t) = e^{tC_+}P_{(0,\gamma)}h_+,$$
$$u_2(t) = e^{tC_+}P_{[\gamma,+\infty)}h_+,$$
$$u_3(t) = e^{tC_-}h_-.$$

First,

$$\|u_3(t)\| = \|e^{-itQ}e^{-tQ_0}e^{-tB}h_-\| \leq \|e^{-tB}h_-\|$$

since e^{-itQ} is unitary and $\|e^{-tQ_0}\| \leq 1$. Next,

$$\|u_2(t)\| = \|e^{itQ}e^{-tB}P_{[\gamma,+\infty)}h_+\| \leq \|e^{-tB}h_+\|.$$

The next estimate is the key one.

For

$$h := P_{(0,\gamma)}(h_+), \tag{3.16}$$

$$\|u_1(t)\| = \|e^{t(-B+Q_0)}h_+\|.$$

We know that $h \in P_{(0,\gamma)}(\mathcal{H})$: assume

$$0 \neq h \in P_{[\delta,\gamma-\delta]}(\mathcal{H}) =: \mathcal{H}_\delta \tag{3.17}$$

for some $\delta > 0$. Let

$$Q_{0\delta} = Q_0P_{[\delta,\gamma-\delta]}.$$

Then, since $F(x) > x$ on $[\delta, \gamma - \delta]$, $F(x) - x \geq \epsilon$ on $[\delta, \gamma - \delta]$ for some $\epsilon > 0$. Thus

$$Q_{0\delta} \geq \epsilon I.$$

Consequently

$$\|u_1(t)\| \geq e^{\epsilon t}\|e^{-tB}h\|.$$

It follows that for some constant C_0,

$$\|u_2(t)\| + \|u_3(t)\| \leq C_0e^{-\epsilon t}\|u_1(t)\|.$$

Thus

$$u(t) = u_1(t)(1 + O(e^{-\epsilon t})). \tag{3.18}$$

We must show that this holds with u_1 replaced by v.

The unique solution of (3.7) is

$$v(t) = e^{-\frac{t}{2}B^{-1}S^2}h.$$

Note that h as defined by (3.8) is $P_\Gamma h_+$ where h_+ is as in (3.15). To compare u_1 with v, we need Taylor's formula with integral remainder, which for $g \in C^3[0,l]$ and for some $l > 0$ says that

$$g(x) = g(0) + g'(0)x + \frac{g''(0)}{2}x^2 + \frac{1}{2}\int_0^x (x-y)^2 g'''(y)dy.$$

Applying this to

$$g(x) = 1 - (1-x)^{\frac{1}{2}}, \quad 0 < x < 1, \tag{3.19}$$

yields

$$g(L)f = \frac{1}{2}Lf + \frac{1}{8}L^2 f + Rf \tag{3.20}$$

where R is a bounded operator commuting with L and satisfying

$$R = R^* \geq 0.$$

Consequently

$$\begin{aligned} \|u_1(t) - v(t)\| &= \|e^{-tB(-I+B^{-1}Q_0)}h - e^{-\frac{t}{2}B^{-1}S^2}h\| \\ &= \|e^{-tB[I-(B^{-2}Q_0^2)^{\frac{1}{2}}]}h - e^{-\frac{t}{2}B^{-1}S^2}h\| \\ &= \|e^{-\frac{t}{2}B^{-1}S^2}\{e^{-\frac{t}{8}L^2}e^{-tR} - I\}h\| \end{aligned} \tag{3.21}$$

by (3.19), (3.20) with $L = B^{-2}(B^2 - S^2)P_{(0,\gamma)} = (I - B^{-2}S^{-2})P_{(0,\gamma)}$.

We have

$$\zeta_1 I \leq R \leq \zeta_2 I$$

on $\mathcal{H}_\delta$ for some constants $0 < \zeta_1 < \zeta_2 < +\infty$. Furthermore, we also have

$$\zeta_3 I \leq L \leq \zeta_4 I$$

on $\mathcal{H}_\delta$ for some positive constants ζ_3, ζ_4. It now follows from (3.20) that

$$\|u_1(t) - v(t)\| = \|e^{-\frac{t}{2}B^{-1}S^2}(I - e^{-\frac{t}{8}L}e^{-tR}h)\|$$

and

$$\|e^{-\frac{t}{8}L}e^{-tR}h\| \leq e^{-t\zeta_5}\|h\|$$

where

$$\zeta_5 = \frac{\zeta_3}{8} + \zeta_1 > 0.$$

Consequently

$$\|u(t) - v(t)\| \leq \|v(t)\| O(e^{-t\zeta_5}).$$

Combining this inequality with (3.18) yields the desired asymptotic relation

$$\frac{\|u(t) - v(t)\|}{\|v(t)\|} = o(e^{-t\epsilon_\delta})$$

for some $\epsilon_\delta > 0$.

Now let $0 \neq h \in P_{(0,\gamma)}(\mathcal{H})$. We must show that

$$\frac{\|u(t) - v(t)\|}{\|v(t)\|} \to 0, \quad \text{and } t \to +\infty. \tag{3.22}$$

We proceed by contradiction. Suppose (3.22) fails to hold for some $h \neq 0$ in $P_{(0,\gamma)}(\mathcal{H})$. Then, there exists $\epsilon_1 > 0$ and $t_n \to +\infty$ such that

$$\frac{\|u(t_n) - v(t_n)\|}{\|v(t_n)\|} \geq \epsilon_1 \tag{3.23}$$

for all $n \in \mathbb{N}$. Choose $\delta > 0$ and $\tilde{h} \in \mathcal{H}_\delta = P_{[\delta,\gamma-\delta]}(\mathcal{H})$ (depending on ϵ_1) such that

$$\|h - \tilde{h}\| < \frac{\epsilon_1}{4}$$

and let $\tilde{f}$, $\tilde{g}$ be the corresponding initial data. Note that

$$P_{[\gamma,+\infty)} l = P_{[\gamma,+\infty)} \tilde{l}$$

for $l = f, g$, and f and g are modified only on the subspace $P_\Lambda(\mathcal{H})$

$$\Lambda := [\delta - \delta_1, \delta + \delta_1] \cup [\gamma - \delta - \delta_1, \gamma - \delta + \delta_1],$$

for some $\delta_1 > 0$ which can be chosen to be arbitrarily small. In particular, given $\epsilon_2 > 0$ we may choose $\tilde{f}, \tilde{g}$ as above and additionally satisfying

$$\|f - \tilde{f}\| + \|g - \tilde{g}\| < \epsilon_2,$$

$$\frac{\|\tilde{h}\|}{\|h\|} \in [1 - \epsilon_2, 1 + \epsilon_2].$$

It follows that

$$\|u(t) - \tilde{u}(t)\|,\ \|v(t) - \tilde{v}(t)\| < \frac{\epsilon_1}{4}$$

for all $t > 0$. Consequently

$$\begin{aligned} \frac{\|u(t) - v(t)\|}{\|v(t)\|} &\leq \frac{\|\tilde{u}(t) - \tilde{v}(t)\|}{\|\tilde{v}(t)\|}\left(\frac{1+\epsilon_2}{1-\epsilon_2}\right) + \frac{\epsilon_1}{4} \\ &\leq \tau_0 e^{-\epsilon_3 t}\left(\frac{1+\epsilon_2}{1-\epsilon_2}\right) + \frac{\epsilon_1}{4} \to \frac{\epsilon_1}{4}, \end{aligned} \tag{3.24}$$

as $t \to +\infty$, since $0 \neq \tilde{h} \in \mathcal{H}_\delta$, and τ_0, ϵ_3 are positive constants depending on δ. But (3.24) contradicts (3.23) for $t = t_n$ with n large enough. It follows that (3.22) holds. This completes the proof of Theorem 3.1. □

We remark that, in general, there does not exist a rate of convergence in (3.22) which works for all $0 \neq h \in \mathcal{H}_\gamma$. This follows from a careful examination of [**1**] and [**2**, Section 3].

4. Examples

Example 4.1. This is the simplest example. Take

$$S^2 := -\Delta + w^2 I \quad \text{on } L^2(\mathbb{R}^N)$$

for $w > 0$. Similarly we can define

$$S_k^2 := S^{2k} = (-\Delta + w^2 I)^k \quad \text{on } L^2(\mathbb{R}^N),\ k \in \mathbb{N}.$$

Theorem 3.1 applies (with $k \in \mathbb{N}$, $w > 0$ fixed) if we take

$$B = a S_k^\alpha = a S^{\alpha k}$$

for $a > 0$, $\alpha \in [0, 1)$. The resulting damped wave equation is

$$u_{tt} + 2a(-\Delta + w^2)^{\frac{\alpha k}{2}} u_t + (-\Delta + w^2)^k u = 0. \tag{4.1}$$

This is a pseudo differential equation unless $\frac{\alpha k}{2} \in \mathbb{N}_0 = \{0, 1, 2, ...\}$, in which case it is a partial differential equation. The simplest case of this is $k = 3$, $\alpha = \frac{2}{3}$, in which case (4.1) reduces to

$$u_{tt} + 2a(-\Delta u_t + w^2 u_t) + (-\Delta + w^2)^3 u = 0.$$

As noted earlier, γ is given by

$$\gamma = a^{\frac{1}{1-\alpha}}$$

in the case of (4.1), for all $k \in \mathbb{N}$.

Example 4.2. We start with a brief summary of the "Wentzell Laplacian" discussed in [**2**] (cf. also [**3**]). Let Ω be an unbounded domain in $\mathbb{R}^N$ with nonempty boundary $\partial\Omega$, such that for every $R > 0$ there exists a ball $B(x_R, R)$ in Ω. Let $A(x)$ be an $N \times N$ matrix for $x \in \overline{\Omega}$ such that $A(x)$ is real, Hermitian and

$$\alpha_1 |\xi|^2 \leq A(x)\xi \cdot \xi \leq \alpha_2 |\xi|^2 \tag{4.2}$$

for all $x \in \overline{\Omega}, \xi \in \mathbb{R}^N$, where

$$0 < \alpha_1 \leq \alpha_2 < \infty$$

are constants. Similarly let $B(x)$ for $x \in \partial\Omega$ be a real Hermitian $(N-1) \times (N-1)$ matrix satisfying (4.2) with the same α_1, α_2. Assume $\partial\Omega, A, B$, and γ, β (introduced below) are sufficiently smooth. Define distributional differential operators on Ω (resp. $\partial\Omega$) by

$$Lu_1 = \nabla \cdot (A(x)\nabla u_1),$$
$$L_\partial u_2 = \nabla_\tau \cdot (B(x)\nabla_\tau u_2)$$

for u_1 (resp. u_2) defined on Ω (resp. $\partial\Omega$). Here ∇_τ is the tangential gradient on $\partial\Omega$. The wave equation (without damping) we consider is

$$u_{tt} = Lu \quad \text{in } \Omega, \tag{4.3}$$

$$Lu + \beta\partial_\nu^A u + \gamma u - q\beta L_\partial u = 0 \quad \text{on } \partial\Omega. \tag{4.4}$$

Here the conormal derivative term is

$$\partial_\nu^A u = (A\nabla u) \cdot \nu$$

at $x \in \partial\Omega$, where ν is the unit outer normal to $\partial\Omega$ at x; $\beta, \gamma \in C^1(\partial\Omega, \mathbb{R}), \beta > 0, \gamma \geq 0$, β, $\frac{1}{\beta}$ and γ are bounded, and $q \in [0, +\infty)$.

In [**2**], it is explained how the problem (4.3), (4.4) can be rewritten as

$$u'' + S^2 u = 0$$

$$u(0) = f, \quad u'(0) = g$$

where the Hilbert space is

$$\mathcal{H} = L^2(\Omega, dx) \oplus L^2\left(\partial\Omega, \frac{d\Gamma}{\beta}\right).$$

Vectors in $\mathcal{H}$ are represented by $u = \begin{pmatrix} u_1 \\ u_2 \end{pmatrix}$ with $u_1 \in L^2(\Omega, dx)$ and $u_2 \in L^2\left(\partial\Omega, \frac{d\Gamma}{\beta}\right)$. The norm in $\mathcal{H}$ is given by

$$\|u\|_{\mathcal{H}} = \{\|u_1\|^2_{L^2(\Omega)} + \|u_2\|^2_{L^2(\partial\Omega, \frac{d\Gamma}{\beta})}\}^{\frac{1}{2}}.$$

Here we write $d\Gamma$ rather than the usual dS for the element of surface measure, since the letter S already is being used to denote the basic operator.

The operator S^2 has the matrix representation

$$S^2 = \begin{pmatrix} -L & 0 \\ \beta\partial_\nu^A & \gamma - q\beta L_\partial \end{pmatrix}.$$

In [**2**] it was shown that $S = [S^2]^{\frac{1}{2}}$, with a suitable domain, satisfies

$$S = S^* \geq 0,\ 0 = \inf \sigma(S),\ 0 \notin \sigma_p(S).$$

Furthermore, for all $u \in D(S)$, we have $u = \begin{pmatrix} u_1 \\ u_2 \end{pmatrix}$, where $u_2 = \mathrm{tr}(u_1)$, the trace of u_1. As in Example 4.1, Theorem 3.1 applies to

$$u'' + 2aS^{\frac{\alpha k}{2}}u' + S^{2k}u = 0, \tag{4.5}$$
$$u(0) = f,\ u'(0) = g, \quad k \in \mathbb{N}.$$

Again, this is a partial differential equation only when $\frac{\alpha k}{2} \in \mathbb{N}$. If $k = 3$ and $\alpha = \frac{2}{3}$, the corresponding parabolic problem is

$$v' + \frac{1}{2a}S^4 v = 0, \quad v(0) = h.$$

The boundary conditions associated with (4.5) are

$$Lw + \beta\partial_\nu^A w + \gamma w - qL_\partial w = 0 \quad \text{on } \partial\Omega$$

for $w = S^{2j}u,\ j = 0, 1, ..., k-1$.

Example 4.3 The simplest example of unidirectional waves in one dimension are described by the equation (for $t, x \in \mathbb{R}$)

$$u_t = cu_x + bu_{xxx} =: Mu, \tag{4.6}$$

where $(b, c) \in \mathbb{R}^2 \setminus \{(0,0)\}$. The most common case is $c \neq 0, b = 0$, in which case the corresponding equation for bidirectional waves is

$$\left(\frac{\partial}{\partial t} - M\right)\left(\frac{\partial}{\partial t} + M\right)u = u_{tt} - c^2 u_{xx} = 0.$$

The case of $b \neq 0$ is the Airy equation, and (4.6) is the linearization of the KdV equation

$$u_t = cu_x + bu_{xxx} + c_1 u u_x.$$

For $c = 0 \neq b$, the bidirectional version of (4.6) is

$$\left(\frac{\partial}{\partial t} - M\right)\left(\frac{\partial}{\partial t} + M\right)u = u_{tt} - b^2 u_{xxxxxx} = 0.$$

Now, let $\mathcal{H} = L^2(\mathbb{R})$, $D = \frac{d}{dx}$ and $T = -D^2 = T^* \geq 0$. Let

$$S^2 = T^3 + a_0 T^2 + a_1 T,$$

where $a_0,\ a_1 \in [0, +\infty)$. Consider

$$u_{tt} - 2au_{xxt} - u_{xxxxxx} + a_0 u_{xxxx} - a_1 u_{xx} = 0,$$
$$u(x, 0) = f(x),\ u_t(x, 0) = g(x).$$

In this case,

$$B = aT = F(S) = F(T^3 + a_0 T^2 + a_1 T).$$

For $x > 0$, we want to consider the function

$$G(x) = \frac{1}{a}(x^3 + a_0 x^2 + a_1 x),$$

so that $G(x) - x$ is negative in $(0, \gamma)$ and positive on (γ, ∞) for some $\gamma > 0$. But

$$\frac{d}{dx}\left(\frac{G(x) - x}{x}\right) = \frac{d}{dx}\left(\frac{1}{a}(x^2 + a_0 x + (a_1 - a))\right) = \frac{1}{a}(2x + a_0) > 0,$$

and $G(x) = x$ for $x \neq 0$ if and only if

$$x = \frac{1}{2}(-a_0 \pm \sqrt{a_0^2 - 4(a_1 - a)}\ ).$$

Thus we get exactly one positive root if and only if

$$a > a_1 + \frac{a_0^2}{4}, \tag{4.7}$$

which we assume. It is now elementary to check that $B = F(S)$ and F satisfies the assumptions of Theorem 3.1. In this case

$$\gamma = \frac{1}{2}(-a_0 + \sqrt{a_0^2 - 4(a_1 - a)}\ ).$$

References

[1] Ted Clarke, Eugene C. Eckstein, and Jerome A. Goldstein, *Asymptotic analysis of the abstract telegraph equation*, Differential Integral Equations **21** (2008), no. 5-6, 433–442. MR2483262 (2010b:35310)

[2] Ted Clarke, Gisèle Ruiz Goldstein, Jerome A. Goldstein, and Silvia Romanelli, *The Wentzell telegraph equation: asymptotics and continuous dependence on the boundary conditions*, Commun. Appl. Anal. **15** (2011), no. 2-4, 313–324. MR2867354

[3] Giuseppe M. Coclite, Angelo Favini, Ciprian G. Gal, Gisèle Ruiz Goldstein, Jerome A. Goldstein, Enrico Obrecht, and Silvia Romanelli, *The role of Wentzell boundary conditions in linear and nonlinear analysis*, Advances in nonlinear analysis: theory, methods and applications, Math. Probl. Eng. Aerosp. Sci., vol. 3, Camb. Sci. Publ., Cambridge, 2009, pp. 279–291. MR2777403

[4] Eugene C. Eckstein, Jerome A. Goldstein, and Mark Leggas, *The mathematics of suspensions: Kac walks and asymptotic analyticity*, Proceedings of the Fourth Mississippi State Conference on Difference Equations and Computational Simulations (1999), Electron. J. Differ. Equ. Conf., vol. 3, Southwest Texas State Univ., San Marcos, TX, 2000, 39–50. MR1784255 (2001h:76099)

[5] Fritz Gesztesy, Jerome A. Goldstein, Helge Holden, and Gerald Teschl, *Abstract wave equations and associated Dirac-type operators*, Ann. Mat. Pura Appl. **191** (2012), no. 4, 631–676, DOI 10.1007/s10231-011-0200-7. MR2993967

[6] Jerome A. Goldstein, *Semigroups of linear operators and applications*, Oxford Mathematical Monographs, The Clarendon Press Oxford University Press, New York, 1985. MR790497 (87c:47056)

[7] Jerome A. Goldstein, Guillermo Reyes, *Asymptotic equipartition of operator weighted energy in damped wave equations*, Asymptotic Anal. 81 (2013), 171–187.

[8] Sidney Goldstein, *On diffusion by discontinuous movements, and on the telegraph equation*, Quart. J. Mech. Appl. Math. **4** (1951), 129–156. MR0047963 (13,960b)

[9] Mark Kac, *A stochastic model related to the telegrapher's equation*, Rocky Mountain J. Math. **4** (1974), 497–509. Reprinting of an article published in 1956; Papers arising from a Conference on Stochastic Differential Equations (Univ. Alberta, Edmonton, Alta., 1972). MR0510166 (58 #23185)

[10] Tosio Kato, *Perturbation theory for linear operators*, 2nd ed., Springer-Verlag, Berlin, 1976. Grundlehren der Mathematischen Wissenschaften, Band 132. MR0407617 (53 #11389)

[11] Peter D. Lax, *Functional analysis*, Pure and Applied Mathematics (New York), Wiley-Interscience [John Wiley & Sons], New York, 2002. MR1892228 (2003a:47001)

[12] Geoffrey I. Taylor, *Diffusion by continuous movements*, Proc. London Math. Soc. (1922), **S2-20**, no. 1, 196–212, DOI 10.1112/plms/s2-20.1.196. MR1577363

Dipartimento di Matematica, Università di Bari "Aldo Moro", Via E. Orabona 4, 70125 Bari - Italy
E-mail address: genni.fragnelli@uniba.it

Department of Mathematical Sciences, University of Memphis, Memphis, Tennessee 38152
E-mail address: ggoldste@memphis.edu

Department of Mathematical Sciences, University of Memphis, Memphis, Tennessee 38152
E-mail address: jgoldste@memphis.edu

Dipartimento di Matematica, Università di Bari "Aldo Moro", Via E. Orabona 4, 70125 Bari, Italy
E-mail address: silvia.romanelli@uniba.it

Proceedings of Symposia in Pure Mathematics
Volume **87**, 2013
http://dx.doi.org/10.1090/pspum/087/01453

On rates in Euler's formula for C_0-semigroups

Alexander Gomilko and Yuri Tomilov

To Fritz Gesztesy on the occasion of his sixtieth anniversary, with admiration

ABSTRACT. By functional calculus methods, we obtain optimal convergence rates in Euler's approximation formula for C_0-semigroups restricted to ranges of generalized Stieltjes functions. Our results include a number of partial cases studied in the literature and cannot essentially be improved.

1. Introduction

Let $-A$ be the generator of a bounded C_0-semigroup $(e^{-tA})_{t\geq 0}$ on a Banach space X. Then the abstract Cauchy problem

$$\begin{cases} x'(t) = -Ax(t), & t \geq 0, \\ x(0) = x_0, & x_0 \in X, \end{cases} \tag{1.1}$$

is well-posed and all its mild solutions are given by the formula

$$x(t) = e^{-tA}x_0, \qquad t \geq 0, \quad x_0 \in X.$$

However, even if A is bounded, the exponential function e^{-tA} can hardly be given in an explicit form. Thus it is of importance for applications to find approximation formulas for e^{-tA} amenable for the purposes of numerical analysis, e.g. formulas involving rational functions of A. Starting from the pioneering works of Hersh and Kato [**10**] and P. Brenner and V. Thomée [**2**], the methods of Hille-Phillips functional calculus have played an important role in the theory of rational approximations of C_0-semigroups, see e.g. [**15**, Introduction and Chapter 1] for a survey.

In this paper, we extend further the functional calculus approach by replacing the "conventional" Hille-Phillips functional calculus by the extended Hille-Phillips functional calculus and then restricting ourselves to the important part of the extended Hille-Phillips calculus given by generalized Stieltjes functions. This approach proved to be quite successful in dealing with rates in mean ergodic theorems for continuous and discrete operator semigroups, see [**6**], [**7**] and [**8**]. To demonstrate the power of our approach we consider the simplest semigroup approximation known as Euler's exponential formula or the Post-Widder inversion formula. The

---

2010 *Mathematics Subject Classification.* Primary 47A60, 65J08, 47D03; Secondary 46N40, 65M12.

Key words and phrases. Euler approximation, C_0-semigroup, functional calculus, convergence rate.

This work was completed with the support of the NCN grant DEC-2011/03/B/ST1/00407.

approximation arises when the abstract Cauchy problem (1.1) is time-discretized by the so-called Euler backward method and it can be defined as

$$E_{n,t}(A)x := \left(1+\frac{t}{n}A\right)^{-n} x, \qquad x \in X, \quad n \in \mathbb{N}, \quad t>0.$$

It is well known that for every $x \in X$,

$$e^{-tA}x = \lim_{n\to\infty} E_{n,t}(A)x \tag{1.2}$$

uniformly for t in compact intervals of positive semi-axis. Thus a natural question is whether it is possible to quantify the convergence in (1.2). It is easy show that, in general, there is no rate of convergence in (1.2) uniform with respect to all elements $x \in X$. However such a rate does exist when the elements are taken from the domain of an appropriate function of A, e.g. a power function. The next theorem surveys known results on the rates of convergence of Euler's formula in this case. Denote

$$\Delta_{n,t}(A) := E_{n,t}(A) - e^{-tA}, \qquad n \in \mathbb{N}, \quad t>0.$$

THEOREM 1.1. *Let $-A$ be the generator of a bounded C_0-semigroup $(e^{-tA})_{t\ge 0}$ on a Banach space X.*

(i) [**2**, Theorem 4] *There exists $c>0$ such that for any $n \in \mathbb{N}$ and $t>0$,*

$$\|\Delta_{n,t}(A)x\| \le c\left(\frac{t}{\sqrt{n}}\right)^2 \|A^2x\|, \qquad x \in \mathrm{dom}\,(A^2);$$

(ii) [**4**, Theorem 1.7] *There exists $c>0$ such that for any $n \in \mathbb{N}$ and $t>0$,*

$$\|\Delta_{n,t}(A)x\| \le c\frac{t}{\sqrt{n}} \|Ax\|, \qquad x \in \mathrm{dom}\,(A);$$

(iii) [**16**, Corollary 4.4] *There exists $c>0$ such that for any $n \in \mathbb{N}$, $t>0$ and $0<\alpha\le 2$,*

$$\|\Delta_{n,t}(A)x\| \le c\left(\frac{t}{\sqrt{n}}\right)^\alpha \|x\|_{\alpha,2,\infty}, \quad t\ge 0, \quad n\in\mathbb{N}, \quad x \in X_{\alpha,2,\infty}, \tag{1.3}$$

where the Banach space $X_{\alpha,2,\infty}$ (called a Favard space) is defined as

$$X_{\alpha,2,\infty} := \left\{x \in X : \|x\|_{\alpha,2,\infty} := \|x\| + \sup_{t>0} \frac{\|(e^{-tA}-I)^2x\|}{t^\alpha} < \infty\right\}.$$

Some comments concerning the last result are in order. Note that by [**14**, Theorem 4.3] (see also [**18**, Theorem 11.3.5]) if $\alpha \in (0,2)$ then $X_{\alpha,2,\infty}$ coincides with Komatsu's (Banach) space $D^\alpha_\infty = D^\alpha_{\infty,2}$,

$$D^\alpha_\infty := \left\{x \in X : \|x\|_{D^\alpha_\infty} := \|x\| + \sup_{\lambda>0} \lambda^\alpha\|[A(A+\lambda)^{-1}]^2x\| < \infty\right\}, \tag{1.4}$$

in the sense that $X_{\alpha,2,\infty} = D^\alpha_\infty$ as sets and the norms are equivalent. On the other hand, $\mathrm{dom}\,(A^2) \subset X_{2,2,\infty}$ and, by [**14**, Proposition 2.8], $\mathrm{dom}\,(A^\alpha)$ is embedded continuously in D^α_∞, $\alpha \in (0,2)$. However, there are examples (see e.g. [**13**, p. 340]) showing the the inclusion $\mathrm{dom}\,(A^\alpha) \subset D^\alpha_\infty$ is in general strict.

Thus, (1.3) implies that there exists $c>0$ such that for any $\alpha \in (0,2]$, $n \in \mathbb{N}$, and $t>0$,

$$\|\Delta_{n,t}(A)x\| \le c\left(\frac{t}{\sqrt{n}}\right)^\alpha \|x\|_{\mathrm{dom}\,(A^\alpha)}, \quad x \in \mathrm{dom}\,(A^\alpha), \tag{1.5}$$

where

$$\|x\|_{\mathrm{dom}\,(A^\alpha)} := \|x\| + \|A^\alpha x\|, \qquad x \in \mathrm{dom}\,(A^\alpha).$$

(Remark that [**16**, Corollary 4.4] only states that (1.3) implies (1.5) for $\alpha = 1, 2$.)

We should also emphasize that the results mentioned in Theorem 1.1 are in fact partial cases of more general statements on convergence rates for A-stable and stable rational approximations of e^{-At} obtained in [**2**], [**4**] and [**16**]. (For similar results see also [**5**].) In this paper we consider a very particular case of Euler's approximation but our results are more general and complete (see also a remark at the end of this section). The main problem addressed in this paper is the characterization of decay rates for $\Delta_{n,t}(A)x, x \in \mathrm{ran}\,(f(A))$, where f is a generalized Stieltjes function of the class $\mathcal{S}_2$.

In particular, we extend Theorem 1.1 substantially by replacing power functions with reciprocals of generalized Stieltjes functions. As a corollary, we are also able to improve Theorem 1.1 by showing that there exists $c > 0$ such that for any $n \in \mathbb{N}$, $t > 0$, and $\alpha \in (0, 2]$,

$$\|\Delta_{n,t}(A)x\| \le c\left(\frac{t}{\sqrt{n}}\right)^\alpha \|A^\alpha x\|, \qquad x \in \mathrm{dom}\,(A^\alpha).$$

This result does not hold for $\alpha > 2$ as it is explained in Remark 6.7. We also show that (1.3) is an easy consequence of our main result and so it is possible to avoid interpolation theory used in [**16**]. Moreover, we prove that our estimates of convergence rates are optimal.

We believe that our method will be fruitful for more general rational approximations as well. However, being confined by space limits, we present only its sample which nevertheless is significant enough to be of value as for semigroup theory so for numerical analysis.

2. Preliminaries and notations

The following elementary integrals will be used frequently throughout the paper:

$$\frac{1}{(n-1)!}\int_0^\infty s^{n-1}e^{-s}\,ds = 1, \tag{2.1}$$

$$\int_0^\infty s^{n-1}e^{-s}(1-s/n)\,ds = 0, \tag{2.2}$$

$$\frac{1}{(n-1)!}\int_0^\infty s^{n-1}e^{-s}(1-s/n)^2\,ds = \frac{1}{n}, \tag{2.3}$$

where $n \in \mathbb{N}$.

To simplify our presentation we introduce the next notation:

$$e_t(z) := e^{-tz}, \quad r(z) := \frac{1}{1+z}, \quad r_{n,t}(z) := r^n(zt/n),$$

$$\Delta_{n,t}(z) := r_{n,t}(z) - e_t(z), \quad n \in \mathbb{N}, \quad t > 0, \quad h := \frac{t}{n}.$$

Thus, in particular, by (2.1) we have for $z \in \mathbb{C}_+$:

$$(n-1)!\, r_{n,t}(z) = \frac{1}{h^n}\int_0^\infty s^{n-1}e^{-h^{-1}s}e^{-zs}\,ds = \int_0^\infty s^{n-1}e^{-s}e^{-zst/n}\,ds. \tag{2.4}$$

For a closed linear operator A on a complex Banach space X we denote by $\mathrm{dom}\,(A)$, $\mathrm{ran}\,(A)$ and $\sigma(A)$ the *domain*, the *range*, and the *spectrum* of A, respectively, and let $\overline{\mathrm{ran}}\,(A)$ stand for the norm-closure of the range. The space of bounded linear operators on X is denoted by $\mathrm{L}(X)$. Finally, we let

$$\mathbb{C}_+ = \{z \in \mathbb{C} : \mathrm{Re}\, z > 0\}, \quad \mathbb{R}_+ = [0, \infty).$$

3. Functional calculus

In this subsection we recall definition and basic properties of functional calculus useful for the sequel.

Let $\mathrm{M}(\mathbb{R}_+)$ be a Banach algebra of bounded Radon measures on $\mathbb{R}_+$. Define the *Laplace transform* of $\mu \in \mathrm{M}(\mathbb{R}_+)$ as

$$(\mathcal{L}\mu)(z) := \int_0^\infty e^{-sz}\,\mu(ds), \qquad z \in \mathbb{C}_+.$$

Note that the space

$$\mathrm{A}^1_+(\mathbb{C}_+) := \{\mathcal{L}\mu : \mu \in \mathrm{M}(\mathbb{R}_+)\}$$

is a commutative Banach algebra with pointwise multiplication and with the norm

$$\|\mathcal{L}\mu\|_{\mathrm{A}^1_+(\mathbb{C}_+)} := \|\mu\|_{\mathrm{M}(\mathbb{R}_+)} = |\mu|(\mathbb{R}_+), \tag{3.1}$$

where $|\mu|(\mathbb{R}_+)$ stands for the total variation of μ on $\mathbb{R}_+$. Moreover, the Laplace transform

$$\mathcal{L} : \mathrm{M}(\mathbb{R}_+) \mapsto \mathrm{A}^1_+(\mathbb{C}_+)$$

is an isometric isomorphism.

Let $-A$ be the generator of a bounded C_0-semigroup $(e^{-tA})_{t\ge 0}$ on a Banach space X. Then the mapping

$$\begin{aligned} \mathrm{A}^1_+(\mathbb{C}_+) &\mapsto \mathrm{L}(X), \\ H(\mathcal{L}\mu)x &:= \int_0^\infty e^{-sA}x\,\mu(ds), \quad x \in X, \end{aligned}$$

defines a continuous algebra homomorphism such that

$$\|H(\mathcal{L}\mu)\| \le \sup_{t\ge 0} \|e^{-tA}\| |\mu|(\mathbb{R}_+). \tag{3.2}$$

The homomorphism is called the *Hille-Phillips* (HP-) functional calculus for A, and we set

$$g(A) = H(\mathcal{L}\mu) \qquad \text{if} \quad g = \mathcal{L}\mu.$$

Basic properties of the Hille-Phillips functional calculus can be found in [**11**, Chapter XV].

The HP-calculus has an automatic extension to a function class much larger then $\mathrm{A}^1_+(\mathbb{C}_+)$. Let us recall how this extension is constructed: if $f : \mathbb{C}_+ \to \mathbb{C}$ is holomorphic such that there exists $e \in \mathrm{A}^1_+(\mathbb{C}_+)$ with $ef \in \mathrm{A}^1_+(\mathbb{C}_+)$ and the operator $e(A)$ is injective, then one defines

$$\begin{aligned} \mathrm{dom}\,(f(A)) :=& \{x \in X : (ef)(A)x \in \mathrm{ran}\,(e(A))\} \\ f(A) :=& e(A)^{-1}\,(ef)(A). \end{aligned}$$

Such f is called *regularizable*, and e is called a *regularizer* for f. This definition of $f(A)$ does not depend on the choice of e and $f(A)$ is a closed operator on X. The set of all regularizable functions f constitute an algebra $\mathcal{A}$ depending on A (see e.g. [**9**, p. 4-5] and [**3**, p. 246-249]). We call the mapping

$$\mathcal{A} \ni f \mapsto f(A)$$

the *extended Hille–Phillips calculus* for A. The next *product rule* of the extended Hille-Phillips calculus (see e.g. [**9**, Chapter 1]) will be crucial for the sequel: *if f is regularizable and $g \in \mathrm{A}^1_+(\mathbb{C}_+)$, then*

$$g(A)f(A) \subset f(A)g(A) = (fg)(A), \tag{3.3}$$

where products of operators have their natural domains. From (3.3) it follows that if f is regularizable and e is a regularizer, then

$$\operatorname{ran}(e(A)) \subset \operatorname{dom}(f(A)). \tag{3.4}$$

4. Generalized Stieltjes functions

Our considerations will rely on the notion of generalized Stieltjes function. We say that a function $f : (0,\infty) \mapsto [0,\infty)$ is *generalized Stieltjes* of order $\alpha > 0$ if it can be written as

$$f(z) = a + \int_0^\infty \frac{\mu(d\tau)}{(z+\tau)^\alpha}, \qquad z > 0,$$

where $a \geq 0$ and μ is a positive Radon measure on $[0,\infty)$ satisfying

$$\int_0^\infty \frac{\mu(d\tau)}{(1+\tau)^\alpha} < \infty.$$

Observe that if f is generalized Stieltjes (of any positive order), then f admits an (unique) analytic extension to $\mathbb{C} \setminus (-\infty, 0]$ which will be identified with f and denoted by the same symbol. The class of generalized Stieltjes functions of order α will be denoted by $\mathcal{S}_\alpha$. In this terminology, Stieltjes functions constitute precisely the class $\mathcal{S}_1$ of generalized Stieltjes functions of order 1, and we will write $\mathcal{S}$ in place of $\mathcal{S}_1$ to denote the class of Stieltjes functions thus using an established notation. Note that $\mathcal{S} \subset \mathcal{S}_2$, and moreover $\mathcal{S} \cdot \mathcal{S} \subset \mathcal{S}_2$. Since for every $\alpha \in (0,2]$ one has $z^{-\alpha} \in \mathcal{S}_\alpha$ and $\mathcal{S}_\alpha \subset \mathcal{S}_2$, it clearly follows that $z^{-\alpha} \in \mathcal{S}_2$ for every $\alpha \in (0,2]$. For these as well as many other properties of generalized Stieltjes functions see [**12**]. A very informative discussion of Stieltjes functions is contained in [**20**, Chapter 2].

We will also need a subclass $\tilde{\mathcal{S}}_2$ of $\mathcal{S}_2$ consisting of products of Stieltjes functions:

$$\tilde{\mathcal{S}}_2 := \{f = f_1 \cdot f_2 : \; f_1, f_2 \in \mathcal{S}\}.$$

Observe that the implication $f_1, f_2 \in \mathcal{S} \Rightarrow f_1^{1/2} \cdot f_2^{1/2} \in \mathcal{S}$ (see [**20**, Proposition 7.10]) yields

$$\tilde{\mathcal{S}}_2 = \{f = f_0^2 : \; f_0 \in \mathcal{S}\}.$$

We can *define* the class of complete Bernstein functions $\mathcal{CBF}$ as $\mathcal{CBF} := \{zf : f \in \mathcal{S}\}$. An important link between the classes of Stieltjes and complete Bernstein functions is provided by the fact that $f \in \mathcal{CBF}, f \neq 0$, if and only if $1/f \in \mathcal{S}$, see e.g. [**20**, Theorem 7.3].

Let now $-A$ be the generator of a bounded C_0-semigroup on a Banach space X. By [**7**, Lemma 2.5] any complete Bernstein function is regularizable by $1/(1+z) \in \mathrm{A}^1_+(\mathbb{C}_+)$. Thus if A is injective then every $f \in \mathcal{S}$ is regularizable by $z/(1+z) \in \mathrm{A}^1_+(\mathbb{C}_+)$. The next proposition shows, in particular, that functions from $\mathcal{S}_2$ (and then from $\tilde{\mathcal{S}}_2$) are regularizable as well and identifies cores of the corresponding operators. To deal with densely defined operators we assume below that the range of A dense. Note that under this condition, for every $x \in X$,

$$\lambda(\lambda + A)^{-1}x \to 0, \qquad \lambda \to 0+,$$

and therefore A is also injective (see e.g. [**1**, p. 261]).

PROPOSITION 4.1. *Let $-A$ be the generator of a bounded C_0-semigroup on a Banach space X, and $\overline{\mathrm{ran}}(A) = X$.*

(i) *If $f \in \mathcal{S}_2$, then f is regularizable by $e(z) = z^2/(1+z)^2 \in \mathrm{A}^1_+(\mathbb{C}_+)$, and thus belongs to the extended Hille-Phillips calculus. Moreover,* ran (A^2) *is a core for $f(A)$.*

(ii) *If $f \in \tilde{\mathcal{S}}_2, f \neq 0$, then $1/f$ is regularizable by $e(z) = 1/(1+z)^2 \in \mathrm{A}^1_+(\mathbb{C}_+)$. Hence $1/f$ belongs to the extended Hille-Phillips calculus and, moreover,* dom (A^2) *is a core for $(1/f)(A)$.*

PROOF. To prove (i) note that, since $A^2(I+A)^{-2}$ is injective, a holomorphic function $f : \mathbb{C}_+ \to \mathbb{C}$ is regularizable by e if and only if $ef \in \mathrm{A}^1_+(\mathbb{C}_+)$. Without loss of generality, we can assume that $f \in \mathcal{S}_2$ is of the form

$$f(z) = \int_0^\infty \frac{\mu(d\tau)}{(z+\tau)^2}, \qquad z \in \mathbb{C}_+. \tag{4.1}$$

Then

$$f(z) = \int_0^\infty e^{-zs} s \int_0^\infty e^{-s\tau}\, \mu(d\tau)\, ds, \qquad z \in \mathbb{C}_+. \tag{4.2}$$

Since

$$\begin{aligned} f(z) &= \int_0^\infty e^{-zs} s \int_0^\infty e^{-st}\mu(dt)\, ds \\ &= \int_0^1 e^{-zs} s \int_0^\infty e^{-st}\mu(dt)\, ds + \int_1^\infty e^{-zs} s \int_0^\infty e^{-st}\mu(dt)\, ds \end{aligned}$$

it is enough to prove that the second term above is regularizable by e.

To this aim note that

$$\begin{aligned} &\int_1^\infty e^{-zs} s \int_0^\infty e^{-st}\mu(dt)\, ds \\ =&\frac{e^{-z}}{z}\int_0^\infty e^{-t}\mu(dt) + \frac{1}{z}\int_1^\infty e^{-sz}\int_0^\infty (1-st)e^{-st}\mu(dt)\, ds \\ =&\frac{e^{-z}}{z}\int_0^\infty e^{-t}\mu(dt) + \frac{e^{-z}}{z^2}\int_0^\infty (1-t)e^{-t}\mu(dt) \\ +&\frac{1}{z^2}\int_1^\infty e^{-zs}\int_0^\infty t(st-2)e^{-st}\mu(dt)\, ds. \end{aligned}$$

The first two terms in the last display are clearly regularizable by e. Let us show that the third term is regularizable by e too. Since

$$\int_1^\infty \left| \int_0^\infty t(ts-2)e^{-st}\,\mu(dt)\right| ds \le \int_0^\infty \int_1^\infty t(ts+2)e^{-st}\, ds\, \mu(dt),$$

and

$$\int_1^\infty t(ts+2)e^{-st}\, ds = (t+3)e^{-t},$$

it follows that there exists $c > 0$ such that

$$\int_1^\infty \left| \int_0^\infty t(ts-2)e^{-st}\,\mu(dt)\right| ds \le \int_0^\infty (t+3)e^{-t}\,\mu(dt)$$
$$\le c\int_0^\infty \frac{\mu(dt)}{(1+t)^2} < \infty.$$

Thus, the function

$$s \mapsto \int_0^\infty t(st-2)e^{-st}\mu(dt)$$

is integrable on $[1,\infty)$. This shows that

$$z \mapsto \int_1^\infty e^{-zs}s\int_0^\infty e^{-st}\mu(dt)\,ds$$

is regularizable by e, and yields $ef \in \mathrm{A}^1_+(\mathbb{C}_+)$.

Moreover, by (3.4) we have

$$\mathrm{ran}\,(A^2) = \mathrm{ran}\,(A^2(I+A)^{-2}) \subset \mathrm{dom}\,(f(A)).$$

To prove that $\mathrm{ran}\,(A^2)$ is a a core for $f(A)$ note that if $e_\epsilon(A) = A^2(\epsilon + A)^{-2}$, $\epsilon > 0$, then $e_\epsilon(A)x \to x$ for every $x \in X$ as $\epsilon \to 0$. Since $e_\epsilon(z) \in \mathrm{A}^1_+(\mathbb{C}_+)$ for each $\epsilon > 0$, the product rule (3.3) implies that if $x \in \mathrm{dom}\,(f(A))$ and $f(A)x = y$ then $f(A)e_\epsilon(A)x = e_\epsilon(A)y$. As $\mathrm{ran}\,(e_\epsilon(A)) = \mathrm{ran}\,(A^2)$, $\epsilon > 0$, the statement follows.

Let us prove (ii) now. Set $g(z) = 1/f(z)$, $z > 0$. As the reciprocal of a nonzero complete Bernstein function is a Stieltjes function, g is a product of two complete Bernstein functions. Then $(1+z)^{-2}g \in \mathrm{A}^1_+(\mathbb{C}_+)$ and, since $(1+A)^2$ is injective, the function g is regularizable by $1/(1+z)^2$. Hence (3.4) yields

$$\mathrm{dom}\,(A^2) = \mathrm{ran}\,((1+A)^{-2}) \subset \mathrm{dom}\,(g(A)).$$

Furthermore, if $e_\epsilon(A) = (1+\epsilon A)^{-2}$, $\epsilon > 0$, then $e_\epsilon(A)x \to x$ for every $x \in X$ as $\epsilon \to 0$. Arguing as in the proof of (i), we infer that $\mathrm{dom}\,(A^2)$ is a core for $g(A)$. □

REMARK 4.2. Let $f \in \mathcal{S}_2$ be of the form (4.1). Using (4.2) and

$$\left(\frac{z}{1+z}\right)^2 = 1 + \int_0^\infty (t-2)e^{-t}e^{-zt}\,dt,$$

by simple transformations, we obtain

$$\frac{z^2}{(1+z)^2}f(z) = \int_0^\infty e^{-zt}r(t)\,dt, \quad z \in \mathbb{C}_+,$$

where

$$r(t) := \int_0^\infty r_0(t,\tau)\,\mu(d\tau), \quad t \ge 0,$$
$$r_0(t,\tau) := te^{-t\tau} + e^{-t}\int_0^t e^{-s\tau}e^s s(t-s-2)\,ds$$
$$= \begin{cases} \frac{(-2+(1-\tau)\tau t)\tau e^{-\tau t}+(t+(2-t)\tau)e^{-t}}{(1-\tau)^3}, & \tau \ne 1, \\ te^{-t}(t^2/6 - t + 1), & \tau = 1. \end{cases}$$

Moreover, it is not to hard to show that

$$\int_0^\infty |r(t)|\,dt \le \int_0^\infty \int_0^\infty |r_0(t,\tau)|\,dt\,\mu(d\tau)$$
$$\le c\int_0^\infty \frac{\mu(d\tau)}{(1+\tau)^2} < \infty,$$

for some constant $c>0$. This leads to an alternative proof of Proposition 4.1, (i).

For $n\in\mathbb{N}$ and $t>0$, denote

$$(4.3)\quad (n-1)!\,L_{n,t}[m] := \int_0^\infty s^{n-1}e^{-s}\int_0^{t|1-s/n|} m(v)\,dv\,ds$$
$$+\int_0^\infty \left|\int_0^{n(v/t+1)} s^{n-1}e^{-s}\,[m(v+t-st/n)-m(v)]\,ds\right|\,dv.$$

LEMMA 4.3. *Let m be a positive measurable function on $[0,\infty)$ such that its Laplace transform $(\mathcal{L}m)(z)$ exists for every $z\in\mathbb{C}_+$. Then for any $n\in\mathbb{N}$ and $t>0$,*

$$(4.4)\qquad \Delta_{n,t}(z)(\mathcal{L}m)(z) = \int_0^\infty e^{-sz}q_{n,t}(s)\,ds,\quad z\in\mathbb{C}_+,$$

where $q_{n,t}(s)$ is a measurable on $[0,\infty)$ and

$$\int_0^\infty |q_{n,t}(s)|\,ds \le L_{n,t}[m].$$

PROOF. By (2.4) we have for every $z\in\mathbb{C}_+$

$$r_{n,t}(z)(\mathcal{L}m)(z) = \frac{1}{(n-1)!h^n}\int_0^\infty e^{-uz}\left(\int_0^u s^{n-1}e^{-h^{-1}s}\,m(u-s)\,ds\right)\,du.$$

Hence after a change of variable $s\mapsto st/n$

$$r_{n,t}(z)(\mathcal{L}m)(z) = \frac{1}{(n-1)!}\int_0^\infty e^{-uz}\left(\int_0^{nu/t} s^{n-1}e^{-s}\,m(u-st/n)\,ds\right)\,du.$$

On the other hand,

$$e^{-tz}(\mathcal{L}m)(z) = \int_0^\infty e^{-(u+t)z}m(u)\,du = \int_t^\infty e^{-uz}\,m(u-t)\,du,\quad z\in\mathbb{C}_+.$$

Then the above two formulas yield (4.4) with

$$(4.5)\qquad q_{n,t}(u) := \frac{1}{(n-1)!}\int_0^{nu/t} s^{n-1}e^{-s}\,m(u-st/n)\,ds - \chi(u-t)m(u-t),$$

where $\chi(\cdot)$ is the characteristic function of $[0,\infty)$.

Taking into account (2.1) we transform (4.5) further to the form

$$q_{n,t}(u) = \frac{\chi(t-u)}{(n-1)!}\int_0^{nu/t} s^{n-1}e^{-s}\,m(u-st/n)\,ds$$
$$-\frac{\chi(u-t)}{(n-1)!}\int_{nu/t}^\infty s^{n-1}e^{-s}m(u-t)\,ds$$
$$+\frac{\chi(u-t)}{(n-1)!}\int_0^{nu/t} s^{n-1}e^{-s}\,[m(u-st/n)-m(u-t)]\,ds.$$

Hence for any $n \in \mathbb{N}$ and $t > 0$,

$$
\begin{aligned}
\int_0^\infty |q_{n,t}(u)|\, du \le& \frac{1}{(n-1)!}\int_0^t \int_0^{nu/t} s^{n-1}e^{-s}\, m(u-st/n)\, ds\, du\\
+&\frac{1}{(n-1)!}\int_t^\infty \int_{nu/t}^\infty s^{n-1}e^{-s} m(u-t)\, ds\, du\\
+&\frac{1}{(n-1)!}\int_t^\infty \left|\int_0^{nu/t} s^{n-1}e^{-s}\left[m(u-st/n)-m(u-t)\right] ds\right| du\\
=&\frac{1}{(n-1)!}\int_0^\infty s^{n-1}e^{-s}\int_0^{t|1-s/n|} m(v)\, dv\, ds\\
+&\frac{1}{(n-1)!}\int_0^\infty \left|\int_0^{n(v/t+1)} s^{n-1}e^{-s}\left[m(v+t-st/n)-m(v)\right] ds\right| dv.
\end{aligned}
$$

The proof is complete. □

Let us illustrate Lemma 4.3 with several examples.

EXAMPLE 4.4. *a)* Let

$$f_1(z) := \frac{1}{z} = \int_0^\infty e^{-zv}\, dv, \qquad z \in \mathbb{C}_+.$$

Then $f_1 = \mathcal{L}m$ with $m(v) \equiv 1$ for $v \ge 0$, and using (2.1) and (2.3), we have

$$
\begin{aligned}
L_{n,t}[m] =& \frac{1}{(n-1)!}\int_0^\infty s^{n-1}e^{-s} t|1-s/n|\, ds\\
\le& \frac{t}{(n-1)!}\left(\int_0^\infty s^{n-1}e^{-s}\, ds\right)^{1/2}\left(\int_0^\infty s^{n-1}e^{-s}(1-s/n)^2\, ds\right)^{1/2}\\
=& \frac{t}{\sqrt{n}}, \qquad n \in \mathbb{N},\ t > 0.
\end{aligned}
$$

Hence by Lemma 4.3,

$$\|\Delta_{n,t} f_1\|_{\mathrm{A}^1_+(\mathbb{C}_+)} \le \frac{t}{\sqrt{n}}, \qquad n \in \mathbb{N},\ t > 0.$$

b) Let

$$f_2(z) := \frac{1}{z^2} = \int_0^\infty e^{-zv} v\, dv, \qquad z \in \mathbb{C}_+,$$

so that $f_2 = \mathcal{L}m$ with $m(v) = v$ for $v \ge 0$. By (2.2) we have

$$\left|\int_0^{n(v/t+1)} s^{n-1}e^{-s}\,(1-s/n)\, ds\right| = \int_{n(v/t+1)}^\infty s^{n-1}e^{-s}\,(s/n-1)\, ds, \tag{4.6}$$

and then (see (2.3))

$$\begin{aligned} L_{n,t}[m] =& \frac{t^2}{2(n-1)!}\int_0^\infty s^{n-1}e^{-s}(1-s/n)^2\,ds \\ &+\frac{t}{(n-1)!}\int_0^\infty\int_{n(v/t+1)}^\infty s^{n-1}e^{-s}\,(s/n-1)\,ds\,dv \\ \le& \frac{3t^2}{2(n-1)!}\int_0^\infty s^{n-1}e^{-s}\,(1-s/n)^2\,ds\,dv \\ =& \frac{3t^2}{2n}, \qquad n\in\mathbb{N},\ \ t>0. \end{aligned}$$

So, in this case, by Lemma 4.3,

$$\|\Delta_{n,t}f_2\|_{\mathrm{A}^1_+(\mathbb{C}_+)} \le \frac{3t^2}{2n}, \qquad n\in\mathbb{N}, \quad t>0.$$

c) Let

$$f_3(z) := \left(1+\frac{\lambda}{z}\right)^2, \qquad \lambda>0, \quad z\in\mathbb{C}_+.$$

Instead of identifying m here we use the previous two examples. Observe that by (2.4) we have $\Delta_{n,t}\in \mathrm{A}^1_+(\mathbb{C}_+)$, and

$$\|\Delta_{n,t}\|_{\mathrm{A}^1_+(\mathbb{C}_+)} \le 2, \qquad n\in\mathbb{N}, \quad t>0. \tag{4.7}$$

By (4.7) and Examples 4.4 a), b),

$$\begin{aligned} \|\Delta_{n,t}f_3\|_{\mathrm{A}^1_+(\mathbb{C}_+)} \le& 2+2\lambda\|\Delta_{n,t}f_1\|_{\mathrm{A}^1_+(\mathbb{C}_+)}+\lambda^2\|\Delta_{n,t}f_2\|_{\mathrm{A}^1_+(\mathbb{C}_+)} \\ \le& 2+\frac{2\lambda t}{\sqrt{n}}+\frac{3\lambda^2t^2}{2n}\le 2\left(1+\frac{\lambda t}{\sqrt{n}}\right)^2, \qquad n\in\mathbb{N},\ \ t>0. \end{aligned}$$

The following technical lemma is crucial in the proof of the main result of this section, Theorem 4.6. We shift its proof to Appendix to clarify our presentation.

LEMMA 4.5. *Let $\tau\ge 0$, $t>0$, and $n\in\mathbb{N}$ be fixed. If*

$$Q^{(1)}_{n,t}(\tau) := \frac{1}{\tau^2}\int_0^\infty s^{n-1}e^{-s}\left(1-(1+\tau t|1-s/n|)e^{-\tau t|1-s/n|}\right)ds, \tag{4.8}$$

and

$$Q^{(2)}_{n,t}(\tau) := \int_0^\infty e^{-\tau v}\psi_{n,t}(v)\,dv, \tag{4.9}$$

$$\psi_{n,t}(v) := \left|\int_0^{n(v/t+1)} s^{n-1}e^{-s}\left((v+t-st/n)e^{-\tau t(1-s/n)}-v\right)ds\right|,$$

then

$$Q_{n,t}(\tau) := \frac{Q^{(1)}_{n,t}(\tau)+Q^{(2)}_{n,t}(\tau)}{(n-1)!} \le \frac{12}{(\sqrt{n}/t+\tau)^2}. \tag{4.10}$$

Lemma 4.5 implies the following key estimate for $\mathrm{A}^1_+(\mathbb{C}_+)$-norms of $\Delta_{n,t}f$ when $f\in\mathcal{S}_2$.

THEOREM 4.6. *Let $f\in\mathcal{S}_2$. Then $\Delta_{n,t}f\in \mathrm{A}^1_+(\mathbb{C}_+)$, and*

$$\|\Delta_{n,t}f\|_{\mathrm{A}^1_+(\mathbb{C}_+)} \le 12f(\sqrt{n}/t), \qquad n\in\mathbb{N}, \quad t>0. \tag{4.11}$$

PROOF. According to (4.7) it suffices to consider f of the form (4.1). Then

$$\Delta_{n,t}(z)f(z) = \int_0^\infty \frac{\Delta_{n,t}(z)}{(z+\tau)^2}\,\mu(d\tau), \qquad z \in \mathbb{C}_+. \tag{4.12}$$

For fixed $\tau \geq 0$ define

$$\varphi_\tau(z) := \frac{1}{(z+\tau)^2} = \int_0^\infty e^{-zv} m(v,\tau)\,du, \quad m(v,\tau) := ve^{-\tau v}, \quad z \in \mathbb{C}_+.$$

Using Lemma 4.3 with $m(v,\tau) = ve^{-\tau v}$ and noting that

$$\int_0^w ve^{-v}dv = 1 - (1+w)e^{-w},$$

we obtain

$$L_{n,t}[m(\cdot,\tau)] = Q_{n,t}(\tau), \quad n \in \mathbb{N}, \quad t > 0, \quad \tau \geq 0,$$

where $L_{n,t}[m(\cdot,\tau)]$ is defined by (4.3) and $Q_{n,t}(\tau)$ is given by (4.10). So, by Lemmas 4.3 and 4.5 we have

$$\|\Delta_{n,t}\varphi_\tau\|_{\mathrm{A}^1_+(\mathbb{C}_+)} \leq Q_{n,t}(\tau) \leq \frac{12}{(\sqrt{n}/t+\tau)^2}. \tag{4.13}$$

Observe further that, in view of (4.7), $\tau \mapsto \Delta_{n,t}\varphi_\tau$ is a continuous, $\mathrm{A}^1_+(\mathbb{C}_+)$-valued function on $(0,\infty)$. Moreover, by (4.13), $\tau \mapsto \|\Delta_{n,t}\varphi_\tau\|_{\mathrm{A}^1_+(\mathbb{C}_+)}$ is Lebesgue integrable on $[0,\infty)$ for any $t > 0$ and $n \in \mathbb{N}$. Thus the $\mathrm{A}^1_+(\mathbb{C}_+)$-valued Bochner integral

$$\int_0^\infty \Delta_{n,t}\varphi_\tau\,\mu(d\tau)$$

is well-defined. Since point evaluations are bounded functionals on $\mathrm{A}^1_+(\mathbb{C}_+)$ and separate elements of $\mathrm{A}^1_+(\mathbb{C}_+)$, (4.12) implies that the integral coincides with $\Delta_{n,t}f$. Then by (4.12), (4.13), and a standard inequality for Bochner integrals (see e.g. [**11**, Theorem 3.7.6]) we obtain for any $n \in \mathbb{N}$ and $t > 0$:

$$\|\Delta_{n,t}f\|_{\mathrm{A}^1_+(\mathbb{C}_+)} \leq \int_0^\infty \|\Delta_{n,t}\varphi_\tau\|_{\mathrm{A}^1_+(\mathbb{C}_+)}\,\mu(d\tau) \leq 12f(\sqrt{n}/t).$$

□

5. Main results

Theorem 4.6 and (3.2) imply immediately the following statement.

THEOREM 5.1. *Let $-A$ be the generator of a C_0-semigroup $(e^{-tA})_{t\geq 0}$ on a Banach space X, and let $\overline{\mathrm{ran}}\,(A) = X$. Assume that*

$$M := \sup_{t\geq 0} \|e^{-tA}\| < \infty.$$

If $f \in \mathcal{S}_2$, then for any $x \in X$,

$$\|f(A)\Delta_{n,t}(A)x\| \leq 12M\|x\|f(\sqrt{n}/t), \qquad n \in \mathbb{N}, \quad t > 0, \tag{5.1}$$

and for any $x = f(A)y$, $y \in \mathrm{dom}\,(f(A))$,

$$\|\Delta_{n,t}(A)x\| \leq 12M\|y\|f(\sqrt{n}/t), \qquad n \in \mathbb{N}, \quad t > 0. \tag{5.2}$$

In particular, if $f(z) = z^{-\alpha}$, $\alpha \in (0,2]$, then for every $x \in \mathrm{dom}\,(A^\alpha)$,

$$\|\Delta_{n,t}(A)x\| \leq 12M\|A^\alpha x\|\left(\frac{t}{\sqrt{n}}\right)^\alpha, \qquad n \in \mathbb{N}, \quad t > 0. \tag{5.3}$$

Let us consider now the case when $-A$ generates a bounded C_0-semigroup but the range of A may not be dense, so that A may not be injective. We will need the next approximation result.

THEOREM 5.2. *Let $-A$ be the generator of a bounded C_0-semigroup on a Banach space X, and let $f \in \tilde{S}_2, f \neq 0$. If $g = 1/f$, then for every $x \in \operatorname{dom}(A^2)$,*

$$\lim_{\delta\to 0+} g(A+\delta)x = g(A)x. \tag{5.4}$$

PROOF. Recall that by Proposition 4.1 $\operatorname{dom}(A^2)$ is a core for $g(A+\delta)$, $\delta \geq 0$. Let $x \in \operatorname{dom}(A^2)$. Then for every $\delta > 0$ there exists $y_\delta \in X$ such that $x = (A+1+\delta)^{-2}y_\delta$. Note that

$$g(A+\delta)x = g(A+\delta)(A+1+\delta)^{-2}y_\delta = [g(z+\delta)\cdot(z+1+\delta)^{-2}](A)y_\delta.$$

Moreover $g(z)/(z+1)^2 \in \mathrm{A}^1_+(\mathbb{C}_+)$ and hence

$$\lim_{\delta\to 0+} \frac{g(\cdot+\delta)}{(\cdot+1+\delta)^2} = \frac{g(\cdot)}{(\cdot+1)^2}.$$

in the Banach algebra $\mathrm{A}^1_+(\mathbb{C}_+)$. Since

$$\lim_{\delta\to 0+} y_\delta = \lim_{\delta\to 0+} (A+1+\delta)^2 x = (A+1)^2 x,$$

we have

$$\begin{aligned}\lim_{\delta\to 0+} g(A+\delta)x =&[g(z)\cdot(z+1)^{-2}](A)(A+1)^2x\\ =&g(A)(A+1)^{-2}(A+1)^2x\\ =&g(A)x.\end{aligned}$$

□

Theorem 5.2 allows us to adopt Theorem 5.1 to the case when the range of the generator may not be dense.

COROLLARY 5.3. *Let $-A$ be the generator of a C_0-semigroup $(e^{-tA})_{t\geq 0}$ on a Banach space X. Assume that*

$$M := \sup_{t\geq 0} \|e^{-tA}\| < \infty.$$

If $g = 1/f$, where $f \in \tilde{S}_2, f \neq 0$, then for every $x \in \operatorname{dom}(g(A))$,

$$\|\Delta_{n,t}(A)x\| \leq 12M\frac{\|g(A)x\|}{g(\sqrt{n}/t)}, \qquad n \in \mathbb{N}, \quad t > 0.$$

In particular, if $g(z) = z^\alpha$, $\alpha \in (0,2]$, then (5.3) holds.

PROOF. Note that for any $\delta > 0$ one has $\operatorname{ran}(A+\delta) = X$. Thus $f(A+\delta)$ is well defined and bounded on X, moreover

$$f(\delta + A) = f_\delta(A),$$

where $f_\delta(z) := f(z+\delta), z > 0$. For $f \in \tilde{S}_2, f \neq 0$, define

$$g(z) := \frac{1}{f(z)}, \quad g_\delta(z) := \frac{1}{f_\delta(z)}, \quad \delta > 0.$$

Then, by the product rule (3.3),

$$f_\delta(A)g_\delta(A)x = g_\delta(A)f_\delta(A)x = x, \quad x \in \operatorname{dom}(A^2). \tag{5.5}$$

From Theorem 5.1 and (5.5) it follows that if $\delta > 0$ and $x \in \operatorname{dom}(A^2)$ then for any $n \in \mathbb{N}$ and $t > 0$,

$$\|\Delta_{n,t}(A)x\| \le 12Mf_\delta(\sqrt{n}/t)\|g_\delta(A)x\| \le 12Mf(\sqrt{n}/t)\|g_\delta(A)x\|.$$

Let $\delta \to 0+$ in the above inequality. Since $\operatorname{dom}(A^2)$ is core for $g(A)$, (5.4) implies the statement. □

We finish this section with the estimate of convergence rate in Euler's formula for Komatsu's spaces D_∞^α (defined in Introduction).

THEOREM 5.4. *Let $-A$ be the generator of a C_0-semigroup $(e^{-tA})_{t\ge 0}$ on a Banach space X. Suppose that*

$$M := \sup_{t\ge 0} \|e^{-tA}\| < \infty.$$

For any $\alpha \in (0,2]$ and $x \in D_\infty^\alpha$,

$$\|\Delta_{n,t}(A)x\| \le 8M\left(\frac{t}{\sqrt{n}}\right)^\alpha \|x\|_{D_\infty^\alpha}, \qquad n \in \mathbb{N}, \quad t > 0. \tag{5.6}$$

PROOF. For fixed $\lambda > 0$ define

$$f_\lambda(z) := \left(1 + \frac{\lambda}{z}\right)^2, \quad g_\lambda(z) := 1/f_\lambda(z) = \left(\frac{z}{\lambda + z}\right)^2, \quad z > 0.$$

Note that $f_\lambda \in \tilde{\mathcal{S}}_2$. Using Corollary 5.3 with $g = g_\lambda$ and taking into account Example 4.4, c), we have for any $x \in X$ and $n \in \mathbb{N}$

$$\begin{aligned}\|\Delta_{n,t}(A)x\| \le& 2M\left(1 + \frac{\lambda t}{\sqrt{n}}\right)^2 \|[A(A+\lambda)^{-1}]^2 x\| \\ =& 2M\lambda^\alpha \|[A(A+\lambda)^{-1}]^2 x\| \frac{(\lambda t/\sqrt{n} + 1)^2}{\lambda^\alpha}, \quad t > 0.\end{aligned} \tag{5.7}$$

Setting now $\lambda = \sqrt{n}/t$ in (5.7) it follows that

$$\|\Delta_{n,t}(A)x\| \le 8M\left(\frac{t}{\sqrt{n}}\right)^\alpha \sup_{s>0}\left(s^\alpha \|[A(A+s)^{-1}]^2 x\|\right), \quad x \in D_\infty^\alpha,$$

and (5.6) holds. □

6. Optimality of rates

In this section we show that our estimates for convergence rates in Euler's approximation formula are in a sense optimal. We will need the next estimate for Stieltjes functions proved independently in many papers. It seems the earliest reference is [**19**, Lemma 2].

LEMMA 6.1. *If $f \in \mathcal{S}$, then*

$$f(s) \le \sqrt{2}|f(\pm is)|, \quad s > 0. \tag{6.1}$$

It will also be convenient to single out an auxiliary inequality involving $\Delta_{n,t}$.

LEMMA 6.2. *For any $n \in \mathbb{N}$ and $t > 0$,*

$$|\Delta_{n,t}(\pm i\sqrt{n}/t)| \ge 1 - \frac{1}{\sqrt{2}}. \tag{6.2}$$

PROOF. For any $n \in \mathbb{N}$ and $t > 0$,

$$|\Delta_{n,t}(\pm i\sqrt{n}/t)| = \left|\frac{1}{(1+i/\sqrt{n})^n} - e^{-i\sqrt{n}}\right| \geq 1 - |1+i/\sqrt{n}|^{-n}.$$

Since

$$\log(1+t) \geq t\log 2, \quad t \in [0,1],$$

it follows that

$$|1+i/\sqrt{n}|^n = e^{(n/2)\log(1+1/n)} \geq e^{(1/2)\log 2} = \sqrt{2}, \quad n \in \mathbb{N},$$

and this yields (6.2). □

The result below shows that Theorem 5.1 and Corollary 5.3 are sharp if the spectrum of the generator is large enough.

THEOREM 6.3. *Let $-A$ be the generator of a bounded C_0-semigroup $(e^{-tA})_{t\geq 0}$ on a Banach space X. Suppose that $\overline{\operatorname{ran}}\,(A) = X$ and*

$$\{|s| : s \in \mathbb{R},\, is \in \sigma(A)\} = \mathbb{R}_+. \tag{6.3}$$

If $f \in \tilde{\mathcal{S}}_2$, then

$$\|f(A)\Delta_{n,t}(A)\| \geq cf(\sqrt{n}/t), \;\; n \in \mathbb{N}, \;\; t > 0, \quad c = \frac{1}{2}\left(1 - \frac{1}{\sqrt{2}}\right). \tag{6.4}$$

In particular,

$$\|A^{-\alpha}\Delta_{n,t}(A)\| \geq c(t/\sqrt{n})^{\alpha}, \;\; n \in \mathbb{N}, \;\; t > 0,$$

for any $\alpha \in (0,2]$.

PROOF. Let $n \in \mathbb{N}$ and $t > 0$ be such that $i\sqrt{n}/t \in \sigma(A)$. By the spectral inclusion theorem for the Hille-Phillips functional calculus (see e.g. [**11**, Theorem 16.3.5] or [**7**, Theorem 2.2]) we obtain for every $t > 0$:

$$\begin{aligned} \|f(A)\Delta_{n,t}(A)\| =& \|(f\cdot\Delta_{n,t})(A)\| \geq \sup_{\lambda\in\sigma(A)} |(f\cdot\Delta_{n,t})(\lambda)| \\ \geq& |(f\cdot\Delta_{n,t})(i\sqrt{n}/t)| = |\Delta_{n,t}(i\sqrt{n}/t)|\cdot|f(i\sqrt{n}/t)|. \end{aligned} \tag{6.5}$$

Then, by Lemmas 6.1 and 6.2, (6.5) implies (6.4).

If $-i\sqrt{n}/t \in \sigma(A)$ then, by Lemmas 6.1 and 6.2, the argument completely analogous to the above gives the same estimate (6.4). □

The assumptions of Theorem 6.3 can trivially be satisfied as the next simple example shows.

EXAMPLE 6.4. Let $X = L^2(\mathbb{R}_+)$. Define

$$(Au)(s) := isu(s), \qquad u \in L^2(\mathbb{R}_+),$$

with the maximal domain. Then $-A$ generates a C_0-semigroup $(e^{-At})_{t\geq 0}$ given by $(e^{-At}u)(s) = e^{-ist}u(s)$, $t \geq 0$, on X, and $\sigma(A) = i\mathbb{R}_+$. Thus, A satisfies the conditions of Theorem 6.3.

The following statement complementing Theorem 6.3 can be proved in the same way as Theorem 6.3.

THEOREM 6.5. *Let $-A$ be the generator of a bounded C_0-semigroup $(e^{-tA})_{t\geq 0}$ on a Banach space X. Suppose that $\overline{\operatorname{ran}}(A) = X$ and $\sigma(A)\cap i\mathbb{R}$ has an accumulation point at infinity. If $f \in \tilde{\mathcal{S}}_2$, then*

$$\limsup_{\sqrt{n}/t\to\infty} \frac{\|f(A)\Delta_{n,t}(A)\|}{f(\sqrt{n}/t)} > 0.$$

Finally, we show that Theorem 5.1 and Corollary 5.3 are sharp in a slightly weaker sense than in Theorem 6.3 but with no restriction on the spectrum of the generator.

COROLLARY 6.6. *Let A and f satisfy the assumptions Theorem 6.5. Suppose in addition that*

$$\lim_{\tau\to\infty} \tau^2 f(\tau) = \infty.$$

Then, whenever $\epsilon : (0,\infty) \mapsto (0,\infty)$ is a decreasing function with $\lim_{\tau\to\infty} \epsilon(\tau) = 0$, there exists $y \in \operatorname{ran}(f(A))$ such that

$$\limsup_{\sqrt{n}/t\to\infty} \frac{\|\Delta_{n,t}(A)y\|}{\epsilon(\sqrt{n}/t)f(\sqrt{n}/t)} = \infty. \tag{6.6}$$

PROOF. By the Theorem 6.5 we have

$$\limsup_{\sqrt{n}/t\to\infty} \frac{\|f(A)\Delta_{n,t}(A)\|}{\epsilon(\sqrt{n}/t)f(\sqrt{n}/t)} = \infty,$$

Since $\operatorname{ran}(A^2) \subset \operatorname{dom}(f(A))$, the product rule (3.3) implies that $f(A)\Delta_{n,t}(A)$ is similar to its restriction to $\operatorname{dom}(A^{-2}) = \operatorname{ran}(A^2)$ by means of the isomorphism $A^2(I+A)^{-2} : X \mapsto \operatorname{ran}(A^2)$. Then the uniform boundedness principle yields $x \in \operatorname{ran}(A^2) \subset \operatorname{dom}(f(A))$ such that

$$\limsup_{\sqrt{n}/t\to\infty} \frac{\|f(A)\Delta_{n,t}(A)x\|_{\operatorname{dom}(A^{-2})}}{\epsilon(\sqrt{n}/t)f(\sqrt{n}/t)} = \infty.$$

On the other hand, setting $y = f(A)x$, and using Example 4.4 *b*), we obtain

$$\begin{aligned} \|f(A)\Delta_{n,t}(A)x\|_{\operatorname{dom}(A^{-2})} &= \|\Delta_{n,t}(A)f(A)x\| + \|A^{-2}\Delta_{n,t}(A)f(A)x\| \\ &= \|\Delta_{n,t}(A)y\| + \|A^{-2}\Delta_{n,t}(A)y\| \\ &\leq \|\Delta_{n,t}(A)y\| + \frac{3M}{2}\left(\frac{t}{\sqrt{n}}\right)^2 \|y\|. \end{aligned} \tag{6.7}$$

Since $\tau^2 f(\tau) \to \infty$, $\tau \to \infty$, we may replace $\epsilon(\tau)$ by $\max\{\epsilon(\tau), (\tau^2 f(\tau))^{-1}\}$, $\tau \geq 1$, and suppose without loss of generality that

$$\beta := \inf_{\tau\geq 1} \epsilon(\tau)\tau^2 f(\tau) > 0.$$

Hence, in view of

$$\sup_{\sqrt{n}/t\geq 1} \frac{1}{\epsilon(\sqrt{n}/t)(\sqrt{n}/t)^2 f(\sqrt{n}/t)} \leq \frac{1}{\beta} < \infty,$$

the statement follows from (6.7). □

REMARK 6.7. Note that Theorem 4.6 does hot hold for wider classes $\mathcal{S}_\alpha$, $\alpha > 2$, of generalized Stieltjes functions. Indeed, note that

$$\lim_{z\to 0+} \frac{\Delta_{n,t}(z)}{z^2} = \lim_{z\to 0+} \frac{r_{n,t}(z) - e_t(z)}{z^2} = \frac{t^2}{2n} \neq 0, \quad \text{if} \quad t > 0.$$

Hence if $f \in \mathcal{S}_\alpha$, $\alpha > 2$, is such that $\lim_{z\to 0+} z^2 f(z) = \infty$, and $\sigma(A)$ has accumulation point at zero, then $f(A)\Delta_{n,t}(A)$ is not bounded. (Otherwise, the inequality

$$\|f(A)\Delta_{n,t}(A)\| \geq \sup_{\lambda\in\sigma(A)} |(f\cdot\Delta_{n,t})(\lambda)|,$$

leads to a contradiction.) Thus, (5.1) and (5.3) are not true in this case.

7. Appendix

PROOF OF LEMMA 4.5. It will be convenient to denote

$$w := w_{n,t}(s,\tau) = \tau t|1 - s/n|. \tag{7.1}$$

Using (4.9) we obtain

$$\begin{aligned}
&Q^{(2)}_{n,t}(\tau)\\
\leq &\int_0^\infty e^{-\tau v} \int_0^{n(v/t+1)} s^{n-1}e^{-s} \left|(v + t - st/n)e^{-\tau t(1-s/n)} - v\right| ds\, dv\\
= &\frac{1}{\tau^2}\int_0^n s^{n-1}e^{-s}\int_0^\infty e^{-v}\left|(v+w)e^{-w} - v\right| dv\, ds\\
+ &\frac{1}{\tau^2}\int_n^\infty s^{n-1}e^{-s}\int_w^\infty e^{-v}\left|(v-w)e^{w} - v\right| dv\, ds\\
= &\frac{1}{\tau^2}\int_0^\infty s^{n-1}e^{-s}\int_0^\infty e^{-v}|(1-e^{-w})v + we^{-w}|\, dv\, ds,
\end{aligned}$$

and therefore

$$Q^{(2)}_{n,t}(\tau) \leq \frac{1}{\tau^2}\int_0^\infty s^{n-1}e^{-s}[(1-e^{-w}) + we^{-w}]\, ds. \tag{7.2}$$

Then, by (4.8) and (7.2),

$$Q_{n,t}(\tau) = \frac{Q^{(1)}_{n,t}(\tau) + Q^{(2)}_{n,t}(\tau)}{(n-1)!} \leq \frac{2}{(n-1)!\tau^2}\int_0^\infty s^{n-1}e^{-s}[1-e^{-w}]\, ds \leq \frac{2}{\tau^2}. \tag{7.3}$$

Now let us prove that

$$Q_{n,t}(\tau) \leq \frac{3t^2}{n}, \quad t \geq 0, \quad n\in\mathbb{N}, \quad \tau > 0. \tag{7.4}$$

Define

$$q_{n,t}(v,s,\tau) := \left|v[e^{-\tau t(1-s/n)} - 1 + \tau t(1-s/n)] + t(1-s/n)(e^{-\tau t(1-s/n)} - 1)\right|.$$

Then using (4.6) we have

$$\begin{aligned}Q^{(2)}_{n,t}(\tau) \le &\int_0^\infty e^{-\tau v}\int_0^{n(v/t+1)} s^{n-1}e^{-s}\, q_{n,t}(v,s,\tau)\,ds\,dv\\ &+t\int_0^\infty e^{-\tau v}|1-\tau v|\int_{n(v/t+1)}^\infty s^{n-1}e^{-s}\,(s/n-1)ds\,dv\\ =&\frac{1}{\tau^2}\int_0^n s^{n-1}e^{-s}\int_0^\infty e^{-v}\tau q_{n,t}(v/\tau,s,\tau)\,dv\,ds\\ &+\frac{1}{\tau^2}\int_n^\infty s^{n-1}e^{-s}\int_w^\infty e^{-v}\tau q_{n,t}(v/\tau,s,\tau)\,dv\,ds\\ &+\frac{1}{\tau^2}\int_n^\infty s^{n-1}e^{-s}wu(w)\,ds,\end{aligned}$$

where

$$u(w) := \int_0^w |1-v|e^{-v}\,dv \le 1-e^{-w}.$$

For $s \le n$ we have

$$\begin{aligned}\tau q_{n,t}(v/\tau,s,\tau) =&\left|v[e^{-w}-1+w]+w(e^{-w}-1)\right|\\ \le& v(e^{-w}-1+w)+w(1-e^{-w}),\end{aligned}$$

and similarly if $s \ge n$ then

$$\begin{aligned}\tau q_{n,t}(v/\tau,s,\tau) =&|v[e^{w}-1-w]-w(e^{w}-1)|\\ \le& v(e^{w}-1-w)+w(e^{w}-1).\end{aligned}$$

So,

$$\begin{aligned}&\tau^2 Q^{(2)}_{n,t}(\tau)\\ \le&\int_0^n s^{n-1}e^{-s}\int_0^\infty e^{-v}(v[e^{-w}-1+w]+w(1-e^{-w}))\,dv\,ds\\ &+\int_n^\infty s^{n-1}e^{-s}\int_w^\infty e^{-v}\,|v[e^{w}-1-w]+w(e^{w}-1)|\,dv\,ds\\ &+\int_n^\infty s^{n-1}e^{-s}w(1-e^{-w})\,ds\\ =&\int_0^n s^{n-1}e^{-s}[2w-1+(1-w)e^{-w}]\,ds\\ &+\int_n^\infty s^{n-1}e^{-s}[1+3w-(1+4w+w^2)e^{-w}]\,ds.\end{aligned}\tag{7.5}$$

Then, by (4.8) and (7.5), we infer that

$$\begin{aligned}\tau^2(Q^{(1)}_{n,t}(\tau)+Q^{(2)}_{n,t}(\tau)) \le& 2\int_0^n s^{n-1}e^{-s}w(1-e^{-w})\,ds\\ &+\int_n^\infty s^{n-1}e^{-s}[2+3w-(w^2+5w+2)e^{-w}]\,ds.\end{aligned}$$

Using now elementary inequalities

$$w(1-e^{-w}) \le w^2, \quad 2+3w-(w^2+5w+2)e^{-w} \le 3w^2, \quad w \ge 0,$$

where $w^2 = \tau^2 t^2 (1 - s/n)^2$ (see (7.1)), we obtain that

$$Q_{n,t}(\tau) \le \frac{3t^2}{(n-1)!} \int_0^\infty s^{n-1}(1 - s/n)^2\, ds = \frac{3t^2}{n}, \tag{7.6}$$

i.e. (7.4) holds.

Hence, from (7.3), (7.6) and the inequality

$$\min\left\{\frac{1}{a^2}, \frac{1}{b^2}\right\} \le \frac{4}{(a+b)^2}, \quad a, b > 0,$$

it follows that

$$Q_{n,t}(\tau) \le 3 \min\left\{\frac{1}{\tau^2}, \frac{t^2}{n}\right\} \le \frac{12}{(\sqrt{n}/t + \tau)^2}, \qquad n \in \mathbb{N},\ \tau \ge 0,\ t > 0.$$

References

[1] Wolfgang Arendt, Charles J. K. Batty, Matthias Hieber, and Frank Neubrander, *Vector-valued Laplace transforms and Cauchy problems*, 2nd ed., Monographs in Mathematics, vol. 96, Birkhäuser/Springer Basel AG, Basel, 2011. MR2798103 (2012b:47109)

[2] Philip Brenner and Vidar Thomée, *On rational approximations of semigroups*, SIAM J. Numer. Anal. **16** (1979), no. 4, 683–694, DOI 10.1137/0716051. MR537280 (80j:47052)

[3] Ralph deLaubenfels, *Automatic extensions of functional calculi*, Studia Math. **114** (1995), no. 3, 237–259. MR1338830 (96f:47029)

[4] *Evolution equations: applications to physics, industry, life sciences and economics*, Progress in Nonlinear Differential Equations and their Applications, 55, Birkhäuser Verlag, Basel, 2003. Edited by Mimmo Iannelli and Günter Lumer. MR2013176 (2004j:34002)

[5] Hassan Emamirad and Arnaud Rougirel, *A functional calculus approach for the rational approximation with nonuniform partitions*, Discrete Contin. Dyn. Syst. **22** (2008), no. 4, 955–972, DOI 10.3934/dcds.2008.22.955. MR2434977 (2009g:47104)

[6] Alexander Gomilko, Markus Haase, and Yuri Tomilov, *On rates in mean ergodic theorems*, Math. Res. Lett. **18** (2011), no. 2, 201–213. MR2784667 (2012e:47028)

[7] Alexander Gomilko, Markus Haase, and Yuri Tomilov, *Bernstein functions and rates in mean ergodic theorems for operator semigroups*, J. Anal. Math. **118** (2012), 545–576, DOI 10.1007/s11854-012-0044-0. MR3000691

[8] A. Gomilko and Yu. Tomilov, *What does a rate in a mean ergodic theorem imply?*, submitted.

[9] Markus Haase, *The functional calculus for sectorial operators*, Operator Theory: Advances and Applications, vol. 169, Birkhäuser Verlag, Basel, 2006. MR2244037 (2007j:47030)

[10] Reuben Hersh and Tosio Kato, *High-accuracy stable difference schemes for well-posed initial value problems*, SIAM J. Numer. Anal. **16** (1979), no. 4, 670–682, DOI 10.1137/0716050. MR537279 (80h:65036)

[11] Einar Hille and Ralph S. Phillips, *Functional analysis and semi-groups*, American Mathematical Society Colloquium Publications, vol. 31, American Mathematical Society, Providence, R. I., 1957. rev. ed. MR0089373 (19,664d)

[12] D. Karp and E. Prilepkina, *Generalized Stieltjes transforms: basic aspects,* arXiv:1111.4271v2.

[13] Hikosaburo Komatsu, *Fractional powers of operators*, Pacific J. Math. **19** (1966), 285–346. MR0201985 (34 #1862)

[14] Hikosaburo Komatsu, *Fractional powers of operators. II. Interpolation spaces*, Pacific J. Math. **21** (1967), 89–111. MR0206716 (34 #6533)

[15] Mihaly Kovacs, *On qualitative properties and convergence of time-discretization methods for semigroups*, ProQuest LLC, Ann Arbor, MI, 2004. Thesis (Ph.D.)–Louisiana State University and Agricultural & Mechanical College. MR2706249

[16] Mihály Kovács, *On the convergence of rational approximations of semigroups on intermediate spaces*, Math. Comp. **76** (2007), no. 257, 273–286, DOI 10.1090/S0025-5718-06-01905-3. MR2261021 (2007k:47070)

[17] Mihály Kovács and Frank Neubrander, *On the inverse Laplace-Stieltjes transform of A-stable rational functions*, New Zealand J. Math. **36** (2007), 41–56. MR2406045 (2009b:44003)

[18] Celso Martínez Carracedo and Miguel Sanz Alix, *The theory of fractional powers of operators*, North-Holland Mathematics Studies, vol. 187, North-Holland Publishing Co., Amsterdam, 2001. MR1850825 (2002k:47038)

[19] E. I. Pustyl′nik, *Some inequalities in the theory of selfadjoint operators*, Izv. Vyssh. Uchebn. Zaved. Mat. **6** (1979), 52–57 (Russian). MR556594 (81b:46084)

[20] René L. Schilling, Renming Song, and Zoran Vondraček, *Bernstein functions*, de Gruyter Studies in Mathematics, vol. 37, Walter de Gruyter & Co., Berlin, 2010. Theory and applications. MR2598208 (2011d:60060)

FACULTY OF MATHEMATICS AND COMPUTER SCIENCE, NICOLAS COPERNICUS UNIVERSITY, UL. CHOPINA 12/18, 87-100 TORUŃ, POLAND

E-mail address: `gomilko@mat.umk.pl`

FACULTY OF MATHEMATICS AND COMPUTER SCIENCE, NICOLAS COPERNICUS UNIVERSITY, UL. CHOPINA 12/18, 87-100 TORUŃ, POLAND — AND — INSTITUTE OF MATHEMATICS, POLISH ACADEMY OF SCIENCES, ŚNIADECKICH 8, 00-956 WARSZAWA, POLAND

E-mail address: `tomilov@mat.umk.pl`

Proceedings of Symposia in Pure Mathematics
Volume **87**, 2013
http://dx.doi.org/10.1090/pspum/087/01442

Construction of a Noncommutative Quantum Field Theory

Harald Grosse and Raimar Wulkenhaar

ABSTRACT. We review our recent successful attempt to construct the planar sector of a nonlocal scalar field model in four dimensional Euclidean deformed space-time, which needs 4 (instead of 3) relevant/marginal operators in the defining Lagrangian. As we have shown earlier, this model is renormalizable up to all orders in pertubation theory. In addition a new fixed point appears, at which the beta function for the coupling constant vanishes. This way, we were able to tame the Landau ghost.

We next discuss Ward identities and Schwinger-Dyson equations and derive integral equations for the renormalized N-point functions. They are the starting point of a nonperturbative construction of the model.

Dear Fritz! I (H.G.) almost cannot believe, that you become 60! I still remember the time, when you came from Graz to Vienna in the early 80's. I enjoyed our long standing interactions, our discussions on spectral concentration, how we handled the non-relativistic limit of the Dirac equation and especially our treatment of index problems and their connection to scattering theory. The last subject became of particular interests through the developments connected to noncommutative geometry and we enjoyed a recent Workshop at ESI on that subject together.

Here I review another outcome of using ideas from noncommutative geometry. I hope you will enjoy reading that a four dimensional quantum field theory model can be constructed on such a deformed space.

I wish you many new results for your interesting work and many happy years to come and hope for your visits to Vienna.

1. Introduction

Our present fundamental physics rests on two pillars: Quantum Field Theory and General Relativity. One of the main question in this area of physics concerns the matching of these two concepts.

In addition we hope to improve quantum field theory models by adding "gravity" effects. Constructive methods led years ago to many beautiful ideas and results, but the main goal to construct a mathematical consistent model of a four dimensional local quantum field theory has not been reached.

The requirements of local quantum field theory are easy to state and consists of quantum mechanical and relativity properties. States are supposed to be represented by vectors of a separable Hilbert space. Field operators are operator valued distribution, which should be smeared with smooth test functions in four coordinates and leads to $\Phi(f)$ acting on a

---

2010 *Mathematics Subject Classification.* Primary .

dense domain of the Hilbert space,

$$\Phi(f) = \int d^4x \Phi(x) f^*(x). \tag{1}$$

The ground or vacuum state is unique (up to a phase) and cyclic. Space-time translations should be symmetries: This implies that the common spectrum of the energy-momentum operator $\sigma(P_\mu)$ lies in the closed forward light cone. The ground state is translation invariant. As for the relativistic properties one likes fields to transform covariantly under a unitary representation of the Poincaré group. One of the most essential postulate concerns **miscroscopic Causality** or locality. If the supports of the smearing functions f and g are space-like separated, then the field operators commute (for Bosons) or anticommute (for Fermions).

Typically one defines the expectation value of the product of smeared field operators called Wightman functions:

$$W_N(f_1 \otimes ... \otimes f_N) := \langle \Omega | \phi(f_1) \cdots \phi(f_N) | \Omega \rangle \tag{2}$$

It is not difficult to rephrase the requirements for the Wightman functions. For many purposes it is easier to go over to Euclidean Schwinger functions obtained by using analyticity of Wightman functions in the coordinate difference variables, implied by the support properties of the Wightman distributions. One has to go over to the so called extended permuted tube.

The formal definition of Schwinger functions reads:

$$S_N(z_1, ..., z_N) = \int \Phi(z_1)...\Phi(z_N) d\nu(\Phi), \qquad d\nu = \frac{1}{Z} e^{-\int L_{int}(\Phi)} d\mu(\Phi), \tag{3}$$

where $d\mu$ is the Gaussian measure corresponding to free fields with two point correlation: $\langle \phi(x_1)\phi(x_2) \rangle = C(x_1, x_2)$, or its Fourier transform: $\tilde{C}(p_1, p_2) = \delta(p_1 - p_2) \frac{1}{p_1^2 + m^2}$, ϕ above is a stochastic variable.

As for interacting fields we have to rely on **(renormalized) pertubation expansions.** We have to put first cut-offs and to expand the interacting part:

$$S_N(x_1 \dots x_N) = \sum_n \frac{(-\lambda)^n}{n!} \int d\mu(\phi) \prod_j^N \phi(x_j) \left(\int dx \, \frac{\phi^4(x)}{4!} \right)^n \tag{4}$$

$$= \sum_{\text{graph } \Gamma_N} \frac{(-\lambda)^n}{Sym_{\Gamma_N}(G)} \int_V \prod_{l \in \Gamma_N} C_\kappa(x_l - y_l) \sim \Lambda^{\omega_D(G)} \tag{5}$$

As a result we may collect contributions to the same Feynman diagram and evaluate the degree of divergence, which is given by $\omega_D(G) = (D-4)n + D - \frac{D-2}{2} N$, $\omega_2(G) = 2 - 2n$, $\omega_4(G) = 4 - N$, where n denotes the order of the graph, or the number of vertices, N the number of external lines, l the number of internal lines. Note that there are $(4n+N)!!$ number of Feynman graphs. Use Stirling formula and the factor $\frac{1}{n!}$ from the exponential, the large order behavior $K^n n!$ for the contributions result, which indicates that a naïve convergence is questionable.

Renormalization If one imposes a finite number of renormalization conditions (here we need 3 conditions to fix a, m and λ), for example:

$$G_2(p^2 = 0) = \frac{1}{m_{phys}^2}, \; \frac{d}{dp^2} G_2(p^2 = 0) = -\frac{a^2}{m_{phys}^4}, G_4(p^2 = 0) = \lambda_{phys} \tag{6}$$

and no new interactions are generated order by order in pertubation theory, we call the model to be renormalizable (this is implied by the BPHZ Theorem for the scalar Φ^4 model).

The program of constructing a nontrivial interacting models was successfully done only in $D = 2, 3$ space-time dimensions. As for $D = 4$ dimensions we have to rely on renormalized pertubation theory and follow the renormalization group flow. But in addition we may add "Gravity" effects, or quantize Space-Time: This led to our program of merging **general relativity ideas with quantum physics through noncommutative geometry**.

Space-Time structure That one should limit localisation in space-time follows from a very simple old argument due to Wheeler and others:

In order to localize two events, which are a distance D apart, one has to do a scattering experiment with particles whose energy hc/λ exceed hc/D. Multiplying these quantities times G/c^4 yields the Schwarzschild radius of the appropriate energy lump. It is natural to require that this radius should be smaller than the distance between the events one started with, since otherwise the scattered particles will be captured by the black hole, which is formed. Putting both inequalities together gives a lower bound to the distance of localizability of events of the order of the Planck length.

$$D \geq R_{ss} = G/c^4 hc/\lambda \geq G/c^4 hc/D \tag{7}$$

which implies that $D \geq l_p =$ Planck length. Early ideas of modifying space-time were phrased already by Schrödinger and Heisenberg, but Snyder in 1947 was the first to formulate a deformed space-time geometry. Such ideas became popular after 1986, when Alain Connes published his work on Noncommutative Geometry. On of us (H. G.) started in 1992 (in work together with J. Madore) to use noncommutative manifolds (algebras) as a natural cut off for quantized field theory models. Doplicher, Fredenhagen and Roberts used the Wheeler argument in 1994 to formulate uncertainty relations for deformed fields and formulated deformed free fields. Filk in 1995 was the first to elaborate on Feynman rules for models defined over deformed space-time, and finally they became popular due to the work of Schomerus (1999), who observed, that such models may result from string theory after taking the zero slope limit.

Ideas: Algebra, fields, diff. calculus,...

Typically one first refers to the Gelfand - Naimark theorem, which states that the algebra of continuous functions over a manifold is isomorphic to a commutative C^* algebra. Next one studies deformations of such algebras, through associative nonlocal star products. Especially simple is the Moyal space. One may start from the algebra of smooth functions over D-dimensional Euclidean space, and define the $\star$-product as $(a \star b)(x) = \int d^D y d^D k a(x+\frac{1}{2}\Theta \cdot k) b(x+y)\, e^{iky}$ where $\Theta = -\Theta^T \in M_D(\mathbb{R})$

Fields are sections of bundles, according to the Serre Swan theorem, they can be identified as projective modules over the algebra A. A very essential requirement concerns the differential calculus, which we would like to keep. Next question results:

Can we make sense of renormalisation in Noncommutative Geometric Models?

As a first step we intend to construct simple quantum field theory models on simple noncommutative geometries, e.g. the Moyal space. Of course, this way we obtain models with non-local interactions.

The naïve application of this procedure to the ϕ^4-action (ϕ-real, Euclidean space) leads on Moyal plane to the action:

$$S = \int d^4x \Big(\frac{1}{2} \partial_\mu \phi \star \partial^\mu \phi + \frac{m^2}{2} \phi \star \phi + \frac{\lambda}{4} \phi \star \phi \star \phi \star \phi \Big)(x) \tag{8}$$

The Feynman rules can be obtained easily. Since we obtain only cyclic invariance at the Vertex, Graphs are best drawn as Ribbon Graphs on Riemann surfaces with a certain genus and a certain number of boundary components. We obtain planar regular contribution and non-planar graphs. The planar graphs still reveal UV divergences, the nonplanar ones are finite for generic momenta. On the other hand for exceptional momenta (if sums of incoming or outgoing momenta vanish) the contributions develop an IR singularity, which spoils **Renormalizability!** In our previous work [**1**] we realized that the UV/IR-mixing problem can be solved by adding a fourth relevant/marginal operator to the Lagrangian
Theorem: The quantum field theory defined by the action

$$S = \int d^4x \Big(\frac{1}{2}\phi \star (\Delta + \Omega^2 \tilde{x}^2 + \mu^2)\phi + \frac{\lambda}{4!}\phi \star \phi \star \phi \star \phi\Big)(x) \tag{9}$$

is **perturbatively renormalisable to all orders in** λ**.**

The additional oscillator potential $\Omega^2\tilde{x}^2$ implements mixing between large and small distance scales and results from the renormalisation proof. Maja Buric and Michael Wohlgenannt [**2**] found an interesting interpretation of this additional term: It results as the coupling of the scalar field to the scalar curvature within the truncation procedure.

Here, $\star$ refers to the Moyal product parametrised by the antisymmetric 4×4-matrix Θ, and $\tilde{x} = 2\Theta^{-1}x$. The model is covariant under the Langmann-Szabo duality transformation [**3**] and becomes self-dual at $\Omega = 1$. Certain variants have also been treated, see [**4**] for a review. Evaluation of the β-functions for the coupling constants Ω, λ in first order of perturbation theory leads to a coupled dynamical system which indicates a fixed-point at $\Omega = 1$, while λ remains bounded [**5**, **6**]. The vanishing of the β-function at $\Omega = 1$ was next proven in [**7**] at three-loop order and finally in [**8**] to all orders of perturbation theory. It implies that there is no infinite renormalisation of λ, which makes the nonperturbative construction simpler. The Landau ghost problem is solved. The vanishing of the β-function to all orders has been obtained using a Ward identity [**8**]. We extend this work and derive an integral equation for the planar sector of the two-point function alone by using the Ward identity and Schwinger-Dyson equations. Usually, Schwinger-Dyson equations couple the two-point function to the four-point function. In our model, we show that the Ward identity allows to express the four-point function in terms of the two-point function, resulting in an equation for the two-point function alone. This is achieved in the first step for the bare two-point function. We are able to perform the mass and wavefunction renormalisation directly in the integral equation, giving a *self-consistent non-linear equation for the renormalised two-point function alone*. Higher n-point functions fulfil a *linear* (inhomogeneous) Schwinger-Dyson equation, with the inhomogeneity given by m-point functions with $m < n$. This means that solving our equation for the two-point function leads to a non-perturbative construction of the planar sector of this interacting quantum field theory in four dimensions. Recently we reduced the question of solving this model to solving one nonlinear integral equation in one variable [**11**]. Of course, the next question concerns the nonplanar sector of this model. We know the appropriate Ward identities, an extension of the reviewed ideas to this sector is under discussion.

In the case of the Φ^4 model with negavite coupling constant, it was possible to sum up the planar graphs, but the nonplanar graphs cannot be summed up, due to lack of stability, see [**12**] and [**13**]. In the present model, we have a positive coupling constant and stability is not a problem. Nevertheless the construction of the full model is still a hard task. A new summation technique has been invented recently for such a situation [**14**]. It has been applied to the two-dimensional model already [**15**].

We hope that a detailed analysis of our model will help for a non-perturbative treatment of more realistic quantum field theories. We expect that we can learn much about non-perturbative renormalization of Euclidean quantum field theories in four dimensions from this almost solvable model.

2. Matrix Model

It is convenient to write the action (9) in the matrix base of the Moyal space, see **[1, 9]**. It simplifies enormously at the self-duality point $\Omega = 1$. We write down the resulting action functionals for the *bare* quantities, which involves the bare mass μ_{bare} and the wave function renormalisation $\phi \mapsto Z^{\frac{1}{2}}\phi$. For simplicity we fix the length scale to $\theta = 4$. This gives

$$S = \sum_{m,n\in\mathbb{N}^2_\Lambda} \frac{1}{2}\phi_{mn}H_{mn}\phi_{nm} + V(\phi), \tag{10}$$

$$H_{mn} = Z\big(\mu^2_{bare} + |m| + |n|\big)\;, \qquad V(\phi) = \frac{Z^2\lambda}{4}\sum_{m,n,k,l\in\mathbb{N}^2_\Lambda} \phi_{mn}\phi_{nk}\phi_{kl}\phi_{lm}\;, \tag{11}$$

It is already used that this model has no renormalisation of the coupling constant **[8]**. All summation indices $m, n, \ldots$ belong to $\mathbb{N}^2$, with $|m| := m_1 + m_2$. The symbol $\mathbb{N}^2_\Lambda$ refers to a cut-off in the matrix size. The scalar field is real, $\phi_{mn} = \overline{\phi_{nm}}$.

3. Ward Identity

The key step in the proof **[8]** that the β-function vanishes is the discovery of a Ward identity induced by inner automorphisms $\phi \mapsto U\phi U^\dagger$. Inserting into the connected graphs the special insertion vertex

$$V^{ins}_{ab} := \sum_n (H_{an} - H_{nb})\phi_{bn}\phi_{na} \tag{12}$$

is the same as the difference of graphs with external indices b and a, respectively, $Z(|a| - |b|)G^{ins}_{[ab]\ldots} = G_{b\ldots} - G_{a\ldots}$:

We write Feynman graphs in the self-dual ϕ^4_4-model as ribbon graphs on a genus-g Riemann surface with B external faces. Adding for each external face an external vertex to get a closed surface, the matrix index is constant at every face. Inserting the special vertex V^{ins}_{ab} leads, however, to an index jump from a to b in an external face which meets an external vertex. The corresponding external sources at the jumped face are thus J_{na} and J_{bm} for some other indices m, n. According to the Ward identity, this is the same as the difference between the graphs with face index b and a, respectively:

(13) $Z(|a| - |b|)$ 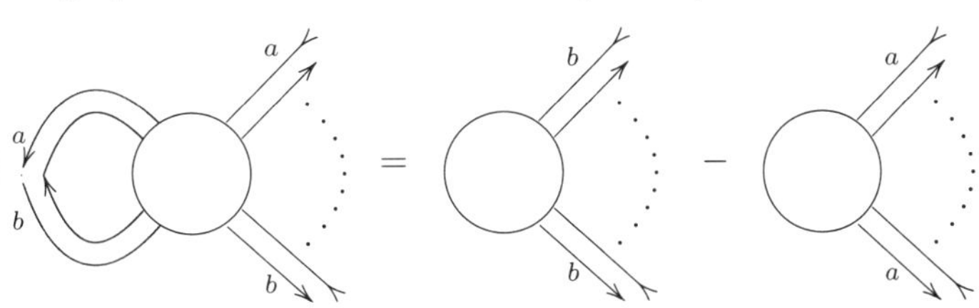

$$Z(|a| - |b|)G^{ins}_{[ab]\ldots} = G_{b\ldots} - G_{a\ldots}\;. \tag{14}$$

The dots in (14) stand for the remaining face indices. We have used $H_{an} - H_{nb} = Z(|a| - |b|)$.

4. Schwinger-Dyson equation

The Schwinger-Dyson equation for the one-particle irreducible two-point function Γ^{ab} reads

(15) $\Gamma_{ab} =$

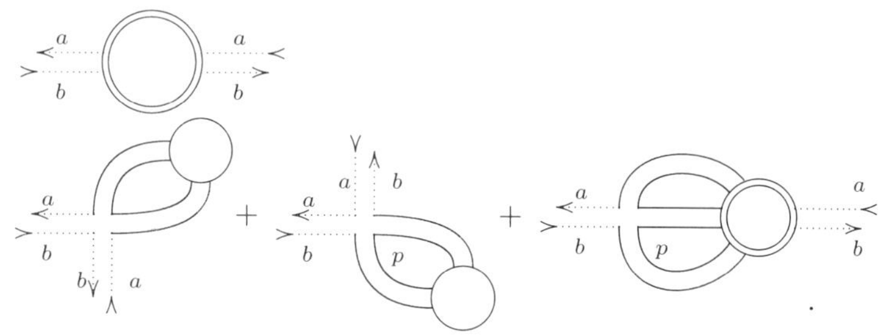

The sum of the last two graphs can be reexpressed in terms of the two-point function with insertion vertex. Adding the left tadpole and using the Ward identity yields

$$\begin{aligned}\Gamma_{ab} &= Z^2\lambda\sum_p\Big(G_{ap} + G_{ab}^{-1}G^{ins}_{[ap]b}\Big) = Z^2\lambda\sum_p\Big(G_{ap} - G_{ab}^{-1}\frac{G_{bp}-G_{ba}}{Z(|p|-|a|)}\Big)\\ &= Z^2\lambda\sum_p\Big(\frac{1}{H_{ap}-\Gamma_{ap}} + \frac{1}{H_{bp}-\Gamma_{bp}} - \frac{1}{H_{bp}-\Gamma_{bp}}\frac{(\Gamma_{bp}-\Gamma_{ab})}{Z(|p|-|a|)}\Big)\,. \end{aligned}\tag{16}$$

This is a closed equation for the two-point function alone. It involves the divergent quantities Γ_{bp} and Z, μ_{bare}.

5. Renormalization

Introducing the renormalised planar two-point function Γ^{ren}_{ab} by Taylor expansion $\Gamma_{ab} = Z\mu^2_{bare} - \mu^2 + (Z-1)(|a|+|b|) + \Gamma^{ren}_{ab}$ and imposing the renormalization condition $\Gamma^{ren}_{00} = 0$ and $(\partial\Gamma^{ren})_{00} = 0$, we obtain a coupled system of equations for Γ^{ren}_{ab}, Z and μ_{bare}. It leads to a closed equation for the renormalised function Γ^{ren}_{ab} alone, which is further analysed in the integral representation.

We replace the indices in $a, b, \ldots \mathbb{N}$ by continuous variables in $\mathbb{R}_+$. Equation (16) depends only on the length $|a| = a_1 + a_2$ of indices. The Taylor expansion respects this feature, so that we replace $\sum_{p\in\mathbb{N}^2_\Lambda}$ by $\int_0^\Lambda |p|\, dp$. After a convenient change of variables $|a| =: \mu^2\frac{\alpha}{1-\alpha}$, $|p| =: \mu^2\frac{\rho}{1-\rho}$ and

$$\Gamma^{ren}_{ab} =: \mu^2\frac{1-\alpha\beta}{(1-\alpha)(1-\beta)}\Big(1 - \frac{1}{G_{\alpha\beta}}\Big)\,,\tag{17}$$

and using an identity resulting from the symmetry $G_{0\alpha} = G_{\alpha 0}$, we arrive at [**10**]:

THEOREM 1. *The renormalised planar connected two-point function $G_{\alpha\beta}$ of the self-dual noncommutative ϕ^4_4-theory satisfies the integral equation*

$$(18)\quad \begin{aligned} G_{\alpha\beta} = 1 + \lambda\Big(& \frac{1-\alpha}{1-\alpha\beta}(\mathcal{M}_\beta - \mathcal{L}_\beta - \beta\mathcal{Y}) + \frac{1-\beta}{1-\alpha\beta}(\mathcal{M}_\alpha - \mathcal{L}_\alpha - \alpha\mathcal{Y}) \\ & + \frac{1-\beta}{1-\alpha\beta}\Big(\frac{G_{\alpha\beta}}{G_{0\alpha}} - 1\Big)(\mathcal{M}_\alpha - \mathcal{L}_\alpha + \alpha\mathcal{N}_{\alpha 0}) \\ & - \frac{\alpha(1-\beta)}{1-\alpha\beta}(\mathcal{L}_\beta + \mathcal{N}_{\alpha\beta} - \mathcal{N}_{\alpha 0}) + \frac{(1-\alpha)(1-\beta)}{1-\alpha\beta}(G_{\alpha\beta} - 1)\mathcal{Y}\Big), \end{aligned}$$

where $\alpha, \beta \in [0, 1)$,

$$\mathcal{L}_\alpha := \int_0^1 d\rho\, \frac{G_{\alpha\rho} - G_{0\rho}}{1-\rho}\,, \quad \mathcal{M}_\alpha := \int_0^1 d\rho\, \frac{\alpha\, G_{\alpha\rho}}{1-\alpha\rho}\,, \quad \mathcal{N}_{\alpha\beta} := \int_0^1 d\rho\, \frac{G_{\rho\beta} - G_{\alpha\beta}}{\rho - \alpha}\,,$$

and $\mathcal{Y} = \lim_{\alpha\to 0} \frac{\mathcal{M}_\alpha - \mathcal{L}_\alpha}{\alpha}$.

Recently we related the construction of this noncommutative quantum field theory to the problem of solving a nonlinear integral equation in one variable [**11**], which we review next.

6. Nonperturbative Construction of this model

We rewrite equation (18) in terms of $D_{\alpha\beta} := \frac{\alpha(1-\beta)}{\beta(1-\alpha)}\Big(\frac{(1-\beta)}{1-\alpha\beta}G_{\alpha\beta} - G_{\alpha 0}\Big)$ and obtain after simple manipulations the integral equation

$$(19)\quad \frac{\beta(1-\alpha)}{\alpha(1-\beta)} + \frac{1 + \lambda\mathcal{Y} + \lambda\pi\alpha\mathcal{H}_\alpha\big[G_{\bullet 0}\big]}{\alpha G_{\alpha 0}} D_{\alpha\beta} - \lambda\pi\mathcal{H}_\alpha\big[D_{\bullet\beta}\big] = -G_{\alpha 0},$$

which is of the Carleman type. $\mathcal{Y}$ is defined as

$$(20)\quad -\lambda\pi\mathcal{H}_0[D_{\bullet 0}] = \frac{\lambda\mathcal{Y}}{1+\lambda\mathcal{Y}}.$$

Here we assume that $D_{\alpha\beta}$ is Hölder continuous. The finite Hilbert transform is given by $\mathcal{H}_\alpha[f(\bullet)] := \frac{1}{\pi}\lim_{\epsilon\to 0}\Big(\int_0^{\alpha-\epsilon} + \int_{\alpha+\epsilon}^1\Big)\frac{f(\rho)}{\rho-\alpha}$.
Equation (19) is a singular linear integral equation of the Carleman type. We quote its solution [Carleman 1922, Tricomi 1957]

Theorem: The singular linear integral equation $a(x)y(x) - \lambda\pi\mathcal{H}_x[y] = f(x)$, $x \in [-1, 1]$ is for $a(x)$ continuous and Hölder continuous near ± 1 and $f \in L^p$ is solved by

$$(21)\quad y(x) = \frac{\sin(\theta(x))}{\lambda\pi}\Big(f(x)\cos(\theta(x)) + e^{\mathcal{H}_x[\theta]}\mathcal{H}_x\Big[e^{-\mathcal{H}_\bullet[\theta]}f(\bullet)\sin(\theta(\bullet))\Big] + \frac{Ce^{\mathcal{H}_x[\theta]}}{1-x}\Big)$$

$$(22)\quad \theta(x) = \operatorname*{arctan}_{[0,\,\pi]}\Big(\frac{\lambda\pi}{a(x)}\Big)\,,\ \sin(\theta(x)) = \frac{|\lambda\pi|}{\sqrt{(a(x))^2 + (\lambda\pi)^2}}$$

where C is an arbitrary constant. We assume first: $C = 0$
We apply the solution of the Carleman equation to (19) and obtain for its solution:

$$(23)\quad \frac{(1-\beta)}{1-\alpha\beta}\frac{G_{\alpha\beta}}{1+\lambda\mathcal{Y}} = \frac{\sin(\theta_\beta(\alpha))}{|\lambda|\pi\alpha}e^{\mathcal{H}_\alpha[\theta_\beta(\bullet)] - \mathcal{H}_0[\theta_0(\bullet)] + \mathcal{H}_1[\theta_0(\bullet) - \theta_\beta(\bullet)]}$$

$$\frac{\lambda\mathcal{Y}}{1+\lambda\mathcal{Y}} = \int_0^1 d\rho\, \frac{\sin^2(\theta_0(\rho))}{\lambda\pi^2\rho^2}, \tag{24}$$

$$\theta_\beta(\alpha) = \underset{[0,\pi]}{\arctan}\left(\frac{\lambda\pi\alpha}{\frac{\beta(1-\alpha)}{1-\beta} + \frac{1+\lambda\mathcal{Y}+\lambda\pi\alpha\mathcal{H}_\alpha[G_{\bullet 0}]}{G_{\alpha 0}}}\right) \tag{25}$$

For the proof one uses the Carleman equation $\lambda\pi\cot\theta_0(\alpha)G_{\alpha 0} - \lambda\pi\mathcal{H}_\alpha[G_{\bullet 0}] = \frac{1+\lambda\mathcal{Y}}{\alpha}$ and Tricomi's identity $e^{-\mathcal{H}_\alpha[\theta_\beta]}\cos(\theta_\beta(\alpha)) + \mathcal{H}_\alpha\left[e^{-\mathcal{H}_\bullet[\theta_\beta]}\sin(\theta_\beta(\bullet))\right] = 1$.
The Carleman equation computes $G_{\alpha\beta}$, as a consequence it implies that $G_{\alpha\beta} \geq 0$! Therefore $G_{0\beta}$ can be evaluated and this implies also a self-consistency equation for $G_{\beta 0}$, since symmetry forces $G_{\beta 0} = G_{0\beta}$.
This leads to the Master equation, whose solution determines the theory completely:

$$G_{\beta 0} = \frac{1+\lambda\mathcal{Y}}{1+(1-\beta)\lambda\mathcal{Y}} exp(-\lambda\int_0^{\frac{\beta}{1-\beta}} dt \int_0^1 \frac{d\rho}{(\lambda\pi\rho)^2 + (t(1-\rho) + \frac{1+\lambda\mathcal{Y}+\lambda\pi\rho\mathcal{H}_\rho[G_{\bullet 0}]}{G_{\rho 0}})^2}) \tag{26}$$

) provided it exists. Of course, together with $\lambda\mathcal{Y}$, which has to be determined from equation (24). Up to now, we deduced various non-perturbative results from this system of equations and used computer calculations for the visualization of the solution of (26). As expected, there is a big difference between the case $\lambda > 0$ and $\lambda < 0$. For positive $\lambda > 0$ we deduce, that $\frac{(1+(1-\beta)\lambda\mathcal{Y})}{1+\lambda\mathcal{Y}}G_{\beta 0} \in \mathcal{C}^1([0,1[)$, is monotonously decreasing and positive. Therefore the limiting value G_{10} exists and $G_{\beta 0} \in \mathcal{C}[0,1]$. For $\lambda < 0$ $\frac{(1+(1-\beta)\lambda\mathcal{Y})}{1+\lambda\mathcal{Y}}G_{\beta 0} \in \mathcal{C}^1([0,1[)$ is monotonously increasing and positive, therefore $G_{\beta 0}$ is unbounded at $\beta = 1$.
Let $\lambda > 0$, $G = TG$ be the master equation and F be within the Hölder class of index λ. Recall $Z^{-1}(G) = 1 + \lambda\mathcal{Y}_G - \lambda\int_0^1 d\rho\,\frac{G_{\rho 0}}{1-\rho}$. We can prove, that if $F(1) \neq 0$, then $(TF)(1) = 0$, if $Z^{-1}(F) \geq \delta > 0$, then $(TF)(1) \geq \epsilon > 0$. If $Z^{-1}(F) < 0$, then $(TF)(1) = 0$. As a consequence we deduce, that $G_{10} = 0$ and $Z^{-1}(G) \leq 0$. But this means, that if

$$G_{\alpha 0} = 0 \quad \Rightarrow \quad 1 + \lambda\mathcal{Y} + \lambda\pi\alpha\mathcal{H}_\alpha\big[G_{\bullet 0}\big] = 0$$

For $\alpha = 1$ this means $Z^{-1}(G) = 0$.

7. Four-point Schwinger-Dyson equation

The knowledge of the two-point function allows a successive construction of the whole theory. As an example we mention the planar connected four-point function G_{abcd}. Following the a-face in direction of an arrow, there is a distinguished vertex at which the first ab-line starts. For this vertex there are two possibilities for the matrix index of the diagonally opposite corner to the a-face: either c or a summation vertex p:

(27)

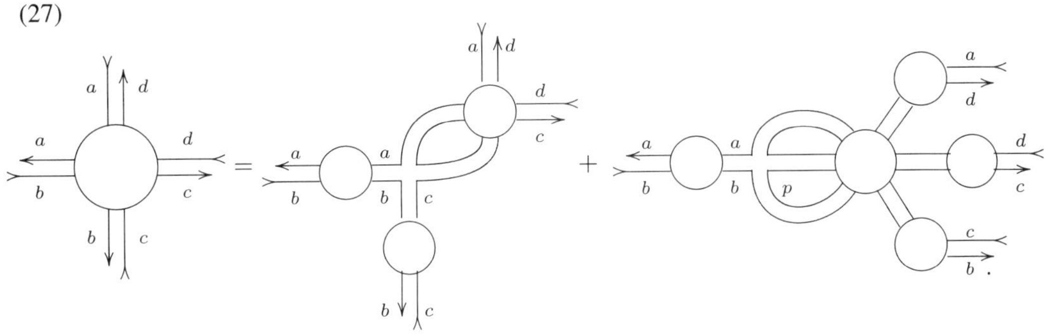

We write the first contribution as a product of the vertex $Z^2\lambda$, the left connected two-point function, the downward two-point function and an insertion, which is reexpressed by means of the Ward-identity. After amputation of the external two-point functions we obtain the Schwinger-Dyson equation for the *renormalised* 1PI four-point function $G_{abcd} = G_{ab}G_{bc}G_{cd}G_{da}\Gamma^{ren}_{abcd}$ as follows:

(28)

$$\Gamma^{ren}_{abcd} = Z\lambda\frac{1}{|a|-|c|}\Big(\frac{1}{G_{ad}} - \frac{1}{G_{cd}}\Big) + Z\lambda\sum_p \frac{1}{|a|-|p|}G_{pb}\Big(\frac{G_{dp}}{G_{ad}}\Gamma^{ren}_{pbcd} - \Gamma^{ren}_{abcd}\Big)\,.$$

We introduce the 1PI function and pass to the integral representation and to the variables α and β and find for $\Gamma_{\alpha\beta\gamma\delta} := \Gamma^{ren}_{abcd}$ an integral equation, which manipulated appropriately allows again to take the limit $\xi \to 1$ after insertion of the expression for the wave function renormalisation constant.

THEOREM 2. *The renormalised planar 1PI four-point function $\Gamma_{\alpha\beta\gamma\delta}$ of self-dual noncommutative ϕ^4_4-theory (with continuous indices $\alpha, \beta, \gamma, \delta \in [0,1)$) satisfies the integral equation*

$$\Gamma_{\alpha\beta\gamma\delta} = \lambda\cdot\frac{\Big(1 - \dfrac{(1-\alpha)(1-\gamma\delta)(G_{\alpha\delta}-G_{\gamma\delta})}{G_{\gamma\delta}(1-\delta)(\alpha-\gamma)} + \displaystyle\int_0^1 \rho\,d\rho\frac{(1-\beta)(1-\alpha\delta)G_{\beta\rho}G_{\delta\rho}}{(1-\beta\rho)(1-\delta\rho)}\frac{\Gamma_{\rho\beta\gamma\delta}-\Gamma_{\alpha\beta\gamma\delta}}{\rho-\alpha}\Big)}{G_{\alpha\delta} + \lambda\Big((\mathcal{M}_\beta - \mathcal{L}_\beta - \mathcal{Y})G_{\alpha\delta} + \displaystyle\int_0^1 d\rho\frac{G_{\alpha\delta}G_{\beta\rho}(1-\beta)}{(1-\delta\rho)(1-\beta\rho)} + \int_0^1 \rho\,d\rho\frac{(1-\beta)(1-\alpha\delta)G_{\beta\rho}}{(1-\beta\rho)(1-\delta\rho)}\frac{(G_{\rho\delta}-G_{\alpha\delta})}{(\rho-\alpha)}\Big)}\,. \tag{29}$$

In lowest order we find

$$\Gamma_{\alpha\beta\gamma\delta} = \lambda - \lambda^2\Big(\frac{(1-\gamma)(I_\alpha-\alpha)-(1-\alpha)(I_\gamma-\gamma)}{\alpha-\gamma} + \frac{(1-\delta)(I_\beta-\beta)-(1-\beta)(I_\delta-\delta)}{\beta-\delta}\Big) + \mathcal{O}(\lambda^3)\,. \tag{30}$$

Note that $\Gamma_{\alpha\beta\gamma\delta}$ is cyclic in the four indices, and that $\Gamma_{0000} = \lambda + \mathcal{O}(\lambda^3)$.
In our recent work, we have been able to solve equation (29) in terms of the two point function and a remarkable simple expression results:

$$\Gamma_{\alpha\beta\gamma\delta} = \frac{\lambda}{(\alpha-\gamma)(\beta-\delta)}\Big(\frac{(1-\alpha\delta)}{G_{\alpha\delta}}\frac{(1-\gamma\beta)}{G_{\gamma\beta}} - \frac{(1-\alpha\beta)}{G_{\alpha\beta}}\frac{(1-\gamma\delta)}{G_{\gamma\delta}}\Big)\,. \tag{31}$$

It was now possible to evaluate the effective coupling in terms of the bare coupling constant. Although the scale is changes ba an infinite amount, a finite coupling constant renormalization results.

8. Conclusions

A remarkable result concerns the appearance of the nontrivial fixed point at $\Omega = 1$, proven to all orders in pertubation theory. We used Ward identities and Schwinger-Dyson equations to deduce integral equations for the renormalized N-point functions. We reduced the construction of this nontrivial noncommutative quantum field theory to solving one nonlinear integral equation for a function of one variable. A survey of this construction is given in [**11**].

We believe, that the first nontrivial four dimensional quantum field theory model, where one is able to sum up the Feynman pertubation expansion, will allow to learn a lot about renormalization.

The zero of the beta function occurs in the one-loop calculation for the degenerate model too. There are attempts to deduce implications for cosmology from space-time noncommutativity. But, of course, there is, up to now, no effect known, which allows a check by experiments in the near future.

9. Acknowledgment

I (H.G.) would like to thank Helge Holden, Barry Simon and Gerald Teschl for the kind invitation to contribute to this volume for my "old friend" Fritz Gesztesy.

References

[1] Harald Grosse and Raimar Wulkenhaar, *Renormalisation of ϕ^4-theory on noncommutative $\mathbb{R}^4$ in the matrix base*, Comm. Math. Phys. **256** (2005), no. 2, 305–374, DOI 10.1007/s00220-004-1285-2. MR2160797 (2006g:81133)

[2] Maja Burić and Michael Wohlgenannt, *Geometry of the Grosse-Wulkenhaar model*, J. High Energy Phys. **3** (2010), 053, 17, DOI 10.1007/JHEP03(2010)053. MR2653469 (2011j:81168)

[3] Edwin Langmann and Richard J. Szabo, *Duality in scalar field theory on noncommutative phase spaces*, Phys. Lett. B **533** (2002), no. 1-2, 168–177, DOI 10.1016/S0370-2693(02)01650-7. MR1911944 (2003d:81305)

[4] Vincent Rivasseau, *Non-commutative renormalization*, Quantum spaces, Prog. Math. Phys., vol. 53, Birkhäuser, Basel, 2007, pp. 19–107, DOI 10.1007/978-3-7643-8522-4_2. MR2382235 (2009e:81143)

[5] H. Grosse and R. Wulkenhaar, *The β-function in duality-covariant non-commutative ϕ^4-theory*, Eur. Phys. J. C Part. Fields **35** (2004), no. 2, 277–282, DOI 10.1140/epjc/s2004-01853-x. MR2075076 (2005b:81213)

[6] Harald Grosse and Raimar Wulkenhaar, *Renormalisation of ϕ^4-theory on non-commutative $\mathbb{R}^4$ to all orders*, Lett. Math. Phys. **71** (2005), no. 1, 13–26, DOI 10.1007/s11005-004-5116-3. MR2136734 (2006d:81172)

[7] M. Disertori and V. Rivasseau, "Two and three loops beta function of non commutative ϕ_4^4 theory," Eur. Phys. J. C **50** (2007) 661 [arXiv:hep-th/0610224].

[8] Margherita Disertori, Razvan Gurau, Jacques Magnen, and Vincent Rivasseau, *Vanishing of beta function of non-commutative Φ_4^4 theory to all orders*, Phys. Lett. B **649** (2007), no. 1, 95–102, DOI 10.1016/j.physletb.2007.04.007. MR2308801 (2008f:81261)

[9] Harald Grosse and Raimar Wulkenhaar, *Renormalisation of ϕ^4-theory on noncommutative $\mathbb{R}^2$ in the matrix base*, J. High Energy Phys. **12** (2003), 019, 26 pp. (electronic), DOI 10.1088/1126-6708/2003/12/019. MR2041157 (2004m:81162)

[10] H. Grosse and R. Wulkenhaar, "Progress in solving a noncommutative quantum field theory in four dimensions," arXiv:0909.1389 [hep-th].

[11] H. Grosse and R. Wulkenhaar, "Self-dual noncommutative ϕ^4-theory in four dimensions is a non-perturbatively solvable and non-trivial quantum field theory," arXiv:1205.0465 [math-ph].

[12] Gerard 't Hooft, *Rigorous construction of planar diagram field theories in four-dimensional Euclidean space*, Comm. Math. Phys. **88** (1983), no. 1, 1–25. MR691195 (84h:81065)

[13] Vincent Rivasseau, *Construction and Borel summability of planar 4-dimensional Euclidean field theory*, Comm. Math. Phys. **95** (1984), no. 4, 445–486. MR767190 (86a:81046)
[14] Vincent Rivasseau, *Constructive matrix theory*, J. High Energy Phys. **9** (2007), 008, 13, DOI 10.1088/1126-6708/2007/09/008. MR2342423 (2008k:81187)
[15] Z. Wang, "Constructive Renormalization of 2-dimensional Grosse-Wulkenhaar Model," arXiv:1205.0196 [hep-th].

FACULTY OF PHYSICS, UNIVERSITY OF VIENNA
E-mail address: harald.grosse@univie.ac.at

MATHEMATISCHES INSTITUT DER WESTFÄLISCHEN WILHELMS-UNIVERSITÄT, MÜNSTER, GERMANY
E-mail address: raimar@math.uni-muenster.de

Proceedings of Symposia in Pure Mathematics
Volume **87**, 2013
http://dx.doi.org/10.1090/pspum/087/01434

Periodic conservative solutions for the two-component Camassa–Holm system

Katrin Grunert, Helge Holden, and Xavier Raynaud

Dedicated with admiration to Fritz Gesztesy on the occasion of his sixtieth anniversary

Abstract. We construct a global continuous semigroup of weak periodic conservative solutions to the two-component Camassa–Holm system, $u_t - u_{txx} + \kappa u_x + 3uu_x - 2u_x u_{xx} - uu_{xxx} + \eta\rho\rho_x = 0$ and $\rho_t + (u\rho)_x = 0$, for initial data $(u, \rho)|_{t=0}$ in $H^1_{\rm per} \times L^2_{\rm per}$. It is necessary to augment the system with an associated energy to identify the conservative solution. We study the stability of these periodic solutions by constructing a Lipschitz metric. Moreover, it is proved that if the density ρ is bounded away from zero, the solution is smooth. Furthermore, it is shown that given a sequence ρ_0^n of initial values for the densities that tend to zero, then the associated solutions u^n will approach the global conservative weak solution of the Camassa–Holm equation. Finally it is established how the characteristics govern the smoothness of the solution.

1. Introduction

In this paper we analyze periodic and conservative weak global solutions of the two-component Camassa–Holm (2CH) system which reads (with $\kappa \in \mathbb{R}$ and $\eta \in (0, \infty)$)

$$u_t - u_{txx} + \kappa u_x + 3uu_x - 2u_x u_{xx} - uu_{xxx} + \eta\rho\rho_x = 0, \tag{1.1a}$$

$$\rho_t + (u\rho)_x = 0. \tag{1.1b}$$

The special case when ρ vanishes identically reduces the system to the celebrated and well-studied Camassa–Holm (CH) equation, first studied in the seminal paper [**4**]. The present system was first introduced by Olver and Rosenau in [**23**, Eq. (43)], and derived in the context of water waves in [**6**], showing η positive and ρ nonnegative to be the physically relevant case. Conservative solutions on the full line for the 2CH system have been studied, see, e.g., [**12**]. However, periodic and conservative solutions for the 2CH system have not been analyzed so far, and this paper aims to fill that gap. It offers some technical challenges that will be described below.

2010 *Mathematics Subject Classification.* Primary 35Q53, 35B35; Secondary 35B20.
Key words and phrases. Two-component Camassa–Holm system, periodic and conservative solutions.
Research supported in part by the Research Council of Norway, and the Austrian Science Fund (FWF) under Grant No. J3147.

The 2CH system can suitably be rewritten as

$$u_t + uu_x + P_x = 0, \tag{1.2a}$$

$$\rho_t + (u\rho)_x = 0, \tag{1.2b}$$

where P is implicitly defined by

$$P - P_{xx} = u^2 + \kappa u + \frac{1}{2}u_x^2 + \eta\frac{1}{2}\rho^2. \tag{1.3}$$

The reason for the intense study of the CH equation is its surprisingly rich structure. In the context of the present paper, the focus is on the wellposedness of global weak solutions of the Cauchy problem. There is an intrinsic dichotomy in the solution that appears after wave breaking, namely between solutions characterized either by conservation or dissipation of the associated energy. The two classes of solutions are for obvious reasons denoted conservative and dissipative, respectively. The fundamental nature of the problem can be understood by the following pregnant example, for simplicity presented here on the full line, rather than the periodic case. The CH equation with $\kappa = 0$ has as special solutions so-called multipeakons given by

$$u(t,x) = \sum_{i=1}^{n} p_i(t)e^{-|x-q_i(t)|},$$

where the $(p_i(t), q_i(t))$ satisfy the explicit system of ordinary differential equations

$$\dot{q}_i = \sum_{j=1}^{n} p_j e^{-|q_i-q_j|}, \quad \dot{p}_i = \sum_{j=1}^{n} p_i p_j \operatorname{sgn}(q_i - q_j)e^{-|q_i-q_j|}.$$

In the special case of $n = 2$ and $p_1 = -p_2$ and $q_1 = -q_2 < 0$ at $t = 0$, the solution consists of two "peaks", denoted peakons, that approach each other. At time $t = t_c$ the two peakons annihilate each other, an example of wave breaking, and the solution satisfies $u = u_x = 0$ pointwise at that time. For positive time two possibilities exist; one is to let the solution remain equal to zero (the dissipative solution), and other one being that that two peakons reemerge (the conservative solution). A more careful analysis reveals that the $H^1(\mathbb{R})$ norm of u remains finite, while u_x becomes singular, at $t = t_c$, and there is an accumulation of energy in the form of a Dirac delta-function at the point of annihilation. The consequences for the wellposedness of the Cauchy problem are severe. The continuation of the solution past wave breaking has been studied, see [**1, 2, 19, 20**]. The method to handle the dichotomy is by reformulating the equation in Lagrangian variables, and analyze carefully the behavior in those variables. We will detail this construction later in the introduction.

The 2CH system has, in spite of its brief history, been studied extensively, and it is not possible to include a complete list of references here. However, we mention [**12, 24**], where a similar approach to the present one, has been employed. The case with $\eta = -1$ has been discussed in [**7**]; our approach does not extend to the case of η negative. In [**14**] it is shown that if the initial density $\rho_0 > 0$, then the solution exists globally and this result is extended here to a local result, Theorem 4.4, where we show how the characteristics govern the local smoothness. For other related results pertaining to the present system, please see [**14–16**]. There exists other two-component generalizations of the CH equation than the one studied here; see, e.g., [**5, 8, 13, 17, 22**].

We now turn to the discussion of the present paper. For simplicity we assume that $\eta = 1$ and $\kappa = 0$. We first make a change from Eulerian to Lagrangian variables and introduce a new energy variable. The change of variables, which we now will detail, is related to the one used in [**19**] and, in particular, [**10**]. Assume that $(u, \rho) = (u(x,t), \rho(x,t))$ is a solution of (1.1), and define the characteristics $y = y(t,\xi)$ by

$$y_t(t,\xi) = u(t, y(t,\xi))$$

and the Lagrangian velocity by

$$U(t,\xi) = u(t, y(t,\xi)).$$

By introducing the Lagrangian energy density ν and density r by

$$\begin{aligned}
\nu(t,\xi) &= u^2(t,y(t,\xi))y_\xi(t,\xi) + u_x^2(t,y(t,\xi))y_\xi(t,\xi) + \rho^2(t,y(t,\xi))y_\xi(t,\xi),\\
r(t,\xi) &= \rho(t,y(t,\xi))y_\xi(t,\xi),
\end{aligned}$$

we find that the system can be rewritten as (introducing $\zeta(t,\xi) = y(t,\xi) - \xi$ for technical reasons)

$$\zeta_t = U, \tag{1.4a}$$

$$U_t = -Q, \tag{1.4b}$$

$$\nu_t = -2QUy_\xi + (3U^2 - 2P)U_\xi, \tag{1.4c}$$

$$r_t = 0, \tag{1.4d}$$

where the functions P and Q are explicitly given by (2.4) and (2.5), respectively. We then establish the existence of a unique global solution for this system (see Theorem 2.3), and we show that the solutions form a continuous semigroup in an appropriate norm. In order to solve the Cauchy problem (1.2) we have to choose the initial data appropriately. To accommodate for the possible concentration of energy we augment the natural initial data u_0 and ρ_0 with a nonnegative Radon measure μ_0 such that the absolutely continuous part $\mu_{0,\mathrm{ac}}$ equals $\mu_{0,\mathrm{ac}} = (u_0^2 + u_{0,x}^2 + \rho_0^2)\,dx$. The precise translation of these initial data is given in Definition 2.5. One then solves the system in Lagrangian coordinates. The translation back to Eulerian variables is described in Theorem 2.9. However, there is an intrinsic problem in this latter translation if one wants a continuous semigroup. This is due to the problem of *relabeling*; to each solution in Eulerian variables there exist several distinct solutions in Lagrangian variables as there are additional degrees of freedom in the Lagrangian variables. In order to resolve this issue to get a continuous semigroup, one has to identify Lagrangian functions corresponding to one and the same Eulerian solution. This is treated in Theorem 2.10. The main existence theorem, Theorem 4.2, states that for $u_0 \in H^1_{\mathrm{per}}$ and $\rho_0 \in L^2_{\mathrm{per}}$ and μ_0 a nonnegative Radon measure with absolutely continuous part $\mu_{0,\mathrm{ac}}$ such that $\mu_{0,\mathrm{ac}} = (u_0^2 + u_{0,x}^2 + \rho_0^2)\,dx$, there exists a continuous semigroup T_t such that (u,ρ), where $(u,\rho,\mu)(t) = T_t(u_0,\rho_0,\mu_0)$, is a weak global and conservative solution of the 2CH system. In addition, the measure μ satisfies

$$\mu_t + (u\mu)_x = (u^3 - 2Pu)_x,$$

weakly. Furthermore, for almost all times the measure μ is absolutely continuous and $\mu = (u^2 + u_x^2 + \rho^2)\,dx$.

The solution so constructed is not Lipschitz continuous in any of the natural norms, say H^1 or L^p. Thus it is an intricate problem to identify a metric that deems the solution Lipschitz continuous, see [**9**, **11**]. For a discussion of Lipschitz metrics in the setting of the Hunter–Saxton equation and relevant examples from ordinary differential equations, see [**3**]. The metric we construct here has to distinguish between conservative and dissipative solutions, and it is closely connected with the construction of the semigroup in Lagrangian variables. We commence by defining a metric in Lagrangian coordinates. To that end, let

$$J(X_\alpha, X_\beta) = \inf_{f,g\in G} \|X_\alpha \bullet f - X_\beta \bullet g\|_E \,. \tag{1.5}$$

Here G contains the labels used for the relabeling, see Definition 2.6, and $X \bullet f$ denotes the solution X with label f. The function J is invariant with respect to relabeling, yet it is not a metric as it does not satisfy the triangle inequality. Introduce $d(X_\alpha, X_\beta)$ by

$$d(X_\alpha, X_\beta) = \inf \sum_{i=1}^{N} J(X_{n-1}, X_n), \quad X_\alpha, X_\beta \in \mathcal{F}, \tag{1.6}$$

where the infimum is taken over all finite sequences $\{X_n\}_{n=0}^N \in \mathcal{F}$ satisfying $X_0 = X_\alpha$ and $X_N = X_\beta$. This will be proved to be a Lipschitz metric in Lagrangian variables. Next the metric is transformed into Eulerian variables, and Theorem 4.3 identifies a metric, denoted $d_{\mathcal{D}^M}$, such that the solution is Lipschitz continuous.

Due to the non-local nature of P in (1.3), see (4.9), information travels with infinite speed. Yet, we show in Theorem 4.4 that regularity is a local property in the following precise sense. A solution is said to be p-regular, with $p \geq 1$ if

$$u_0 \in W^{p,\infty}(x_0, x_1), \quad \rho_0 \in W^{p-1,\infty}(x_0, x_1), \quad \text{and } \mu_0 = \mu_{0,ac} \text{ on } (x_0, x_1),$$

and that

$$\rho_0(x)^2 \geq c > 0$$

for $x \in (x_0, x_1)$. If the initial data (u_0, ρ_0, μ_0) is p-regular, then the solution $(u, \rho, \mu)(t, \cdot\,)$, for $t \in \mathbb{R}_+ = [0, \infty)$, remains p-regular on the interval $(y(t, \xi_0), y(t, \xi_1))$, where ξ_0 and ξ_1 satisfy $y(0, \xi_0) = x_0$ and $y(0, \xi_1) = x_1$ and are defined as

$$\xi_0 = \sup\{\xi \in \mathbb{R} \mid y(0, \xi) \leq x_0\} \text{ and } \xi_1 = \inf\{\xi \in \mathbb{R} \mid y(0, \xi) \geq x_1\}.$$

It is interesting to consider how the standard CH equation is obtained when the density ρ vanishes since the CH equation formally is obtained when ρ is identically zero in the 2CH system. In order to analyze the behavior of the solution, we need to have a sufficiently strong stability result. Consider a sequence of initial data $(u_0^n, \rho_0^n, \mu_0^n)$ such that $u_0^n \to u_0$ in H^1_{per}, $\rho_0^n \to 0$ in L^2_{per} with $\rho_0^n \geq d_n > 0$ for all n. Assume that the initial measure is absolutely continuous, that is, $\mu_0^n = \mu_{0,\text{ac}}^n = ((u_{0,x}^n)^2 + (\rho_0^n)^2)\, dx$. Then we show in Theorem 4.6 that the sequence $u^n(t)$ converges in L^∞_{per} to the weak, conservative global solution of the Camassa–Holm equation with initial data u_0. To illustrate this result we have plotted, in Figure 1, a peakon anti-peakon solution u of the Camassa–Holm equation (that is, with ρ identically zero) which enjoys wave breaking. In addition, we have plotted the corresponding energy function $u^2 + u_x^2$. A closer analysis reveals that at the time $t = t_c$ of wave breaking, all the energy is concentrated at one point, which can be described as (a multiple of) a Dirac delta function. In contrast to that, Figure 2 shows that if we choose as initial condition the same peakon anti-peakon function u_0 together with $\rho_0(x) = 0.5$, then no wave breaking takes place, but at the time

t_c where $u(t_c, x) \approx 0$, a considerable part of the energy is transferred from $u^2 + u_x^2$ to ρ^2, while the total energy $\int_0^1 (u^2 + u_x^2 + \rho^2)dx$ remains constant.

Personal note by HH. We started our collaboration almost 30 years ago, on solvable models in quantum mechanics. We have shared the intense joy of discovery as well as long and often strenuous computations. Fritz's ability to get computations right never ceases to amaze me. He is willing to spend whatever time is required to get the details right. His care for detail and accurate references to the pertinent literature represent professionalism and scholarship the way it should be, but rarely is. As G.-C. Rota states[1] *"Sitzfleisch" (the ability to spend endless hours at a desk doing gruesome work) is considered by mathematicians to be a better gauge of success than any other definition of talent.* Fritz has a lot of Sitzfleisch!

2. Eulerian and Lagrangian variables

The two-component Camassa–Holm (2CH) system with $\kappa \in \mathbb{R}$ and $\eta \in (0, \infty)$ reads

$$u_t - u_{txx} + \kappa u_x + 3uu_x - 2u_x u_{xx} - uu_{xxx} + \eta\rho\rho_x = 0, \tag{2.1a}$$

$$\rho_t + (u\rho)_x = 0. \tag{2.1b}$$

If (u, ρ) is a solution of (2.1) then the pair (v, τ), given by $v(t, x) = u(t, x - \alpha t) + \alpha$ and $\tau(t, x) = \sqrt{\beta}\rho(t, x)$, is a solutions to the 2CH system with κ and η replaced by $\kappa - 2\alpha$ and $\frac{\eta}{\beta}$, respectively. Thus we can assume without loss of generality that $\kappa = 0$ and $\eta = 1$. Moreover, we will only consider the Cauchy problem for initial data, and hence also of solutions, of period unity, i.e., $u(t, x+1) = u(t, x)$ and $\rho(t, x+1) = \rho(t, x)$. All our results carry over with only slight modifications to the case of a general period.

To any pair (u_0, ρ_0) in $H^1_{\mathrm{per}} \times L^2_{\mathrm{per}}$ we can introduce the corresponding Lagrangian coordinates $(y(0, \xi), U(0, \xi), \nu(0, \xi), r(0, \xi))$ and describe their time evolution using the weak formulation of the 2CH system. Namely, the characteristics $y(t, \xi)$ are defined as solutions of

$$y_t(t, \xi) = u(t, y(t, \xi))$$

for a given $y(0, \xi)$ such that $y(0, \xi+1) = y(0, \xi)+1$. The Lagrangian velocity $U(t, \xi)$ defined as

$$U(t, \xi) = u(t, y(t, \xi)).$$

The energy derivative reads

$$\nu(t, \xi) = (u^2 + u_x^2 + \rho^2)(t, y(t, \xi))y_\xi(t, \xi)$$

together with the energy $h(t) = \int_0^1 \nu(t, \xi)d\xi$, and, finally,

$$r(t, \xi) = \rho(t, y(t, \xi))y_\xi(t, \xi)$$

is the Lagrangian density.

Rewriting the 2CH system as

$$u_t + uu_x + P_x = 0, \tag{2.2a}$$

$$\rho_t + (u\rho)_x = 0, \tag{2.2b}$$

[1]In *Indiscrete Thoughts*, Birkhauser, Boston, 1997, p. 64.

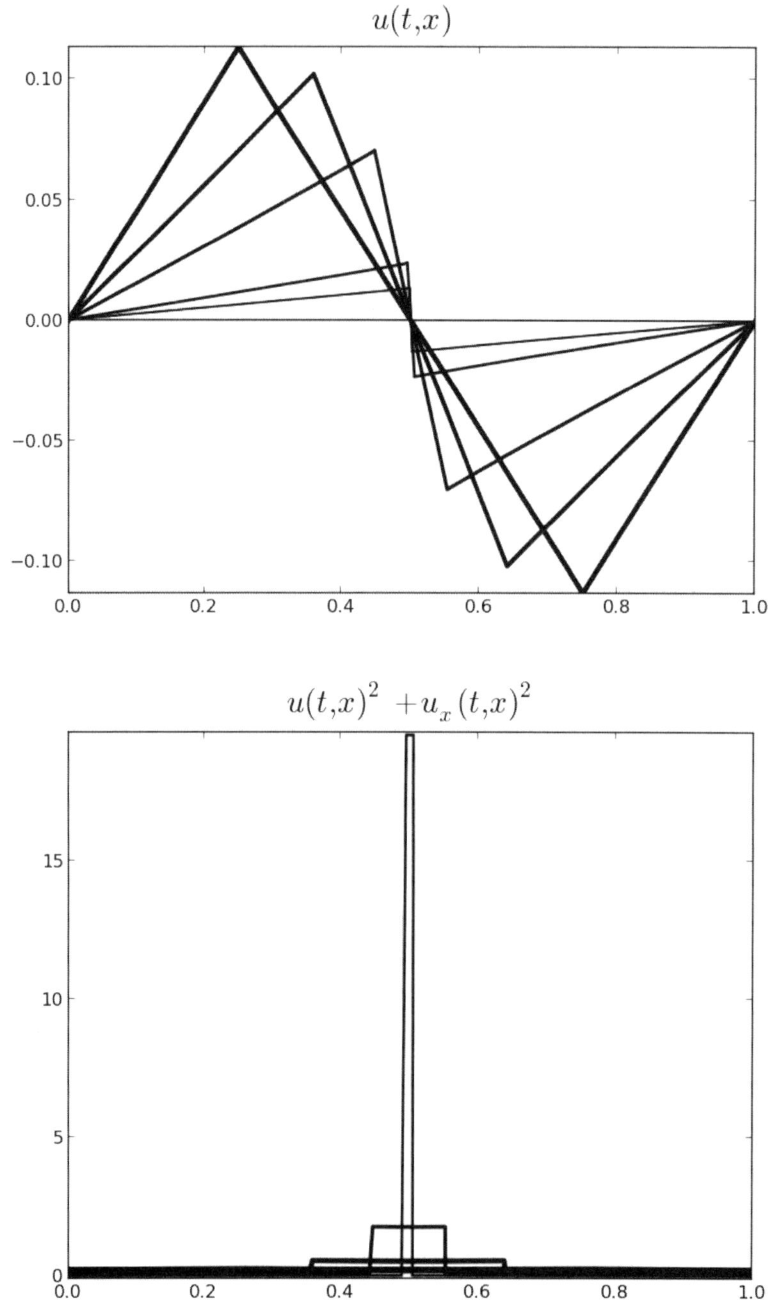

FIGURE 1. Plot of a peakon anti-peakon solution u of the CH equation with ρ identically zero at all times (the thinner the curve is, the later time it represents). In this case, we obtain a conservative solution of the scalar Camassa–Holm equation. We observe that the total energy $u^2 + u_x^2$ converges to a multiple of a Dirac delta function at $t = t_c$.

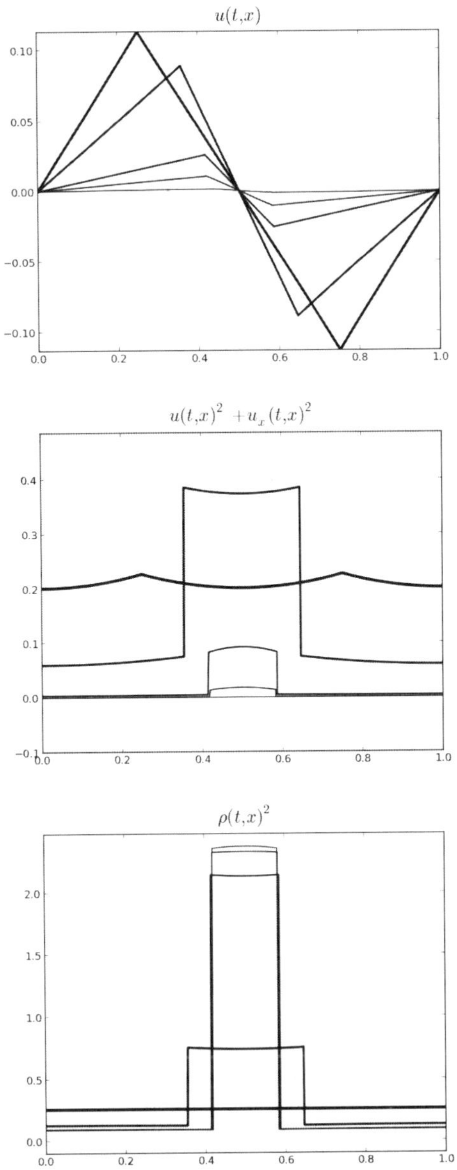

FIGURE 2. Here we employ the same initial condition as in Figure 1 for u while $\rho_0(x) = 0.5$. The total energy $\int_0^1 (u^2 + u_x^2 + \rho^2)\, dx$ is preserved. We observe first a concentration of the part of the energy given by $u^2 + u_x^2$. However, as we get closer to t_c, there is a transfer of energy from $u^2 + u_x^2$ to ρ^2.

where $P = P(t, x)$ implicitly is given as the solution of $P - P_{xx} = u^2 + \frac{1}{2}u_x^2 + \frac{1}{2}\rho^2$, enables us to derive how (y, U, ν, r) change with respect to time. In particular, direct computations yield, after setting $y(t,\xi)=\xi+\zeta(t,\xi)$, that

$$\zeta_t = U, \tag{2.3a}$$

$$U_t = -Q, \tag{2.3b}$$

$$\nu_t = -2QUy_\xi + (3U^2-2P)U_\xi, \tag{2.3c}$$

$$r_t = 0, \tag{2.3d}$$

where

$$P(t,\xi) = \frac{1}{2(e-1)}\int_0^1 \cosh(y(t,\xi)-y(t,\eta))(U^2y_\xi+\nu)(t,\eta)\,d\eta \\ + \frac{1}{4}\int_0^1 \exp\big(-\operatorname{sgn}(\xi-\eta)(y(t,\xi)-y(t,\eta))\big)(U^2y_\xi+\nu)(t,\eta)\,d\eta, \tag{2.4}$$

and

(2.5)

$$Q(t,\xi) = \frac{1}{2(e-1)}\int_0^1 \sinh(y(t,\xi)-y(t,\eta))(U^2y_\xi+\nu)(t,\eta)\,d\eta \\ - \frac{1}{4}\int_0^1 \operatorname{sgn}(\xi-\eta)\exp\big(-\operatorname{sgn}(\xi-\eta)(y(t,\xi)-y(t,\eta))\big)(U^2y_\xi+\nu)(t,\eta)\,d\eta.$$

First, we will consider this system of ordinary differential equations in the Banach space $E = W^{1,1}_{\text{per}}\times W^{1,1}_{\text{per}}\times L^1_{\text{per}}\times L^1_{\text{per}}$, where

$$W^{1,1}_{\text{per}} = \{f\in W^{1,1}_{\text{loc}}(\mathbb{R}) \mid f(\xi+1)=f(\xi) \text{ for all } \xi\in\mathbb{R}\}, \tag{2.6a}$$

$$L^1_{\text{per}} = \{f\in L^1_{\text{loc}}(\mathbb{R}) \mid f(\xi+1)=f(\xi) \text{ for all } \xi\in\mathbb{R}\}, \tag{2.6b}$$

and the corresponding norms are given by

$$\|f\|_{W^{1,1}_{\text{per}}} = \|f\|_{L^\infty([0,1])} + \|f_\xi\|_{L^1([0,1])}, \text{ and } \|f\|_{L^1_{\text{per}}} = \|f\|_{L^1([0,1])},$$
$$\|(y,U,\nu,r)\|_E = \|y-\text{id}\|_{W^{1,1}_{\text{per}}} + \|U\|_{W^{1,1}_{\text{per}}} + \|\nu\|_{L^1_{\text{per}}} + \|r\|_{L^1_{\text{per}}}.$$

The existence and uniqueness of short time solutions of (2.3), will follow from a contraction argument once we can show that the right-hand side of (2.3) is Lipschitz continuous on bounded sets. Note that this is the case if and only if the same holds for P and Q. The latter statement has been proved in [**11**, Lemma 2.1], and we state the result here for completeness.

LEMMA 2.1. *For any $X=(y,U,\nu,r)$ in E, we define the maps $\mathcal{Q}$ and $\mathcal{P}$ as $\mathcal{Q}(X)=Q$ and $\mathcal{P}(X)=P$ where P and Q are given by (2.4) and (2.5), respectively. Then, $\mathcal{P}$ and $\mathcal{Q}$ are Lipschitz maps on bounded sets from E to $W^{1,1}_{\text{per}}$. More precisely, we have the following bounds. Let*

$$B_M = \{X=(y,U,\nu,r)\in E \mid \|U\|_{W^{1,1}_{\text{per}}} + \|y_\xi\|_{L^1_{\text{per}}} + \|\nu\|_{L^1_{\text{per}}} \le M\}. \tag{2.7}$$

Then for any $X,\tilde{X}\in B_M$, we have

$$\left\|\mathcal{Q}(X)-\mathcal{Q}(\tilde{X})\right\|_{W^{1,1}_{\text{per}}} \le C_M\left\|X-\tilde{X}\right\|_E, \tag{2.8}$$

and

$$\left\|\mathcal{P}(X)-\mathcal{P}(\tilde{X})\right\|_{W^{1,1}_{\text{per}}} \le C_M\left\|X-\tilde{X}\right\|_E, \tag{2.9}$$

where the constant C_M only depends on the value of M.

To establish the global existence of solutions, we have to impose more conditions on our initial data and solutions in Lagrangian coordinates.

DEFINITION 2.2. *The set $\mathcal{F}$ is composed of all $(y, U, \nu, r) \in E$ such that*

$$(y, U) \in W^{1,\infty}_{\text{loc}}(\mathbb{R}) \times W^{1,\infty}_{\text{loc}}(\mathbb{R}),\ (\nu, r) \in L^\infty(\mathbb{R}) \times L^\infty(\mathbb{R}), \tag{2.10a}$$

$$y_\xi \geq 0,\ \nu \geq 0,\ y_\xi + \nu \geq c \text{ almost everywhere, for some constant } c > 0, \tag{2.10b}$$

$$y_\xi \nu = y_\xi^2 U^2 + U_\xi^2 + r^2 \text{ almost everywhere.} \tag{2.10c}$$

The set $\mathcal{F}$ is preserved with respect to time and plays a special role when proving the global existence of solutions. In particular, for $X(t) \in \mathcal{F}$, we have $h(t) = h(0)$ for all $t \in \mathbb{R}$ which implies that $\|X(t)\|_E$ cannot blow up within a finite time interval. Note that the first three equations in (2.3) are independent of r and coincide with the system considered in [**11**]. Moreover, the last variable r is preserved with respect to time. Hence, by following closely the proofs of [**11**, Lemma 2.3, Theorem 2.4], we get the global existence of solutions.

THEOREM 2.3. *For any $\bar{X} = (\bar{y}, \bar{U}, \bar{\nu}, \bar{r}) \in \mathcal{F}$, the system (2.3) admits a unique global solution $X(t) = (y(t), U(t), \nu(t), r(t))$ in $C^1(\mathbb{R}_+, E)$ with initial data $\bar{X} = (\bar{y}, \bar{U}, \bar{\nu}, \bar{r})$. We have $X(t) \in \mathcal{F}$ for all times. Let the mapping $S\colon \mathcal{F} \times \mathbb{R}_+ \to \mathcal{F}$ be defined as*

$$S_t(X) = X(t).$$

Given $M > 0$ and $T > 0$, we define B_M as before, that is,

$$B_M = \{X = (y, U, \nu, r) \in E \mid \|U\|_{W^{1,1}_{\text{per}}} + \|y_\xi\|_{L^1_{\text{per}}} + \|\nu\|_{L^1_{\text{per}}} \leq M\}. \tag{2.11}$$

Then there exists a constant C_M which depends only on M and T such that, for any two elements X_α and X_β in B_M, we have

$$\|S_t X_\alpha - S_t X_\beta\|_E \leq C_M \|X_\alpha - X_\beta\|_E \tag{2.12}$$

for any $t \in [0, T]$.

So far we have proved that there exist global, unique solutions to the 2CH system in Lagrangian coordinates. However, we still have to show that the assumptions are sufficiently general to accommodate rather general initial data in Eulerian coordinates. In particular, we must admit initial data (in Eulerian coordinates) that consists not only of the functions u_0 and ρ_0 but also of a positive, periodic Radon measure. This is necessary due to the fact that when wave breaking occurs, energy is concentrated at sets of measure zero. More precisely, we define the set of Eulerian coordinates as follows.

DEFINITION 2.4. *The set $\mathcal{D}$ of possible initial data consists of all triplets (u, ρ, μ) such that $u \in H^1_{\text{per}}$, $\rho \in L^2_{\text{per}}$, and μ is a positive, periodic Radon measure whose absolute continuous part, μ_{ac}, satisfies*

$$\mu_{\text{ac}} = (u^2 + u_x^2 + \rho^2)dx. \tag{2.13}$$

Having identified our set of Eulerian coordinates we can map them to the corresponding set of Lagrangian coordinates, using the mapping $\tilde{L}$.

DEFINITION 2.5. *For any* (u, ρ, μ) *in* $\mathcal{D}$, *let*

$$\begin{aligned} h &= \mu([0,1)), \\ y(\xi) &= \sup\{y \mid F_\mu(y) + y < (1+h)\xi\}, \\ \nu(\xi) &= (1+h) - y_\xi(\xi), \\ U(\xi) &= u \circ y(\xi), \\ r(\xi) &= \rho \circ y(\xi) y_\xi(\xi) \end{aligned} \tag{2.14}$$

where

$$F_\mu(x) = \begin{cases} \mu([0,x)) & \text{if } x > 0, \\ 0 & \text{if } x = 0, \\ -\mu([x,0)) & \text{if } x < 0. \end{cases} \tag{2.15}$$

Then $(y, U, \nu, r) \in \mathcal{F}$. *We define* $\tilde{L}(u, \mu) = (y, U, \nu, r)$. *The functions* P *and* Q *are given by* (2.4) *and* (2.5), *respectively.*

That this definition is well-posed follows after some slight modifications as in [**11**].

However, notice that we have three Eulerian coordinates in contrast to four Lagrangian coordinates, and hence there can at best be a one-to-one correspondence between triplets in Eulerian coordinates and *equivalence classes* in Lagrangian coordinates. When defining equivalence classes, relabeling functions will play a key role, and we will see why we had to impose (2.10b) in the definition of $\mathcal{F}$. Therefore we will now focus on the set G of relabeling functions.

DEFINITION 2.6. *Let* G *be the set of all functions* f *such that* f *is invertible,*

$$f \in W^{1,\infty}_{\text{loc}}(\mathbb{R}), \quad f(\xi+1) = f(\xi) + 1 \text{ for all } \xi \in \mathbb{R}, \text{ and} \tag{2.16}$$

$$f - \text{id} \text{ and } f^{-1} - \text{id} \text{ both belong to } W^{1,\infty}_{\text{per}}. \tag{2.17}$$

One of the main reasons for the choice of G is that any $f \in G$ satisfies

$$\frac{1}{1+\alpha} \leq f_\xi \leq 1 + \alpha,$$

for some constant $\alpha > 0$ according to [**21**, Lemma 3.2]. This allows us, following the same lines as in [**11**, Definition 3.2, Proposition 3.3] to define a group action of G on $\mathcal{F}$.

DEFINITION 2.7. *We define the map* $\Phi\colon G \times \mathcal{F} \to \mathcal{F}$ *as follows*

$$\begin{cases} \bar{y} = y \circ f, \\ \bar{U} = U \circ f, \\ \bar{\nu} = \nu \circ f f_\xi, \\ \bar{r} = r \circ f f_\xi, \end{cases}$$

where $(\bar{y}, \bar{U}, \bar{\nu}, \bar{r}) = \Phi(f, (y, U, \nu, r))$. *We denote* $(\bar{y}, \bar{U}, \bar{\nu}, \bar{r}) = (y, U, \nu, r) \bullet f$.

Using Φ, we can identify a subset of $\mathcal{F}$ which contains one element of each equivalence class. We introduce $\mathcal{F}_0 \subset \mathcal{F}$,

$$\mathcal{F}_0 = \{X = (y, U, \nu, r) \in \mathcal{F} \mid y_\xi + \nu = 1 + h\},$$

where $h = \|\nu\|_{L^1_{\mathrm{per}}}$. In addition, let $\mathcal{H} \subset \mathcal{F}_0$ be defined as follows

$$\mathcal{H} = \{(y, U, \nu, r) \in \mathcal{F}_0 \mid \int_0^1 y(\xi) d\xi = 0\}.$$

We can then associate to any element $X \in \mathcal{F}$ a unique element $\bar{X} \in \mathcal{H}$. This means there is a bijection between $\mathcal{H}$ and $\mathcal{F}/G$. Indeed, let $\Pi_1 \colon \mathcal{F} \to \mathcal{F}_0$ be given by

$$\Pi_1(X) = X \bullet f^{-1},$$

with $f(\xi) = \frac{1}{1+h}(y(\xi) + \int_0^\xi \nu(\eta) d\eta) \in G$ for $X = (y, U, \nu, r) \in \mathcal{F}$ due to (2.10b) and $\Pi_2 \colon \mathcal{F}_0 \to \mathcal{H}$ by

$$\Pi_2(x) = X(\xi - a),$$

where $a = \int_0^1 y(\xi) d\xi$ for $X = (y, U, \nu, r)$. Then

$$\Pi = \Pi_2 \circ \Pi_1$$

is a projection from $\mathcal{F}$ to $\mathcal{H}$, and, since $\Pi(X)$ is unique, $\mathcal{F}/G$ and $\mathcal{H}$ are in bijection. We can now redefine our mapping from Eulerian to Lagrangian coordinates such that any triplet $(u, \rho, \mu) \in \mathcal{D}$ is mapped to the corresponding element $(y, U, \nu, r) \in \mathcal{H}$ by applying $\tilde{L}$ followed by Π.

THEOREM 2.8. *For any $(u, \rho, \mu) \in \mathcal{D}$ let $X = (y, U, \nu, r) \in \mathcal{H}$ be given by $X = L(u, \rho, \mu) = \Pi \circ \tilde{L}(u, \rho, \mu)$. Then $L \colon \mathcal{D} \to \mathcal{H}$.*

Furthermore, since $r_t = 0$, [**11**, Lemma 3.5] implies directly that S_t is equivariant, i.e.,

$$S_t(X \bullet f) = S_t(X) \bullet f, \tag{2.18}$$

for $X \in \mathcal{F}$ and $f \in G$. In particular, we can define the semigroup $\bar{S}_t$ on $\mathcal{H}$ as

$$\bar{S}_t = \Pi \circ S_t. \tag{2.19}$$

The final and last step is to go back from Lagrangian to Eulerian coordinates, which is a generalization of [**21**, Theorem 3.11] and the adaptation of [**11**, Theorem 4.10] to the periodic case.

THEOREM 2.9. *Let $X \in \mathcal{F}$, then the periodic measure[2] $y_\#(r d\xi)$ is absolutely continuous and (u, ρ, μ) given by*

$$\begin{aligned} u(x) &= U(\xi) \text{ for any } \xi \text{ such that } x = y(\xi), \\ \mu &= y_\#(\nu d\xi), \\ \rho(x) dx &= y_\#(r d\xi), \end{aligned} \tag{2.20}$$

belongs to $\mathcal{D}$. We denote by M the mapping from $\mathcal{F}$ to $\mathcal{D}$ which to any $X \in \mathcal{F}$ associates the element $(u, \rho, \mu) \in \mathcal{D}$ given by (2.20).

Note that the mapping M is independent of the representative in every equivalence class we choose, i.e.,

$$M = M \circ \Pi. \tag{2.21}$$

In order to be able to get back and forth between Eulerian and Lagranian coordinates at any possible time it is left to clarify the relation between L and M.

---

[2]The push-forward of a measure ν by a measurable function f is the measure $f_\#\nu$ defined as $f_\#\nu(B) = \nu(f^{-1}(B))$ for any Borel set B.

THEOREM 2.10. *The maps $L : \mathcal{D} \to \mathcal{H}$ and $M : \mathcal{H} \to \mathcal{D}$ are invertible. We have*

$$L \circ M = \Pi, \quad \text{and} \quad M \circ L = \mathrm{id}, \tag{2.22}$$

where the mapping $\Pi : \mathcal{F} \to \mathcal{H}$ is a projection which associates to any element $X \in \mathcal{F}$ a unique element $\tilde{X} \in \mathcal{H}$, which means, in particular, that $\mathcal{F}/G$ and $\mathcal{H}$ are in bijection.

The proof follows the same lines as [**21**, Theorem 3.12], and we therefore do not present it here. We will see later that the last theorem together with (2.19) allows us to define a semigroup of solutions. To obtain a continuous semigroup we have to study the stability of solutions in Lagrangian coordinates, which is the aim of the next section.

3. Lipschitz metric

We will now construct a Lipschitz metric in Lagrangian coordinates which will be invariant under relabeling. It will be quite similar to the one in [**11**] due to the fact that the first three equations in (2.3) are independent of r and coincide with the system considered in [**11**] and because $r(t) = r(0)$ for all $t \in \mathbb{R}$.

Let X_α, $X_\beta \in \mathcal{F}$. We introduce the function $J(X_\alpha, X_\beta)$ by

$$J(X_\alpha, X_\beta) = \inf_{f,g \in G} \|X_\alpha \bullet f - X_\beta \bullet g\|_E\,, \tag{3.1}$$

which is invariant with respect to relabeling. That means, for any $X_\alpha, X_\beta \in \mathcal{F}$ and $f, g \in G$, we have

$$J(X_\alpha \bullet f, X_\beta \bullet g) = J(X_\alpha, X_\beta). \tag{3.2}$$

Note that the mapping J does not define a metric, since it does not satisfy the triangle inequality, which is the reason why we introduce the following mapping d.

Let $d(X_\alpha, X_\beta)$ be defined by

$$d(X_\alpha, X_\beta) = \inf \sum_{n=1}^{N} J(X_{n-1}, X_n), \quad X_\alpha, X_\beta \in \mathcal{F}, \tag{3.3}$$

where the infimum is taken over all finite sequences $\{X_n\}_{n=0}^N \in \mathcal{F}$ satisfying $X_0 = X_\alpha$ and $X_N = X_\beta$. In particular, d is relabeling invariant, that means for any $X_\alpha, X_\beta \in \mathcal{F}$ and $f, g \in G$, we have

$$d(X_\alpha \bullet f, X_\beta \bullet g) = d(X_\alpha, X_\beta). \tag{3.4}$$

In order to prove that d is a Lipschitz metric on bounded sets, we have to choose one element in each equivalence class, and we will apply (2.12). One problem we are facing in that context is that the constant on the right-hand side of (2.12) depends on the set B_M we choose, but B_M is not preserved by the time evolution while it is invariant with respect to relabeling. Hence we will try to find a suitable set, which is invariant with respect to time and relabeling and is in some sense equivalent to B_M. To that end we define the subsets of bounded energy $\mathcal{F}^M$ of $\mathcal{F}_0$ by

$$\mathcal{F}^M = \{X = (y, U, \nu, r) \in \mathcal{F} \mid h = \|\nu\|_{L^1_{\mathrm{per}}} \leq M\}$$

and let $\mathcal{H}^M = \mathcal{H} \cap \mathcal{F}^M$. The important property of the set $\mathcal{F}^M$ is that it is preserved both by the flow and relabeling. In particular, we have that

$$B_M \cap \mathcal{H} \subset \mathcal{H}^M \subset B_{\bar{M}} \cap \mathcal{H} \tag{3.5}$$

for $\bar{M} = 6(1+M)$ and hence the sets $B_M \cap \mathcal{H}$ and $\mathcal{H}^M$ are in this sense equivalent.

DEFINITION 3.1. *Let d_M be the metric on $\mathcal{H}^M$ which is defined, for any $X_\alpha, X_\beta \in \mathcal{H}^M$, as*

$$d_M(X_\alpha, X_\beta) = \inf \sum_{n=1}^{N} J(X_{n-1}, X_n) \tag{3.6}$$

where the infimum is taken over all finite sequences $\{X_n\}_{n=0}^N \in \mathcal{H}^M$ which satisfy $X_0 = X_\alpha$ and $X_N = X_\beta$.

By definition d_M is relabeling invariant and the triangle inequality is satisfied. In this way we obtain a metric which in addition can be compared with other norms on $\mathcal{H}^M$ (cf. [**11**, Lemma 4.3]).

LEMMA 3.2. *The mapping $d_M \colon \mathcal{H}^M \times \mathcal{H}^M \to \mathbb{R}_+$ is a metric on $\mathcal{H}^M$. Moreover, given X_α, $X_\beta \in \mathcal{H}^M$, define $R_\alpha(\xi) = \int_0^\xi r_\alpha(\eta)d\eta$ and $R_\beta(\xi) = \int_0^\xi r_\beta(\eta)d\eta$ for $\xi \in [0,1]$. Then we have*

$$\|y_\alpha - y_\beta\|_{L^\infty} + \|U_\alpha - U_\beta\|_{L^\infty} + |h_\alpha - h_\beta| + \|R_\alpha - R_\beta\|_{L^\infty} \leq C_M d_M(X_\alpha, X_\beta) \tag{3.7}$$

and

$$d(X_\alpha, X_\beta) \leq \|X_\alpha - X_\beta\|_E, \tag{3.8}$$

where C_M denotes some fixed constant which depends only on M.

To show that we not only obtained a relabeling invariant metric but in fact a Lipschitz metric, we combine all results we obtained so far as in [**11**, Theorem 4.6]. This yields the following Lipschitz stability theorem for $\bar{S}_t$.

THEOREM 3.3. *Given $T > 0$ and $M > 0$, there exists a constant C_M which depends only on M and T such that, for any $X_\alpha, X_\beta \in \mathcal{H}^M$ and $t \in [0,T]$, we have*

$$d_M(\bar{S}_t X_\alpha, \bar{S}_t X_\beta) \leq C_M d_M(X_\alpha, X_\beta). \tag{3.9}$$

4. Global weak solutions

It is left to check that we obtain a global weak solution of the 2CH system by solving (2.3) and using the maps between Eulerian and Lagrangian coordinates. In the case of conservative solutions we have that for any triplet $(u(t,x), \rho(t,x), \mu(t,x))$ in Eulerian coordinates, the function $P(t,x)$ is given by

$$\begin{aligned} P(t,x) &= \frac{1}{2(e-1)}\int_0^1 \cosh(x-z)u^2(t,z)dz + \frac{1}{2(e-1)}\int_0^1 \cosh(x-z)d\mu(t,z) \\ &\quad + \frac{1}{4}\int_0^1 e^{-|x-z|}u^2(t,z)dz + \frac{1}{4}\int_0^1 e^{-|x-z|}d\mu(t,z). \end{aligned} \tag{4.1}$$

Applying the mapping L maps $P(t,x)$ to $P(t,\xi)$ given by (2.4) and $P_x(t,x)$ to $Q(t,\xi)$ given by (2.5). Since the set of times where wave breaking occurs has measure zero, $\mu = \mu_{ac} = (u^2 + u_x^2 + \rho^2)dx$ for almost all times, $P(t,x)$ defined by (4.1) coincides for almost all times and all $x \in \mathbb{R}$ with the solution of $P - P_{xx} = u^2 + \frac{1}{2}u_x^2 + \frac{1}{2}\rho^2$.

DEFINITION 4.1. *Let $u\colon \mathbb{R}_+ \times \mathbb{R} \to \mathbb{R}$ and $\rho\colon \mathbb{R}_+ \times \mathbb{R} \to \mathbb{R}$. Assume that u and ρ satisfy*

(i) $u \in L^\infty([0,\infty), H^1_{\mathrm{per}})$, $\rho \in L^\infty([0,\infty), L^2_{\mathrm{per}})$,
(ii) the equations

$$\iint_{\mathbb{R}_+\times[0,1]} \Big(- u(t,x)\phi_t(t,x) + (u(t,x)u_x(t,x) + P_x(t,x))\phi(t,x)\Big) dx dt = \int_{[0,1]} u(0,x)\phi(0,x)dx, \tag{4.2}$$

$$\iint_{\mathbb{R}_+\times[0,1]} \Big((P(t,x) - u^2(t,x) - \frac{1}{2}u_x^2(t,x))\phi(t,x) + P_x(t,x)\phi_x(t,x)\Big) dx dt = 0, \tag{4.3}$$

and

$$\iint_{\mathbb{R}_+\times[0,1]} \Big(- \rho(t,x)\phi_t(t,x) - u(t,x)\rho(t,x)\phi_x(t,x)\Big) dx dt = \int_{[0,1]} \rho(0,x)\phi(0,x)dx, \tag{4.4}$$

hold for all spatial periodic functions $\phi \in C_0^\infty([0,\infty),\mathbb{R})$. Then we say that (u,ρ) is a global weak solution of the two-component Camassa–Holm system.

If (u,ρ) in addition satisfies

$$(u^2 + u_x^2 + \rho^2)_t + (u(u^2 + u_x^2 + \rho^2))_x - (u^3 - 2Pu)_x = 0$$

in the sense that

$$\begin{aligned}\iint_{\mathbb{R}_+\times[0,1]} \Big[&(u^2(t,x) + u_x^2(t,x) + \rho^2(t,x))\phi_t(t,x)\\ &+ (u(t,x)(u^2(t,x) + u_x^2(t,x) + \rho^2(t,x)))\phi_x(t,x)\\ &- (u^3(t,x) - 2P(t,x)u(t,x))\phi_x(t,x)\Big] dx dt = 0,\end{aligned} \tag{4.5}$$

for any spatial periodic function $\phi \in C_0^\infty((0,\infty)\times\mathbb{R})$, we say that (u,ρ) is a weak global conservative solution of the two-component Camassa–Holm system.

Introduce the mapping T_t from $\mathcal{D}$ to $\mathcal{D}$ by

$$T_t = M\bar{S}_t L. \tag{4.6}$$

Then one can check that for any $(u_0,\rho_0,\mu_0) \in \mathcal{D}$ such that μ_0 is purely absolutely continuous, the pair $(u(t,x),\rho(t,x))$ given by $(u,\rho,\mu)(t) = T_t(u_0,\rho_0,\mu_0)$ satisfies (4.2)–(4.4).

THEOREM 4.2. *Given any initial condition $(u_0,\rho_0) \in H^1_{\mathrm{per}} \times L^2_{\mathrm{per}}$, we define $\mu_0 = (u_0^2 + u_{0,x}^2 + \rho_0^2)dx$, and we denote $(u,\rho,\mu)(t) = T_t(u_0,\rho_0,(u_0^2 + u_{0,x}^2 + \rho_0^2)dx)$. Then (u,ρ) is a periodic and global weak solution of the 2CH system and μ satisfies weakly*

$$\mu_t + (u\mu)_x = (u^3 - 2Pu)_x.$$

Moreover, $\mu(t)$ consists of an absolutely continuous and a singular part, that means

$$\int_{[0,1]} d\mu(t,x) = \int_{[0,1]} (u^2 + u_x^2 + \rho^2)(t,x)\, dx + \int_{[0,1]} d\mu_{\mathrm{sing}}(t,x).$$

In particular, $\mathrm{supp}(\mu_{\mathrm{sing}}(t,\cdot))$ coincides with the set of points where wave breaking occurs at time t.

Note that since we are looking for global weak solutions for initial data in $H^1_{\rm per}\times L^2_{\rm per}$ it is no restriction to assume that μ is purely absolutely continuous initially while we in general will not have that μ remains purely absolutely continuous at any later time. In particular, if μ is not absolutely continuous at a particular time, we know how much and where the energy has concentrated, and this energy must be given back to the solution in order to obtain conservative solutions. Therefore the measure plays an important role.
It is also possible to define global weak solutions for initial data where the measure μ_0 is not purely absolutely continuous by defining $P(0,x)$ using (4.1), which is then mapped to (2.4) by applying (2.14) directly. Moreover, $P(0,\xi)$ can be mapped back to $P(0,x)$ via M. In addition, this point of view allows us to jump between Eulerian and Lagrangian coordinates at any time.

Moreover, one can show that the sets $\mathcal{F}^M$ and $\mathcal{H}^M$ in Lagrangian coordinates correspond to the set $\mathcal{D}^M$ in Eulerian coordinates. Given $M>0$, we define

$$\mathcal{D}^M=\{(u,\rho,\mu)\in\mathcal{D}\mid\mu([0,1))\leq M\}. \tag{4.7}$$

Thus it is natural to define a Lipschitz metric on the sets of bounded energy in Eulerian coordinates as follows,

$$d_{\mathcal{D}^M}((u,\rho,\mu),(\tilde{u},\tilde{\rho},\tilde{\mu}))=d_M(L(u,\rho,\mu),L(\tilde{u},\tilde{\rho},\tilde{\mu})). \tag{4.8}$$

In particular, we have the following result.

THEOREM 4.3. *The semigroup $(T_t,d_{\mathcal{D}})$, which corresponds to solutions of the 2CH system, is a continuous semigroup on $\mathcal{D}$ with respect to the metric $d_{\mathcal{D}}$. The semigroup is Lipschitz continuous on sets of bounded energy, that is, given $M>0$ and a time interval $[0,T]$, there exists a constant C which only depends on M and T such that, for any (u,ρ,μ) and $(\tilde{u},\tilde{\rho},\tilde{\mu})$ in $\mathcal{D}^M$, we have*

$$d_{\mathcal{D}^M}(T_t(u,\rho,\mu),T_t(\tilde{u},\tilde{\rho},\tilde{\mu}))\leq Cd_{\mathcal{D}^M}((u,\rho,\mu),(\tilde{u},\tilde{\rho},\tilde{\mu}))$$

for all $t\in[0,T]$.

Last, but not least, we want to investigate the regularity of solutions and the connection of the topology in $\mathcal{D}$ with other topologies. Due to the global interaction term given for almost all times by

$$\begin{aligned}P(t,x)=&\ \frac{1}{2(e-1)}\int_0^1\cosh(x-z)(2u^2+u_x^2+\rho^2)(t,z)dz\\&+\frac{1}{4}\int_0^1 e^{-|x-z|}(2u^2+u_x^2+\rho^2)(t,z)dz,\end{aligned} \tag{4.9}$$

the 2CH system has an infinite speed of propagation [**18**]. However, the system remains essentially hyperbolic in nature, and we prove that singularities travel with finite speed. In [**12**, Theorem 6.1] we showed that the local regularity of a solution depends on the regularity of the initial data and that $\rho_0(x)^2$ can be bounded from below by a strictly positive constant. Since this result is a local result, it carries over to the periodic case and we state it here for the sake of completeness.

THEOREM 4.4. *We consider initial data $(u_0,\rho_0,\mu_0)\in\mathcal{D}$. Furthermore, we assume that there exists an interval (x_0,x_1) such that (u_0,ρ_0,μ_0) is p-regular, with $p\geq 1$, in the sense that*

$$u_0\in W^{p,\infty}(x_0,x_1),\quad \rho_0\in W^{p-1,\infty}(x_0,x_1),\quad \text{and } \mu_0=\mu_{0,ac} \text{ on } (x_0,x_1),$$

and that

$$\rho_0(x)^2 \geq c > 0$$

for $x \in (x_0, x_1)$. *Then for any* $t \in \mathbb{R}_+$, $(u, \rho, \mu)(t, \cdot)$ *is p-regular on the interval* $(y(t, \xi_0), y(t, \xi_1))$, *where* ξ_0 *and* ξ_1 *satisfy* $y(0, \xi_0) = x_0$ *and* $y(0, \xi_1) = x_1$ *and are defined as*

$$\xi_0 = \sup\{\xi \in \mathbb{R} \mid y(0, \xi) \leq x_0\} \text{ and } \xi_1 = \inf\{\xi \in \mathbb{R} \mid y(0, \xi) \geq x_1\}.$$

In other words, we see that the regularity is preserved between characteristics. As an immediate consequence we obtain the following result.

THEOREM 4.5. *If the initial data* $(u_0, \rho_0, \mu_0) \in \mathcal{D}$ *satisfies* $u_0, \rho_0 \in C^\infty(\mathbb{R})$, μ_0 *is absolutely continuous and* $\rho_0^2(x) \geq d > 0$ *for all* $x \in \mathbb{R}$, *then* $u, \rho \in C^\infty(\mathbb{R} \times \mathbb{R})$ *is the unique classical solution to* (2.1) *with* $\kappa = 0$ *and* $\eta = 1$.

In particular this result implies that if $\rho_0^2(x) \geq c$ for some positive constant $c > 0$, no wave breaking occurs. Hence if we can compare the topology on $\mathcal{D}$ with standard topologies we have a chance to approximate conservative solutions of the CH equation which enjoy wave breaking by global smooth solutions of the 2CH system. Indeed, the mapping

$$(u, \rho) \mapsto (u, \rho, (u^2 + u_x^2 + \rho^2)dx), \tag{4.10}$$

is continuous from $H^1_{\mathrm{per}} \times L^2_{\mathrm{per}}$ to $\mathcal{D}$. This means, given a sequence $(u_n, \rho_n) \in H^1_{\mathrm{per}} \times L^2_{\mathrm{per}}$ converging to $(u, \rho) \in H^1_{\mathrm{per}} \times L^2_{\mathrm{per}}$, then $(u_n, \rho_n, (u_n^2 + u_{n,x}^2 + \rho_n^2)dx)$ converges to $(u, \rho, (u^2 + u_x^2 + \rho^2)dx)$ in $\mathcal{D}$.
Conversely if (u_n, ρ_n, μ_n) is a sequence in $\mathcal{D}$ which converges to $(u, \rho, \mu) \in \mathcal{D}$, then

$$u_n \to u \text{ in } L^\infty_{\mathrm{per}}, \quad \rho_n \overset{*}{\rightharpoonup} \rho, \text{ and } \mu_n \overset{*}{\rightharpoonup} \mu. \tag{4.11}$$

Putting now everything together we have the following result.

THEOREM 4.6. *Let* $u_0 \in H^1_{\mathrm{per}}$. *We consider the approximating sequence of initial data* $(u_0^n, \rho_0^n, \mu_0^n) \in \mathcal{D}$ *given by* $u_0^n \in C^\infty(\mathbb{R})$ *with* $\lim_{n\to\infty} u_0^n = u_0$ *in* H^1_{per}, $\rho_0^n \in C^\infty(\mathbb{R})$ *with* $\lim_{n\to\infty} \rho_0^n = 0$ *in* L^2_{per}, $(\rho_0^n)^2 \geq d_n$ *for some constant* $d_n > 0$ *and for all* n *and* $\mu_0^n = ((u_0^n)^2 + (u_{0,x}^n)^2 + (\rho_0^n)^2)dx$. *We denote by* (u^n, ρ^n) *the unique classical solution to* (2.1), *with* $\kappa = 0$ *and* $\eta = 1$, *in* $C^\infty(\mathbb{R}_+ \times \mathbb{R}) \times C^\infty(\mathbb{R}_+ \times \mathbb{R})$ *with* $(u^n, \rho^n)|_{t=0} = (u_0^n, \rho_0^n)$. *Then for every* $t \in \mathbb{R}_+$, *the sequence* $u^n(t, \cdot)$ *converges to* $u(t, \cdot)$ *in* $L^\infty(\mathbb{R})$, *where* u *is the conservative solution of the Camassa–Holm equation*

$$u_t - u_{txx} + 3uu_x - 2u_x u_{xx} - uu_{xxx} = 0, \tag{4.12}$$

with initial data $u_0 \in H^1_{\mathrm{per}}$.

References

[1] Alberto Bressan and Adrian Constantin, *Global conservative solutions of the Camassa-Holm equation*, Arch. Ration. Mech. Anal. **183** (2007), no. 2, 215–239, DOI 10.1007/s00205-006-0010-z. MR2278406 (2007j:35183)

[2] Alberto Bressan and Adrian Constantin, *Global dissipative solutions of the Camassa-Holm equation*, Anal. Appl. (Singap.) **5** (2007), no. 1, 1–27, DOI 10.1142/S0219530507000857. MR2288533 (2007k:35394)

[3] Alberto Bressan, Helge Holden, and Xavier Raynaud, *Lipschitz metric for the Hunter-Saxton equation*, J. Math. Pures Appl. (9) **94** (2010), no. 1, 68–92, DOI 10.1016/j.matpur.2010.02.005. MR2653980 (2011f:35294)

[4] Roberto Camassa and Darryl D. Holm, *An integrable shallow water equation with peaked solitons*, Phys. Rev. Lett. **71** (1993), no. 11, 1661–1664, DOI 10.1103/PhysRevLett.71.1661. MR1234453 (94f:35121)

[5] Robin Ming Chen and Yue Liu, *Wave breaking and global existence for a generalized two-component Camassa-Holm system*, Int. Math. Res. Not. IMRN **6** (2011), 1381–1416, DOI 10.1093/imrn/rnq118. MR2806508 (2012k:35459)

[6] Adrian Constantin and Rossen I. Ivanov, *On an integrable two-component Camassa-Holm shallow water system*, Phys. Lett. A **372** (2008), no. 48, 7129–7132, DOI 10.1016/j.physleta.2008.10.050. MR2474608 (2009m:35418)

[7] Joachim Escher, Olaf Lechtenfeld, and Zhaoyang Yin, *Well-posedness and blow-up phenomena for the 2-component Camassa-Holm equation*, Discrete Contin. Dyn. Syst. **19** (2007), no. 3, 493–513, DOI 10.3934/dcds.2007.19.493. MR2335761 (2008j:35154)

[8] Ying Fu and Changzheng Qu, *Well posedness and blow-up solution for a new coupled Camassa-Holm equations with peakons*, J. Math. Phys. **50** (2009), no. 1, 012906, 25, DOI 10.1063/1.3064810. MR2492603 (2010b:35397)

[9] Katrin Grunert, Helge Holden, and Xavier Raynaud, *Lipschitz metric for the Camassa–Holm equation*, Discrete Contin. Dyn. Syst. **33** (2013), 2809–2827.

[10] Katrin Grunert, Helge Holden, and Xavier Raynaud, *Global conservative solutions to the Camassa-Holm equation for initial data with nonvanishing asymptotics*, Discrete Contin. Dyn. Syst. **32** (2012), no. 12, 4209–4227, DOI 10.3934/dcds.2012.32.4209. MR2966743

[11] Katrin Grunert, Helge Holden, and Xavier Raynaud, *Lipschitz metric for the periodic Camassa-Holm equation*, J. Differential Equations **250** (2011), no. 3, 1460–1492, DOI 10.1016/j.jde.2010.07.006. MR2737213 (2012d:35319)

[12] Katrin Grunert, Helge Holden, and Xavier Raynaud. Global solutions for the two-component Camassa–Holm system. *Comm. Partial Differential Equations* 37: 2245–2271, 2012.

[13] Chunxia Guan and Zhaoyang Yin, *Global weak solutions for a modified two-component Camassa-Holm equation*, Ann. Inst. H. Poincaré Anal. Non Linéaire **28** (2011), no. 4, 623–641, DOI 10.1016/j.anihpc.2011.04.003 (English, with English and French summaries). MR2823888 (2012k:35062)

[14] Chunxia Guan and Zhaoyang Yin, *Global existence and blow-up phenomena for an integrable two-component Camassa-Holm shallow water system*, J. Differential Equations **248** (2010), no. 8, 2003–2014, DOI 10.1016/j.jde.2009.08.002. MR2595712 (2011b:35072)

[15] Guilong Gui and Yue Liu, *On the Cauchy problem for the two-component Camassa-Holm system*, Math. Z. **268** (2011), no. 1-2, 45–66, DOI 10.1007/s00209-009-0660-2. MR2805424 (2012g:35183)

[16] Guilong Gui and Yue Liu, *On the global existence and wave-breaking criteria for the two-component Camassa-Holm system*, J. Funct. Anal. **258** (2010), no. 12, 4251–4278, DOI 10.1016/j.jfa.2010.02.008. MR2609545 (2011i:35215)

[17] Zhengguang Guo and Yong Zhou, *On solutions to a two-component generalized Camassa-Holm equation*, Stud. Appl. Math. **124** (2010), no. 3, 307–322, DOI 10.1111/j.1467-9590.2009.00472.x. MR2655943 (2011d:35431)

[18] David Henry, *Infinite propagation speed for a two component Camassa-Holm equation*, Discrete Contin. Dyn. Syst. Ser. B **12** (2009), no. 3, 597–606, DOI 10.3934/dcdsb.2009.12.597. MR2525158 (2010g:35266)

[19] Helge Holden and Xavier Raynaud, *Global conservative solutions of the Camassa-Holm equation—a Lagrangian point of view*, Comm. Partial Differential Equations **32** (2007), no. 10-12, 1511–1549, DOI 10.1080/03605300601088674. MR2372478 (2010a:35220)

[20] Helge Holden and Xavier Raynaud, *Dissipative solutions for the Camassa-Holm equation*, Discrete Contin. Dyn. Syst. **24** (2009), no. 4, 1047–1112, DOI 10.3934/dcds.2009.24.1047. MR2505693 (2010k:35408)

[21] Helge Holden and Xavier Raynaud, *Periodic conservative solutions of the Camassa-Holm equation*, Ann. Inst. Fourier (Grenoble) **58** (2008), no. 3, 945–988 (English, with English and French summaries). MR2427516 (2009e:35233)

[22] Pavel Aleksandrovich Kuz′min, *On two-component generalizations of the Camassa-Holm equation*, Mat. Zametki **81** (2007), no. 1, 149–152, DOI 10.1134/S0001434607010142 (Russian); English transl., Math. Notes **81** (2007), no. 1-2, 130–134. MR2333873 (2008d:37134)

[23] Peter J. Olver and Philip Rosenau, *Tri-Hamiltonian duality between solitons and solitary-wave solutions having compact support*, Phys. Rev. E (3) **53** (1996), no. 2, 1900–1906, DOI 10.1103/PhysRevE.53.1900. MR1401317 (97c:35172)

[24] Yujuan Wang, Jincun Huang, and Lele Chen, *Global conservative solutions of the two-component Camassa-Holm shallow water system*, Int. J. Nonlinear Sci. **9** (2010), no. 3, 379–384. MR2670950

(K. Grunert) Department of Mathematical Sciences, Norwegian University of Science and Technology, NO-7491 Trondheim, Norway

E-mail address: katring@math.ntnu.no

URL: http://www.math.ntnu.no/~katring/

(H. Holden) Department of Mathematical Sciences, Norwegian University of Science and Technology, NO-7491 Trondheim, Norway — and — Centre of Mathematics for Applications, University of Oslo, NO-0316 Oslo, Norway

E-mail address: holden@math.ntnu.no

URL: http://www.math.ntnu.no/~holden/

(X. Raynaud) Centre of Mathematics for Applications, University of Oslo, NO-0316 Oslo, Norway

E-mail address: xavierra@cma.uio.no

URL: http://folk.uio.no/xavierra/

Proceedings of Symposia in Pure Mathematics
Volume **87**, 2013
http://dx.doi.org/10.1090/pspum/087/01426

A Minimal Uncertainty Product for One Dimensional Semiclassical Wave Packets

George A. Hagedorn

Happy 60th birthday, Mr. Fritz!

ABSTRACT. Although real, normalized Gaussian wave packets minimize the product of position and momentum uncertainties, generic complex normalized Gaussian wave packets do not. We prove they minimize an alternative product of uncertainties that correspond to variables that are phase space rotations of position and momentum.

1. Introduction

In studying small $\hbar$ asymptotics of solutions to the time–dependent Schrödinger equation, semiclassical wave packets have proven very useful. (See *e.g.*, [**3**] and its references, or more modern works, such as [**1**], [**2**], or [**4**].) In the case of one degree of freedom, for fixed, allowed values of the parameters A, B, $\hbar$, a, and η, these wave packets are an orthonormal basis of $L^2(\mathbb{R},\, dx)$ that we denote by $\{\, \varphi_k(A,\, B,\, \hbar,\, a,\, \eta,\, x)\,\}$, where $k = 0,\, 1,\, \dots$.

By making a proper choice of the parameters, one can write any normalized, one dimensional complex Gaussian wave packet as $\varphi_0(A,\, B,\, \hbar,\, a,\, \eta,\, x)$. Two of the restrictions on the parameters are the conditions $\mathrm{Re}\,\overline{A}B = 1$ and $\hbar > 0$. By a fairly straightforward calculation, one can prove that the usual position, momentum uncertainty product in the state $\varphi_0(A,\, B,\, \hbar,\, a,\, \eta,\, x)$ satisfies $\Delta x\, \Delta p = \frac{\hbar}{2}\, |A|\, |B|$. When the complex phases of A and B are the same, $\varphi_0(A,\, B,\, \hbar,\, a,\, \eta,\, x)$ is a phase times a real Gaussian, and it is a standard result that this product takes its minimal value $\Delta x\, \Delta p = \frac{\hbar}{2}$. When the phases of A and B are different, $|A|\, |B| > 1$, and consequently, $\Delta x\, \Delta p > \frac{\hbar}{2}$. As usual, x and p are the position and momentum operators, and we define the uncertainty for a self-adjoint observable X in the state ψ to be $\Delta X = \sqrt{\langle\, \psi,\, X^2\, \psi\,\rangle - \langle\, \psi,\, X\, \psi\,\rangle^2}$.

1991 *Mathematics Subject Classification.* Primary 81Q20, 81S30.
Key words and phrases. Quantum Mechanics, Semiclassical Wave Packets.
The author was supported in part by NSF Grant #DMS–1210928.

The goal of this paper is to prove an alternative minimum uncertainty product result for the state $\varphi_0(A,\, B,\, \hbar,\, a,\, \eta,\, x)$. We define two "rotated" operators

$$\begin{aligned} \alpha &= \cos(\theta)\, x + \sin(\theta)\, p \qquad \text{and} \\ \beta &= -\sin(\theta)\, x + \cos(\theta)\, p. \end{aligned}$$

We show that in any normalized state,

$$\Delta\alpha\, \Delta\beta \geq \frac{\hbar}{2},$$

and we show in Theorem 5.2 that by choosing

$$\theta = \frac{1}{2} \arctan\left(\frac{2\, \mathrm{Im}\,(B\overline{A})}{|B|^2 - |A|^2} \right),$$

one has

$$\Delta\alpha\, \Delta\beta = \frac{\hbar}{2}$$

in the state $\varphi_0(A,\, B,\, \hbar,\, a,\, \eta,\, x)$. So, general semiclassical wave packets satisfy a minimal uncertainty relation. It is not the usual relation, but the product for the rotated operators.

REMARK 1.1. Employing a sort of microlocal intuition, we often like to think heuristically of supports of quantum states in phase space. Weyl asymptotics and Bohr–Sommerfeld rules suggest that a normalized state should occupy a phase space area of $2\pi\hbar$. From this viewpoint, we envision the usual, frequency ω harmonic oscillator ground state as having phase space support of the interior of an ellipse that is centered at the origin and has semiaxes $\sqrt{2\,\hbar/\omega}$ in the x direction and $\sqrt{2\,\hbar\omega}$ in the p direction.

On this intuitive level, the region of phase space corresponding to $\varphi_0(A,\, B,\, \hbar,\, a,\, \eta,\, x)$ is such an ellipse that has been rotated through the angle θ and then translated so its center is at $(a,\, \eta)$.

REMARK 1.2. The situation in more than one dimension is significantly more complicated. Although we believe an analogous result must be true, and we have received some preliminary calculations from Vidian Rousse [**5**], we have not seen a complete argument. In n dimensions, one would like to find an orthogonal, symplectic matrix depending on A and B that would define rotated operators α_j and β_j for $j = 1,\, 2,\, \cdots,\, n$, such that in the state $\varphi_0(A,\, B,\, \hbar,\, a,\, \eta,\, x)$,

$$\Delta\alpha_j\, \Delta\beta_j = \frac{\hbar}{2}$$

for each j.

We have tried to generalize the one dimensional proof given in Section 5 to n dimensions. The proof relies on finding the minimum of a function of n^2 variables. (The symplectic orthogonal group is isomorphic to the n^2 dimensional real Lie group $U(n)$.) Using the second derivative test to separate minima, maxima, and saddles becomes exceptionally complicated.

2. Observations about the Angle of Rotation

The particular value of θ may seem rather bizarre, but it is natural. The wave packet $\varphi_0(A,\, B,\, \hbar,\, a,\, \eta,\, x)$ is the ground state of a Hamiltonian that is quadratic in x and p. That Hamiltonian is explicitly

$$H \;=\; \frac{1}{2}\,(x \quad p)\begin{pmatrix} |B|^2 & \mathrm{Im}\,(B\overline{A}) \\ \mathrm{Im}\,(B\overline{A}) & |A|^2 \end{pmatrix}\begin{pmatrix} x \\ p \end{pmatrix}.$$

In terms of the raising and lowering operators we use below,

$$H \;=\; \frac{\hbar}{2}\,(\mathcal{A}^*\mathcal{A} \,+\, \mathcal{A}\mathcal{A}^*)\,.$$

A rotation through angle θ diagonalizes the real symmetric matrix

$$\frac{1}{2}\begin{pmatrix} |B|^2 & \mathrm{Im}\,(B\overline{A}) \\ \mathrm{Im}\,(B\overline{A}) & |A|^2 \end{pmatrix},$$

whose eigenvalues are

$$\frac{1}{4}\left\{(|A|^2+|B|^2) \;\pm\; \sqrt{(|A|^2-|B|^2)^2 \,+\, 4\,(\mathrm{Im}\,(B\overline{A}))^2}\right\}.$$

The product of these eigenvalues is the determinant of the matrix for H. The calculation is a bit tedious, but for allowed values of the parameters, the product is $1/4$.

When H is associated with a diagonal matrix, $\theta = 0$, and one is essentially back to considering a standard frequency ω harmonic oscillator Hamiltonian, where $A = \omega^{-1/2}$ and $B = \omega^{1/2}$. In this case, the standard uncertainty product is $\hbar/2$. When considering the matrix associated with H, the value of θ is very natural.

The vectors $\varphi_k(A,\, B,\, \hbar,\, a,\, \eta,\, x)$ also diagonalize the Hamiltonians

$$H_1 \;=\; \hbar\,\mathcal{A}\mathcal{A}^* \qquad \text{and} \qquad H_2 \;=\; \hbar\,\mathcal{A}^*\mathcal{A}.$$

The matrices associated with these quadratic Hamiltonians are

$$\frac{1}{2}\begin{pmatrix} |B|^2 & -\,i\,B\overline{A} \\ i\,A\overline{B} & |A|^2 \end{pmatrix} \qquad \text{and} \qquad \frac{1}{2}\begin{pmatrix} |B|^2 & i\,A\overline{B} \\ -\,i\,B\overline{A} & |A|^2 \end{pmatrix},$$

respectively. In general, these matrices are not real symmetric and do not have real eigenvectors. However, the quadratic forms in the classical variables x and p are the same as the one for H. So, they lead to the same value of θ.

3. Preliminary Comments about Uncertainty Products

We wish to begin with the standard argument for the Heisenberg uncertainty relation

$$\Delta x\,\Delta p \;\geq\; \frac{\hbar}{2}. \tag{3.1}$$

This is a consequence of the more general result that for any two self-adjoint operators X and Y, and any normalized state ψ, we have

$$\Delta X\,\Delta Y \;\geq\; \frac{1}{2}\;|\,\langle\,\psi,\,[X,\,Y]\,\psi\,\rangle\,|\,. \tag{3.2}$$

(One can take both sides as infinite if ψ is not in the appropriate domains.)

The formal argument for proving this is to note that the square of the norm of

$$\{\,(X\,-\,\langle\psi,\,X\,\psi\rangle)\,+\,i\,\lambda\,(Y\,-\,\langle\psi,\,Y\,\psi\rangle)\,\}\;\psi$$

must be positive. By explicit calculation, this leads to the inequality

$$\Delta X^2\,+\,\lambda^2\,\Delta Y^2\,+\,\lambda\,\langle\,\psi,\,i\,[X,\,Y]\,\psi\,\rangle\;\geq\;0.$$

The left hand side is minimized by taking $\lambda\;=\;-\,\dfrac{\langle\,\psi,\,i\,[X,\,Y]\,\psi\,\rangle}{2\,\Delta Y^2}$. Since the inequality is true for this value for λ, we see that

$$\Delta X^2\;-\;\frac{\langle\,\psi,\,i\,[X,\,Y]\,\psi\,\rangle^2}{4\,\Delta Y^2}\;\geq\;0,$$

and inequality (3.2) follows immediately.

Since $[x,\,p]\;=\;i\,\hbar$, inequality (3.2) implies inequality (3.1). We then note that for any value of θ, we have

$$\begin{aligned}[\alpha,\,\beta]\;&=\;[x\,\cos(\theta)\,+\,p\,\sin(\theta),\,-\,x\,\sin(\theta)\,+\,p\,\cos(\theta)]\\ &=\;\cos^2(\theta)\,[x,\,p]\;-\;\sin^2(\theta)\,[p,\,x]\\ &=\;i\,\hbar.\end{aligned}$$

So, by another application of inequality (3.2), we obtain the uncertainty relation that for any normalized state ψ,

$$\Delta\alpha\,\Delta\beta\;\geq\;\frac{\hbar}{2}. \tag{3.3}$$

Our main result is that one actually has equality when $\psi\;=\;\varphi_0(A,\,B,\,\hbar,\,a,\,\eta,\,x)$ and $\theta\;=\;\dfrac{1}{2}\,\arctan\left(\dfrac{2\,\mathrm{Im}\,(B\overline{A})}{|B|^2-|A|^2}\right)$.

4. One Dimensional Semiclassical Wave Packets

The one dimensional semiclassical wave packets are most easily defined by using raising and lowering operators [**3**]: The number $a\in\mathbb{R}$ denotes the mean position of the wave packets. The number $\eta\in\mathbb{R}$ denotes the mean momentum. We assume the semiclassical parameter $\hbar$ is positive, and we choose any two complex numbers A and B that satisfy $\overline{A}B\,+\,\overline{B}A\;=\;2$. We then define

$$\varphi_0(A,\,B,\,\hbar,\,a,\,\eta,\,x)\;=\;\pi^{-1/4}\,\hbar^{-1/4}\,A^{-1/2}\,\exp\left\{\,-\,\frac{B\,(x-a)^2}{2\,A\,\hbar}\;+\;i\,\eta\,(x-a)/\hbar\right\}.$$

(The square root $A^{-1/2}$ can take either sign. In applications, the sign is determined by an initial choice and continuity in time.) We note that this vector is normalized because $\overline{A}B\,+\,\overline{B}A\;=\;2$. Also, any complex, normalized Gaussian can be written this way.

In analogy with the usual harmonic oscillator, we define raising and lowering operators by

$$\mathcal{A}(A,\,B,\,\hbar,\,a,\,\eta)^*\;=\;\frac{1}{\sqrt{2\,\hbar}}\;\left[\,\overline{B}\,(x-a)\;-\;i\,\overline{A}\,(p-\eta)\,\right]$$

and

$$\mathcal{A}(A,\,B,\,\hbar,\,a,\,\eta)\;=\;\frac{1}{\sqrt{2\,\hbar}}\;\left[\,B\,(x-a)\,+\,i\,A\,(p-\eta)\,\right].$$

Using the raising operator inductively, starting from $k = 0$, we define

$$\varphi_{k+1}(A,\,B,\,\hbar,\,a,\,\eta,\,x) \;=\; \frac{1}{\sqrt{k+1}}\;\mathcal{A}(A,\,B,\,\hbar,\,a,\,\eta)^*\;\varphi_k(A,\,B,\,\hbar,\,a,\,\eta,\,x).$$

Then, for the lowering operator, we also have

$$\mathcal{A}(A,\,B,\,\hbar,\,a,\,\eta)\;\varphi_0(A,\,B,\,\hbar,\,a,\,\eta,\,x) \;=\; 0$$

and

$$\mathcal{A}(A,\,B,\,\hbar,\,a,\,\eta)\;\varphi_k(A,\,B,\,\hbar,\,a,\,\eta,\,x) \;=\; \sqrt{k}\;\varphi_{k-1}(A,\,B,\,\hbar,\,a,\,\eta,\,x),$$

for $k = 1,\,2,\,\dots$.

By using the raising and lowering operators, it is quite easy [**3**] to prove that $\{\,\varphi_k(A,\,B,\,\hbar,\,a,\,\eta,\,x)\,\}$ is an orthonormal basis of $L^2(\mathbb{R},\,dx)$, and that in the state $\varphi_k(A,\,B,\,\hbar,\,a,\,\eta,\,x)$,

$$\Delta x \;=\; \left(\frac{\hbar}{2}\right)^{1/2}\;|A|\;\sqrt{2k+1},$$

and

$$\Delta p \;=\; \left(\frac{\hbar}{2}\right)^{1/2}\;|B|\;\sqrt{2k+1}.$$

The proof we present below for φ_0 generalizes to show that with $\theta = \dfrac{1}{2}\arctan\left(\dfrac{2\,\mathrm{Im}\,(B\overline{A})}{|B|^2-|A|^2}\right)$, in the state $\varphi_k(A,\,B,\,\hbar,\,a,\,\eta,\,x)$ we have

$$\Delta\alpha\,\Delta\beta \;=\; \frac{\hbar}{2}\;(2k+1).$$

Our proof of the rotated uncertainty relation (3.3) will make use of the raising and lowering operators to represent $(x-a)$ and $(p-\eta)$. From the definitions above, one easily sees that

$$(x-a) \;=\; \sqrt{\frac{\hbar}{2}}\;\left\{\,A\,\mathcal{A}(A,\,B,\,\hbar,\,a,\,\eta)^* \;+\; \overline{A}\,\mathcal{A}(A,\,B,\,\hbar,\,a,\,\eta)\,\right\} \tag{4.1}$$

and

$$(p-\eta) \;=\; i\,\sqrt{\frac{\hbar}{2}}\;\left\{\,B\,\mathcal{A}(A,\,B,\,\hbar,\,a,\,\eta)^* \;-\; \overline{B}\,\mathcal{A}(A,\,B,\,\hbar,\,a,\,\eta)\,\right\}. \tag{4.2}$$

5. The Rotated Uncertainty Product

We begin with a technical lemma that is proved by simple calculation.

LEMMA 5.1. *If $A(t)$ and $B(t)$ satisfy*

$$\dot{A}(t) \;=\; i\,B(t) \qquad \textit{and} \tag{5.1}$$

$$\dot{B}(t) \;=\; i\,A(t), \tag{5.2}$$

then we have the following time derivatives:

$$\frac{d}{dt}\,\left(|A(t)|^2\,|B(t)|^2\right) \;=\; 2\,\left(|A(t)|^2-|B(t)|^2\right)\;Im\,(B(t)\overline{A(t)}), \tag{5.3}$$

$$\frac{d}{dt}\,\left(Im\,(B(t)\overline{A(t)})\right) \;=\; |A(t)|^2-|B(t)|^2, \qquad \textit{and} \tag{5.4}$$

$$\frac{d}{dt}\,\left(|A(t)|^2-|B(t)|^2\right) \;=\; -\,4\,Im\,(B(t)\overline{A(t)}). \tag{5.5}$$

We now state and prove our main result:

THEOREM 5.2. *Let A, B, $\hbar$, a, and η be any allowed values of the parameters. If we choose*

$$\theta = \frac{1}{2} \arctan \left(\frac{2\, Im(B\overline{A})}{|B|^2 - |A|^2} \right),$$

then the state $\varphi_0(A,\, B,\, \hbar,\, a,\, \eta,\, x)$ minimizes the uncertainty product for α and β. I.e.,

$$\Delta\alpha\, \Delta\beta \;=\; \frac{\hbar}{2}.$$

PROOF. We can prove this by an explicit, but tedious calculation or by the following more appealing argument.

First we note that the values of a and η are irrelevant, so we can set them both to zero. Second we note that clockwise rotations of phase space are generated by the standard, frequency 1, classical harmonic oscillator Hamiltonian $(p^2 + x^2)/2$. If we propagate with its quantum analog, the state $\varphi_0(A(0),\, B(0),\, \hbar,\, 0,\, 0,\, x)$ evolves to a new Gaussian $\varphi_0(A(t),\, B(t),\, \hbar,\, 0,\, 0,\, x)$, where [**3**] $A(t)$ and $B(t)$ are given in equations (5.7) below.

Applying the "counterclockwise" rotation to the quantum operators x and p to obtain α and β is equivalent to keeping the original operators x and p, but rotating the state φ_0 in the clockwise direction. Thus, proving the theorem is equivalent to showing that $\Delta x\, \Delta p \;=\; \hbar/2$ in the state $\varphi_0(A(\theta),\, B(\theta),\, \hbar,\, 0,\, 0,\, x)$.

Since $\Delta x\, \Delta p \;=\; \dfrac{\hbar}{2}\, |A(\theta)|\, |B(\theta)|$, it suffices to show that $|A(\theta)|\, |B(\theta)| \;=\; 1$.

We can find the minimum of $|A(t)|\, |B(t)|$ by setting the derivative of $f(t) \;=\; |A(t)|^2\, |B(t)|^2$ to zero. By formula (5.3), this requires

$$\dot{f}(t) \;=\; 2\, \left(|A(t)|^2 - |B(t)|^2\right)\, \mathrm{Im}\, (B(t)\overline{A(t)}) \;=\; 0.$$

Thus, we must have $\mathrm{Im}\, (B(t)\overline{A(t)}) \;=\; 0$ or $|A(t)|^2 - |B(t)|^2 \;=\; 0$.

To obtain a contradiction, suppose a relative minimum occurs with $\mathrm{Im}(B(t)\overline{A(t)}) \neq 0$. Then we must have $|A(t)|^2 - |B(t)|^2 \;=\; 0$.

From (5.4) and (5.5), we see that the second derivative of f is

$$\ddot{f}(t) \;=\; 2\, \left(|A(t)|^2 - |B(t)|^2\right)^2 \;-\; 8\, \left(\mathrm{Im}\, (B(t)\overline{A(t)})\right)^2.$$

From our assumptions above, this quantity is strictly negative, and we have found a maximum of f instead of a minimum. Thus, at any minimum of f, we must have

$$\mathrm{Im}\, (B(t)\overline{A(t)}) \;=\; 0. \tag{5.6}$$

Since $\mathrm{Re}\, (B(t)\overline{A(t)}) \;=\; 1$, this condition forces $|A(t)|\, |B(t)| \;=\; 1$, and hence our desired result $\Delta x\, \Delta p \;=\; \dfrac{\hbar}{2}$.

We next note that

$$\begin{aligned} A(t) &\;=\; A(0)\, \cos(t) \;+\; i\, B(0)\, \sin(t) \\ B(t) &\;=\; i\, A(0)\, \sin(t) \;+\; B(0)\, \cos(t). \end{aligned} \tag{5.7}$$

We obtain these relations by explicitly solving the linear system of ordinary differential equations (2.16) of [**3**] for this easy special case.

From these relations, we see that

$$\begin{aligned}
&\operatorname{Im}(B(t)\,\overline{A(t)}) \\
=\ &\operatorname{Im}\{\,(i\,A(0)\,\sin(t) + B(0)\,\cos(t))\left(\overline{A(0)}\,\cos(t) - i\,\overline{B(0)}\,\sin(t)\right) \\
=\ &\operatorname{Im}\left\{i\,(|A(0)|^2 - |B(0)|^2)\cos(t)\sin(t) + B(0)\overline{A(0)}\cos^2(t) + A(0)\overline{B(0)}\sin^2(t)\right\} \\
=\ &(|A(0)|^2 - |B(0)|^2)\,\cos(t)\,\sin(t) + \operatorname{Im}(B(0)\overline{A(0)})\,(\cos^2(t) - \sin^2(t)).
\end{aligned}$$

So, equation (5.6) is equivalent to

$$(|A(0)|^2 - |B(0)|^2)\,\sin(2t) + 2\operatorname{Im}(B(0)\overline{A(0)})\,\cos(2t) = 0,$$

which is satisfied if we choose $t = \theta$. This proves the theorem. □

REMARK 5.3. Our comments about the choice of θ in Section 2 relied on operators that were quadratic in the raising and lowering operators. In the spirit of the proof of Theorem 5.2, we can make similar comments that just involve the lowering operator.

Let $U(t)$ denote the propagator for the standard, frequency 1, quantum harmonic oscillator. Choosing $a = 0$ and $\eta = 0$, we have $\mathcal{A}\,\varphi_0(A,\,B,\,\hbar,\,0,\,0,\,x) = 0$. Thus,

$$\begin{aligned}
0 &= U(t)\,\mathcal{A}\,\varphi_0(A,\,B,\,\hbar,\,0,\,0,\,x) \\
&= U(t)\,\mathcal{A}\,U(t)^{-1}\,U(t)\,\varphi_0(A,\,B,\,\hbar,\,0,\,0,\,x) \\
&= [B(x\,\cos(t) + p\,\sin(t)) + iA(-x\,\sin(t) + p\,\cos(t))]\,U(t)\,\varphi_0(A,\,B,\,\hbar,\,0,\,0,\,x) \\
&= [(B\cos(t) - iA\sin(t))x + i(A\cos(t) - iB\sin(t))\,p]\,U(t)\,\varphi_0(A,\,B,\,\hbar,\,0,\,0,\,x).
\end{aligned}$$

(5.8)

If we choose $t = \theta$, the coefficients of x and p inside the square brackets have the same complex phase. This is equivalent to

$$[\,\gamma\,x + i\,\delta\,p\,]\;U(t)\,\varphi_0(A,\,B,\,\hbar,\,0,\,0,\,x) = 0,$$

where γ and δ are real. This implies $U(t)\,\varphi_0(A,\,B,\,\hbar,\,0,\,0,\,x)$ is a phase times a real Gaussian, and that it consequently minimizes $\Delta x\,\Delta p$. Undoing the rotation of phase space shows that $\varphi_0(A,\,B,\,\hbar,\,0,\,0,\,x)$ minimizes the uncertainty product of $\Delta\alpha\,\Delta\beta$.

Requiring the coefficients in (5.8) to have the same phases is equivalent to solving (5.6) and equivalent to solving the eigenvalue problem of Section 2.

REMARK 5.4. If one is not interested in the value of θ, but only its existence, the portion of the proof after equation (5.6) can be replaced by the following:

It suffices to prove that equation (5.6) is satisfied for some t. All rotations of $\mathbb{R}^2$ are symplectic, and the rotation group $SO(2)$ is compact. So, there exists a value of t at which $g(t) = \left(\operatorname{Im}(B(t)\overline{A(t)})\right)^2$ takes its minimum. At that minimum, $\dot{g}(t) = 2\,g(t)\,(|A(t)|^2 - |B(t)|^2)$ must be zero. So, $g(t) = 0$ or $(|A(t)|^2 - |B(t)|^2) = 0$. If $g(t) \neq 0$, we must have $(|A(t)|^2 - |B(t)|^2) = 0$, and in that case, it follows from explicit calculation that $\ddot{g}(t) = -8\,g(t)^2 < 0$. This cannot happen at a minimum, so we conclude that $g(t)$ must be zero.

References

[1] Erwan Faou, Vasile Gradinaru, and Christian Lubich, *Computing semiclassical quantum dynamics with Hagedorn wavepackets*, SIAM J. Sci. Comput. **31** (2009), no. 4, 3027–3041, DOI 10.1137/080729724. MR2520310 (2010f:65200)

[2] Gradinaru, V. and Hagedorn, G. A., *A Time–Splitting for the Semiclassical Schrödinger Equation.* (2012 preprint).

[3] George A. Hagedorn, *Raising and lowering operators for semiclassical wave packets*, Ann. Physics **269** (1998), no. 1, 77–104, DOI 10.1006/aphy.1998.5843. MR1650826 (99f:81052)

[4] George A. Hagedorn and Alain Joye, *Exponentially accurate semiclassical dynamics: propagation, localization, Ehrenfest times, scattering, and more general states*, Ann. Henri Poincaré **1** (2000), no. 5, 837–883, DOI 10.1007/PL00001017. MR1806980 (2001k:81066)

[5] Rousse, V., (personal communication).

Department of Mathematics and Center for Statistical Mechanics, Mathematical Physics, and Theoretical Chemistry, Virginia Polytechnic Institute and State University, Blacksburg, Virginia 24061–0123

E-mail address: `hagedorn@math.vt.edu`

Proceedings of Symposia in Pure Mathematics
Volume **87**, 2013
http://dx.doi.org/10.1090/pspum/087/01433

On a transformation of Bohl and its discrete analogue

Evans M. Harrell II and Manwah Lilian Wong

Abstract. Fritz Gesztesy's varied and prolific career has produced many transformational contributions to the spectral theory of one-dimensional Schrödinger equations. He has often done this by revisiting the insights of great mathematical analysts of the past, connecting them in new ways, and reinventing them in a thoroughly modern context.

In this short note we recall and relate some classic transformations that figure among Fritz Gestesy's favorite tools of spectral theory, and indeed thereby make connections among some of his favorite scholars of the past, Bohl, Darboux, and Green. After doing this in the context of one-dimensional Schrödinger equations on the line, we obtain some novel analogues for discrete one-dimensional Schrödinger equations.

Dem einzigartigen Fritz gewidmet.

1. Introduction

In 1906 [**3**], Bohl introduced a nonlinear transformation for solutions of Sturm-Liouville equations, which is an exact, albeit implicit, counterpart to the Liouville-Green aproximation [**23**]. Bohl used the transformation as a tool in oscillation theory, and this has continued to be the main use of the Bohl transformation in the hands of later authors. Notably, Ráb [**24**] used the Bohl transformation to prove necessary and sufficient conditions for oscillation of solutions, and showed that it is an effective foundation for Sturm-Liouville oscillation theory. Willett's lecture in [**29**] provides a clear description in English of the Bohl transformation in oscillation theory, including the contributions of Ráb, while Reid's monograph [**27**] compares and contrasts it with the Prüfer transformation. See also [**13, 16–18, 26**].

In [**10**] §4, Davies and Harrell introduced a non-oscillatory variant of the Bohl transformation to connect the notions of Liouville-Green approximation, Green functions, and Agmon metrics for exponential decay of solutions. Some spectral bounds were derived as consequences. This analysis was extended in a series of articles by Chernyavskaya and Shuster (e.g., [**4, 5, 8**]), to address questions of solvability, regularity, estimates of Green functions, and asymptotics in Sturm-Liouville theory.

In this note we begin with a largely expository treatment of the classic Bohl transformation, concentrating for simplicity on the situation where all coefficients are real and regular and the Sturm-Liouville equation is in the standard form of the one-dimensional Schrödinger equation. Then in the last section we show how

2010 *Mathematics Subject Classification.* Primary .

the technique can be adapted to the case of a discrete Schrödinger equation on the integers.

2. The interplay of the Bohl and Green functions

Let V be real-valued and continuous, and consider a solution basis for the Sturm-Liouville equation

$$-u'' + V(x)u = 0; \tag{2.1}$$

we may normalize the basis $u_{1,2}(x)$ so that $W[u_1, u_2] := u_1 u_2' - u_2 u_1' = 1$. (There is no assumption of an eigenvalue 0. For our purposes a possible nonzero spectral parameter has simply been incorporated into V.) The Bohl transformation maps this solution basis onto a second solution basis with remarkable properties, some of which are collected in a nutshell version in this section.

DEFINITION 2.1. Given a solution basis $\{u_{1,2}(x)\}$ of (2.1), chosen so that the Wronskian $W[u_1, u_2] := u_1 u_2' - u_2 u_1' = 1$, we define the *diagonal function* by

$$Z[u_1, u_2](x) := (u_1(x)u_2(x))^{1/2}. \tag{2.2}$$

The *Bohl transformation* of $\{u_{1,2}(x)\}$ is an equivalent solution basis of (2.1), defined in terms of $\{u_{1,2}(x)\}$ by

$$\mathcal{B} : \{u_1(x), u_2(x)\} \to \left\{\phi^{\pm}(x) := Z(x) \exp\left(\pm \int^x \frac{1}{2Z^2(t)} dt\right)\right\}. \tag{2.3}$$

REMARK 1. a) The choice of the complex phase of the square root in (2.2) is unimportant, but should be continuous in x. For brevity we write $Z(x)$ for $Z[u_1, u_2](x)$ when the dependence on $u_{1,2}$ is clear. The reason for calling it the diagonal function is that, as will be seen below,

$$Z^2(x) = G_0(x, x),$$

where $G_0(x, x)$ is the diagonal of a certain Green function $G_0(x, y)$ for (2.1). Among the useful properties of the function Z is that it solves the *diagonal differential equation*

$$\mathcal{J}[Z] := -Z'' + V(x)Z - \frac{1}{4Z^3} = 0, \tag{2.4}$$

[**10**, **17**, **18**, **29**].

b). In fact, with the oscillatory situation in mind Bohl originally wrote the solution basis in the Liouville-Green form

$$\frac{1}{\sqrt{R}} \sin\left(\int^x R(t)dt\right)$$

and

$$\frac{1}{\sqrt{R}} \cos\left(\int^x R(t)dt\right),$$

which is equivalent to (2.3) under the identification $2Z^2 \to i/R$ and some harmless linear combinations.

c). We recall that Gesztesy and Simon [**12**] have made connections between the Krein spectral shift function, the related Xi function, and the diagonal of the Green function.

Calling upon [**10**] §4, we collect some facts, which are verifiable directly:

THEOREM 2.1.

(1) *If* $u_1(x)$ *and* $u_2(x)$ *are solutions to* (2.1) *such that* $W[u_1, u_2] = 1$, *and* $u_1(x)u_2(x)$ *does not vanish on the interval* (a, b), *then* $Z[u_1, u_2](x)$ *satisfies* (2.4) *on* (a, b).

(2) *If* Z *is a nonvanishing solution of* (2.4) *on* (a, b), *then* $\phi^{\pm}(x)$ *as defined in* (2.3) *provide a pair of independent solutions of* (2.1) *on* (a, b). *In particular, each* $\phi^{\pm}(x)$ *is a linear combination of* $\{u_{1,2}\}$ *and vice versa.*

(3) *If* $x_{>,<} := \max(x, y)$, *resp.* $\min(x, y)$, *then*

$$G_0(x, y) := Z(x)Z(y) \exp\left(-\int_{x_<}^{x_>} \frac{1}{2Z^2(t)} dt\right) \tag{2.5}$$

is a Green function for (2.1), *in the sense that*

$$\left(-\frac{\partial^2}{\partial x^2} + V(x)\right) G_0(x, y) = \delta(x - y). \tag{2.6}$$

As an integral kernel, G_0 defines the inverse of a particular realization of $-\frac{d^2}{dx^2} + V$, but not *a priori* one for which the domain of definition includes u_1 or u_2, because of a possible mismatch of boundary conditions at finite points. This issue is not important for questions of oscillation or asymptotic behavior at infinity, but another concern remains, namely the possibility that Z vanishes, which would invalidate the transformation. Because of this we recall that in the absence of imposed finite boundary conditions, complex solutions can always be used to prevent Z from vanishing:

LEMMA 2.2. *Suppose that* u *is a solution of* (2.1) *on a finite or infinite interval* (a, b), *and that at some* $x_0 \in (a, b)$, $\mathrm{Re}(u(x_0))\mathrm{Im}(u(x_0)) \neq 0$ *and* $u'(x_0)/u(x_0) \notin \mathbb{R}$. *Then* u *does not vanish on* (a, b).

PROOF. Because V is real-valued, $\mathrm{Re}\, u$ and $\mathrm{Im}\, u$ each satisfy (2.1), and it therefore suffices to show that they are independent.

Letting $\alpha = u'(x_0)/u(x_0)$, a calculation shows that.

$$\frac{\mathrm{Re}\, u'(x_0)}{\mathrm{Re}\, u(x_0)} = \mathrm{Re}\, \alpha - \mathrm{Im}\, \alpha \frac{\mathrm{Im}\, u(x_0)}{\mathrm{Re}\, u(x_0)},$$

$$\frac{\mathrm{Im}\, u'(x_0)}{\mathrm{Im}\, u(x_0)} = \mathrm{Re}\, \alpha + \mathrm{Im}\, \alpha \frac{\mathrm{Re}\, u(x_0)}{\mathrm{Im}\, u(x_0)}.$$

It follows that

$$\frac{\mathrm{Re}\, u'(x_0)}{\mathrm{Re}\, u(x_0)} - \frac{\mathrm{Im}\, u'(x_0)}{\mathrm{Im}\, u(x_0)} = -\mathrm{Im}\, \alpha \left(\frac{\mathrm{Im}\, u(x_0)}{\mathrm{Re}\, u(x_0)} + \frac{\mathrm{Re}\, u(x_0)}{\mathrm{Im}\, u(x_0)}\right),$$

and therefore

$$W[\mathrm{Im}\, u, \mathrm{Re}\, u] = -\mathrm{Im}\, \alpha \left((\mathrm{Re}\, u(x_0))^2 + (\mathrm{Im}\, u(x_0))^2\right) \neq 0.$$

□

This standard lemma implies that given any two linearly independent solutions of (2.1) it is always possible to find a pair of complex-valued linearly independent combinations that are non-vanishing on (a, b). The Wronskian of the new pair may be set to 1 by multiplying one solution by an appropriate constant, justifying the conclusions of Theorem 2.1. It is of some use to consider a particular Z determined as follows.

By construction any solution u as set forth in Lemma 2.2 will be linearly independent of its complex conjugate. We may therefore choose a complex number α so that $W[u, \alpha^2 \overline{u}] = 1$, which ensures that

$$Z(x) = \alpha|u(x)|. \tag{2.7}$$

Indeed, $\arg(\alpha)$ is restricted to the values $k\pi/4$ for integer k, as can be seen for example from (2.4), which implies that

$$-|u|'' + V|u| = \frac{1}{4\alpha^4|u|^3} \in \mathbb{R}. \tag{2.8}$$

(In passing we note the implication that the expression on the left does not change sign.) In fact, there are only two truly distinct cases for $\arg(\alpha)$, *viz.*, 0 and $\frac{\pi}{4}$, due to the simple scalings in (2.4) and Theorem 2.1 when Z is replaced by iZ. If $\alpha > 0$, i.e., $Z(x) > 0$, then the solutions $\phi^{\pm}$ do not change sign, which corresponds to the case of *disconjugacy* for the ODE (2.1) in the classical theory [**20, 26**]. This situation was the focus of [**10**], in which some spectral bounds were derived and it was argued that $1/2Z^2$ defines an Agmon metric.

Otherwise, it may be assumed without of loss of generality that $\arg(\alpha) = \frac{\pi}{4}$, for which solutions may oscillate, in that their arguments increase or decrease by $n\pi$ for $n > 1$ as $x \to \infty$. A central question of Sturmian theory is whether solutions oscillate infinitely often, or only finitely often. When the increase in the argument of a solution is infinite, the equation (2.1) is said to be *oscillatory*. An approach to oscillation theory, equivalent to that of [**24, 29**] but bringing out the role of Green functions, can be based on the following version of a result of Gagliardo, as cited in [**29**]:

COROLLARY 2.3 (Cf. [**29**], Corollary 3.2.). *Suppose that* (2.1) *holds on an infinite interval* (a, ∞), *and let* $G_B(x, y)$ *be the Green function constructed according to the prescription leading to* (2.7). *Then either* $G_B(x, y) \in \mathbb{R}$ *for all* x, y, *in which case the solution basis* $\phi^{\pm}$ *is nonoscillatory, or else: The phase of* $\phi^{\pm}$ *has only finite increase on* (a, ∞) *iff* $1/G(x, x) \in L^1(a, \infty)$.

As has been known since the work of von Neumann and Wigner, it is possible for eigenvalues to be embedded in the continuous spectrum of Sturm-Liouville equations [**22, 25**]. This phenomenon requires oscillatory solutions to be square-integrable, and can thus be related to the Bohl transformation as follows:

COROLLARY 2.4 (Ráb). *An oscillatory solution of* (2.1), *written in the form*

$$Z(x) \exp\left(\pm \frac{i}{2} \int_a^x \frac{1}{|Z(t)|^2} dt \right),$$

exists and is square integrable if and only if the nonlinear equation

$$-w''(x) + V(x)w(x) = -\frac{1}{4w^3(x)}$$

has a square-integrable solution.

We close this section with a Darboux-type factorization [**9, 14**], the novel feature of which is the role played by the diagonal function and the Bohl solution basis (2.3). For any complex valued, nonvanishing function $Z(x) \in AC^1[(a, b)]$, define

$$\mathcal{D}^{\pm}[Z] := \frac{d}{dx} - \frac{Z'}{Z} \mp \frac{1}{2Z^2}.$$

It is immediate to see that

$$\mathcal{D}^{\pm}[Z]\phi^{\pm} = 0,$$

where $\phi^{\pm}$ are defined in terms of Z by (2.3). A further calculation reveals that

$$\left(\mathcal{D}^{\pm}[Z] - 2\frac{d}{dx}\right)\mathcal{D}^{\pm}[Z] = -\frac{d^2}{dx^2} + \frac{Z''}{Z} + \frac{1}{4Z^4} = -\frac{d^2}{dx^2} + V(x), \tag{2.9}$$

provided that Z satisfies the diagonal differential equation (2.4). An alternative way to express the two factorizations in (2.9) is that

$$-\frac{d^2}{dx^2} + V(x) = \left(\mathcal{D}\left[\overline{Z}\right]^{\mp}\right)^{*} \mathcal{D}[Z]^{\pm},$$

where * designates the formal adjoint operation.

3. The discrete form of the Bohl $\mathcal{J}[Z]$

In this section we show that most of the transformations and relationships presented in the first section have counterparts for discrete one-dimensional Schrödinger equations. (Part of the material in this section has appeared in a preprint [**19**], which has been expanded and divided for publication as two articles.) some details are rather different from the continuous case, making it uncertain how far the analogy goes, especially in the oscillatory case. A full-fledged oscillation theory for discrete problems based on an analogue of the Bohl transformation and its connection to Green functions would be an interesting next project.

Let Δ denote the discrete second-difference operator on the positive integers. We standardize the Laplacian such that $(\Delta f)_n := f_{n+1} + f_{n-1} - 2f_n$ for $f = (f_n) \in \ell^2(\mathbb{N})$, and consider equations of the form

$$(-\Delta + V)u = 0, \tag{3.1}$$

where the potential-energy function V is a diagonal operator with real values V_n.

Eq. (3.1) and its solutions share many of the properties of classical Sturm-Liouville equations, as is laid out for example in [**1**]. For our purposes we recall that: The solution space is two-dimensional, and the Wronskian of any two solutions

$$W[u^{(1)}, u^{(2)}] := u_n^{(1)} u_{n+1}^{(2)} - u_{n+1}^{(1)} u_n^{(2)} \tag{3.2}$$

is constant. A *Green matrix* as a solution of

$$(-\Delta + V)G = I, \tag{3.3}$$

where I is the identity matrix, and every Green matrix can be written as the sum of a vector in the null space of $(-\Delta + V)$ and the particular Green matrix

$$G_{m,n}^{(p)} := \frac{u^{(1)}_{\max(m,n)} u^{(2)}_{\min(m,n)}}{W[u^{(1)}, u^{(2)}]}, \tag{3.4}$$

provided that $\{u^{(1)}, u^{(2)}\}$ are linearly independent.

A feature of the discrete Schrödinger equation (3.1) that is not shared by (2.1) is an invariance under the transformation

$$\begin{aligned} u_n &\to (-1)^n u_n \\ V_n &\to -4 - V_n, \end{aligned} \tag{3.5}$$

as can be easily checked. Among other things, this implies that any fact proved under the assumption, for example, that $V_n > 0$ has a counterpart for $V_n < -4$, with systematic sign changes.

Our goal in this section is to present an analogue of the Bohl transformation for the discrete Schrödinger equation (3.1). In particular, we offer a discrete version of some of the results of [**10**], §4, and show in particular that the diagonal elements G_{nn} of the Green matrix allow the full solution space to be recovered formulaically. We build on some earlier steps in this direction by Chernyavskaya and Shuster [**6**, **7**]. As in [**10**] we furthermore point out connections between the diagonal of the Green matrix and an Agmon distance for (3.1).

In the discrete situation the use of exponentials of integrals is not the most natural, so we instead seek to represent a pair of solutions in the forms

$$\varphi_n^+ = z_n \prod_{\ell=1}^{n} S_\ell, \qquad \varphi_n^- = z_n \left(\prod_{\ell=1}^{n} S_\ell \right)^{-1}. \tag{3.6}$$

Since the product of these two solutions is the diagonal of a Green matrix, up to a constant multiple, this suggests that if we begin by selecting a Green matrix such that G_{nn} is nonvanishing, then we can directly define

$$z_n := (G_{nn})^{1/2}.$$

It remains to work out the most convenient form of $S_\ell^{\pm 1}$. If the Wronskian is scaled so that $W[\varphi^-, \varphi^+] = 1$, then substitution of the ansatz (3.6) leads after a calculation to

$$S_n - \frac{1}{S_n} = \frac{1}{z_n z_{n-1}}. \tag{3.7}$$

Here we pause to observe two ambiguities in relating $\varphi_n^\pm$ to the potential V. The first is that, due to the invariance (3.5), if

$$G_{mn} = \psi^+_{\min(m,n)} \psi^-_{\max(m,n)} \tag{3.8}$$

is the Green matrix for some potential function V_n, then the same diagonal elements G_{nn} also belong to the Green matrix for an equation of type (3.1) but with potential function $\widetilde{V}_n = -4 - V_n$. Secondly, (3.7) is equivalent to a quadratic expression for S_n, and therefore the solution is generally nonunique. These ambiguities are avoided when the Schrödinger operator $H = -\Delta + V$ in (3.1) is positive, so $G_{nn} > 0$ and by convention $z_n > 0$. We can then fix S_n as the larger root of (3.7).

Accordingly, in this situation we simply define

$$S_n^{[z]} := \frac{1 + \sqrt{1 + 4 z_n^2 z_{n-1}^2}}{2 z_n z_{n-1}}. \tag{3.9}$$

A pair of functions $\varphi_n^{\pm}$ can now be defined by the ansatz (3.6), i.e., when expressed in terms of z_n,

$$\varphi_n^+ := z_n \prod_{k=m+1}^{n} \left(\frac{1+\sqrt{1+4z_n^2 z_{n-1}^2}}{2z_n z_{n-1}} \right), \tag{3.10}$$

$$\varphi_n^- = z_n \left(\prod_{k=m+1}^{n} \left(\frac{1+\sqrt{1+4z_n^2 z_{n-1}^2}}{2z_n z_{n-1}} \right) \right)^{-1}.$$

Remarkably, with this definition, both φ^+ and φ^- solve a single equation of the form (3.1), where the potential function V_n is determined from z_n via

$$V_n^{[z]} := \frac{\Delta \varphi_n^+}{\varphi_n^+} = \frac{1+\sqrt{1+4z_n^2 z_{n+1}^2}}{2z_n^2} + \frac{2z_{n-1}^2}{1+\sqrt{1+4z_n^2 z_{n-1}^2}} - 2$$

$$= \frac{z_{n+1}}{z_n} S_{n+1}^{[z]} + \frac{z_{n-1}}{z_n S_n^{[z]}} - 2, \tag{3.11}$$

provided that $V_n > -2$. (Else a different root must be chosen in (3.9).) To see that $\varphi_n^{\pm}$ solve the same discrete Schrödinger equation, let us separately calculate

$$\frac{\Delta \varphi_n^-}{\varphi_n^-} = \frac{z_{n+1}}{z_n S_{n+1}^{[z]}} + \frac{z_{n-1}}{z_n} S_n^{[z]} - 2, \tag{3.12}$$

and note that since $S_n^{[z]}$ has been chosen to satisfy (3.9), the difference between these last two expressions is

$$\frac{1}{z_n^2} - \frac{1}{z_n^2} = 0.$$

This leads to a theorem in the spirit of [**10**].

THEOREM 3.1. *Suppose that* (3.1) *has two independent positive solutions for* $m \le n \le N$, *with* $N \ge M+2$, *and denote the associated Green matrix* G_{mn}. *Since* $G_{nn} > 0$ *for* $m \le n \le N$, *we may define* $z_n := \sqrt{G_{nn}} > 0$. *In terms of* z_n, *determine* $S_n^{[z]}$ *and* $\varphi_n^{\pm}$ *according to* (3.9) *and* (3.10). *Then*

(1) $\varphi_n^{\pm}$ *is an independent pair of solutions of* (3.1) *for* $m < n \le N$.
(2) $G_{nm} = z_n z_m \prod_{\ell=m+1}^{n} \frac{1}{S_\ell^{[z]}}$, $M < m < n \le N$.
(3) *The potential function is determined from* G_{nn} *by a nonlinear difference equation,*

$$\frac{1}{2}\left(\sqrt{1+4G_{n\,n}G_{n+1\,n+1}} + \sqrt{1+4G_{n\,n}G_{n-1\,n-1}}\right) = (V_n+2)G_{nn}. \tag{3.13}$$

REMARK 2. In what follows we are mainly concerned with what happens when $N \to \infty$. In that case the assumption that there are two positive solutions is a question of disconjugacy in the theory of ordinary differential equations, cf. [**1, 20**]. If, for example, $V_n > 0$ for $n \ge N_0$, then it is not difficult to show that no solution can change sign more than once, and that therefore the positivity assumption is satisfied for n sufficiently large. As will be seen in the proof, a necessary condition for the assumption is that $V_n > -2$.

Per the symmetry remarked upon in (3.5), an alternative to positivity is the assumption that there are two solutions $\psi_n^{\pm}$ such that $(-1)^n\psi_n^{\pm} > 0$. A sufficient condition for this is that $V_n < -4$ and a necessary condition is that $V_n < -2$.

PROOF. The essential calculation was provided in the discussion before the statement of the theorem. Given that the Wronskian of φ^- and φ^+ is 1, these two functions are linearly independent and therefore a basis for the solution space of

$$(-\Delta + V_n^{[z]})\varphi = 0,$$

$V_n^{[z]}$ being defined by (3.11). Moreover,

$$G_{mn} = \varphi^+_{\min(m,n)}\varphi^-_{\max(m,n)}$$

is a Green function for $-\Delta + V_n^{[z]}$.

Hence the crux is to show that $V_n^{[z]}$ is the same as the original V_n of (3.1). Because $S_n^{[z]}$ was defined such that

$$S_n^{[z]} - \frac{1}{S_n^{[z]}} = \frac{1}{z_n z_{n-1}},$$

we may rewrite (3.11) as

$$V_n^{[z]} + 2 = \frac{1}{2z_n^2}\left(\sqrt{1 + 4z_n^2 z_{n+1}^2} + \sqrt{1 + 4z_n^2 z_{n-1}^2}\right). \tag{3.14}$$

From the definition of z_n and the assumptions of the theorem, we know that for some independent set of positive solutions $\psi_n^{\pm}$ of (3.1), with Wronskian 1, $z_n^2 = \psi_n^+\psi_n^-$. Therefore

$$\begin{aligned} 4z_n^2 z_{n\pm1}^2 &= 4(\psi_n^+\psi_{n\pm1}^-)(\psi_n^-\psi_{n\pm1}^+) \\ &= (\psi_n^+\psi_{n\pm1}^- + \psi_n^-\psi_{n\pm1}^+)^2 - (\psi_n^+\psi_{n\pm1}^- - \psi_n^-\psi_{n\pm1}^+)^2 \\ &= (\psi_n^+\psi_{n\pm1}^- + \psi_n^-\psi_{n\pm1}^+)^2 - 1. \end{aligned}$$

Hence (3.14) yields

$$\begin{aligned} V_n^{[z]} + 2 &= \frac{1}{2\psi_n^+\psi_n^-}\left(\psi_n^+\psi_{n+1}^- + \psi_n^-\psi_{n+1}^+ + \psi_n^+\psi_{n-1}^- + \psi_n^-\psi_{n-1}^+\right) \\ &= \frac{1}{2\psi_n^+\psi_n^-}\left(\psi_n^+ V_n\psi_n^- + \psi_n^- V_n\psi_n^+\right) \\ &= V_n + 2, \end{aligned}$$

as claimed, and establishes (3.13). □

It may well be asked at this stage why we have restricted ourselves to the situation where $G_{nn} > 0$, for at the formal level the calculations given above remain valid without assuming positivity. In the discrete setting, continuity is not available to connect the values of a solution φ_n as n varies, and hence without an assumption such as positivity, there is a degree of indeterminateness in defining solutions by a prescription such as (3.9). For *some* choice of phases in (3.14), it will still be true that $V_n^{[z]}$ as defined there coincides with V_n, but the implicit nature of these choices of square root is problematic. Possibly a suitable canonical choice of phase or ideas from Teschl's oscillation theory for Jacobi operators [**28**] could help avoid implicit definitions, and we hope to elaborate this point in future work.

Returning to the case where $G_{nn} > 0$, Formula (3.10) suggests that S_n can be related to an Agmon distance [**2, 21**], that is, a metric $d_A(m,n)$ on the positive integer lattice such that every ℓ^2 solution ϕ^- of (3.1) satisfies a bound of the form

$$e^{d_A(0,n)}\phi_n^- \in \ell^\infty,$$

and that as a consequence ϕ_n^- decays rapidly as $n \to \infty$. Thus if z_n is bounded we expect an Agmon distance to be something like $\sum_{\ell=m+1}^{n} \ln S_\ell^{[z]}$, assuming $n > m$. (We write the Agmon distance in this way because the triangle inequality is an equality on the integer lattice, which implies that any metric takes the form of a sum of quantities defined at values of ℓ from $m+1$ to n.) In Agmon's theory, however, it is desirable that the distance function be a quantity that can be calculated directly from the potential alone (or at least dominated by some such expression). As we shall now see, understanding the diagonal of the Green matrix allows the derivation of Agmonish bounds. We begin by showing that G_{nn} is comparable to $(V_n+2)^{-1}$ in a precise sense.

LEMMA 3.2. *Suppose that* $\liminf_{n\to\infty} V_n > C > 0$, *and let* G_{mn} *be any positive Green matrix for* (3.1) *on the positive integers. Define*

$$K_A := \sqrt{1 + \left(\frac{2}{C(C+2)}\right)^2} + \frac{2}{C(C+2)}.$$

Then for n sufficiently large,

$$\frac{1}{V_n+2} \le G_{nn} \le \frac{K_A}{V_n+2}. \tag{3.15}$$

Consequently,

$$\begin{aligned}\frac{\sqrt{(V_n+2)(V_{n-1}+2)} + \sqrt{4+(V_n+2)(V_{n-1}+2)}}{2K_A} &\le S_n^{[z]} \\ \le \frac{\sqrt{(V_n+2)(V_{n-1}+2)} + \sqrt{4+(V_n+2)(V_{n-1}+2)}}{2}&.\end{aligned} \tag{3.16}$$

REMARK 3. The upper bound is of the same form as a semiclassical upper bound proved in [**19**]. To simplify it, K_A could be replaced in these inequalities by

$$\sqrt{1+\frac{4}{C^2}} > K_A$$

(see proof).

PROOF. The lower bound on G_{nn} is immediate from Statement (3) of Theorem 3.1, the left member of which is larger than 1.

The upper bound in (3.15) requires a spectral estimate. The Green matrix G_{mn} is the kernel of the resolvent operator of a self-adjoint realization of $-\Delta + V$ on $\ell^2([N,\infty))$ for some N, where the boundary condition at $n = N, N+1$ is that satisfied by φ_n^+. Since $-\Delta > 0$ on this space (as an operator), $\inf \operatorname{sp}(-\Delta+V) > C$, and hence, by the spectral mapping theorem, $\|(-\Delta+V)^{-1}\|_{\text{op}} < C^{-1}$. Since $G_{nn} = \langle e_n, (-\Delta+V)^{-1}e_n\rangle$, where $\{e_n\}$ designate the standard unit vectors in ℓ^2, it follows that $G_{nn} < C^{-1}$. Inserting this into (3.11) would already imply (3.15) with K_A replaced by $\sqrt{1+4/C^2}$. To improve the constant, replace only the terms $G_{n\pm1\,n\pm1}$ in (3.14) by $1/C$, getting

$$(V_n+2) \le \frac{\sqrt{1+\frac{4G_{nn}}{C}}}{G_{nn}}. \tag{3.17}$$

Since

$$\frac{\sqrt{1+xy}}{x}$$

is a decreasing function of x when $x, y > 0$, an upper bound on G_{nn} is the larger root of the case of equality in (3.17) (which is effectively a quadratic). The claimed upper bound with the constant K_A results by keeping one factor V_n+2 in the solution of the quadratic, replacing the others by $C+2$.
The bounds on $S_n^{[z]}$ result from inserting the bounds on G_{nn} into (3.9) and collecting terms. □

We can now state some Agmonish bounds.

COROLLARY 3.3. *Suppose that* $\liminf_{n\to\infty} V_n > C > 0$ *and fix a positive integer* m. *Then the subdominant (i.e., eventually decreasing) solution* ϕ^- *of* (3.1) *satisfies*
(a)

$$\left(\prod_{\ell=m}^{n} \frac{V_\ell+2}{K_A}\right)\phi_n^- \in \ell^\infty.$$

(b) If, in addition, $n(V_{n+1}-V_n) \in \ell^1$, *then*

$$\left(\prod_{\ell=m}^{n} \frac{V_\ell+2+\sqrt{V_\ell(V_\ell+4)}}{2}\right)\phi_n^- \in \ell^\infty.$$

PROOF. The ansatz (3.9) allows an identification of ϕ^- with a constant multiple of φ^-, in the representation (3.10). Because z_n is bounded, so is

$$\left(\prod_{\ell}^{n} S_\ell^{[z]}\right)\varphi_n^-.$$

We then use the lower bound on $S_\ell^{[z]}$ from the lemma, but simplify by dropping the 4, which allows the product to telescope in a pleasing way, producing (a).

For (b) we note that the additional assumption on V_n allows us to conclude that φ is well-approximated by a Liouville-Green expression in [**19**], Theorem 4.1, which is a bounded quantity times the reciprocal of the expression in parentheses. □

Thus when $\liminf_{n\to\infty} V_n > 0$, a suitable Agmon distance $d_A(m,n)$ for (3.1) is given by

$$\sum_{\ell=m+1}^{n} (\ln(V_l+2) - \ln K_A),$$

or by

$$\sum_{\ell=m+1}^{n} \ln\frac{V_\ell+2+\sqrt{V_\ell(V_\ell+4)}}{2},$$

provided that $n(V_{n+1}-V_n) \in \ell^1$.

We close with a Darboux-type factorization for a generic discrete Schrödinger equation (3.1). A Darboux-type factorization for general Jacobi operators was previously considered by Gesztesy and Teschl in [**15**]. As in §2 the novel feature of our factorization is that it is constructed using the diagonal of the Green matrix.

To this end, choose a Green matrix such that G_{nn} is nonvanishing for a range of values of n, and define a solution φ_n^+ according to (3.9). The phase of the square roots is chosen (if necessary) to ensure that $V_n^{[z]}$ from (3.14) equals V_n.

THEOREM 3.4. *Given a Green matrix such that the diagonal G_{kk} is nonvanishing for $n-1 \le k \le n+2$, and choosing the phase of the square roots as described above,*

$$-\Delta + V_n = \mathcal{R}\left[-\nabla^+ - 1 + \frac{2G_{nn}}{1+(1+4G_{nn}G_{n+1\,n+1})^{1/2}}\right]\left[\nabla^+ + 1 - \frac{1+(1+4G_{nn}G_{n+1\,n+1})^{1/2}}{2G_{nn}}\right],$$

where $\mathcal{R}$ is the shift operator such that $[\mathcal{R}f]_n = f_{n-1}$ and the right-difference operator is defined by $[\nabla^+ f]_n := f_{n+1} - f_n$.

REMARK 4. As with (2.9), there is a second factorization, with shifts and differences reversed, and $n+1$ replaced by $n-1$.

PROOF. Writing

$$Q_n := 1 - \frac{1+(1+4G_{nn}G_{n+1\,n+1})^{1/2}}{2G_{nn}} = 1 - \frac{z_{n+1}S_{n+1}^{[z]}}{z_n},$$

with $S_k^{[z]}$ defined in (3.9), we first note that, by a simple calculation,

$$\left[\nabla^+ + Q_n\right]\varphi^+ = 0. \tag{3.18}$$

This motivates calculating

$$H = \left[-\nabla^+ + \frac{Q_n}{1-Q_n}\right]\left[\nabla^+ + Q_n\right],$$

which is well-defined because $Q_n \neq 1$, owing to (3.9) with z_k nonvanishing. The left factor was chosen to produce a convenient cancellation, ensuring that H has the form of a discrete Schrödinger equation, with a shifted index:

$$(Hf)_n = (-\Delta f)_{n+1} + \left(\frac{Q_n}{1-Q_n} - Q_{n+1}\right) f_{n+1} + 0 \cdot f_n.$$

We now verify that when the index is shifted back, the potential term is indeed V_n:

$$\begin{aligned}\left(\frac{Q_{n-1}}{1-Q_{n-1}} - Q_n\right) &= \left(\frac{z_{n-1}}{z_n S_n^{[z]}} - 1\right) - \left(1 - \frac{z_{n+1}S_{n+1}^{[z]}}{z_n}\right)\\ &= -2 + \frac{z_{n-1}}{z_n S_n^{[z]}} + \frac{z_{n+1}S_{n+1}^{[z]}}{z_n},\end{aligned}$$

which reduces to V_n according to (3.11). □

References

[1] R. P. Agarwal, *Difference equations and inequalities*, 2nd ed., Monographs and Textbooks in Pure and Applied Mathematics, vol. 228, Marcel Dekker Inc., New York, 2000. Theory, methods, and applications. MR1740241 (2001f:39001)

[2] S. Agmon, *Lectures on exponential decay of solutions of second-order elliptic equations: bounds on eigenfunctions of N-body Schrödinger operators*, Mathematical Notes, vol. 29, Princeton University Press, Princeton, NJ, 1982. MR745286 (85f:35019)

[3] P. Bohl, *Über eine Differentialgleichung der Störungstheorie*, J. reine angew. Math. **131** (1906) 268–321.

[4] N. Chernyavskaya, *On a problem of Hartman and Wintner*, Proc. Roy. Soc. Edinburgh Sect. A **128** (1998), no. 5, 1007–1022, DOI 10.1017/S0308210500030031. MR1642124 (99h:34077)

[5] N. Chernyavskaya and L. Shuster, *Estimates for the Green function of a general Sturm-Liouville operator and their applications*, Proc. Amer. Math. Soc. **127** (1999), no. 5, 1413–1426, DOI 10.1090/S0002-9939-99-05049-2. MR1625725 (99h:34040)

[6] N. A. Chernyavskaya and L. A. Shuster, *Regularity of the inversion problem for the Sturm-Liouville difference equation. I. Representation of the Davies-Harrell type for the Green difference function*, J. Math. Anal. Appl. **246** (2000), no. 1, 150–163, DOI 10.1006/jmaa.2000.6773. MR1761155 (2001g:39010)

[7] N. A. Chernyavskaya and L. A. Shuster, *Regularity of the inversion problem for the Sturm-Liouville difference equation. I. Representation of the Davies-Harrell type for the Green difference function*, J. Math. Anal. Appl. **246** (2000), no. 1, 150–163, DOI 10.1006/jmaa.2000.6773. MR1761155 (2001g:39010)

[8] N. A. Chernyavskaya and L. A. Shuster, *Davies-Harrell representations, Otelbaev's inequalities and properties of solutions of Riccati equations*, J. Math. Anal. Appl. **334** (2007), no. 2, 998–1021, DOI 10.1016/j.jmaa.2006.10.092. MR2338644 (2008d:34087)

[9] G. Darboux, *Sur une proposition relative aux équations linéaires*, C. R. Acad. Sci. (Paris) **94** (1882), 1456–1459.

[10] E. B. Davies and E. M. Harrell II, *Conformally flat Riemannian metrics, Schrödinger operators, and semiclassical approximation*, J. Differential Equations **66** (1987), no. 2, 165–188, DOI 10.1016/0022-0396(87)90030-1. MR871993 (88a:35061)

[11] F. Gesztesy, Y. Latushkin, and K. A. Makarov, *Evans functions, Jost functions, and Fredholm determinants*, Arch. Ration. Mech. Anal. **186** (2007), no. 3, 361–421, DOI 10.1007/s00205-007-0071-7. MR2350362 (2008k:34209)

[12] F. Gesztesy and B. Simon, *The xi function*, Acta Math. **176** (1996), no. 1, 49–71, DOI 10.1007/BF02547335. MR1395669 (97e:47078)

[13] F. Gesztesy, B. Simon, and G. Teschl, *Zeros of the Wronskian and renormalized oscillation theory*, Amer. J. Math. **118** (1996), no. 3, 571–594. MR1393260 (97g:34105)

[14] F. Gesztesy and G. Teschl, *On the double commutation method*, Proc. Amer. Math. Soc. **124** (1996), no. 6, 1831–1840, DOI 10.1090/S0002-9939-96-03299-6. MR1322925 (96h:34171)

[15] F. Gesztesy and G. Teschl, *Commutation methods for Jacobi operators*, J. Differential Equations **128** (1996), no. 1, 252–299, DOI 10.1006/jdeq.1996.0095. MR1392402 (97i:47079)

[16] F. Gesztesy and M. Ünal, *Perturbative oscillation criteria and Hardy-type inequalities*, Math. Nachr. **189** (1998), 121–144, DOI 10.1002/mana.19981890108. MR1492926 (99a:34069)

[17] S. Goff and D. F. St. Mary, *The Bohl transformation for second order linear differential systems*, J. Math. Anal. Appl. **140** (1989), no. 1, 95–106, DOI 10.1016/0022-247X(89)90096-6. MR997845 (90m:34087)

[18] S. Goff and D. F. St. Mary, *The Bohl transformation and oscillation of linear differential systems*, SIAM J. Math. Anal. **20** (1989), no. 1, 215–221, DOI 10.1137/0520017. MR977500 (89k:34046)

[19] E. M. Harrell II and M. L. Wong, *On the behavior at infinity of solutions to difference equations in Schrödinger form*. arXiv:1109.4691.

[20] P. Hartman, *Ordinary differential equations*, S. M. Hartman, Baltimore, Md., 1973. Corrected reprint. MR0344555 (49 #9294)

[21] P. D. Hislop and I. M. Sigal, *Introduction to spectral theory*, Applied Mathematical Sciences, vol. 113, Springer-Verlag, New York, 1996. With applications to Schrödinger operators. MR1361167 (98h:47003)

[22] J. v. Neumann and E. Wigner, *Über merkwürdige diskrete Eigenwerte*, Phys. Zeitschr. **30** (1929) 465–467.
[23] F. W. J. Olver, *Asymptotics and special functions*, Academic Press [A subsidiary of Harcourt Brace Jovanovich, Publishers], New York-London, 1974. Computer Science and Applied Mathematics. MR0435697 (55 #8655)
[24] M. Ráb, *Kriterien für die Oszillation der Lösungen der Differentialgleichung* $[p(x)y']' + q(x)y = 0$, Časopis Pěst. Mat. 84 (1959), 335-370; erratum **85** (1959), 91 (German, with Czech and Russian summaries). MR0114964 (22 #5773)
[25] M. Reed and B. Simon, *Methods of modern mathematical physics. IV. Analysis of operators*, Academic Press [Harcourt Brace Jovanovich Publishers], New York, 1978. MR0493421 (58 #12429c)
[26] W. T. Reid, *Ordinary differential equations*, John Wiley & Sons Inc., New York, 1971. MR0273082 (42 #7963)
[27] W. T. Reid, *Sturmian theory for ordinary differential equations*, Applied Mathematical Sciences, vol. 31, Springer-Verlag, New York, 1980. With a preface by John Burns. MR606199 (82f:34002)
[28] G. Teschl, *Oscillation theory and renormalized oscillation theory for Jacobi operators*, J. Differential Equations **129** (1996), no. 2, 532–558, DOI 10.1006/jdeq.1996.0126. MR1404392 (98m:47053)
[29] D. Willett, *Classification of second order linear differential equations with respect to oscillation*, Advances in Math. **3** (1969), 594–623 (1969). MR0280800 (43 #6519)

SCHOOL OF MATHEMATICS, GEORGIA INSTITUTE OF TECHNOLOGY, ATLANTA, GEORGIA 30332-0610

E-mail address: harrell@math.gatech.edu

SCHOOL OF MATHEMATICS, GEORGIA INSTITUTE OF TECHNOLOGY, ATLANTA, GEORGIA 30332-0610

E-mail address: wong@math.gatech.edu

Proceedings of Symposia in Pure Mathematics
Volume **87**, 2013
http://dx.doi.org/10.1090/pspum/087/01436

The Morse and Maslov indices for matrix Hill's equations

Christopher K. R. T. Jones, Yuri Latushkin, and Robert Marangell

To Fritz Gesztesy on the occasion of his 60-th birthday with best wishes

ABSTRACT. For Hill's equations with matrix valued periodic potential, we discuss relations between the Morse index, counting the number of unstable eigenvalues, and the Maslov index, counting the number of signed intersections of a path in the space of Lagrangian planes with a fixed plane. We adapt to the one dimensional periodic setting the strategy of a recent paper by J. Deng and C. Jones relating the Morse and Maslov indices for multidimensional elliptic eigenvalue problems.

1. Introduction

Various results on Hill's equation are among many fundamental contributions made by Fritz Gesztesy in mathematical physics and analysis, see, for example, [**GW96**, **GT09**]. In the current paper, we discuss a symplectic approach to counting positive θ-eigenvalues for Hill's equations with matrix valued periodic potentials, that is, the values of λ for which there exists a nontrivial solution of the eigenvalue problem

$$Hy := y'' + V(x)y = \lambda y, \quad y = (y_1(x), \dots, y_n(x))^\top, \tag{1.1}$$

that satisfies the boundary conditions

$$y(L) = e^{i\theta} y(-L), \quad y'(L) = e^{i\theta} y'(-L). \tag{1.2}$$

Here, $x \in \mathbb{R}$, $y_i : \mathbb{R} \to \mathbb{C}$, $\theta \in [0, 2\pi]$, and $V(x)$ is an $n \times n$ symmetric matrix whose entires are real valued piecewise continuous periodic functions of period $2L$. We will denote by H_θ the differential operator in $L^2([-L, L])$ associated with the eigenvalue problem (1.1), (1.2).

A great deal of attention is devoted to Schrödinger operators with periodic potentials, see, e.g., [**MW**, **ReSi78**, **Kr97**] and the bibliography therein. In the current paper, our main concern is the *Morse index*, $\mathrm{Mor}(H_\theta)$, a ubiquitous number that appears in many areas, from variational calculus [**B56**, **D76**, **M63**] to stability of traveling waves [**J88**, **SS08**], and which is defined as the dimension of the spectral

2010 *Mathematics Subject Classification.* Primary 53D12, 34L40; Secondary 37J25, 70H12.

Key words and phrases. Schrödinger equation, Hamiltonian systems, periodic potentials, eigenvalues, stability, differential operators, discrete spectrum.

Partially supported by the grants NSF DMS-0754705, DMS-1067929, DMS-0410267 and ONR N00014-05-1-0791, and by the Research Council and the Research Board of the University of Missouri.

subspace of a self-ajoint operator corresponding to its positive (unstable) discrete eigenvalues. We will relate it to the *Maslov index*, $\mathrm{Mas}(\Gamma, \mathcal{X})$, which is defined as the signed number of intersections of a curve Γ in the set of Lagrangian planes with a given subvariety, called the *train* of a fixed Lagrangian plane $\mathcal{X}$ (see [**Ar67**, **Ar85**, **J88**, **RS93**, **RS95**] as well as more recent papers [**F04**, **FJN03**, **O90**] and the bibliography therein for a discussion of this beautiful subject).

One of the main motivations for studying the Maslov index in the context of second order differential operators was a generalization in [**Ar85**] (for the case of matrix valued potentials) of the classical Sturm oscillation theorems; in connection with the latter we mention [**GST96**], [**G07**, Chapter 3] and the bibliography therein. That the Morse and Maslov indices for periodic problems are related is of course well known (see, e.g., the classical sources [**D76**, **CZ84**], an excellent book [**A01**] which has a detailed bibliography, and the important recent work done in [**CDB06**, **CDB09**, **CDB11**]). However, all literature that we know deals only with the case of periodic eigenvalues corresponding to the particular case of $\theta = 0$ or $\theta = 2\pi$ (but also see [**SB**]).

More importantly, in the present paper we use a novel approach of determining the Maslov index borrowed from a recent paper [**DJ11**] where the relations between the Morse and Maslov indices have been established in the *multidimensional* situation, in particular, for elliptic problems in a star-shaped domain $\mathcal{D}$ in $\mathbb{R}^d$ containing zero. The main idea in [**DJ11**] is to consider a family of "shrinking" domains $\mathcal{D}_s$ parametrized by $s \in (0, 1]$ and such that a point $\mathbf{x} \in \partial\mathcal{D}$ if and only if $s\mathbf{x} \in \partial\mathcal{D}_s$. Rescaling the original elliptic equation for λ-eigenfunctions from $\mathcal{D}_s$ to $\mathcal{D}$, one then defines a trace map ϕ_s^λ acting from the Sobolev space $H^1(\mathcal{D})$ into the trace space $H^{1/2}(\partial\mathcal{D}) \times H^{-1/2}(\partial\mathcal{D})$. It maps a weak solution of the eigenvalue equation with no boundary conditions at all into a vector function on the boundary whose components are the Dirichlet and Neumann traces of the solution. Using Green's formulas, one defines a symplectic structure in the trace space so that if $\mathcal{Y}_{s,\lambda}$ denotes the set of the weak solutions then $\phi_s^\lambda(\mathcal{Y}_{s,\lambda})$ forms a curve in the set of Fredholm Lagrangian planes. The boundary conditions define a plane, and an intersection of the curve with the train of the plane defined via the boundary conditions corresponds to an eigenvalue of the elliptic operator at hand, eventually leading to a formula relating the Morse and Maslov indices.

In the current paper, for the boundary value problem (1.1), (1.2) on $[-L, L]$, following the strategy in [**DJ11**], we consider a family, parametrized by $s \in (0, L]$, of boundary value problems for (1.1) on the segments $[-s, s]$ with the boundary conditions

$$y(s) = e^{i\theta} y(-s), \quad y'(s) = e^{i\theta} y'(-s). \tag{1.3}$$

By changing s and λ and using the traces of solutions of the differential equation (1.1) at the boundary of the segment $[-s, s]$, we construct a path in the set of finite dimensional Lagrangian planes. The construction of the path is the first crucial ingredient of the current paper. The second key point is to utilize and further develop an idea from [**Ga93**] to augment the first order system corresponding to (1.1) by considering a supplementary linear complex $(2n \times 2n)$ first order ODE system with the coefficient $\frac{i\theta}{2s} I_{2n}$ whose solutions automatically satisfy the boundary conditions (1.3). This allows one to replace the $e^{i\theta}$-periodic boundary conditions in (1.2), (1.3) by certain "Dirichlet-type" boundary conditions for the augmented system.

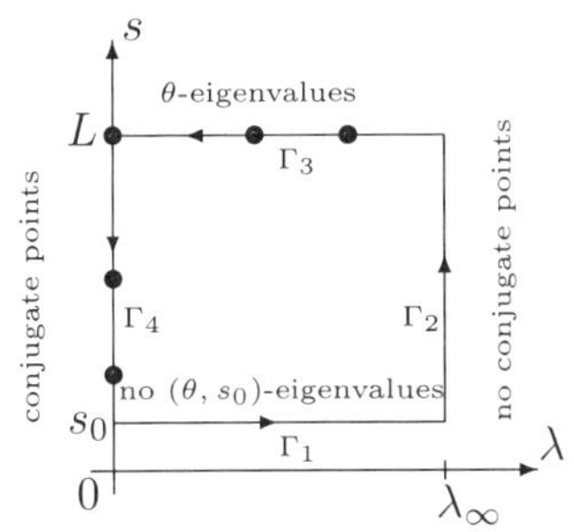

FIGURE 1. $\lambda = 0$ is a θ-eigenvalue, $\theta \in (0, 2\pi)$, and s_0 is small enough

Our plan then is to re-write the eigenvalue equation (1.1) as a complex $(2n \times 2n)$ first order system, separate the real and imaginary parts of the solutions in the eigenvalue equation and the boundary conditions, thus arriving at a $(4n \times 4n)$ real system, consider the augmented $(8n \times 8n)$ real system, and then to define a trace map, Φ_s^λ, for each $s \in (0, L]$ and $\lambda \in \mathbb{R}$, that maps a solution $(\mathbf{p}, \mathbf{w})^\top$ of the augmented system on $[-L, L]$ with no boundary conditions at all into its trace $(\mathbf{p}(-s), \mathbf{w}(-s), \mathbf{p}(s), \mathbf{w}(s))^\top \in \mathbb{R}^{16n}$ on the boundary of of the segment $[-s, s]$. This leads to the critical observation (see Proposition 3.7 below) that if $Y_{s,\lambda}$ denotes the set of the solutions of the augmented system then λ is an eigenvalue of (1.1), (1.3) on $[-s, s]$ if and only if the plane $\Phi_s^\lambda(Y_{s,\lambda})$ intersects the plane $X \times X$ in $\mathbb{R}^{16}$ consisting of vectors whose respective $\mathbf{p}(\pm s)$- and $\mathbf{w}(\pm s)$-components are equal; here and below we denote $\mathbf{p} = (p, q)^\top$, $\mathbf{w} = (w, z)^\top$, and use notation

$$X = \{(p, q, w, z)^\top \in \mathbb{R}^{8n} \,|\, p = w, q = z\}. \tag{1.4}$$

Thus, the "Dirichlet-type" boundary condition $\Phi_s^\lambda\big((\mathbf{p}, \mathbf{w})^\top\big) \in X \times X$ replaces the $e^{i\theta}$-periodic boundary condition (1.3).

There is a natural symplectic structure in $\mathbb{R}^{16n}$ such that the planes $\Phi_s^\lambda(Y_{s,\lambda})$ and $X \times X$ in $\mathbb{R}^{16n}$ are Lagrangian (see Theorem 3.4). Thus, one can consider crossings with the train of $X \times X$ of the Lagrangian curve $\Gamma = \Gamma_1 \cup \Gamma_2 \cup \Gamma_3 \cup \Gamma_4$ formed by $\Phi_s^\lambda(Y_{s,\lambda})$ when (λ, s) runs along the boundary of the square $[s_0, L] \times [0, \lambda_\infty]$, for a small $s_0 > 0$ and a large λ_∞, where Γ_j, $j = 1, 2, 3, 4$, correspond to the four sides of the square, see Figure 1. We stress that Γ depends on the choice of s_0 and λ_∞ while the location of the crossings of course depends on θ; we sometimes write $\Gamma_{(\theta, s_0)}$ and $\Gamma_{j,(\theta,s_0)}$.

A homotopy argument implies that the Maslov index $\mathrm{Mas}(\Gamma, X \times X)$ of the entire curve Γ is equal to zero (see Corollary 3.9). By general properties of the Maslov index one infers $\mathrm{Mas}(\Gamma, X \times X) = \sum_{j=1}^4 \mathrm{Mas}(\Gamma_j, X \times X)$. For $\theta \in (0, 2\pi)$ one can show that there are no crossings along Γ_1 (when $s = s_0$ and $\lambda \in [0, \lambda_\infty]$) provided s_0 is chosen small enough (see Lemma 3.12). For $\theta = 0$ or $\theta = 2\pi$, assuming that the potential V is continuous at the point $x = 0$, and $s_0 > 0$ is small enough, one can show that the number of crossings along Γ_1 is equal to the number $\mathrm{Mor}(V(0))$ of positive eigenvalues of the matrix $V(0)$ (Lemma 4.3). Since the spectrum of the operator H_θ is bounded from above, there are no crossings along Γ_2 (when $\lambda = \lambda_\infty$ and $s \in [s_0, L]$) provided λ_∞ is chosen large enough, see Lemma 3.12.

The crossings of the curve Γ_3 (when $s = L$ and $\lambda \in [0, \lambda_\infty]$), correspond to the θ-eigenvalues of (1.1), (1.2). A local computation shows that all crossings along Γ_1 have the same signs and all crossings along Γ_3 have the same signs, see Lemma 4.1. This important monotonicity property of the Maslov index implies that the Morse index $\mathrm{Mor}(H_\theta)$ is equal to the number of crossings along Γ_3 counting their multiplicities.

It turns out that the crossings along Γ_4 (when $\lambda = 0$ and $s \in [s_0, L]$) correspond to conjugate points of the Hill's equation on $[-L, L]$, that is, to the points s where the number $e^{i\theta}$ is an eigenvalue of the propagator of this equation transforming the value of its solution at the point $-s$ into the value at the point s (Proposition 3.7). Thus, $\mathrm{Mas}(\Gamma_4, X \times X)$ can be viewed as the Maslov index of the boundary value problem (1.1), (1.2) for the Hill equation. Yet another local computation shows that all crossings along Γ_4 have the same sign provided that, in addition, the potential is sign definite (see Lemma 4.2).

Since $\mathrm{Mas}(\Gamma, X \times X) = 0$, we therefore arrive at the desired formula

$$\mathrm{Mor}(H_\theta) = \begin{cases} -\mathrm{Mas}(\Gamma_4, X \times X), & \text{if } \theta \in (0, 2\pi), \\ & \text{for small } s_0 = s_0(\theta) > 0, \\ -\mathrm{Mas}(\Gamma_4, X \times X) + \mathrm{Mor}(V(0)), & \text{if } \theta = 0 \text{ or } \theta = 2\pi, \\ & \text{for small } s_0 > 0, \end{cases} \tag{1.5}$$

relating the Maslov index of the boundary value problem for the Hill equation and the Morse index of the corresponding differential operator (see Theorem 4.4 summarizing our results). For instance, for a fixed $s_0 > 0$, when θ changes from a positive value to zero, the crossings move from Γ_4 to Γ_1 through the left bottom corner of the square in Figure 1, thus keeping the proper balance in formula (1.5).

The paper is organized as follows. In Section 2 we set up the stage and introduce the augmented system for the Hill equation (1.1). After a brief reminder of basics on the Maslov index, in Section 3 we introduce an appropriate Lagrangian structure, and relate the crossings of the path Φ_s^λ to the eigenvalues of differential operators. In Section 4 we prove monotonicity of the crossings, and summarize the main results of the paper. Finally, in Section 5 we conduct several numerical experiments calculating the Maslov and Morse indices for a particular Mathieu equation.

Notations. We denote by I_n and 0_n the $n \times n$ identity and zero matrix. For an $n \times m$ matrix $A = (a_{ij})_{i=1,j=1}^{n,m}$ and a $k \times \ell$ matrix $B = (b_{ij})_{i=1,j=1}^{k,\ell}$, we denote by $A \otimes B$ the Kronecker product, that is, the $nk \times m\ell$ matrix composed of $k \times \ell$ blocks $a_{ij}B$, $i = 1, \dots n$, $j = 1, \dots m$. We let $\langle \cdot, \cdot \rangle_{\mathbb{R}^n}$ denote the real scalar product in the space $\mathbb{R}^n$ of $n \times 1$ vectors, and let $\top$ denote transposition. We denote by $A \oplus B$ the matrix $\begin{pmatrix} A & 0 \\ 0 & B \end{pmatrix}$ and use notation $J = \begin{pmatrix} 0 & 1 \\ -1 & 0 \end{pmatrix}$ for the standard symplectic matrix. When $a = (a_i)_{i=1}^n \in \mathbb{R}^n$ and $b = (b_j)_{j=1}^m \in \mathbb{R}^m$ are $(n \times 1)$ and $(m \times 1)$ column vectors, we use notation $(a, b)^\top$ for the $(n + m) \times 1$ column vector with the entries $a_1, \dots, a_n, b_1, \dots, b_m$ (just avoiding the use of $(a^\top, b^\top)^\top$). We denote by $\mathcal{B}(\mathcal{X})$ the set of linear bounded operators on a Hilbert space $\mathcal{X}$ and by $\mathrm{Spec}(T) = \mathrm{Spec}(T; \mathcal{X})$ the spectrum of an operator on $\mathcal{X}$.

Acknowledgment. We thank Konstantin Makarov and Holger Dullin for their valuable suggestions.

2. Hill's equation and an augmented equation

We start with the eigenvalue problem (1.1), where we consider $\lambda \in \mathbb{R}$, and consider complex valued solutions to (1.1). Setting

$$\begin{aligned} p_i &:= (\mathrm{Re}(y_i), \mathrm{Im}(y_i))^\top \in \mathbb{R}^2, \quad p := (p_1, \dots, p_n)^\top \in \mathbb{R}^{2n}, \\ q_i &:= (\mathrm{Re}(y_i'), \mathrm{Im}(y_i'))^\top \in \mathbb{R}^2, \quad q := (q_1, \dots, q_n)^\top \in \mathbb{R}^{2n}, \end{aligned} \tag{2.1}$$

we can write (1.1) as follows:

$$\begin{pmatrix} p \\ q \end{pmatrix}' = \begin{pmatrix} 0_{2n} & I_{2n} \\ \lambda I_{2n} - V(x) \otimes I_2 & 0_{2n} \end{pmatrix} \begin{pmatrix} p \\ q \end{pmatrix}. \tag{2.2}$$

It is sometimes convenient to denote $\mathbf{p} := (p, q)^\top \in \mathbb{R}^{4n}$ and to write (2.2) as

$$\mathbf{p}' = A(x, \lambda)\mathbf{p}, \quad A(x, \lambda) = \begin{pmatrix} 0_{2n} & I_{2n} \\ \lambda I_{2n} - V(x) \otimes I_2 & 0_{2n} \end{pmatrix}. \tag{2.3}$$

We are interested in studying bounded on $\mathbb{R}$ solutions of (1.1). To this end, for each $\theta \in [0, 2\pi]$, we will examine for which λ there exists a nontrivial solution y of (1.1) that satisfies the boundary condition (1.2). In particular, if $\theta = 0$ or $\theta = 2\pi$ we have periodic boundary conditions, and if $\theta = \pi$ we have antiperiodic ones. Equivalently, using (2.1) and writing out (1.2) in real and imaginary parts, we seek a nontrivial solution $\mathbf{p} = (p, q)^\top$ of (2.2) such that the following boundary condition is satisfied:

$$\begin{pmatrix} p(L) \\ q(L) \end{pmatrix} = \begin{pmatrix} I_n \otimes U(\theta) & 0 \\ 0 & I_n \otimes U(\theta) \end{pmatrix} \begin{pmatrix} p(-L) \\ q(-L) \end{pmatrix}, \tag{2.4}$$

where we denote

$$U(\theta) := \begin{pmatrix} \cos\theta & -\sin\theta \\ \sin\theta & \cos\theta \end{pmatrix}. \tag{2.5}$$

In the notation of equation (2.3), condition (2.4) is written as

$$\mathbf{p}(L) = (I_{2n} \otimes U(\theta))\mathbf{p}(-L). \tag{2.6}$$

Since the boundary conditions (1.2) and (1.3) are the same in the case when $\theta = 0$ or $\theta = 2\pi$, out of these two possibilities we will always consider only the former.

We now briefly discuss the spectrum of the operators associated with (1.1). On the space $L^2(\mathbb{R})$ of $(n \times 1)$ complex vector valued functions, or on the space $BUC(\mathbb{R})$ of bounded uniformly continuous complex vector valued functions, one can associate to equation (1.1) a differential operator, H, defined by $H = \frac{d^2}{dx^2} + V(x)$, whose domain is given by the following formula (we recall that the potential V is bounded):

$$\mathrm{dom}(H) = \left\{ y \in L^2(\mathbb{R}) \,\middle|\, y, y' \in AC_{\mathrm{loc}}(\mathbb{R}), y'' \in L^2(\mathbb{R}) \right\} \tag{2.7}$$

(for the space $BUC(\mathbb{R})$ one has to replace the space $L^2(\mathbb{R})$ in (2.7) by $BUC(\mathbb{R})$). There is a standard way, see [**ReSi78**, Section XIII.16], of associating with H a family of operators, H_θ, with $\theta \in [0, 2\pi]$, acting in $L^2([-L, L])$ and induced by the complex boundary conditions (1.2). Indeed, we may identify $L^2(\mathbb{R})$ and $L^2([0, 2\pi]; L^2([-L, L])) = L^2([0, 2\pi] \times [-L, L])$ by introducing, see [**ReSi78**, eqn. (147)], a family of operators $W_\theta : L^2(\mathbb{R}) \to L^2([-L, L])$ by

$$(W_\theta y)(x) = \sum_{n \in \mathbb{Z}} \mathrm{e}^{-in\theta} y(x + 2Ln), \quad x \in [-L, L]. \tag{2.8}$$

Obviously, $(W_\theta y)(L) = \mathrm{e}^{i\theta}(W_\theta y)(-L)$, and analogously for the derivative y', leading to the fact the H is similar to the direct integral, $\oplus\int_0^{2\pi} H_\theta \frac{d\theta}{2\pi}$, of the operators H_θ defined in $L^2([-L,L])$ as follows: $H_\theta = \frac{d^2}{dx^2} + V(x)$ with

$$\begin{aligned}\operatorname{dom}(H_\theta) = \Big\{ y \in L^2([-L,L]) \Big| & y, y' \in AC_{\text{loc}}([-L,L]), \\ & y'' \in L^2([-L,L]) \text{ and the boundary condition (1.2) holds} \Big\}.\end{aligned} \tag{2.9}$$

Similarly, one can introduce the operator H_θ on the space $C([-L,L])$ of continuous functions by replacing $L^2([-L,L])$ in (2.9) by $C([-L,L])$.

For each $\theta \in [0, 2\pi]$, the spectrum $\operatorname{Spec}(H_\theta)$ consists of discrete eigenvalues; when θ varies, they fill up spectral bands with or without spectral gaps between them, thus forming the spectrum $\operatorname{Spec}(H)$, see [**MW**, **ReSi78**] for a detailed exposition.

DEFINITION 2.1. We say that λ is a θ-*eigenvalue* of equation (1.1) if there is a nonzero solution of (2.2) such that the boundary condition (2.4) is satisfied.

For each $\lambda \in \mathbb{R}$, we let $\Psi_A(x,\lambda)$ denote the fundamental matrix solution to equation (2.3) such that $\Psi_A(-L,\lambda) = I_{4n}$ and, for each $s \in (0, L]$, let $M_A(s,\lambda) := \Psi_A(s,\lambda)\Psi_A(-s,\lambda)^{-1}$ denote its propagator so that $\mathbf{p}(s) = M(s,\lambda)\mathbf{p}(-s)$ for a solution of (2.3). In particular, $M_A(L,\lambda) = \Psi_A(L,\lambda)$ denotes the monodromy matrix for (2.3). We recall that in [**Ga93**], λ is said to be a γ-*eigenvalue* if $\gamma \in \{\gamma \in \mathbb{C} : |\gamma| = 1\}$ is an eigenvalue of the monodromy matrix of equation (2.2). We note that our definition of θ-eigenvalue is consistent with the definition of γ-eigenvalue, with $\gamma = e^{i\theta}$, given in [**Ga93**], as the following proposition shows.

PROPOSITION 2.2. *On $L^2(\mathbb{R})$ or $BUC(\mathbb{R})$, the following assertions are equivalent:*

(i) $\lambda \in \operatorname{Spec}(H)$;
(ii) *equation* (2.2) *has a bounded solution on* $\mathbb{R}$;
(iii) $\operatorname{Spec}(M_A(L,\lambda)) \cap \{\gamma \in \mathbb{C} : |\gamma| = 1\} \neq \emptyset$;
(iv) *equation* (2.2) *has a solution on* $[-L, L]$ *satisfying* (2.4) *for a* $\theta \in [0, 2\pi]$;
(v) $\lambda \in \operatorname{Spec}(H_\theta)$ *for a* $\theta \in [0, 2\pi]$.

PROOF. This follows immediately from Proposition 2.1 in [**Ga93**] and its proof and from the results in [**ReSi78**, Section XIII.16]. The equivalence of the last three assertions is also proved in a slightly more general Proposition 3.7 below. □

We will now introduce a family of systems of equations parametrized by $s \in (0, L]$ which augment (2.2). Each system will be a linear constant coefficient system whose solutions satisfy the same boundary condition (2.4) as our original system but with L replaced by s. To this end let us consider the system

$$\begin{pmatrix}\zeta\\ \xi\end{pmatrix}' = \begin{pmatrix}\frac{i\theta}{2s} & 0\\ 0 & \frac{i\theta}{2s}\end{pmatrix}\begin{pmatrix}\zeta\\ \xi\end{pmatrix}, \quad \zeta, \xi : \mathbb{R} \to \mathbb{C}^n, \theta \in [0, 2\pi],\ s \in (0, L]. \tag{2.10}$$

Setting

$$\begin{aligned} w_i &:= (\operatorname{Re}(\zeta_i), \operatorname{Im}(\zeta_i))^\top \in \mathbb{R}^2, \quad w := (w_1, \dots, w_n)^\top \in \mathbb{R}^{2n}, \\ z_i &:= (\operatorname{Re}(\xi_i), \operatorname{Im}(\xi_i))^\top \in \mathbb{R}^2, \quad z := (z_1, \dots, z_n)^\top \in \mathbb{R}^{2n}, \end{aligned} \tag{2.11}$$

we observe that w and z satisfy the following system of ODEs:

$$(2.12)\qquad \begin{pmatrix} w \\ z \end{pmatrix}' = \begin{pmatrix} I_n \otimes \mathfrak{u}(s,\theta) & 0 \\ 0 & I_n \otimes \mathfrak{u}(s,\theta) \end{pmatrix} \begin{pmatrix} w \\ z \end{pmatrix}, \quad \mathfrak{u}(s,\theta) := \begin{pmatrix} 0 & \frac{-\theta}{2s} \\ \frac{\theta}{2s} & 0 \end{pmatrix}.$$

Any solution of (2.10), respectively, (2.12) will automatically satisfy the same boundary conditions as in (1.2), respectively, (2.4), with L replaced by s, that is, the boundary conditions $(\zeta(s), \xi(s))^\top = e^{i\theta}(\zeta(-s), \xi(-s))^\top$, respectively,

$$(2.13)\qquad \begin{pmatrix} w(s) \\ z(s) \end{pmatrix} = \begin{pmatrix} I_n \otimes U(\theta) & 0 \\ 0 & I_n \otimes U(\theta) \end{pmatrix} \begin{pmatrix} w(-s) \\ z(-s) \end{pmatrix}.$$

As before, sometimes it is convenient to write equation (2.12) in a more condensed form denoting $\mathbf{w} := (w, z)^\top$, and writing equation (2.12) as the following equation with x-independent coefficient:

$$(2.14)\qquad \mathbf{w}' = B(s,\theta)\mathbf{w}, \quad B(s,\theta) := I_{2n} \otimes \mathfrak{u}(s,\theta).$$

For each $s \in (0, L]$, we let $\Psi_B(x, s)$ denote the fundamental matrix solution to the equation (2.12) such that $\Psi_B(-L, \lambda) = I_{4n}$, and remark that

$$(2.15)\quad \Psi_B(x,s) = I_{2n} \otimes e^{\mathfrak{u}(s,\theta)(x+L)} = I_{2n} \otimes \begin{pmatrix} \cos\frac{\theta}{2s}(x+L) & -\sin\frac{\theta}{2s}(x+L) \\ \sin\frac{\theta}{2s}(x+L) & \cos\frac{\theta}{2s}(x+L) \end{pmatrix}$$

is an orthogonal matrix: $\Psi_B(x,s)^\top = \big(\Psi_B(x,s)\big)^{-1}$.

It is sometimes convenient to combine (2.3) and (2.14) as follows:

$$(2.16)\quad \begin{pmatrix} \mathbf{p} \\ \mathbf{w} \end{pmatrix}' = \begin{pmatrix} A(x,\lambda) & 0_{4n} \\ 0_{4n} & B(s,\theta) \end{pmatrix} \begin{pmatrix} \mathbf{p} \\ \mathbf{w} \end{pmatrix}, \; x \in [-L, L],\; \theta \in [0, 2\pi],\; s \in (0, L].$$

We will now reformulate the boundary value problems for equations (2.2) and (2.12) with $s = L$ in a way amenable for symplectic analysis. We consider X defined in (1.4) as a $4n$-plane in $\mathbb{R}^{8n}$. We claim that λ is a θ-eigenvalue of equation (1.1) if and only if there is a nonzero solution to the following (augmented) boundary value problem:

$$(2.17)\quad \begin{pmatrix} p \\ q \\ w \\ z \end{pmatrix}' = \begin{pmatrix} 0 & I_{2n} & 0 & 0 \\ \lambda I_{2n} - I_2 \otimes V(x) & 0 & 0 & 0 \\ 0 & 0 & I_n \otimes \mathfrak{u}(L,\theta) & 0 \\ 0 & 0 & 0 & I_n \otimes \mathfrak{u}(L,\theta) \end{pmatrix} \begin{pmatrix} p \\ q \\ w \\ z \end{pmatrix},$$

$$(2.18)\quad \big(p(-L), q(-L), w(-L), z(-L)\big)^\top, \big(p(L), q(L), w(L), z(L)\big)^\top \in X.$$

It is convenient to write (2.17) and (2.18) as follows:

$$(2.19)\qquad \begin{pmatrix} \mathbf{p} \\ \mathbf{w} \end{pmatrix}' = \begin{pmatrix} A(x,\lambda) & 0_{4n} \\ 0_{4n} & B(L,\theta) \end{pmatrix} \begin{pmatrix} \mathbf{p} \\ \mathbf{w} \end{pmatrix},$$

$$(2.20)\qquad \begin{pmatrix} \mathbf{p}(-L) \\ \mathbf{w}(-L) \end{pmatrix}, \begin{pmatrix} \mathbf{p}(L) \\ \mathbf{w}(L) \end{pmatrix} \in X.$$

To justify the claim, we note that if $\mathbf{w}$ satisfies (2.14) with $s = L$ then $\mathbf{w}$ automatically satisfies (2.13) with $s = L$. Thus, if (2.20) holds then $\mathbf{p}$ satisfies (2.6). Conversely, given a $\mathbf{p}$ satisfying (2.6), pick a solution $\mathbf{w}$ of (2.14) with $s = L$ such that $\mathbf{w}(-L) = \mathbf{p}(-L)$. Then (2.20) holds.

3. A symplectic approach to counting eigenvalues

We begin by recalling some notions regarding symplectic structures and the Maslov index; for a detailed exposition see [**Ar67**, **Ar85**, **RS93**, **RS95**] and a review [**F04**], for a brief but extremely informative account see [**FJN03**].

A skew-symmetric non-degenerate quadratic form ω on $\mathbb{R}^{2n}$ is called symplectic. Symplectic forms are in one-to-one correspondence with orthogonal skew-symmetric matrices Ω, such that $\Omega^\top = \Omega^{-1} = -\Omega$, via the relation $\omega(v_1, v_2) = \langle v_1, \Omega v_2\rangle_{\mathbb{R}^{2n}}$, $v_1, v_2 \in \mathbb{R}^{2n}$. A real Lagrangian plane V is an n-dimensional subspace in $\mathbb{R}^{2n}$ such that $\omega(v_1, v_2) = 0$ for all $v_1, v_2 \in V$. The set of all Lagrangian planes in $\mathbb{R}^{2n}$ is denoted by $\Lambda(n)$.

Let $\mathrm{Train}(V)$ denote the train of a Lagrangian plane $V \in \Lambda(n)$, that is the set of all Lagrangian planes whose intersection with V is non trivial. Obviously, $\mathrm{Train}(V) = \cup_{k=1}^n \mathcal{T}_k(V)$ where $\mathcal{T}_k(V) = \big\{V_0 \in \Lambda(n) \,\big|\, \dim(V \cap V_0) = k\big\}$. Each set $\mathcal{T}_k(V)$ is an algebraic submanifold of $\Lambda(n)$ of codimension $k(k+1)/2$. In particular, $\operatorname{codim} \mathcal{T}_1(V) = 1$; moreover, $\mathcal{T}_1(V)$ is two-sidedly imbedded in $\Lambda(n)$, that is, there is a continuous vector field tangent to $\Lambda(n)$ which is transversal to $\mathcal{T}_1(V)$. Hence, one can speak about the positive and negative sides of $\mathcal{T}_1(V)$. Thus, given a smooth closed curve Φ in $\Lambda(n)$ that intersects $\mathrm{Train}\, V$ transversally (and thus in $\mathcal{T}_1(V)$), one can define the *Maslov index* $\mathrm{Mas}(\Phi, V)$ as the signed number of intersections.

We now recall a more detailed definition of the Maslov index as well as how to calculate it from local data. Let $\Phi : [a, b] \to \Lambda(n)$ be a smooth path. A *crossing* is a point $t_0 \in (a, b)$ of intersection of the path $\{\Phi(t) : t \in [a, b]\}$ with $\mathrm{Train}(V)$. Let $t_0 \in (a, b)$ be a crossing for a smooth path Φ, that is, assume that $\Phi(t_0) \cap V \neq \{0\}$. Let $V^\perp$ be a subspace in $\mathbb{R}^{2n}$ transversal to $\Phi(t_0)$. Then $V^\perp$ is transversal to $\Phi(t)$ for all $t \in [t_0 - \varepsilon, t_0 + \varepsilon]$ for $\varepsilon > 0$ small enough. Thus, there exists a smooth family of matrices, $\phi(t)$, viewed as operators from $\Phi(t_0)$ into $V^\perp$, so that $\Phi(t)$ is the graph of $\phi(t)$ for $|t - t_0| \leq \varepsilon$. The bilinear form $Q_{\mathcal{M}} = Q_{\mathcal{M}}(\Phi(t_0), V)$ defined by

$$Q_{\mathcal{M}}(\mathrm{v}, \mathrm{w}) = \frac{d}{dt}\omega(\mathrm{v}, \phi(t)\mathrm{w})\big|_{t=t_0} \quad \text{for } \mathrm{v}, \mathrm{w} \in \Phi(t_0) \cap V, \tag{3.1}$$

is called the *crossing form.*

A crossing is called *regular* if the crossing form is non degenerate. At a regular crossing t_0, denote the signature of the crossing form by $\operatorname{sign} Q_{\mathcal{M}}(\Phi(t_0), V)$. The Maslov index $\mathrm{Mas}(\Phi, V)$ of the path Φ with only regular crossings of $\mathrm{Train}(V)$ is then defined as

$$\begin{aligned} \mathrm{Mas}(\Phi, V) :=\ & \frac{1}{2}\operatorname{sign} Q_{\mathcal{M}}(\Phi(a), V) \\ & + \sum_{t \in (a,b)} \operatorname{sign} Q_{\mathcal{M}}(\Phi(t), V) + \frac{1}{2}\operatorname{sign} Q_{\mathcal{M}}(\Phi(b), V), \end{aligned} \tag{3.2}$$

where the summation above is over all crossings t (one can verify that regular crossings are isolated [**RS93**]). At the endpoints, take the appropriate left or right limit definition of the derivative in (3.1) to compute the bilinear form $Q_{\mathcal{M}}$ (and hence its signature). We remark that now we have a Maslov index even if the crossing does not take place in $\mathcal{T}_1$. It will sometimes be convenient to refer to the absolute value of the local Maslov index of a crossing as the *multiplicity* of the crossing. In the sequel, a curve with only regular crossings will also be called *regular*. From the context it should always be clear whether regular refers to a crossing or to the curve itself.

The important features of the Maslov index for this work are summarized below.

THEOREM 3.1. [**RS93**]

(1) (Naturality) *If T is a symplectic linear transformation then*

$$\mathrm{Mas}(T\Phi(t), TV) = \mathrm{Mas}(\Phi, V).$$

(2) (Catenation) *For $a < c < b$*

$$\mathrm{Mas}(\Phi, V) = \mathrm{Mas}(\Phi(\cdot)\big|_{[a,c]}, V) + \mathrm{Mas}(\Phi(\cdot)\big|_{[c,b]}, V).$$

(3) (Homotopy) *Two paths $\Phi_0, \Phi_1 : [a, b] \to \Lambda(n)$, with $\Phi_0(a) = \Phi_1(a)$ and $\Phi_0(b) = \Phi_1(b)$, are homotopic with fixed endpoints if and only if they have the same Maslov index.*

REMARK 3.2 (The generic case). A crossing t_0 is called *simple* if it is regular and $\Phi(t_0) \in \mathcal{T}_1(V)$. A curve has only simple crossings if and only if it is transverse to every $\mathcal{T}_k(V)$. Suppose that a curve $\Phi : [a, b] \to \Lambda(n)$ with $\Phi(a), \Phi(b) \in \mathcal{T}_0(V) := \big\{V_0 \in \Lambda(n) \,\big|\, \dim(V \cap V_0) = 0\big\}$ has only simple crossings. Then the two-sidedness of $\mathcal{T}_1(V)$ allows one to define m_+ to be the number of crossings by which $\Phi(t)$ passes from the negative side of $\mathcal{T}_1(V)$ to the positive side, and m_- to be the number of crossings from negative to positive. We then have that $\mathrm{Mas}(\Phi, V) = m_+ - m_-$.

REMARK 3.3. We remark that at a regular crossing t_0 the Maslov index of the path $\Phi : [t_0 - \varepsilon, t_0 + \varepsilon] \to \Lambda(n)$, for small enough ε, is equal to the signature of the crossing form at the crossing. In particular, the crossing is called *positive* (respectively *negative*) if the crossing form is positive (negative) definite. In this case the local Maslov index at the crossing is equal to plus (respectively minus) the dimension of the subspace $\Phi(t_0) \cap V$ (i.e. the multiplicity of the crossing is the real dimension of this subspace).

We will now return to the augmented system (2.16). Following [**DJ11**], for each $\lambda \in \mathbb{R}$ and $s \in (0, L]$ we now define the following set of vector valued functions on $[-L, L]$:

$$\begin{aligned} Y_{s,\lambda} = \Big\{(\mathbf{p}, \mathbf{w})^\top \,\Big|\, &\mathbf{p}, \mathbf{w} \in AC_{\mathrm{loc}}([-L, L]), \\ &\text{and } (\mathbf{p}, \mathbf{w})^\top \text{ is a solution of (2.16) on } [-L, L]\Big\}. \end{aligned} \tag{3.3}$$

That is, we consider the ($8n$ dimensional) solution space to the augmented equation (2.16), defined on $[-L, L]$, without any boundary conditions at all. We stress that by solutions $(\mathbf{p}, \mathbf{w})^\top$ of (2.16) on $[-L, L]$ we understand the mild solutions, that is, absolutely continuous vector valued functions such that (2.16) holds for almost all $x \in [-L, L]$; in other words, $\mathbf{p}(x) = \Psi_A(x, \lambda)\mathbf{p}(-L)$ and $\mathbf{w}(x) = \Psi_B(x, s)\mathbf{w}(-L)$, $x \in [-L, L]$, where $\Psi_A(\cdot, \lambda)$ and $\Psi_B(\cdot, s)$ are the fundamental matrix solutions to equations (2.2) and (2.12), respectively.

Next, for each $\lambda \in \mathbb{R}$ and $s \in (0, L]$, let us define the trace map $\Phi_s^\lambda : Y_{s,\lambda} \to \mathbb{R}^{16n}$ by the following formula:

$$\Phi_s^\lambda : (\mathbf{p}, \mathbf{w})^\top \mapsto \big(\mathbf{p}(-s), \mathbf{w}(-s), \mathbf{p}(s), \mathbf{w}(s)\big)^\top \in \mathbb{R}^{16n}. \tag{3.4}$$

We remark that Φ_s^λ can be identified with the following $(16n \times 8n)$ matrix,

$$\widehat{\Phi}_s^\lambda = \begin{pmatrix} \Psi_A(-s,\lambda) & 0_{4n} \\ 0_{4n} & \Psi_B(-s,s) \\ \Psi_A(s,\lambda) & 0_{4n} \\ 0_{4n} & \Psi_B(s,s) \end{pmatrix}, \tag{3.5}$$

since for the solution $(\mathbf{p},\mathbf{w})^\top \in Y_{s,\lambda}$ given by $\mathbf{p}(x) = \Psi_A(x,\lambda)\mathbf{p}(-L)$ and $\mathbf{w}(x) = \Psi_B(x,s)\mathbf{w}(-L)$, clearly, the vector $\Phi_s^\lambda((\mathbf{p},\mathbf{w})^\top) \in \mathbb{R}^{16n}$ is the product of the matrix $\widehat{\Phi}_s^\lambda$ and the vector $(\mathbf{p}(-L),\mathbf{w}(-L))^\top \in \mathbb{R}^{8n}$.

Let us introduce the $(16n \times 16n)$ orthogonal skew-symmetric matrix Ω (and thus a symplectic structure on $\mathbb{R}^{16n}$) by the formula

$$\Omega = (J \otimes I_{2n}) \oplus (I_{2n} \otimes J^T) \oplus (J^T \otimes I_{2n}) \oplus (I_{2n} \otimes J), \quad J = \begin{pmatrix} 0 & 1 \\ -1 & 0 \end{pmatrix}, \tag{3.6}$$

where J is the standard symplectic matrix.

THEOREM 3.4. *For all $s \in (0, L]$ and $\lambda \in \mathbb{R}$ the plane $\Phi_s^\lambda(Y_{s,\lambda})$ belongs to the space $\Lambda(8n)$ of Lagrangian $8n$-planes in $\mathbb{R}^{16n}$, with the Lagrangian structure $\omega(v_1, v_2) = \langle v_1, \Omega v_2\rangle_{\mathbb{R}^{16n}}$ given by Ω defined in (3.6).*

PROOF. Equations (2.3) and (2.14) are Hamiltonian, with the symplectic structure defined by the matrices

$$J_n := J \otimes I_{2n} \quad \text{and} \quad J^{(n)} := I_{2n} \otimes J^\top, \tag{3.7}$$

respectively. In particular,

$$J_n A(x,\lambda) = \big(J_n A(x,\lambda)\big)^\top, \quad J^{(n)} B(s,\theta) = \big(J^{(n)} B(s,\theta)\big)^\top. \tag{3.8}$$

Writing (3.6) as

$$\Omega = J_n \oplus J^{(n)} \oplus (-J_n) \oplus (-J^{(n)}), \tag{3.9}$$

for any two vectors from $\Phi_s^\lambda(Y_{s,\lambda})$, $v_1 = \big(\mathbf{p}_1(-s), \mathbf{w}_1(-s), \mathbf{p}_1(s), \mathbf{w}_1(s)\big)^\top$ and $v_2 = \big(\mathbf{p}_2(-s), \mathbf{w}_2(-s), \mathbf{p}_2(s), \mathbf{w}_2(s)\big)^\top$, we infer:

$$\begin{aligned}
\langle v_1, \Omega v_2\rangle_{\mathbb{R}^{16n}} &= \langle \mathbf{p}_1(-s), J_n \mathbf{p}_2(-s)\rangle_{\mathbb{R}^{4n}} + \langle \mathbf{w}_1(-s), J^{(n)}\mathbf{w}_2(-s)\rangle_{\mathbb{R}^{4n}} \\
&\quad - \langle \mathbf{p}_1(s), J_n \mathbf{p}_2(s)\rangle_{\mathbb{R}^{4n}} - \langle \mathbf{w}_1(s), J^{(n)}\mathbf{w}_2(s)\rangle_{\mathbb{R}^{4n}} \\
&= \int_s^{-s} \frac{d}{dx}\Big(\langle \mathbf{p}_1(x), J_n \mathbf{p}_2(x)\rangle_{\mathbb{R}^{4n}} + \langle \mathbf{w}_1(x), J^{(n)}\mathbf{w}_2(x)\rangle_{\mathbb{R}^{4n}}\Big)\, dx \\
&= \int_s^{-s} \Big(\langle \mathbf{p}_1'(x), J_n \mathbf{p}_2(x)\rangle_{\mathbb{R}^{4n}} + \langle \mathbf{p}_1(x), J_n \mathbf{p}_2'(x)\rangle_{\mathbb{R}^{4n}} \\
&\quad + \langle \mathbf{w}_1'(x), J^{(n)}\mathbf{w}_2(x)\rangle_{\mathbb{R}^{4n}} + \langle \mathbf{w}_1(x), J^{(n)}\mathbf{w}_2'(x)\rangle_{\mathbb{R}^{4n}}\Big)\, dx \\
&= \int_s^{-s} \Big(\langle A(x,\lambda)\mathbf{p}_1(x), J_n \mathbf{p}_2(x)\rangle_{\mathbb{R}^{4n}} + \langle \mathbf{p}_1(x), J_n A(x,\lambda)\mathbf{p}_2(x)\rangle_{\mathbb{R}^{4n}} \\
&\quad + \langle B(s,\theta)\mathbf{w}_1(x), J^{(n)}\mathbf{w}_2(x)\rangle_{\mathbb{R}^{4n}} + \langle \mathbf{w}_1(x), J^{(n)} B(s,\theta)\mathbf{w}_2(x)\rangle_{\mathbb{R}^{4n}}\Big)\, dx \\
&= \int_s^{-s} \Big(- \langle J_n A(x,\lambda)\mathbf{p}_1(x), \mathbf{p}_2(x)\rangle_{\mathbb{R}^{4n}} + \langle \mathbf{p}_1(x), J_n A(x,\lambda)\mathbf{p}_2(x)\rangle_{\mathbb{R}^{4n}} \\
&\quad - \langle J^{(n)} B(s,\theta)\mathbf{w}_1(x), \mathbf{w}_2(x)\rangle_{\mathbb{R}^{4n}} + \langle \mathbf{w}_1(x), J^{(n)} B(s,\theta)\mathbf{w}_2(x)\rangle_{\mathbb{R}^{4n}}\Big)\, dx
\end{aligned}$$

$= 0,$

where in the last two lines we used $(J_n)^\top = -J_n$, $(J^{(n)})^\top = -J^{(n)}$ and (3.8). $\square$

We remark that $X \times X$ with X defined in (1.4) is a Lagrangian plane in $\mathbb{R}^{16n}$ with the same symplectic structure given by Ω (indeed, this was why Ω was chosen in the first place). This can be verified by a straightforward calculation.

DEFINITION 3.5. For a given λ, a point $s \in (0, L]$ is called a *(λ-)conjugate point* if $\Phi_s^\lambda(Y_{s,\lambda}) \in \mathrm{Train}(X \times X)$, where X is defined in (1.4).

The latter inclusion means that there exists a solution of the system of equations (2.2), (2.12) on the segment $[-s, s]$ satisfying the boundary conditions (2.4) with L replaced by s, that is, the boundary condtions

$$\begin{pmatrix} p(s) \\ q(s) \end{pmatrix} = \begin{pmatrix} I_n \otimes U(\theta) & 0 \\ 0 & I_n \otimes U(\theta) \end{pmatrix} \begin{pmatrix} p(-s) \\ q(-s) \end{pmatrix}, \tag{3.10}$$

and the boundary conditions (2.13).

Our next objective is to relate the crossings of the path $\{\Phi_\lambda^s(Y_{s,\lambda})\}$ and eigenvalues of differential operators $H_{\theta,s}$ in $L^2([-s, s])$ introduced as follows, cf. (2.9). For any $s \in (0, L]$ and $\theta \in [0, 2\pi]$, let $H_{\theta,s} = \frac{d^2}{dx^2} + V(x)$ with

$$\begin{aligned} \mathrm{dom}(H_\theta) = \Big\{ y \in L^2([-s, s]) \Big| y, y' \in AC_{\mathrm{loc}}([-s, s]), \\ y'' \in L^2([-s, s]) \text{ and the boundary condition (1.3) holds} \Big\}. \end{aligned} \tag{3.11}$$

In particular, $H_{\theta,L} = H_\theta$. We remark that $y \in \ker\big(H_{\theta,s} - \lambda I_{L^2([-s,s])}\big)$ if and only if the vector valued function $\mathbf{p} = (p, q)^\top$ defined in (2.1) is a solution of (2.3) on $[-s, s]$ that satisfies the boundary conditions (3.10).

DEFINITION 3.6. We say that λ is an (θ, s)-*eigenvalue* of equation (1.1) if there is a nonzero solution of (2.2) such that the boundary conditions (3.10) are satisfied.

Recall that $\Psi_A(x, \lambda)$ is the fundamental matrix solution of the system (2.2), and $M_A(s, \lambda) = \Psi_A(s, \lambda)\Psi_A(-s, \lambda)^{-1}$ is the propagator for $s \in (0, L]$ so that $\mathbf{p}(s) = M(s, \lambda)\mathbf{p}(-s)$ for a solution of (2.3). Also, we recall that the multiplicity of the eigenvalue λ of the operator H_θ is the (complex) dimension of the solution space of the boundary value problem (1.1), (1.2) on $[-L, L]$.

PROPOSITION 3.7. *For any $\lambda \in \mathbb{R}$, $\theta \in [0, 2\pi]$, and $s \in (0, L]$ the following assertions are equivalent:*

(i) $\lambda \in \mathrm{Spec}(H_{\theta,s})$ in $L^2([-s, s])$;

(ii) $e^{i\theta} \in \mathrm{Spec}\big(M_A(s, \lambda)\big)$;

(iii) s is a λ-conjugate point, that is, $\Phi_s^\lambda(Y_{s,\lambda}) \in \mathrm{Train}(X \times X)$.

Moreover, the multiplicity of the eigenvalue λ of the operator $H_{\theta,s}$ is equal to the real dimension of the subspace $\Phi_s^\lambda(Y_{s,\lambda}) \cap (X \times X)$. In particular, λ is a θ-eigenvalue of (1.1) if and only if L is a λ-conjugate point, that is, $\Phi_L^\lambda(Y_{L,\lambda}) \in \mathrm{Train}(X \times X)$, and λ is an (θ, s)-eigenvalue of (1.1) if and only $\Phi_s^\lambda(Y_{s,\lambda}) \in \mathrm{Train}(X \times X)$.

PROOF. (*i*) $\Rightarrow$ (*ii*) Take a nonzero $y \in \ker\big(H_{\theta,s} - \lambda I_{L^2([-s,s])}\big)$ and let $\mathbf{y} = (y, y')^\top$ be the complex valued $(n \times 1)$ solution of the first order system

$$\mathbf{y}' = A^{\mathbb{C}}(x, \lambda)\mathbf{y}, \quad A^{\mathbb{C}}(x, \lambda) = \begin{pmatrix} 0_n & I_n \\ \lambda I_n - V(x) & 0_n \end{pmatrix} \tag{3.12}$$

that satisfies the boundary condition (1.3). Let $\Psi_A^{\mathbb{C}}(x,\lambda)$ denote the fundamental matrix solution to (3.12) such that $\Psi_A^{\mathbb{C}}(-L,\lambda)=I_n$, so that $\mathbf{y}(x)=\Psi_A^{\mathbb{C}}(x,\lambda)\mathbf{y}(-L)$, and let $M_A^{\mathbb{C}}(s,\lambda)=\Psi_A^{\mathbb{C}}(s,\lambda)\Psi_A^{\mathbb{C}}(-s,\lambda)^{-1}$ denote the propagator such that $\mathbf{y}(s)=M_A^{\mathbb{C}}(s,\lambda)\mathbf{y}(-s)$. Due to (1.3), we have $e^{i\theta}\in\operatorname{Spec}\big(M_A^{\mathbb{C}}(s,\lambda)\big)$. Let $T:\mathbb{C}^{2n}\to\mathbb{R}^{4n}$ be the map $\mathbf{y}\mapsto\mathbf{p}=(p,q)^\top$ defined in (2.1). Then $\Psi_A(x,\lambda)=T\Psi_A^{\mathbb{C}}(x,\lambda)T^{-1}$ and $M_A(s,\lambda)=TM_A^{\mathbb{C}}(s,\lambda)T^{-1}$, yielding *(ii)*.

(ii) $\Rightarrow$ *(iii)* For a vector $\mathbf{v}\in\mathbb{C}^{2n}$ satisfying $M_A^{\mathbb{C}}(s,\lambda)\mathbf{v}=e^{i\theta}\mathbf{v}$ let $\mathbf{y}(x)=\Psi_A^{\mathbb{C}}(x,\lambda)\mathbf{v}$ be the solution of (3.12) satisfying (1.3). Using (2.1), construct the solution $\mathbf{p}$ of (2.3) satisfying (3.10), that is, satisfying $\mathbf{p}(s)=(I_{2n}\otimes U(\theta))\mathbf{p}(-s)$. Pick the solution $\mathbf{w}$ of (2.14) such that $\mathbf{w}(-s)=\mathbf{p}(-s)$. Since solutions of (2.14) automatically satisfy (2.13), we have $\mathbf{p}(\pm s)=\mathbf{w}(\pm s)$ and thus $\Phi_s^\lambda(Y_{s,\lambda})\in\operatorname{Train}(X\times X)$.

(iii) $\Rightarrow$ *(i)* Pick a solution $(\mathbf{p},\mathbf{w})^\top$ of (2.16) such that $\Phi_s^\lambda\big((\mathbf{p},\mathbf{w})^\top\big)\in X\times X$; then $\mathbf{p}(\pm s)=\mathbf{w}(\pm s)$. Since $\mathbf{w}$ automatically satisfies (2.13), the boundary condition $\mathbf{p}(s)=(I_{2n}\otimes U(\theta))\mathbf{p}(-s)$ holds. It follows that the solution $\mathbf{y}$ of (3.12) related to $\mathbf{p}=(p,q)^\top$ via (2.1) satisfies the boundary condition (1.3), thus yielding $H_{\theta,s}y=\lambda y$.

To prove the equality of the multiplicity and the dimension of the intersection, we remark that the linear map $y\mapsto(\mathbf{p}(-s),\mathbf{w}(-s),\mathbf{p}(s),\mathbf{w}(s))^\top$ from the finite dimensional space

$$\begin{aligned}\ker(H_{\theta,s}-\lambda I_{L^2([-s,s])})=\Big\{y\in L^2([-s,s])\Big|\,&y,y'\in AC_{\text{loc}}([-s,s]),\\ &\text{and (1.1), (1.3) hold}\Big\}\end{aligned}\tag{3.13}$$

into the finite dimensional space $\Phi_s^\lambda(Y_{s,\lambda})\cap(X\times X)$ has zero kernel, and thus is an isomorphism yielding $\dim_{\mathbb{C}}\ker(H_{\theta,s}-\lambda I_{L^2([-s,s])})=\dim_{\mathbb{R}}\big(\Phi_s^\lambda(Y_{s,\lambda})\cap(X\times X)\big)$. $\square$

Since the boundary value problem on the segment $[-s,s]$ makes sense only for positive s, we may restrict s to $s\in[s_0,L]$ for some $s_0>0$. Since the operator $H_{\theta,s}$ is semibounded from above, for a λ_∞ large enough there are no (θ,s)-eigenvalues with $\lambda\geq\lambda_\infty$. Therefore, we may restrict λ to $\lambda\in[0,\lambda_\infty]$. As we will see in Lemma 3.12, for λ_∞ large enough there are no $s\in[s_0,L]$ such that $\Phi_s^{\lambda_\infty}(Y_{s,\lambda})\in\operatorname{Train}(X\times X)$ provided $\theta\in[0,2\pi]$, and for s_0 small enough there are no $\lambda\in[0,\lambda_\infty]$ such that $\Phi_{s_0}^\lambda(Y_{s_0,\lambda})\in\operatorname{Train}(X\times X)$ provided $\theta\in(0,2\pi)$.

With no loss of generality (by varying θ a little, if needed), we may assume that $\lambda=0$ is *not* a θ-eigenvalue for a given θ, see Figures 1 and 2. (This ensures that all crossings take place away from the upper left corner in Figure 2). This is not actually necessary, but more of a convenience. We can simply use the crossing form calculation at the endpoints if there is a crossing at the upper left corner, taking into account half of the local Maslov index each time.

We also remark that for a fixed λ_∞, and s_0, we can view the map Φ_s^λ as a continuous map from the square $\Phi_s^\lambda:[0,\lambda_\infty]\times[s_0,L]\to\Lambda(8n)$ to the space of Lagrangian planes, see Figure 2. As such, its image must be homotopic to a point, and so we have the following theorem.

THEOREM 3.8. *The homotopy class of the image of the boundary of the square* $[0,\lambda_\infty]\times[s_0,L]$ *under the map* Φ *is zero in* $\pi_1(\Lambda(8n))$.

It is well known that $\pi_1(\Lambda(8n))\approx\mathbb{Z}$, and that the class of a closed curve can be determined by the number of intersections of such a curve (up to homotopy) with the train of a fixed Lagrangian plane (see for example, [**Ar67**], or [**RS93**] and

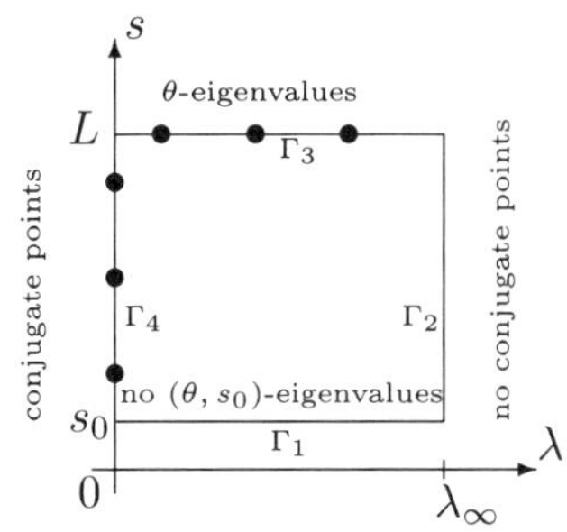

FIGURE 2. $\lambda = 0$ is *not* a θ-eigenvalue, $\theta \in (0, 2\pi)$, and s_0 is small enough

the references therein). Denote by Γ the boundary of the image of $[0, \lambda_\infty] \times [s_0, L]$ under Φ_s^λ. The key idea here is that under the construction given above, we have an eigenvalue interpretation for the intersection of $\Phi_s^\lambda(Y_{s,\lambda})$ with the train of a special plane. Since the signed number of intersections does not change under homotopy, and Γ is homotopic to a point, we have the following result.

COROLLARY 3.9. *As we travel along* Γ, *the signed number of intersections of* Γ *with* $\mathrm{Train}(X \times X)$, *counted with multiplicity, is equal to zero.*

REMARK 3.10. It is convenient for us to break up the curve $\Gamma = \Gamma_{(\theta,s_0)}$ into the four pieces corresponding to the sides of the square from which it comes. Let $\Gamma_1 = \Gamma_{1,(\theta,s_0)}$ denote $\{\Phi_{s_0}^\lambda(Y_{s_0,\lambda}) \big| \lambda \in [0, \lambda_\infty]\}$, let $\Gamma_2 = \Gamma_{2,(\theta,s_0)}$ denote $\{\Phi_s^{\lambda_\infty}(Y_{s,\lambda_\infty}) \big| s \in [s_0, L]\}$, let $\Gamma_3 = \Gamma_{3,(\theta,s_0)}$ denote $\{\Phi_L^\lambda(Y_{L,\lambda}) \big| \lambda \in [\lambda_\infty, 0]\}$, and let $\Gamma_4 = \Gamma_{4,(\theta,s_0)}$ denote $\{\Phi_s^0(Y_{s,0}) \big| s \in [L, s_0]\}$.

Let $A_i = A_{i,(\theta,s_0)}$ denote the Maslov index of each piece of Γ_i, as defined in (3.2), that is,

$$A_i := \mathrm{Mas}(\Gamma_i, X \times X). \tag{3.14}$$

We will also denote by $B_i = B_{i,(\theta,s_0)}$ the following expression:

$$\begin{aligned} B_i := &\left|\frac{1}{2}\mathrm{sign}\ Q_{\mathcal{M}}(\Gamma_i(a_i), X \times X)\right| \\ &+ \sum_{t \in (a_i, b_i)} |\mathrm{sign}\ Q_{\mathcal{M}}(\Gamma_i(t), X \times X)| + \left|\frac{1}{2}\mathrm{sign}\ Q_{\mathcal{M}}(\Gamma_i(b_i), X \times X)\right|, \end{aligned} \tag{3.15}$$

where $\Gamma_i(a_i)$ and $\Gamma_i(b_i)$ denote the endpoints of the curve Γ_i. That is, B_i is the number of crossings along Γ_i each counted regardless of sign, but taking into account the multiplicity of crossings. For instance, if we have three simple crossings on Γ_i with signs $+, -, +$, we would have that $A_i = 1$, while $B_i = 3$. It is also worth noting that in all cases $|A_i| \leq B_i$.

We will now show that $B_2 = 0$ provided λ_∞ is large enough and $\theta \in [0, 2\pi]$, and that $B_1 = 0$ provided $s_0 > 0$ is small enough and $\theta \in (0, 2\pi)$. For $\theta = 0$, see a computation of B_1 in Lemma 4.3. We will repeatedly use the following elementary fact.

THEOREM 3.11. [**K80**, Theorem V.4.10] *Let* $\mathcal{H}$ *be selfadjoint and* $\mathcal{V} \in \mathcal{B}(\mathcal{X})$ *be symmetric operators on a Hilbert space* $\mathcal{X}$. *Then*

$$\mathrm{dist}\left(\mathrm{Spec}(\mathcal{H} + \mathcal{V}), \mathrm{Spec}(\mathcal{H})\right) \leq \|\mathcal{V}\|_{\mathcal{B}(\mathcal{X})}.$$

We recall that V is a bounded matrix valued function on $[-L, L]$ and denote the supremum of its matrix norm by $\|V\|_\infty = \sup_{x\in[-L,L]} \|V(x)\|_{\mathbb{R}^n\times\mathbb{R}^n}$.

LEMMA 3.12.

(i) *Assume that* $\theta \in [0, 2\pi]$. *If* $\lambda_\infty > \|V\|_\infty$ *then* $B_2 = 0$.
(ii) *Assume that* $\theta \in (0, 2\pi)$. *If* $\lambda_\infty > \|V\|_\infty$ *and*

$$s_0 < \frac{1}{2}\min\{\theta, 2\pi - \theta\}\,(\|V\|_\infty + \lambda_\infty)^{-1/2},$$

then $B_1 = 0$.

PROOF. Let $H_{\theta,s}^{(0)} = \frac{d^2}{dx^2}$ with $\mathrm{dom}(H_{\theta,s}^{(0)}) = \mathrm{dom}(H_{\theta,s})$. The eigenvalues of $H_{\theta,s}^{(0)}$ are given by the formula

$$\mu_k = -\Big(\frac{\theta + 2\pi k}{2s}\Big)^2, \; k \in \mathbb{Z}. \tag{3.16}$$

Indeed, inserting the general solution $y(x) = c_1 e^{\sqrt{\mu}x} + c_2 e^{-\sqrt{\mu}x}$ of the equation $y'' = \mu y$ in the boundary conditions (1.3), we obtain the system of equations for c_1, c_2, whose determinant must be equal to zero, yielding (3.16).

For $s \in (0, L]$ and μ_k in (3.16) we denote $\mu(s) = \max_{k\in\mathbb{Z}} \mu_k$. Then (3.16) implies

$$\mu(s) = -\big(\min\{\theta, 2\pi - \theta\}/(2s)\big)^2, \quad \theta \in [0, 2\pi],\, s \in (0, L], \tag{3.17}$$

and

$$\mathrm{Spec}(H_{\theta,s}^{(0)}) \subset \big(-\infty, \mu(s)\big] \subset (-\infty, 0], \text{ for each } s \in (0, L]. \tag{3.18}$$

By Theorem 3.11 we infer:

$$\mathrm{dist}\,\big(\mathrm{Spec}(H_{\theta,s}), \mathrm{Spec}(H_{\theta,s}^{(0)})\big) \le \|V\|_{\mathcal{B}(L^2([-s,s]))} \le \|V\|_\infty. \tag{3.19}$$

This and the second inclusion in (3.18) yield $\mathrm{Spec}(H_{\theta,s}) \subset (-\infty, \|V\|_\infty]$. If s is a conjugation point for a given λ, then there is a solution y of the equation $H_{\theta,s}y = \lambda y$ satisfying (1.3), that is, λ is an eigenvalue of $H_{\theta,s}$. Thus, there are no conjugation points for λ_∞ provided $\lambda_\infty > \|V\|_\infty$, proving assertion (i).

(ii) Assume that $\theta \in (0, 2\pi)$, fix $\lambda_\infty > \|V\|_\infty$, and consider any $\lambda \in [0, \lambda_\infty]$ and $s \in (0, L]$. If y is a solution of the equation $y'' + V(x)y = \lambda y$ for $|x| \le s$ satisfying boundary conditions (1.3) then $z(x) = y(sx/L)$ for $|x| \le L$ satisfies the equation

$$H_{\theta,L}^{(1)} z := z'' + \Big(\big(s/L\big)^2 V(sx/L) - \lambda(s/L)^2\Big) z = 0,\; x \in [-L, L], \tag{3.20}$$

and boundary conditions (1.2). In other words, λ is an eigenvalue of $H_{\theta,s}$ on $L^2([-s, s])$ if and only if zero is an eigenvalue of $H_{\theta,L}^{(1)}$ on $L^2([-L, L])$. Since the potential in $H_{\theta,L}^{(1)}$ is $\big(s/L\big)^2 V(s(\cdot)/L) - \lambda(s/L)^2$, by Theorem 3.11 we infer:

$$\begin{aligned} \mathrm{dist}\,\big(\mathrm{Spec}(H_{\theta,L}^{(1)}), \mathrm{Spec}(H_{\theta,L}^{(0)})\big) &\le \|\big(s/L\big)^2 V(s(\cdot)/L) - \lambda(s/L)^2\|_{\mathcal{B}(L^2([-L,L]))} \\ &\le \big(s/L\big)^2\big(\|V\|_\infty + \lambda_\infty\big). \end{aligned} \tag{3.21}$$

This and the first inclusion in (3.18) with $s = L$ imply

$$\mathrm{Spec}(H_{\theta,L}^{(1)}) \subset \Big(-\infty,\, \big(s/L\big)^2\big(\|V\|_\infty + \lambda_\infty\big) + \mu(L)\Big], \tag{3.22}$$

where $\mu(L) < 0$ due to $\theta \in (0, 2\pi)$. In particular, using (3.16), if

$$s_0^2\big(\|V\|_\infty + \lambda_\infty\big) < (\min\{\theta, 2\pi - \theta\}/2)^2$$

then zero is not an eigenvalue of $H_{\theta,L}^{(1)}$ and thus λ is not an eigenvalue of H_{θ,s_0} on $L^2([-s_0, s_0])$, as needed in (ii). □

Alternatively, one can prove that for $\theta \in (0, 2\pi)$ there are no conjugate points, provided s is sufficiently small, using Proposition 3.7 (ii): Since Spec $\big(M_A(s,\lambda)\big) \to \{1\}$ as $s \to 0$, we infer that $e^{i\theta} \notin$ Spec $\big(M_A(s,\lambda)\big)$ for s small enough.

The periodic case $\theta = 0$ or $\theta = 2\pi$ is somehow special and should be treated separately. Since the periodic boundary conditions (1.2), (1.3) hold when either $\theta = 0$ or $\theta = 2\pi$, we will conclude this section by considering the case $\theta = 0$ (see also Lemma 4.3 for more information regarding this case).

We will begin by constructing the curve $\Gamma = \Gamma_{(0,0)}$ for $\theta = 0$ and $s_0 = 0$ (note that the construction described in Remark 3.10 does not work as (2.14) is not defined for $s = 0$). If $\theta = 0$ then $\mathbf{u}(s, 0) = 0_2$ in (2.12) and $B(s, 0) = 0_{4n}$ in (2.14) for all $s > 0$. Thus, we have $\Psi_B(x, s) = I_{4n}$ for $\theta = 0$ and $s > 0$. Letting $\Psi_B(x, 0) = I_{4n}$ for $s = 0$ and all $x \in [-L, L]$, we can extend $\Psi_B(x, s)$ continuously from $s > 0$ to $s = 0$ although the differential equation (2.14) is not defined for $s = 0$. This allows us to define the curve Γ_1 for $\theta = 0$ and $s_0 = 0$ as follows: Recall that the curve $\Gamma_1 = \Gamma_{1,(\theta,s_0)}$ is defined via (3.5) as the set

$$\Gamma_{1,(\theta,s_0)} = \big\{\widehat{\Psi}_{s_0}^{\lambda}\mathbf{v}\big|\, \lambda \in [0, \lambda_\infty], \mathbf{v} \in \mathbb{R}^{8n}\big\}. \tag{3.23}$$

Setting $\theta = 0$ and passing in (3.5) to the limit yields $\widehat{\Psi}_{s_0}^{\lambda} \to \widehat{\Psi}_0^{\lambda}$ as $s_0 \to 0^+$ uniformly for $\lambda \in [0, \lambda_\infty]$, where we define

$$\widehat{\Phi}_0^{\lambda} = \begin{pmatrix} \Psi_A(0,\lambda) & 0_{4n} \\ 0_{4n} & I_{4n} \\ \Psi_A(0,\lambda) & 0_{4n} \\ 0_{4n} & I_{4n} \end{pmatrix}. \tag{3.24}$$

Letting

$$\Gamma_{1,(0,0)} = \big\{\widehat{\Psi}_0^{\lambda}\mathbf{v}\big|\, \lambda \in [0, \lambda_\infty], \mathbf{v} \in \mathbb{R}^{8n}\big\}, \tag{3.25}$$

we thus introduce the curve $\Gamma_1 = \Gamma_{1,(0,0)}$ for $\theta = 0$ and $s_0 = 0$. This curve is homotopic to the curve $\Gamma_1 = \Gamma_{1,(\theta,s_0)}$ for $\theta > 0$ and $s_0 > 0$ although the endpoints of the two curves are not fixed. A direct computation shows that $\big(\widehat{\Psi}_0^{\lambda}\big)^{\top}\Omega\widehat{\Psi}_0^{\lambda} = 0_{16n}$, and thus $\Gamma_{1,(0,0)}$ is a curve in $\Lambda(8n)$. Clearly, $\Gamma_{1,(0,0)}$ lies in Train$(X \times X)$, and thus is not regular. This makes the computation of Mas$(\Gamma_1, X \times X)$ for $\theta = 0$ and $s_0 = 0$ with this choice of Γ_1 difficult. By appending to $\Gamma_{1,(0,0)}$ the three remaining curves $\Gamma_{j,(0,0)}$, $j = 2, 3, 4$, corresponding to the remaining three sides of the square $[0, L] \times [0, \lambda_\infty]$, we construct the entire curve $\Gamma = \Gamma_{(0,0)}$ for $\theta = 0$ and $s_0 = 0$ which is homotopic to the curve $\Gamma = \Gamma_{(\theta,s_0)}$ for $\theta > 0$ and $s_0 > 0$. We can appeal to a theorem in [**RS93**] which says that every continuous curve is homotopic to a curve with only regular crossings. Thus we can compute the Maslov index of $\Gamma = \Gamma_{(0,0)}$ and verify that it is indeed 0, that is, that Corollary 3.9 holds for $\theta = 0$ and $s_0 = 0$.

One can also define conjugate point as a point s where $e^{i\theta} \in$ Spec $\big(M_A(s,\lambda)\big)$, see Proposition 3.7 (ii). Unlike Definition 3.5, this latter definition is applicable for $s = 0$ as well. But for $\theta = 0$, since $M_A(0,\lambda) = I_{4n}$, we have that $1 = e^{i0} \in$

Spec $\big(M_A(0,\lambda)\big)$, and thus $s=0$ is the conjugate point for all $\lambda\in[0,\lambda_\infty]$. In particular, for $\theta=0$ the curve Γ_2 has a conjugate point at $s=0$.

We summarize the discussion as follows and refer to Lemma 4.3 for more information regarding the case $\theta=0$.

COROLLARY 3.13. *Assume that $\theta=0$ and that Γ is the curve just defined for $s_0=0$ using (3.24), and parametrized by the sides of the square $[0,L]\times[0,\lambda_\infty]$. Then $\mathrm{Mas}(\Gamma, X\times X)=0$. Each point of the curve Γ_1 belongs to $\mathrm{Train}(X\times X)$. The lower endpoints of the curves Γ_2 and Γ_4, and all points of Γ_1 are conjugate points in the sense that $1=e^{i0}\in\mathrm{Spec}\big(M_A(0,\lambda)\big)$ for all $\lambda\in[0,\lambda_\infty]$.*

4. Monotonicity of the Maslov index

We will now establish monotonicity of the Maslov index with respect to the parameter λ and, under some additional assumptions, with respect to the parameter s. Let us begin with λ. We recall from Remark 3.10 that the curve Γ_3 is parametrized by the parameter λ *decaying* from λ_∞ to 0 while the curve Γ_1 is parametrized by the parameter λ *growing* from 0 to λ_∞. The strategy of the proof of the next result follows the proof of [**DJ11**, Lemma 4.7].

LEMMA 4.1. *For any $\theta\in[0,2\pi]$ and any fixed $s\in(0,L]$, each crossing $\lambda_0\in(0,\lambda_\infty)$ of the path $\big\{\Phi_s^\lambda(Y_{s,\lambda})\big\}_{\lambda=\lambda_0-\varepsilon}^{\lambda_0+\varepsilon}$, with $\varepsilon>0$ small enough, is negative. In particular, if $0\notin\mathrm{Spec}(H_\theta)$, then $B_3=A_3$ and if $0\notin\mathrm{Spec}(H_{\theta,s_0})$ then $B_1=-A_1$.*

PROOF. Let $\lambda_0\in(0,\lambda_\infty)$ be a crossing, so that $\Phi_s^{\lambda_0}(Y_{s,\lambda_0})\cap(X\times X)\neq\{0\}$. Let $V^\perp$ be a subspace in $\mathbb{R}^{16n}$ transversal to $\Phi_s^{\lambda_0}(Y_{s,\lambda_0})$. Then $V^\perp$ is transversal to $\Phi_s^\lambda(Y_{s,\lambda})$ for all $\lambda\in[\lambda_0-\varepsilon,\lambda_0+\varepsilon]$ for $\varepsilon>0$ small enough. Thus, there exists a smooth family of matrices, $\phi(\lambda)$, for $\lambda\in[\lambda_0-\varepsilon,\lambda_0+\varepsilon]$, viewed as operators $\phi(\lambda):\Phi_s^{\lambda_0}(Y_{s,\lambda_0})\to V^\perp$, such that $\Phi_s^\lambda(Y_{s,\lambda})$ is the graph of $\phi(\lambda)$. Fix any nonzero $\mathrm{v}\in\Phi_s^{\lambda_0}(Y_{s,\lambda_0})\cap(X\times X)$ and consider the curve $v(\lambda)=\mathrm{v}+\phi(\lambda)\mathrm{v}\in\Phi_s^\lambda(Y_{s,\lambda})$ for $\lambda\in[\lambda_0-\varepsilon,\lambda_0+\varepsilon]$ with $v(\lambda_0)=\mathrm{v}$. By the definition of $Y_{s,\lambda}$, there is a family of solutions $(\mathbf{p}(\cdot,\lambda),\mathbf{w}(\cdot,\lambda))^\top$ of (2.16) such that $v(\lambda)=\Phi_s^\lambda\big((\mathbf{p}(\cdot,\lambda),\mathbf{w}(\cdot,\lambda))^\top\big)$. We claim that

$$\omega\big(v(\lambda_0),\frac{\partial v}{\partial\lambda}(\lambda_0)\big)<0. \tag{4.1}$$

Assuming the claim, we finish the proof as follows: Since for each nonzero $\mathrm{v}\in\Phi_s^{\lambda_0}(Y_{s,\lambda_0})\cap(X\times X)$ the crossing form $Q_{\mathcal{M}}$ satisfies

$$\begin{aligned}Q_{\mathcal{M}}(\mathrm{v},\mathrm{v})&=\frac{d}{d\lambda}\Big|_{\lambda=\lambda_0}\omega(\mathrm{v},\phi(\lambda)\mathrm{v})=\frac{d}{d\lambda}\Big|_{\lambda=\lambda_0}\omega(\mathrm{v},\mathrm{v}+\phi(\lambda)\mathrm{v})\\&=\omega\big(v(\lambda_0),\frac{\partial v}{\partial\lambda}(\lambda_0)\big)<0,\end{aligned}$$

the form is negative definite. Thus, the crossing $\lambda_0\in(0,\lambda_\infty)$ is negative. In particular, taking into account that the path $\Gamma_3=\big\{\Phi_L^\lambda(Y_{L,\lambda})\big\}_{\lambda=\lambda_\infty}^0$ is parametrized by the parameter λ *decaying* from λ_∞ to 0, each crossing λ_0 along Γ_3 is positive. Thus, the Maslov index A_3 of the path Γ_3 is equal to B_3. Taking into account the parametrization of $\Gamma_1=\big\{\Phi_{s_0}^\lambda\big\}_{\lambda=0}^{\lambda_\infty}$, a similar argument yields $A_1=-B_1$.

Starting the proof of claim (4.1), for the solution $\mathbf{p}=\mathbf{p}(x,\lambda)$ we compute the λ-derivative (for brevity, denoted below by dot) in equation (2.3), and obtain the equation

$$\dot{\mathbf{p}}'(x) = A(x,\lambda)\dot{\mathbf{p}}(x) + (\sigma_0 \otimes I_{2n})\mathbf{p}(x); \tag{4.2}$$

here and below we abbreviate $\sigma_0 = \begin{pmatrix} 0 & 0 \\ 1 & 0 \end{pmatrix}$ and recall notations J_n and $J^{(n)}$ in (3.7) and formula (3.9). Computing the scalar product in $\mathbb{R}^{4n}$ of both parts of (4.2) with $J_n\mathbf{p}$, integrating from $-s$ to s, and using the identities

$$\int_{-s}^{s} \langle \dot{\mathbf{p}}'(x),\, J_n\mathbf{p}(x)\rangle_{\mathbb{R}^{4n}}\, dx = \langle \dot{\mathbf{p}},\, J_n\mathbf{p}\rangle_{\mathbb{R}^{4n}}\Big|_{-s}^{s} - \int_{-s}^{s} \langle \dot{\mathbf{p}}(x),\, J_n\mathbf{p}'(x)\rangle_{\mathbb{R}^{4n}}\, dx$$
(integration by parts),

$$\int_{-s}^{s} \langle A(x,\lambda)\dot{\mathbf{p}}(x),\, J_n\mathbf{p}(x)\rangle_{\mathbb{R}^{4n}}\, dx = -\int_{-s}^{s} \langle \dot{\mathbf{p}}(x),\, J_nA(x,\lambda)\mathbf{p}(x)\rangle_{\mathbb{R}^{4n}}\, dx$$
(formulas $J_n^{\top} = -J_n$ and (3.8)),

$$\begin{aligned}\int_{-s}^{s} \langle (\sigma_0 \otimes I_{2n})\mathbf{p}(x),\, J_n\mathbf{p}(x)\rangle_{\mathbb{R}^{4n}}\, dx &= -\int_{-s}^{s} \langle ((J\sigma_0)\otimes I_{2n})\mathbf{p}(x),\, \mathbf{p}(x)\rangle_{\mathbb{R}^{4n}}\, dx \\ &= -\int_{-s}^{s} \|p(x)\|^2_{\mathbb{R}^{2n}}\, dx \qquad (\text{because } \mathbf{p} = (p,q)^{\top}),\end{aligned}$$

and $J_n\mathbf{p}' = JA(x,\lambda)\mathbf{p}$, we arrive at the equality

$$\begin{aligned}\langle \mathbf{p}(-s),\, J_n\dot{\mathbf{p}}(-s)\rangle_{\mathbb{R}^{4n}} + \langle \mathbf{p}(s),\, (-J_n)\dot{\mathbf{p}}(s)\rangle_{\mathbb{R}^{4n}} &= \langle \dot{\mathbf{p}},\, J_n\mathbf{p}\rangle_{\mathbb{R}^{4n}}\Big|_{-s}^{s} \\ &= -\int_{-s}^{s} \|p(x)\|^2_{\mathbb{R}^{2n}}\, dx.\end{aligned} \tag{4.3}$$

A similar argument for $\mathbf{w} = \mathbf{w}(x,\lambda)$ yields

$$\langle \mathbf{w}(-s),\, J^{(n)}\dot{\mathbf{w}}(-s)\rangle_{\mathbb{R}^{4n}} + \langle \mathbf{w}(s),\, (-J^{(n)})\dot{\mathbf{w}}(s)\rangle_{\mathbb{R}^{4n}} = 0. \tag{4.4}$$

Combining (4.3), (4.4) with (3.9) and

$$v(\lambda_0) = \Phi_s^{\lambda_0}\big((\mathbf{p}(\cdot\,,\lambda_0), \mathbf{w}(\cdot\,,\lambda_0))^{\top}\big) = \big(\mathbf{p}(-s,\lambda_0), \mathbf{w}(-s,\lambda_0), \mathbf{p}(s,\lambda_0), \mathbf{w}(s,\lambda_0)\big)^{\top}$$

we infer

$$\begin{aligned}\omega\big(v(\lambda_0),\, \dot{v}(\lambda_0)\big) &= \big\langle v(\lambda_0),\, \Omega\dot{v}(\lambda_0)\big\rangle_{\mathbb{R}^{4n}} \\ &= \langle \mathbf{p}(-s),\, J_n\dot{\mathbf{p}}(-s)\rangle_{\mathbb{R}^{4n}} + \langle \mathbf{p}(s),\, (-J_n)\dot{\mathbf{p}}(s)\rangle_{\mathbb{R}^{4n}} \\ &\quad + \langle \mathbf{w}(-s),\, J^{(n)}\dot{\mathbf{w}}(-s)\rangle_{\mathbb{R}^{4n}} + \langle \mathbf{w}(s),\, (-J^{(n)})\dot{\mathbf{w}}(s)\rangle_{\mathbb{R}^{4n}} \\ &= -\int_{-s}^{s} \|p(x,\lambda_0)\|^2_{\mathbb{R}^{2n}}\, dx < 0,\end{aligned}$$

thus completing the proof of (4.1) and the lemma. □

We will now establish monotonicity of the Maslov index with respect to the parameter s. The strategy of the proof of the next lemma is similar to the proof of Lemma 4.1. In the lemma we formulate a simple sufficient condition for the crossing form to be sign-definite; however, in the course of its proof we give a general formula (4.9). We recall that the curve Γ_4 is parametrized by the parameter s *decaying* from L to s_0.

LEMMA 4.2. *For any $\theta \in [0, 2\pi]$, any fixed $\lambda \in (0, \lambda_\infty)$, and any $s_0 \in (0, L)$, each crossing $s_* \in (s_0, L)$ of the path $\{\Phi_s^\lambda(Y_{s,\lambda})\}_{s=s_*-\varepsilon}^{s_*+\varepsilon}$, with $\varepsilon > 0$ small enough, is positive provided the potential V is continuous at the points $\pm s_*$ and the matrix*

$$\frac{1}{2}\big(V(-s_*) + V(s_*)\big) - \lambda I_n \quad \text{is positive definite.} \tag{4.5}$$

In particular, $B_4 = -A_4$ provided V is continuous and positive definite at each point of $[-L, L]$, and $0 \notin \operatorname{Spec}(H_\theta)$, $0 \notin \operatorname{Spec}(H_{\theta,s_0})$.

PROOF. Let $s_* \in (s_0, L)$ be a crossing, so that $\Phi_{s_*}^\lambda(Y_{s_*,\lambda}) \cap (X \times X) \neq \{0\}$. Let $V^\perp$ be a subspace in $\mathbb{R}^{16n}$ transversal to $\Phi_{s_*}^\lambda(Y_{s_*,\lambda})$. Then $V^\perp$ is transversal to $\Phi_s^\lambda(Y_{s,\lambda})$ for all $s \in [s_* - \varepsilon, s_* + \varepsilon]$ for $\varepsilon > 0$ small enough. Thus, there exists a smooth family of matrices, $\phi(s)$, for $s \in [s_* - \varepsilon, s_* + \varepsilon]$, viewed as operators $\phi(s) : \Phi_{s_*}^\lambda(Y_{s_*,\lambda}) \to V^\perp$, such that $\Phi_s^\lambda(Y_{s,\lambda})$ is the graph of $\phi(s)$. Fix any nonzero $\mathrm{v} \in \Phi_{s_*}^\lambda(Y_{s_*,\lambda}) \cap (X \times X)$ and consider the curve $v(s) = \mathrm{v} + \phi(s)\mathrm{v} \in \Phi_s^\lambda(Y_{s,\lambda})$ for $s \in [s_* - \varepsilon, s_* + \varepsilon]$ with $v(s_*) = \mathrm{v}$. By the definition of $Y_{s,\lambda}$, there is a family of solutions $(\mathbf{p}(\cdot\,, s), \mathbf{w}(\cdot\,, s))^\top$ of (2.16) such that $v(s) = \Phi_s^\lambda\big((\mathbf{p}(\cdot\,, s), \mathbf{w}(\cdot\,, s))^\top\big)$. Denoting by dot the derivative with respect to the variable s, we claim that

$$\omega\big(v(s_*),\, \dot{v}(s_*)\big) > 0 \tag{4.6}$$

provided (4.5) holds. Assuming the claim, we finish the proof as follows: Since for each nonzero $\mathrm{v} \in \Phi_{s_*}^\lambda(Y_{s_*,\lambda}) \cap (X \times X)$ the crossing form $Q_{\mathcal{M}}$ satisfies

$$\begin{aligned} Q_{\mathcal{M}}(\mathrm{v}, \mathrm{v}) &= \frac{d}{ds}\Big|_{s=s_*} \omega(\mathrm{v}, \phi(\lambda)\mathrm{v}) = \frac{d}{ds}\Big|_{s=s_*} \omega(\mathrm{v}, \mathrm{v} + \phi(s)\mathrm{v}) \\ &= \omega\big(v(s_*),\, \dot{v}(s_*)\big) > 0, \end{aligned}$$

the form is positive definite. Thus, the crossing $s_* \in (s_0, L)$ is positive. In particular, taking into account that the path $\Gamma_4 = \{\Phi_s^\lambda(Y_{s,\lambda})\}_{s=L}^{s_0}$ is parametrized by the parameter s *decaying* from L to s_0, each crossing along Γ_4 is negative since the assumptions $0 \notin \operatorname{Spec}(H_\theta)$, $0 \notin \operatorname{Spec}(H_{\theta,s_0})$ and Proposition 3.7 imply that all crossings for $\lambda = 0$ belong to (s_0, L). Thus, the Maslov index A_4 of the path Γ_4 is equal to $-B_4$.

Starting the proof of claim (4.6), we remark that s-derivatives of the solutions $\mathbf{p}(\cdot, s)$ and $\mathbf{w}(\cdot, s)$ of (2.3) and (2.14) satisfy the differential equations

$$\dot{\mathbf{p}}'(x) = A(x, \lambda)\dot{\mathbf{p}}(x), \;\; \dot{\mathbf{w}}'(x) = \dot{B}(s, \theta)\mathbf{w}(x) + B(s, \theta)\dot{\mathbf{w}}(x), \tag{4.7}$$

where $\dot{B}(s, \theta)$ is computed similarly to (2.14), (2.12) but with $\frac{\mp\theta}{2s}$ replaced by $\frac{\pm\theta}{2s^2}$:

$$B(s) = -\frac{\theta}{2s}\big(I_{2n} \otimes J\big), \quad \dot{B}(s) = \frac{\theta}{2s^2}\big(I_{2n} \otimes J\big). \tag{4.8}$$

Clearly, $v(s) = \big(\mathbf{p}(-s, s), \mathbf{w}(-s, s), \mathbf{p}(s, s), \mathbf{w}(s, s)\big)^\top$ yields

$$\begin{aligned} \dot{v}(s) = \big(&-\mathbf{p}'(-s, s) + \dot{\mathbf{p}}(-s, s), -\mathbf{w}'(-s, s) + \dot{\mathbf{w}}(-s, s), \\ &\mathbf{p}'(s, s) + \dot{\mathbf{p}}(s, s), \mathbf{w}'(s, s) + \dot{\mathbf{w}}(s, s)\big)^\top. \end{aligned}$$

Using (3.6), we split the expression for $\omega\big(v(s),\, \dot{v}(s)\big)$ as follows:

$$\begin{aligned} &\langle v(s), \Omega\dot{v}(s)\rangle_{\mathbb{R}^{16n}} \\ &\quad = -\langle \mathbf{p}(-s, s),\, (J \otimes I_{2n})\mathbf{p}'(-s, s)\rangle_{\mathbb{R}^{4n}} + \langle \mathbf{p}(-s, s),\, (J \otimes I_{2n})\dot{\mathbf{p}}(-s, s)\rangle_{\mathbb{R}^{4n}} \\ &\qquad + \langle \mathbf{w}(-s, s),\, (I_{2n} \otimes J)\mathbf{w}'(-s, s)\rangle_{\mathbb{R}^{4n}} - \langle \mathbf{w}(-s, s),\, (I_{2n} \otimes J)\dot{\mathbf{w}}(-s, s)\rangle_{\mathbb{R}^{4n}} \end{aligned}$$

$$
\begin{aligned}
&- \langle \mathbf{p}(s,s), (J\otimes I_{2n})\mathbf{p}'(s,s)\rangle_{\mathbb{R}^{4n}} - \langle \mathbf{p}(s,s), (J\otimes I_{2n})\dot{\mathbf{p}}(s,s)\rangle_{\mathbb{R}^{4n}}\\
&+ \langle \mathbf{w}(s,s), (I_{2n}\otimes J)\mathbf{w}'(s,s)\rangle_{\mathbb{R}^{4n}} + \langle \mathbf{w}(s,s), (I_{2n}\otimes J)\dot{\mathbf{w}}(s,s)\rangle_{\mathbb{R}^{4n}}\\
=&\alpha_1+\alpha_2+\alpha_3+\alpha_4,
\end{aligned}
$$

where, using (4.7) and rearranging terms, the expressions α_j are defined and computed as follows:

$$
\begin{aligned}
\alpha_1 &= -\langle \mathbf{p}(-s,s), (J\otimes I_{2n})A(-s,\lambda)\mathbf{p}(-s,s)\rangle_{\mathbb{R}^{4n}}\\
&\qquad\qquad - \langle \mathbf{p}(s,s), (J\otimes I_{2n})A(s,\lambda)\mathbf{p}(s,s)\rangle_{\mathbb{R}^{4n}};\\
\alpha_2 &= \langle \mathbf{p}(-s,s), (J\otimes I_{2n})\dot{\mathbf{p}}(-s,s)\rangle_{\mathbb{R}^{4n}} - \langle \mathbf{p}(s,s), (J\otimes I_{2n})\dot{\mathbf{p}}(s,s)\rangle_{\mathbb{R}^{4n}}\\
&= -\int_{-s}^{s}\frac{d}{dx}\big(\langle \mathbf{p}(x,s), (J\otimes I_{2n})\dot{\mathbf{p}}(x,s)\rangle_{\mathbb{R}^{4n}}\big)\,dx\\
&= -\int_{-s}^{s}\big(\langle A(x,\lambda)\mathbf{p}(x,s), (J\otimes I_{2n})\dot{\mathbf{p}}(x,s)\rangle_{\mathbb{R}^{4n}}\\
&\qquad\qquad + \langle \mathbf{p}(x,s), (J\otimes I_{2n})A(x,\lambda)\dot{\mathbf{p}}(x,s)\rangle_{\mathbb{R}^{4n}}\big)\,dx \quad \text{(using (4.7))}\\
&= -\int_{-s}^{s}\big(-\langle (J\otimes I_{2n})A(x,\lambda)\mathbf{p}(x,s), \dot{\mathbf{p}}(x,s)\rangle_{\mathbb{R}^{4n}}\\
&\qquad\qquad + \langle \mathbf{p}(x,s), (J\otimes I_{2n})A(x,\lambda)\dot{\mathbf{p}}(x,s)\rangle_{\mathbb{R}^{4n}}\big)\,dx\\
&= 0 \qquad \text{(using (3.8))};\\
\alpha_3 &= \langle \mathbf{w}(-s,s), (I_{2n}\otimes J)\mathbf{w}'(-s,s)\rangle_{\mathbb{R}^{4n}} + \langle \mathbf{w}(s,s), (I_{2n}\otimes J)\mathbf{w}'(s,s)\rangle_{\mathbb{R}^{4n}}\\
&= \langle \mathbf{w}(-s,s), (I_{2n}\otimes J)B(s,\theta)\mathbf{w}(-s,s)\rangle_{\mathbb{R}^{4n}}\\
&\qquad\qquad + \langle \mathbf{w}(s,s), (I_{2n}\otimes J)B(s,\theta)\mathbf{w}(s,s)\rangle_{\mathbb{R}^{4n}}\\
&= \langle \mathbf{w}(-s,s), -\frac{\theta}{2s}(I_{2n}\otimes J)^2\mathbf{w}(-s,s)\rangle_{\mathbb{R}^{4n}}\\
&\qquad\qquad + \langle \mathbf{w}(s,s), -\frac{\theta}{2s}(I_{2n}\otimes J)^2\mathbf{w}(s,s)\rangle_{\mathbb{R}^{4n}} \quad \text{(using (4.8))}\\
&= \frac{\theta}{2s}\big(\|\mathbf{w}(-s,s)\|^2_{\mathbb{R}^{4n}} + \|\mathbf{w}(s,s)\|^2_{\mathbb{R}^{4n}}\big)\\
&= \frac{\theta}{s}\|\mathbf{w}(-s,s)\|^2_{\mathbb{R}^{4n}} \qquad \text{(since } \Psi_B(x,\theta) \text{ is orthogonal)};\\
\alpha_4 &= -\langle \mathbf{w}(-s,s), (I_{2n}\otimes J)\dot{\mathbf{w}}(-s,s)\rangle_{\mathbb{R}^{4n}} + \langle \mathbf{w}(s,s), (I_{2n}\otimes J)\dot{\mathbf{w}}(s,s)\rangle_{\mathbb{R}^{4n}}\\
&= \int_{-s}^{s}\frac{d}{dx}\big(\langle \mathbf{w}(x,s), (I_{2n}\otimes J)\dot{\mathbf{w}}(x,s)\rangle_{\mathbb{R}^{4n}}\big)\,dx\\
&= \int_{-s}^{s}\big(\langle B(s,\theta)\mathbf{w}(x,s), (I_{2n}\otimes J)\dot{\mathbf{w}}(x,s)\rangle_{\mathbb{R}^{4n}}\\
&\quad + \langle \mathbf{w}(x,s), (I_{2n}\otimes J)\big(\dot{B}(s,\theta)\mathbf{w}(x,s) + B(s,\theta)\dot{\mathbf{w}}(x,s)\big)\rangle_{\mathbb{R}^{4n}}\big)\,dx \text{ (by (4.7))}\\
&= \int_{-s}^{s}\langle \mathbf{w}(x,s), (I_{2n}\otimes J)\dot{B}(s,\theta)\mathbf{w}(x,s)\rangle dx \quad \text{(distributing and using (3.8))}\\
&= -\frac{\theta}{2s^2}\int_{-s}^{s}\|\mathbf{w}(x,s)\|^2_{\mathbb{R}^{4n}}\,dx \quad \text{(using (4.8))}\\
&= -\frac{\theta}{2s^2}\int_{-s}^{s}\|\mathbf{w}(-s,s)\|^2_{\mathbb{R}^{4n}}\,dx = -\frac{\theta}{s}\|\mathbf{w}(-s,s)\|^2_{\mathbb{R}^{4n}}
\end{aligned}
$$

(since $\Psi_B(x,\theta)$ is orthogonal). Thus, $\langle v(s),\, \Omega\dot{v}(s)\rangle_{\mathbb{R}^{16n}} = \alpha_1$. After a short calculation using the condition $\mathbf{p}(s_*,s_*) = \big(I_{2n}\otimes U(\theta)\big)\mathbf{p}(-s_*,s_*)$ (which holds since s_* is a conjugation point), the orthogonality of $U(\theta)$, and formulas

$$\begin{aligned}
&\big(J\otimes I_{2n}\big)A(\pm s_*,\lambda) = \big((\lambda I_n - V(\pm s_*))\otimes I_2\big)\oplus\big(-I_{2n}\big),\\
&I_{2n}\otimes U(\theta)^{\pm 1} = \big(I_n\otimes U(\theta)^{\pm 1}\big)\oplus\big(I_n\otimes U(\theta)^{\pm 1}\big),\\
&\big(I_{2n}\otimes U(\theta)^{-1}\big)\big(J\otimes I_{2n}\big)A(s_*,\lambda)\big(I_{2n}\otimes U(\theta)\big)\\
&\qquad\qquad = \big((\lambda I_n - V(s_*))\otimes I_2\big)\oplus\big(-I_{2n}\big),
\end{aligned}$$

we conclude that $\omega\big(v(s_*),\,\dot{v}(s_*)\big) = \alpha_1\Big|_{s=s_*}$ is equal to

$$-\big\langle \mathbf{p}(-s_*,s_*),\, \Big(\big(2\lambda I_{2n} - (V(-s_*)+V(s_*))\otimes I_2\big)\oplus\big(-2I_{2n}\big)\Big)\mathbf{p}(-s_*,s_*)\big\rangle_{\mathbb{R}^{4n}}.$$

Since $\mathbf{p}(-s_*,s_*) = \big(p(-s_*,s_*), q(-s_*,s_*)\big)^{\top}$, we therefore have the following final formula for the crossing form:

$$\begin{aligned}
\omega\big(v(s_*),\,\dot{v}(s_*)\big) &= 2\big\langle p(-s_*,s_*),\, \big(\tfrac{1}{2}\big(V(-s_*)+V(s_*)\big)\otimes I_2 - \lambda I_{2n}\big)p(-s_*,s_*)\big\rangle_{\mathbb{R}^{2n}}\\
&\quad + 2\|q(-s_*,s_*)\|^2_{\mathbb{R}^{2n}}.
\end{aligned}\tag{4.9}$$

In particular, (4.5) implies (4.6). ☐

We will prove next a version of Lemma 3.12 (ii) for $\theta = 0$ or $\theta = 2\pi$. It is interesting to note that although the conclusion of the next lemma concerns the spectrum of the operators $H_{0,s}$, its proof uses topological arguments which led to Corollary 3.9. We recall the notation $\mathrm{Mor}(\mathcal{H}) = \dim(\operatorname{ran}\mathcal{P})$ for the Morse index of an invertible selfadjoint semi-bounded from above operator $\mathcal{H}$; here,

$$\mathcal{P} = (2\pi i)^{-1}\int_\gamma (z-\mathcal{H})^{-1}\,dz \tag{4.10}$$

is the Riesz projection corresponding to the positive part $\mathrm{Spec}(\mathcal{H})\cap(0,+\infty)$ of the spectrum of $\mathcal{H}$, and γ is a smooth curve enclosing this part of the spectrum.

LEMMA 4.3. *Assume that $\theta = 0$ or $\theta = 2\pi$ and that the potential V is continuous at $x = 0$ and the matrix $V(0)$ is invertible. If $\lambda_\infty > \|V\|_\infty$ and $s_0 \in (0,L]$ is sufficiently small then $0\notin \mathrm{Spec}(H_{0,s_0})$ and $B_1 = \mathrm{Mor}(V(0))$; in particular, if $V(0)$ is negative definite then $B_1 = 0$.*

PROOF. Since $H_{0,s} = H_{2\pi,s}$ because the boundary conditions (1.3) are the same for $\theta = 0$ and $\theta = 2\pi$, and taking into account Proposition 3.7, we will consider only the case $\theta = 0$. If $\theta = 0$ and $s > 0$ then $H_{0,s}$ is the operator in $L^2([-s,s])$ defined by $(H_{0,s}y)(x) = y''(x) + V(x)y(x)$, $|x|\le s$, with the domain

$$\begin{aligned}
\mathrm{dom}(H_{0,s}) = \big\{y\in L^2([-s,s])\,\big|\, &y, y'\in AC_{\mathrm{loc}}([-s,s]), y''\in L^2([-s,s]) \text{ and the}\\
&\text{periodic boundary conditions } y(s)=y(-s),\ y'(s)=y'(-s) \text{ hold}\big\}.
\end{aligned}$$

It is convenient to "rescale" the operator $H_{0,s}$ to $L^2([-L,L])$ by introducing the operator $H_0(s)$ in $L^2([-L,L])$ defined by

$$(H_0(s)y)(x) = \big(L/s\big)^2 y''(x) + V\big(sx/L\big)y(x),\quad |x|\le L,$$

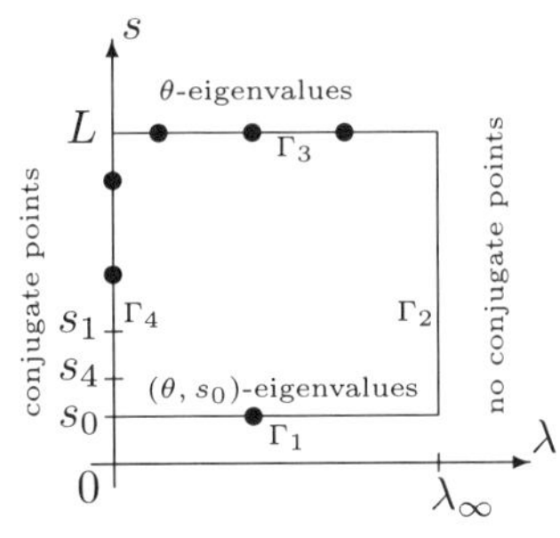

FIGURE 3. $\theta = 0$ and the numbers $s_1 > s_2 > s_3 > s_4 \geq s_0 > 0$ in the proof of Lemma 4.3 are small enough

with the domain

$$\operatorname{dom}(H_0(s)) = \big\{y \in L^2([-L,L]) \big| \, y, y' \in AC_{\rm loc}([-L,L]), y'' \in L^2([-L,L]) \text{ and the periodic boundary conditions } y(L) = y(-L),\ y'(L) = y'(-L) \text{ hold}\big\}.$$

Writing the eigenvalue equation $(H_{0,s}y)(x) = \lambda y(x)$, $|x| \leq s$, at the point $x = s\widehat{x}/L$ for $|\widehat{x}| \leq L$, introducing $z(\widehat{x}) = y(s\widehat{x}/L)$, and passing to the eigenvalue equation $(H_0(s)z)(\widehat{x}) = \lambda z(\widehat{x})$, $|\widehat{x}| \leq L$, we observe that

$$\operatorname{Spec}(H_{0,s}; L^2([-s,s])) = \operatorname{Spec}(H_0(s); L^2([-L,L])) \quad \text{for all } s \in (0,L]. \tag{4.11}$$

In addition to $H_0(s)$, we introduce a constant coefficient operator $H_0^{(0)}(s)$ on $L^2([-L,L])$ defined by $(H_0^{(0)}(s)y)(x) = \big(L/s\big)^2 y''(x) + V(0)y(x)$, $|x| \leq L$, with the domain $\operatorname{dom} H_0^{(0)}(s) = \operatorname{dom} H_0(s)$. Since

$$\|H_0(s) - H_0^{(0)}(s)\|_{\mathcal{B}(L^2([-L,L]))} = \sup_{|x|\leq L} \|V(sx/L) - V(0)\| \to 0 \text{ as } s \to 0 \tag{4.12}$$

by the continuity assumption in the lemma, we can use Theorem 3.11 to conclude that

$$\operatorname{dist}\big(\operatorname{Spec}(H_0(s)), \operatorname{Spec}(H_0^{(0)}(s))\big) \to 0 \text{ as } s \to 0. \tag{4.13}$$

Since the operator $H_0^{(0)}(s)$ is a constant coefficient operator with periodic boundary conditions, passing to the Fourier series $y(x) = \sum_{k\in\mathbb{Z}} y_k e^{i\pi kx/L}$, $|x| \leq L$, we calculate:

$$\operatorname{Spec}(H_0^{(0)}(s)) = \bigcup_{k\in\mathbb{Z}} \Big(- \big(\pi k/s\big)^2 + \operatorname{Spec}(V(0))\Big). \tag{4.14}$$

Let ν_j denote the eigenvalues of the matrix $V(0)$ and let $\varkappa = \operatorname{Mor}(V(0))$ denote the number of the positive eigenvalues counting multiplicities. Since $0 \notin \operatorname{Spec}(V(0))$ by the assumption, we can find a $\delta > 0$, and enumerate the eigenvalues in $\operatorname{Spec}(V(0))$ such that

$$-\|V(0)\| \leq \cdots \leq \nu_{-1} < -\delta < 0 < \delta < \nu_1 \leq \cdots \leq \nu_\varkappa \leq \|V(0)\|.$$

Choose $s_1 \in (0,L)$ so small that $\|V(0)\| + \delta < (\pi/s_1)^2$, see Figure 3. Then, for each $s \in (0, s_1]$, the eigenvalues $\nu_j - (\pi k/s)^2$ of the operator $H_0^{(0)}(s)$ are positioned as follows:

$$0 < \delta < \nu_1 \leq \cdots \leq \nu_\varkappa, \text{ for } j \geq 1 \text{ and } k = 0,$$
$$\ldots \leq \nu_1 - (k\pi/s)^2 \leq \cdots \leq \nu_\varkappa - (k\pi/s)^2 \leq \cdots \leq \nu_1 - (\pi/s)^2 \leq \cdots \leq \nu_\varkappa - (\pi/s)^2$$

$$\le \|V(0)\| - (\pi/s)^2 \le \|V(0)\| - (\pi/s_1)^2 < -\delta < 0, \text{ for } j \ge 1 \text{ and } k \in \mathbb{Z}\setminus\{0\},$$
$$\nu_j - (k\pi/s)^2 < -\delta, \text{ for } j \le -1 \text{ and } k \in \mathbb{Z}.$$

In particular, for each $s \in (0, s_1]$ we have $\mathrm{Spec}(H_0^{(0)}(s)) \cap [-\delta, \delta] = \emptyset$,

$$\mathrm{Spec}(H_0^{(0)}(s)) \cap (0, +\infty) = \{\nu_1, \dots, \nu_\varkappa\} \subset \big(\delta, \|V(0)\|\big), \tag{4.15}$$

and $\mathrm{Mor}(H_0^{(0)}(s)) = \mathrm{Mor}(V(0))$. Using (4.13), (4.15) one can find a sufficiently small $s_2 \in (0, s_1)$ such that for all $s \in (0, s_2]$ one has:

$$0 \notin \mathrm{Spec}(H_0(s)) \text{ and } \mathrm{Spec}(H_0(s)) \cap (0, +\infty) \subset \big(\delta/2, \|V(0)\| + \delta/2\big). \tag{4.16}$$

By (4.11), conclusions (4.16) hold with $\mathrm{Spec}(H_0(s))$ replaced by $\mathrm{Spec}(H_{0,s})$.

Fix any $s_3 \in (0, s_2)$. We now claim that

$$\sup_{s\in(0,s_3]} \mathrm{Mor}(H_0(s)) < \infty. \tag{4.17}$$

Postponing the proof of claim (4.17), we proceed with the proof of the lemma.

Let P_s, respectively, $P_s^{(0)}$ denote the Riesz projection for the operator $H_0(s)$, respectively, $H_0^{(0)}(s)$ that corresponds to the positive part of its spectrum. The Riesz projection is defined by formula (4.10) with $\mathcal{H}$ replaced by $H_0(s)$, respectively, $H_0^{(0)}(s)$, and with γ being the rectangle in the complex plane with the vertices at the points $\pm i\delta$ and $\|V(0)\| + \delta \pm i\delta$. Inclusions (4.15), (4.16) imply, for $s \in (0, s_2]$,

$$\mathrm{dist}\big(\mathrm{Spec}(H_0^{(0)}(s)), \gamma\big) \ge \delta, \ \mathrm{dist}\big(\mathrm{Spec}(H_0(s)), \gamma\big) \ge \delta/2. \tag{4.18}$$

Using (4.18) and that $H_0(s)$ is selfadjoint, for $z \in \gamma$ we arrive at the estimate

$$\|\big(z - H_0(s)\big)^{-1}\| = 1/\,\mathrm{dist}(\mathrm{Spec}(H_0(s)), z) \le 1/\,\mathrm{dist}(\mathrm{Spec}(H_0(s)), \gamma) \le 2/\delta,$$

and a similar estimate for $\|\big(z - H_0(s)\big)^{-1}\|$. Using (4.12) and

$$P_s - P_s^{(0)} = (2\pi i)^{-1} \int_\gamma \Big(\big(z - H_0(s)\big)^{-1}\big(H_0(s) - H_0^{(0)}(s)\big)\big(z - H_0^{(0)}(s)\big)^{-1}\Big)\, dz,$$

we conclude that

$$\|P_s - P_s^{(0)}\|_{\mathcal{B}(L^2([-L,L]))} \to 0 \text{ as } s \to 0. \tag{4.19}$$

Noting that $\mathrm{Mor}(H_0^{(0)}(s)) = \dim(\mathrm{ran} P_s^{(0)}) = \mathrm{tr}(P_s^{(0)}) = \mathrm{Mor}(V(0))$ by (4.15) and that the dimensions $\mathrm{Mor}(H_0(s)) = \dim(\mathrm{ran} P_s) = \mathrm{tr}(P_s)$ are bounded uniformly for $s \in (0, s_3]$ by (4.17), we introduce the subspace $R_s = \mathrm{ran} P_s^{(0)} \oplus \mathrm{ran} P_s$ and denote $R = \sup_{s\in(0,s_3]} \dim R_s < \infty$. Viewing $P_s - P_s^{(0)}$ as a finite dimensional operator in R_s, we infer, using a simple inequality relating trace and norm:

$$\begin{aligned} \big|\mathrm{tr}(P_s) - \mathrm{tr}(P_s^{(0)})\big| &= \big|\mathrm{tr}(P_s - P_s^{(0)})\big| \\ &\le R\|P_s - P_s^{(0)}\|_{\mathcal{B}(R_s)} \le R\|P_s - P_s^{(0)}\|_{\mathcal{B}(L^2([-L,L]))}. \end{aligned} \tag{4.20}$$

We recall the $0 \notin \mathrm{Spec}(H_{0,s})$ and $\mathrm{Mor}(H_{0,s}) = \mathrm{Mor}(H_0(s))$ by (4.16) and (4.11). Thus, using Proposition 3.7, for any $s_0 \in (0, s_3]$, the number $B_1 = \mathrm{Mor}(H_{0,s_0})$ of the crossings along Γ_1 is equal to the Morse index $\mathrm{Mor}(H_0(s_0)) = \mathrm{tr}(P_{s_0})$. In order to establish the required in the lemma equality $B_1 = \mathrm{Mor}(V(0))$, it suffices to show that $\mathrm{tr}(P_{s_0}) = \mathrm{tr}(P_{s_0}^{(0)})$ for all small enough s_0. Indeed, the latter equality implies

$$B_1 = \mathrm{Mor}(H_{0,s_0}) = \mathrm{Mor}(H_0(s_0)) = \mathrm{tr}(P_{s_0}) = \mathrm{tr}(P_{s_0}^{(0)}) = \mathrm{Mor}(V(0)), \tag{4.21}$$

as needed in the lemma. Since the functions $s \mapsto \mathrm{tr}(P_s)$, $s \mapsto \mathrm{tr}(P_s^{(0)})$ take integer values, it suffices to show the existence of a small $s_4 \in (0, s_3)$ such that the right-hand side of (4.20) is smaller than 1 for all $s \in (0, s_4]$. But this follows from (4.19), thus concluding the proof of the lemma.

It remains to prove claim (4.17). This is the part of the proof based on Corollary 3.9. Since $0 \notin \mathrm{Spec}(H_{0,s})$ and $\mathrm{Mor}(H_{0,s}) = \mathrm{Mor}(H_0(s))$ for all $s \in (0, s_2]$ by (4.16) and (4.11), in order to show (4.17) it suffices to check that the number of crossings $B_1 = \mathrm{Mor}(H_{0,s_0})$ along the curve $\Gamma_1 = \Gamma_{1,(0,s_0)}$ is estimated from above by a finite number that does not depend on $s_0 \in (0, s_3]$ (we recall that $s_3 \in (0, s_2)$). Take any $s_0 \in (0, s_3]$ and construct the curve $\Gamma = \Gamma_{1,(0,s_0)} \cup \Gamma_2 \cup \Gamma_3 \cup \Gamma_{4,(0,s_0)}$ as described in Remark 3.10. First, we remark that due to Proposition 3.7 there are no crossings of the portion of the curve $\Gamma_{4,(0,s_0)}$ given by $\{\Phi_s^0(Y_{s,0}))|s \in [s_0, s_2]\}$ since $0 \notin \mathrm{Spec}(H_{0,s})$ for all $s \in [s_0, s_2]$. Second, we remark that with no loss of generality we may assume that the curve $\Gamma_{4,(0,s_0)} \cup \Gamma_3$ is regular. (Indeed, otherwise, use homotopy with the fixed endpoints $\Phi_{s_2}^0(Y_{s_2,0})$ and $\Phi_L^{\lambda_\infty}(Y_{L,\lambda_\infty})$ of the portion of this curve given by $\{\Phi_s^0(Y_{s,0}))|s \in [s_2, L]\} \cup \Gamma_3$ to transform it into a regular curve.) Since the regular crossings are isolated, based on the two remarks just made we conclude that the number B_4 of the crossings of $\Gamma_{4,(0,s_0)}$ is finite and does not depend on s_0. Clearly, B_3 is finite and does not depend on s_0 either. By Corollary 3.9 we know that $0 = A_1 + A_2 + A_3 + A_4$. By Lemma 3.12 (i) we have $A_2 = 0$. By Lemma 4.1 we know that $A_3 = B_3$ and $A_1 = -B_1$. Combining all this, we have the required estimate

$$\mathrm{Mor}(H_{0,s_0}) = B_1 = -A_1 = A_3 + A_4 = B_3 + A_4 \leq B_3 + |A_4| \leq B_3 + B_4,$$

which concludes the proof of claim (4.17) and the lemma. □

We will now summarize the count of eigenvalues and conjugate points via the Morse and Maslov indices. Recall that the number of positive eigenvalues of a self-adjoint operator (counting their multiplicities) is called its *Morse index*, and is denoted by $\mathrm{Mor}(\cdot)$. Also, recall definition (3.2) of the Maslov index. The Maslov index of $\Gamma_4 = \Gamma_{4,(\theta,s_0)}$ is the number $A_4 = A_{4,(\theta,s_0)}$ which can also be thought of as the *Maslov index* of equation (1.1). Also, we recall definition (3.15) of the numbers B_i, and note that the expressions for A_i and B_i do not contain the first and the last terms provided the endpoints of Γ_i are *not* crossings. In this case, we can interpret B_3 and B_4 in terms of the eigenvalues of H_θ and the conjugate points.

THEOREM 4.4. *Let us fix $\theta \in [0, 2\pi]$, and let the numbers $A_i = A_{i,(\theta,s_0)}$ and $B_i = B_{i,(\theta,s_0)}$ be defined in (3.14) and (3.15) for a (small) $s_0 > 0$ and a (large) $\lambda_\infty > 0$. Then the following assertions hold.*

(i) *The Maslov index of the curve Γ is zero for any $s_0 \in (0, L)$ and $\lambda_\infty > 0$.*
(ii) *If $0 \notin \mathrm{Spec}(H_\theta)$ then the Maslov index A_3 of the curve Γ_3 satisfies $A_3 = B_3$.*
(iii) *If $0 \notin \mathrm{Spec}(H_\theta)$ and λ_∞ is large enough then B_3 is equal to the number of the positive λ for which there exists a solution to the original boundary value problem for equation (2.2) on $[-L, L]$ subject to the boundary condition (2.4) counting multiplicities, that is, to the* Morse index *of the operator H_θ in $L^2([-L, L])$:*

$$B_3 = \mathrm{Mor}\left(H_\theta\right).$$

(iv) *If* $0 \notin \operatorname{Spec}(H_\theta)$ *and* $0 \notin \operatorname{Spec}(H_{\theta,s_0})$ *for some* $s_0 > 0$ *then* B_4 *is equal to the number of the conjugate points for* $\lambda = 0$ *counting multiplicities, that is, the number of such* $s \in (s_0, L)$ *for which there exists a nontrivial solution to the boundary value problem for equation (2.2) on* $[-s, s]$ *subject to the boundary condition* (3.10).

(v) *If* $\theta \in (0, 2\pi)$, $\lambda_\infty > 0$ *is large enough and* $s_0 > 0$ *is small enough then* $A_3 = -A_4$. *If, in addition,* $0 \notin \operatorname{Spec}(H_\theta)$ *then the Maslov index and the Morse index are related as follows:*

$$\operatorname{Mor}(H_\theta) = -\operatorname{Mas}(\Gamma_4, X \times X). \tag{4.22}$$

(vi) *If the potential* V *is continuous and positive definite on* $[-L, L]$, *and the assumptions in (iv) hold then* $A_4 = -B_4$. *If, in addition, the assumptions in (v) hold then the Morse index can be computed as follows:*

$$\operatorname{Mor}(H_\theta) = B_4. \tag{4.23}$$

(vii) *If* $\theta = 0$ *or* $\theta = 2\pi$, $\lambda_\infty > 0$ *is large enough and* $s_0 > 0$ *is small enough, the potential* V *is continuous at the point* $x = 0$, *and* $0 \notin \operatorname{Spec}(V(0))$ *then* $A_1 = -B_1 = \operatorname{Mor}(V(0))$. *If, in addition,* $0 \notin \operatorname{Spec}(H_\theta)$ *then*

$$\operatorname{Mor}(H_\theta) = -\operatorname{Mas}(\Gamma_4, X \times X) + \operatorname{Mor}(V(0)). \tag{4.24}$$

Finally, if 0 *is not in the spectrum of the operator* H *in* $L^2(\mathbb{R})$, *then, for all* $\theta \in [0, 2\pi]$, *the Morse index of* H_θ *does not depend on* θ, *and is greater than or is equal to the number of disjoint spectral bands of* H *in the (unstable) right half-line, and is equal to the number of the spectral bands of* H *if they are disjoint.*

PROOF. Assertion (i) is proved in Corollary 3.9. Assertion (ii) follows from Lemma 4.1, while (iii) and (iv) are proved in Proposition 3.7. Assertion (v) follows from (i) and Lemma 3.12 while (vi) follows from Lemma 4.2. Assertion (vii) follows from Lemma 4.3. □

5. The Mathieu equation: an example

Now we will use a well known Mathieu equation, see, e.g. [**BO78**, **JS99**], as an illustration of the phenomena described in Theorem 4.4. This example will also give some indication as to how to handle the loss of regularity of crossings when $\theta = 0$ or $\theta = 2\pi$ and the curve $\Gamma_{(0,0)}$ is constructed as in Corollary 3.13. Specifically, let us consider the Mathieu equation

$$y'' + 3.2\cos(2x)y = \lambda y,\ x \in [-\pi, \pi], \tag{5.1}$$

where we have chosen the usual parameter in the equation to be -1.6 purely for convenience and choose $L = \pi$ (we could of course also choose $L = k\pi/2$, $k \in \mathbb{Z}$).

Letting $\Psi_A^{\mathbb{C}}(x, \lambda)$ be the fundamental solution matrix to the (2×2) first order (complex) system associated with (5.1) such that $\Psi_A^{\mathbb{C}}(-\pi, \lambda) = I_2$, we have then that the propagator for all $s \in [0, \pi]$ is given by $M_A^{\mathbb{C}}(s, \lambda) = \Psi_A^{\mathbb{C}}(s, \lambda)\big(\Psi_A^{\mathbb{C}}(-s, \lambda)\big)^{-1}$. Now for a fixed value of $\theta \in [0, 2\pi]$ we can look for θ-eigenvalues and conjugate points of (5.1). A θ-eigenvalue will occur when $M_A^{\mathbb{C}}(\pi, \lambda)$ has an eigenvalue $e^{i\theta}$, while a conjugate point will be a value of s such that $M_A^{\mathbb{C}}(s, 0)$ has an eigenvalue $e^{i\theta}$. That is, the following two quantities can be computed:

$$B_3 = \{\text{The number of } \lambda \in [0, \infty) \text{ such that } \det\big(M_A^{\mathbb{C}}(\pi, \lambda) - e^{i\theta} I_2\big) = 0\},$$

$$B_4 = \{\text{The number of } s \in [0, \pi] \text{ such that } \det\big(M_A^{\mathbb{C}}(s, 0) - e^{i\theta} I_2\big) = 0\}.$$

The number B_4 here corresponds to the choice $s_0 = 0$. The graphs in Figures 4 and 5 were computed using *Mathematica*'s numerical Mathieu equations, and plot the values of the θ-eigenvalues and the conjugate points versus values of $\theta \in [0, 2\pi]$.

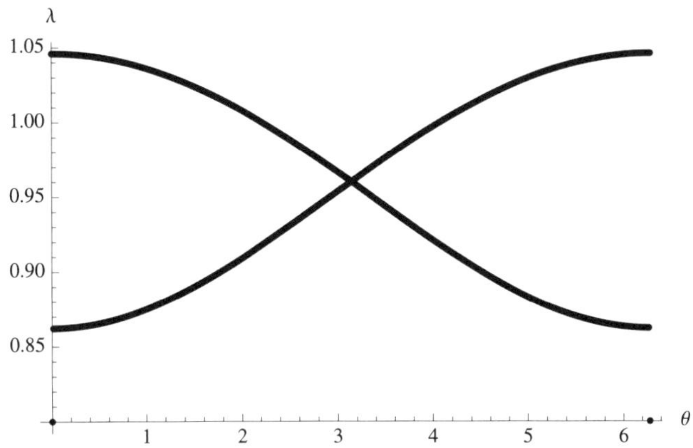

FIGURE 4. A plot of the location of the θ-eigenvalues versus θ in the Mathieu example (where $s = \pi$). It is clear that at $\theta = \pi$ there is a double eigenvalue.

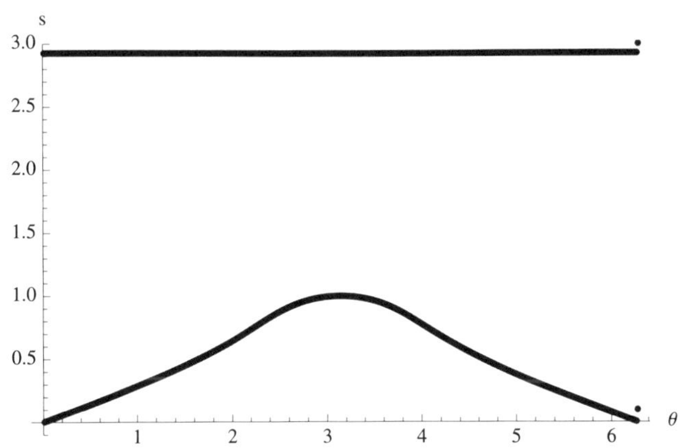

FIGURE 5. A plot of the location of the conjugate points s versus θ in the Mathieu example (where $\lambda = 0$).

The graphs show that in our numerical experiments the quantities B_3 and B_4 are equal for all $\theta \in [0, 2\pi]$. It is worth noting that the multiplicity of the eigenvalue λ when $\theta = \pi$ is two, however this is 'canceled' out by two crossings along Γ_4 - i.e. we have two separate conjugate points, each with multiplicity one, and thus our calculations are in concert with Theorems 3.8 and 4.4.

Theorem 4.4 (ii), (v) tells us that for any $\theta \in (0, 2\pi)$ the number of θ-eigenvalues for $s = \pi$ will be the same as the (signed) count of the number of conjugate points for $\lambda = 0$, that is, that $B_3 = A_3 = -A_4$ as soon as we chose $\lambda_\infty > 0$ large enough and $s_0 > 0$ small enough. We now need to choose a small $s_0 > 0$ as the arguments in Theorem 4.4 (v) involve the curve Γ_1 as defined in Remark 3.10.

We recall that $B_3 = A_3$ due to Lemma 4.1 for any $\theta \in [0, 2\pi]$. Also, for the chosen value 3.2 of the parameter in the Mathieu equation it is known that $\lambda = 0$ is not a θ-eigenvalue for any $\theta \in [0, 2\pi]$, see [**BO78, JS99**].

Lemma 4.2 can be applied for the crossings at the conjugate points $s_* \in (0, \pi)$ such that $3.2 \cos 2s_* > 0$. For any $\theta \in [0, 2\pi]$ the latter inequality certainly holds for the upper conjugate point in Figure 5, and thus the crossing form is positive at this crossing by Lemma 4.2. At the lower conjugate point in Figure 5 the inequality $3.2 \cos 2s_* > 0$ does not hold for θ close to π, and thus one can not use the sufficient condition (4.5) for the crossing form to be positive formulated in Lemma 4.2. However, using the explicit formula for the crossing form in equation (4.9), we computed directly the value of the crossing form at each crossing. As can be seen from Figure 6, the crossing form is positive also for the lower conjugate point. This implies that $B_4 = -A_4$ for the crossings in $(0, \pi)$, and confirms the equality $B_3 = A_3 = B_4 = -A_4$ in yet another way and for all $\theta \in (0, 2\pi)$.

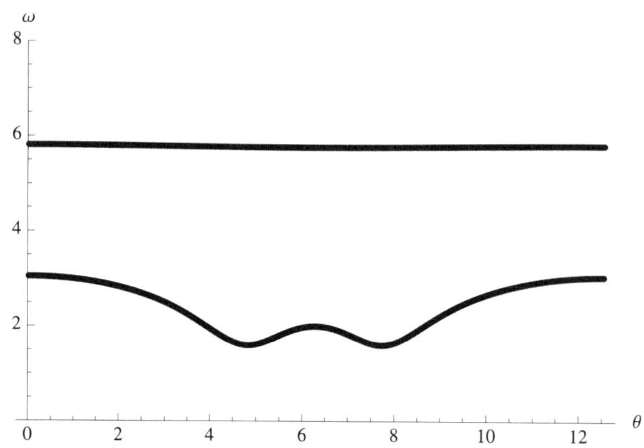

FIGURE 6. A plot of the value $\omega = \omega(v(s_*(\theta)), \dot{v}(s_*(\theta)))$ of the crossing form versus θ at the two conjugate points $s_* = s_*(\theta)$. The crossing form was computed using the right-hand side of equation (4.9), and is positive.

The case as $\theta \to 0$ or $\theta \to 2\pi$ poses more of a problem. As these two possibilities are analogous, we consider, as usual, the case $\theta \to 0$. In this case the hypothesis of Lemma 3.12 (ii) is not satisfied, so we can not expect to have a non-zero lower bound $\lim_{\theta \to 0^+} s_0^{\min}(\theta)$, where $s_0^{\min}(\theta)$ is defined as follows:

$$s_0^{\min}(\theta) := \inf \big\{ s \in (0, L] \big| \text{ for some } \lambda > 0 \text{ on } [-s, s] \text{ there exists a nonzero solution} \\ \text{of the boundary value problem (1.1), (1.3)} \big\}.$$

Indeed, as seen from the plot, $\lim_{\theta \to 0^+} s_0^{\min}(\theta) = 0$ in the Mathieu example (see Figure 5). One can choose, however, a sequence of nonzero θ_ε that converges to zero (see Figure 7). It is worth noting that the θ_ε-eigenvalues stabilize away from zero, even though the lower bound $s_0^{\min}(\theta_\varepsilon)$ tends to zero. This is because we have chosen the parameter q in the Mathieu equation $q = -1.6$ so that 0 was not an eigenvalue for any θ (and in particular for $\theta = 0$). We could have similarly perturbed θ_ε away from zero in the negative direction, and we observe the same behavior. The

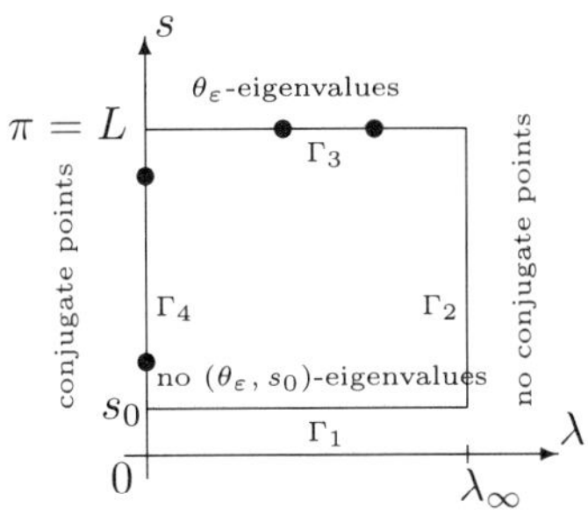

FIGURE 7. We let $\theta_\varepsilon \to 0$ and choose $s_0 = s_0(\theta_\varepsilon) > 0$ sufficiently small. Although $s_0^{\min}(\theta_\varepsilon) \to 0$, for each θ_ε we still have that $\lambda = 0$ is not a θ_ε-eigenvalue, $\lambda = 0$ is not a $(\theta_\varepsilon, s_0)$-eigenvalue, and conclusions of Theorems 3.8 and 4.4 still hold.

numerical calculations can be summarized as follows: For small $\theta_{\pm\varepsilon}$, the values of the θ-eigenvalues are 0.862 and 1.046, each of multiplicity one, while the conjugate points are a small positive number and 2.926, each of multiplicity one.

It is also worth noting that even though the entire boundary curve Γ when $\theta = 0$ has to be defined for $s_0 = 0$ as in Corollary 3.13 because (2.10) is not defined at $s = 0$, the curves Γ_3 and Γ_4 are regular.

Alternatively, if $\theta = 0$, we can pick a small $s_0 > 0$ and define Γ_1 and Γ_4 as described in Remark 3.10. In this case, the curve Γ_4 contains only one conjugate point (the upper conjugate point s_* on the vertical line $\theta = 0$, see Figure 5). Thus, $B_4 = 1$. Since $\cos s_* > 0$ for the upper conjugate point, by Lemma 4.2 we have $A_4 = -B_4 = -1$. Since $\cos 0 > 0$, we have $\mathrm{Mor}(V(0)) = 1$, thus confirming the count in (4.24) since $B_3 = 2$ is the number of θ-eigenvalues when $\theta = 0$.

References

[A01] Alberto Abbondandolo, *Morse theory for Hamiltonian systems*, Chapman & Hall/CRC Research Notes in Mathematics, vol. 425, Chapman & Hall/CRC, Boca Raton, FL, 2001. MR1824111 (2002e:37103)

[Ar67] V. I. Arnol′d, *On a characteristic class entering into conditions of quantization*, Funkcional. Anal. i Priložen. **1** (1967), 1–14 (Russian). MR0211415 (35 #2296)

[Ar85] V. I. Arnol′d, *Sturm theorems and symplectic geometry*, Funktsional. Anal. i Prilozhen. **19** (1985), no. 4, 1–10, 95 (Russian). MR820079 (87j:58033)

[BO78] Carl M. Bender and Steven A. Orszag, *Advanced mathematical methods for scientists and engineers*, McGraw-Hill Book Co., New York, 1978. International Series in Pure and Applied Mathematics. MR538168 (80d:00030)

[B56] Raoul Bott, *On the iteration of closed geodesics and the Sturm intersection theory*, Comm. Pure Appl. Math. **9** (1956), 171–206. MR0090730 (19,859f)

[CDB06] F. Chardard, F. Dias, and T. J. Bridges, *Fast computation of the Maslov index for hyperbolic linear systems with periodic coefficients*, J. Phys. A **39** (2006), no. 47, 14545–14557, DOI 10.1088/0305-4470/39/47/002. MR2277067 (2007k:37081)

[CDB09] Frédéric Chardard, Frédéric Dias, and Thomas J. Bridges, *Computing the Maslov index of solitary waves. I. Hamiltonian systems on a four-dimensional phase space*, Phys. D **238** (2009), no. 18, 1841–1867, DOI 10.1016/j.physd.2009.05.008. MR2598511 (2012b:37160)

[CDB11] Frédéric Chardard, Frédéric Dias, and Thomas J. Bridges, *Computing the Maslov index of solitary waves, Part 2: Phase space with dimension greater than four*, Phys. D **240** (2011), no. 17, 1334–1344, DOI 10.1016/j.physd.2011.05.014. MR2831770 (2012k:37130)

[CZ84] Charles Conley and Eduard Zehnder, *Morse-type index theory for flows and periodic solutions for Hamiltonian equations*, Comm. Pure Appl. Math. **37** (1984), no. 2, 207–253, DOI 10.1002/cpa.3160370204. MR733717 (86b:58021)

[DJ11] Jian Deng and Christopher Jones, *Multi-dimensional Morse index theorems and a symplectic view of elliptic boundary value problems*, Trans. Amer. Math. Soc. **363** (2011), no. 3, 1487–1508, DOI 10.1090/S0002-9947-2010-05129-3. MR2737274 (2011k:35156)

[D76] J. J. Duistermaat, *On the Morse index in variational calculus*, Advances in Math. **21** (1976), no. 2, 173–195. MR0649277 (58 #31190)

[FJN03] Roberta Fabbri, Russell Johnson, and Carmen Núñez, *Rotation number for non-autonomous linear Hamiltonian systems. I. Basic properties*, Z. Angew. Math. Phys. **54** (2003), no. 3, 484–502, DOI 10.1007/s00033-003-1068-1. MR2048667 (2005f:37114)

[F04] Kenro Furutani, *Fredholm-Lagrangian-Grassmannian and the Maslov index*, J. Geom. Phys. **51** (2004), no. 3, 269–331, DOI 10.1016/j.geomphys.2004.04.001. MR2079414 (2005g:53150)

[Ga93] R. A. Gardner, *On the structure of the spectra of periodic travelling waves*, J. Math. Pures Appl. (9) **72** (1993), no. 5, 415–439. MR1239098 (94j:35072)

[G07] Fritz Gesztesy, *Inverse spectral theory as influenced by Barry Simon*, Spectral theory and mathematical physics: a Festschrift in honor of Barry Simon's 60th birthday, Proc. Sympos. Pure Math., vol. 76, Amer. Math. Soc., Providence, RI, 2007, pp. 741–820. MR2307754 (2008d:34010)

[GST96] F. Gesztesy, B. Simon, and G. Teschl, *Zeros of the Wronskian and renormalized oscillation theory*, Amer. J. Math. **118** (1996), no. 3, 571–594. MR1393260 (97g:34105)

[GT09] Fritz Gesztesy and Vadim Tkachenko, *A criterion for Hill operators to be spectral operators of scalar type*, J. Anal. Math. **107** (2009), 287–353, DOI 10.1007/s11854-009-0012-5. MR2496408 (2010d:47063)

[GW96] Fritz Gesztesy and Rudi Weikard, *Picard potentials and Hill's equation on a torus*, Acta Math. **176** (1996), no. 1, 73–107, DOI 10.1007/BF02547336. MR1395670 (97f:14046)

[J88] Christopher K. R. T. Jones, *Instability of standing waves for nonlinear Schrödinger-type equations*, Ergodic Theory Dynam. Systems **8*** (1988), no. Charles Conley Memorial Issue, 119–138, DOI 10.1017/S014338570000938X. MR967634 (90d:35267)

[JS99] D. W. Jordan and P. Smith, *Nonlinear ordinary differential equations*, 3rd ed., Oxford Texts in Applied and Engineering Mathematics, vol. 2, Oxford University Press, Oxford, 1999. An introduction to dynamical systems. MR1743361 (2000j:34001)

[Kr97] Yulia E. Karpeshina, *Perturbation theory for the Schrödinger operator with a periodic potential*, Lecture Notes in Mathematics, vol. 1663, Springer-Verlag, Berlin, 1997. MR1472485 (2000i:35002)

[K80] Tosio Kato, *Perturbation theory for linear operators*, 2nd ed., Springer-Verlag, Berlin, 1976. Grundlehren der Mathematischen Wissenschaften, Band 132. MR0407617 (53 #11389)

[MW] Wilhelm Magnus and Stanley Winkler, *Hill's equation*, Dover Publications Inc., New York, 1979. Corrected reprint of the 1966 edition. MR559928 (80k:34001)

[M63] J. Milnor, *Morse theory*, Based on lecture notes by M. Spivak and R. Wells. Annals of Mathematics Studies, No. 51, Princeton University Press, Princeton, N.J., 1963. MR0163331 (29 #634)

[O90] V. Yu. Ovsienko, *Selfadjoint differential operators and curves on a Lagrangian Grassmannian that are subordinate to a loop*, Mat. Zametki **47** (1990), no. 3, 65–73, 142, DOI 10.1007/BF01138507 (Russian); English transl., Math. Notes **47** (1990), no. 3-4, 270–275. MR1052069 (91h:58044)

[ReSi78] Michael Reed and Barry Simon, *Methods of modern mathematical physics. IV. Analysis of operators*, Academic Press [Harcourt Brace Jovanovich Publishers], New York, 1978. MR0493421 (58 #12429c)

[RS93] Joel Robbin and Dietmar Salamon, *The Maslov index for paths*, Topology **32** (1993), no. 4, 827–844, DOI 10.1016/0040-9383(93)90052-W. MR1241874 (94i:58071)

[RS95] Joel Robbin and Dietmar Salamon, *The spectral flow and the Maslov index*, Bull. London Math. Soc. **27** (1995), no. 1, 1–33, DOI 10.1112/blms/27.1.1. MR1331677 (96d:58021)

[SS08] Björn Sandstede and Arnd Scheel, *Relative Morse indices, Fredholm indices, and group velocities*, Discrete Contin. Dyn. Syst. **20** (2008), no. 1, 139–158. MR2350063 (2009h:35031)

[SB] H. Schulz-Baldes, *Sturm intersection theory for periodic Jacobi matrices and linear Hamiltonian systems*, Linear Algebra Appl. **436** (2012), 498–515.

Mathematics Department, The University of North Carolina at Chapel Hill Chapel Hill, North Carolina 27599

E-mail address: ckrtj@email.unc.edu

Department of Mathematics, The University of Missouri, Columbia, Missouri 65211

E-mail address: latushkiny@missouri.edu

Department of Mathematics and Statistics, The University of Sydney Sydney, NSW 2006, Australia

E-mail address: r.marangell@maths.usyd.edu.au

Proceedings of Symposia in Pure Mathematics
Volume **87**, 2013
http://dx.doi.org/10.1090/pspum/087/01435

1–D Schrödinger operators with local point interactions: a review

Aleksey Kostenko and Mark Malamud

Dedicated with great pleasure to Fritz Gesztesy on the occasion of his 60th birthday.

Abstract. We review recent developments in the theory of 1-D Schrödinger operators with local point interactions on a discrete set. The progress in this area was stimulated by recent advances in the extension theory of symmetric operators and in the theory of ordinary differential operators with distributional coefficients.

Contents

1. Introduction

Schrödinger operators with potentials supported on a finite or a discrete set of points are known as solvable models of quantum mechanics. These models called "solvable" since their resolvents can be computed explicitly in terms of the interaction strengths and the location of the sources. As a consequence the spectrum, the

2010 *Mathematics Subject Classification.* Primary .
The research was funded by the Austrian Science Fund (FWF) under project No. M1309–N13.

eigenfunctions, and further spectral properties can be determined explicitly. Models of this type have been extensively discussed in the physical literature, mainly in atomic, nuclear and solid state physics. A comprehensive treatment of Schrödinger operators with point interactions as well as a detailed list of references can be found in the monograph [**2**] published in 1988. In its second edition [**3**], published in 2005 by the American Mathematical Society, an account on the progress in the field for the period from 1988 until 2005 was summarized by Pavel Exner (see Appendix K "Seize ans après" in [**3**]). Our main aim is to review recent advances in the spectral theory of 1–D Schrödinger operators with local point interactions on a discrete set of points.

Historically, the first influential paper on 1–D Schrödinger operators with point interactions was the paper by Kronig and Penney [**77**]. They considered the Hamiltonian

$$\mathbf{H}_{KP} = -\frac{d^2}{dx^2} + \sum_{k\in\mathbb{Z}} a\delta(x-k), \tag{1.1}$$

where $a \in \mathbb{R}$ is fixed and δ is a Dirac delta-function. This Hamiltonian, now known as "the Kronig–Penney model", describes a nonrelativistic electron moving in a fixed crystal lattice. Our main objects of interest are the following operators $\mathbf{H}_{X,\alpha,q}$ and $\mathbf{H}_{X,\beta,q}$ associated with the formal differential expressions

$$\ell_{X,\alpha,q} := -\frac{\mathrm{d}^2}{\mathrm{d}x^2} + q(x) + \sum_{x_n\in X} \alpha_n\delta_n, \tag{1.2}$$

$$\ell_{X,\beta,q} := -\frac{\mathrm{d}^2}{\mathrm{d}x^2} + q(x) + \sum_{x_n\in X} \beta_n(\cdot,\delta_n')\delta_n', \tag{1.3}$$

where $\delta_n := \delta(x-x_n)$. These operators describe δ- and δ'-interactions, respectively, on a discrete set $X = \{x_n\}_{n\in I} \subset \mathcal{I} = (a,b)$, and the coefficients α_n, $\beta_n \in \mathbb{R}$ are called the strengths of the interaction at the point $x = x_n$. Clearly, (1.1) is a particular case of (1.2) with $\mathcal{I} = \mathbb{R}$, $X = \mathbb{Z}$, $\alpha_n \equiv \alpha$, and $q \equiv 0$.

The spectral properties of Hamiltonians associated with (1.2) and (1.3) are widely studied under the assumption that there is a positive uniform lower bound on the distance between interaction centers,

$$d_* := \inf_{i,j} |x_i - x_j| > 0. \tag{1.4}$$

A comprehensive account on related results can be found in the monograph [**3**]. One of the main reasons for the assumption (1.4) is that without this assumption even in the case $q \equiv 0$ the Hamiltonian (1.2) might be non-self-adjoint, i.e., symmetric with nontrivial deficiency indices (the first example was given by Shubin and Stolz in [**104**]). In the last few years this difficulty has been overcome due to recent advances in the extension theory of symmetric operators and in the theory of ordinary differential operators with distributional coefficients.

We would like to complete the introduction with a few words about Fritz's work on point interactions. It is difficult to overestimate his influence on the theory of Schrödinger operators with point interactions. He is one of the founders and promoters of the spectral theory of Schrödinger operators with point interactions. Under his influence, and with his participation over a long period, the subject has taken its present form.

Roughly speaking, his research in the field can be divided into two parts: (i) the study of Schrödinger operators with infinitely many interaction centers, and (ii) a rigorous definition of various classes of point interactions. His papers [**37**] and [**40**] written jointly with Holden and Šeba, respectively, originated a new concept of relativistic and non-relativistic Hamiltonians with δ'-interactions and had a long-year discussion in both physical and mathematical literature. It is also difficult to overestimate the role of the monograph [**2**], which represents the foundation of a new and rapidly developing branch in the spectral theory of Schrödinger operators.

Happy Birthday, Fritz, and many new important and influential results!

Notation. $\mathbb{N}, \mathbb{C}, \mathbb{R}$ have the usual meaning; $\mathbb{R}_+ = [0, \infty)$.

For a potential q and sequences α and β we set $q^{\pm}(x) := (|q(x)| \pm q(x))/2$, $\alpha_k^{\pm} := (|\alpha_k| \pm \alpha_k)/2$, and $\beta_k^{\pm} := (|\beta_k| \pm \beta_k)/2$.

For a self–adjoint operator $T = T^*$ acting in a Hilbert space $\mathfrak{H}$, $E_T(\cdot)$ denotes the spectral measure, $T^- := TE_T(-\infty, 0)$ and $T^+ := TE_T(0, +\infty)$ are the negative and positive parts of T, respectively, and $\kappa_{\pm}(T) := \dim\big(\operatorname{ran}(T_{\pm})\big)$ (if $\kappa_{\pm}(T) < \infty$, then $\kappa_{\pm}(T)$ is the number of negative/positive eigenvalues of T counting multiplicities). Further, $\sigma(T)$ and $\rho(T)$ are the spectrum and the resolvent set of T, respectively. By $\sigma_{\mathrm{p}}(T)$, $\sigma_{\mathrm{pp}}(T)$, $\sigma_{\mathrm{ess}}(T)$, $\sigma_{\mathrm{ac}}(T)$, and $\sigma_{\mathrm{sc}}(T)$ we denote point, pure point, essential, absolutely continuous and singular continuous spectra of T.

Let X be a discrete subset of $\mathbb{R}_+$, $X = \{x_k\}_1^{\infty}$ and $x_k \uparrow +\infty$. Also we shall use the following Sobolev spaces ($n \in \mathbb{N}$ and $p \in [1, \infty]$)

$$W^{n,p}(\mathbb{R}_+ \setminus X) := \{f \in L^p(\mathbb{R}_+) : f \in W^{n,p}[x_{k-1}, x_k],\ k \in \mathbb{N},\ f^{(n)} \in L^p(\mathbb{R}_+)\},$$
$$W_0^{n,p}(\mathbb{R}_+ \setminus X) := \{f \in W^{n,p}(\mathbb{R}_+) : f(x_k) = \ldots = f^{(n-1)}(x_k) = 0,\ k \in \mathbb{N}\},$$
$$W_{\mathrm{comp}}^{n,p}(\mathbb{R}_+ \setminus X) := W^{n,p}(\mathbb{R}_+ \setminus X) \cap L_{\mathrm{comp}}^p(\mathbb{R}_+).$$

2. Hamiltonians with δ-interactions

2.1. Definition of δ-interactions. There are several ways to associate an operator with the differential expression $\ell_{X,\alpha,q}$. For example, a δ-interaction at a point $x = x_0$ may be defined using the *form method*, that is the operator $-\frac{\mathrm{d}^2}{\mathrm{d}x^2} + \alpha_0\delta(x - x_0)$ is defined as an operator associated in $L^2(\mathbb{R})$ with the quadratic form

$$\mathfrak{t}[f] := \int_{\mathbb{R}} |f'(t)|^2 dt + \alpha_0 |f(x_0)|^2, \qquad \operatorname{dom}(\mathfrak{t}) := W_2^1(\mathbb{R}),$$

which is closed and lower semibounded by the KLMN Theorem (see [**94**, p. 168]). Another way to introduce a local interaction at x_0 is to consider a symmetric operator $\mathbf{H}_{\min} := \mathbf{H}_{\min}^- \oplus \mathbf{H}_{\min}^+$, where $\mathbf{H}_{\min}^-$ and $\mathbf{H}_{\min}^+$ are the minimal operators generated by $-\frac{\mathrm{d}^2}{\mathrm{d}x^2}$ in $L^2(-\infty, x_0)$ and $L^2(x_0, +\infty)$, respectively, and then to consider its extension subject to *the boundary conditions* connecting x_0+ and x_0-:

$$f(x_0+) = f(x_0-), \quad f'(x_0+) - f'(x_0-) = \alpha_0 f(x_0). \tag{2.1}$$

Both these methods have disadvantages if the set X is infinite. The form method works only for the case of lower semibounded operators. If we apply the method of boundary conditions, then the corresponding minimal operator $\mathbf{H}_{\min}$ has infinite deficiency indices and the description of self-adjoint extensions of $\mathbf{H}_{\min}$ is a rather complicated problem in this case.

In what follows, without loss of generality we shall consider $\ell_{X,\alpha,q}$ on the positive semi-axis $\mathbb{R}_+$ assuming that the sequence $X = \{x_k\}_{k=1}^{\infty}$ is strictly increasing,

$0 = x_0 < x_1 < x_2 < .. < x_k < x_{k+1} < ...$, and satisfies $x_k \uparrow +\infty$. We shall define the Hamiltonian with δ-interactions on X as follows: let

$$\mathbf{H}^0_{X,\alpha,q} f := \tau_q f = -f'' + q(x)f, \quad f \in \operatorname{dom}(\mathbf{H}^0_{X,\alpha,q}), \tag{2.2}$$

$$\operatorname{dom}(\mathbf{H}^0_{X,\alpha,q}) := \Big\{ f \in W^{2,1}_{\text{comp}}(\mathbb{R}_+ \setminus X) : f(0) = 0, \tag{2.3}$$
$$\begin{matrix} f(x_k+) = f(x_k-) \\ f'(x_k+) - f'(x_k-) = \alpha_k f(x_k) \end{matrix}, \ \tau_q f \in L^2(\mathbb{R}_+) \Big\}.$$

Clearly, the operator $\mathbf{H}^0_{X,\alpha,q}$ is symmetric. Let us denote its closure by $\mathbf{H}_{X,\alpha,q}$:

$$\mathbf{H}_{X,\alpha,q} := \overline{\mathbf{H}^0_{X,\alpha,q}}. \tag{2.4}$$

If $q = \mathbf{0}$, we set $\mathbf{H}_{X,\alpha} := \mathbf{H}_{X,\alpha,0}$; if either $X = \emptyset$ or $\alpha = \mathbf{0}$, then $\mathbf{H}_{X,\alpha,q}$ will be denoted by $\mathbf{H}_q$.

An alternative approach was proposed by A. Savchuk and A. Shkalikov in [**98**] (see also [**99**])[1]. Namely, they suggested to consider (1.2) with the help of quasi-derivatives. The potential $v(x) = q(x) + \sum_{k=1}^{\infty} \alpha_k \delta(x - x_k)$ is a derivative of the function $V(x) = \int_0^x q(t)dt + \sum_{x_k < x} \alpha_k$ in the sense of distributions. Therefore, we can rewrite the differential expression (1.2) as follows

$$\ell_{X,\alpha,q}\, y = \ell_{V'}\, y := -(y^{[1]})' - V(x)y^{[1]} - V^2(x)y, \quad y^{[1]} := y' - V(x)y, \tag{2.5}$$

and then define the operator $\mathbf{H}^0_{V'}$ by (2.5) on the domain

$$\operatorname{dom}(\mathbf{H}^0_{V'}) = \{ f \in L^2_{\text{comp}}(\mathbb{R}_+) : f, f^{[1]} \in AC_{\text{loc}}(\mathbb{R}_+),\ f(0) = 0,\ \ell_{V'} f \in L^2(\mathbb{R}_+) \}. \tag{2.6}$$

It is straightforward to check that the operators $\mathbf{H}^0_{X,\alpha,q}$ and $\mathbf{H}^0_{V'}$ defined by (2.2)–(2.3) and by (2.5)–(2.6), respectively, coincide (see [**99**]). This definition preserves the main features of the classical Sturm–Liouville theory (for instance, Weyl–Titchmarsh theory [**28**]). Moreover, it allows ones to study the direct and inverse spectral problems for 1–D Schrödinger operators with potentials distributions. We decided not to discuss this topic here in order to keep our review at a reasonable length (however, see remarks at the end of this section).

Let us emphasize that definition (2.5)–(2.6) is applicable only under the assumption that $V \in L^2_{\text{loc}}(\mathbb{R}_+)$, that is, the original potential is a $W^{-1,2}_{\text{loc}}$-distribution (see discussion in [**99**]). Clearly, this excludes the case of δ'-interactions.

Next assuming that $q \in L^\infty(\mathbb{R}_+)$, we specify a description of the domains $\operatorname{dom}(\mathbf{H}_{X,\alpha,q})$ and $\operatorname{dom}(\mathbf{H}^*_{X,\alpha,q})$ equipped with the graph norms of the operators $\mathbf{H}_{X,\alpha,q}$ and $\mathbf{H}^*_{X,\alpha,q}$, respectively.

PROPOSITION 2.1. *Let $q \in L^\infty(\mathbb{R}_+)$. Then:*

(i) *The operator $\mathbf{H}_{X,\alpha,q}$ is symmetric and its adjoint is given by the same differential expression τ_q on the domain*

$$\operatorname{dom}\left(\mathbf{H}^*_{X,\alpha,q}\right) := \left\{ f \in W^{2,2}(\mathbb{R}_+ \backslash X) : f(0) = 0, \ \begin{matrix} f(x_k+) = f(x_k-) \\ f'(x_k+) - f'(x_k-) = \alpha_k f(x_k) \end{matrix} \right\}. \tag{2.7}$$

[1]This regularization method was used in [**14**] in the particular case $q(x) = 1/x$ and then further developed for generic $W^{-1,2}_{\text{loc}}$-distributional potentials in [**98**], [**99**]. For further historical remarks we refer to [**29**], [**30**], [**49**], and [**50**].

(ii) *Assume additionally that* $d_* > 0$,

$$d_* := \inf_k d_k, \quad \text{and} \quad d_k := x_k - x_{k-1}, \quad k \in \mathbb{N}. \tag{2.8}$$

Then the embedding

$$\operatorname{dom}(\mathbf{H}_{X,\alpha,q}) \hookrightarrow \operatorname{dom}(\mathbf{H}^*_{X,\alpha,q}) \hookrightarrow W^{1,2}(\mathbb{R}_+) \tag{2.9}$$

holds and is continuous.

PROOF. (i) follows from integration by parts of the expression $(\mathbf{H}_{X,\alpha,q}f, g)$.

(ii) If $d_* > 0$, then applying the Sobolev embedding theorem to the spaces $W^{2,2}[x_{k-1}, x_k]$, $k \in \mathbb{N}$ (see [**66**, inequality (IV.1.12)] and also the proof of [**74**, Proposition 2.1(ii)]), we conclude that $W^{2,2}(\mathbb{R}_+ \setminus X) = \bigoplus_{k=1}^\infty W^{2,2}[x_{k-1}, x_k]$ is continuously embedded into $W^{1,2}(\mathbb{R}_+ \setminus X) = \bigoplus_{k=1}^\infty W^{1,2}[x_{k-1}, x_k]$. The description (2.7) of $\operatorname{dom}(\mathbf{H}^*_{X,\alpha,q})$ completes the proof. □

REMARK 2.2. *Let us stress that in the case* $d_* = 0$ *the embedding* (2.9) *depends on* α *and might be false (see Example* 2.3 *and also Remark* 2.16*(iii) below).*

EXAMPLE 2.3. *Let* $X = \{x_k\}_{k=1}^\infty$ *be such that* $d_{2k-1} = d_{2k} = \frac{1}{k}$, $k \in \mathbb{N}$. *Let also* $\alpha_k = \frac{2}{d_k}$, $k \in \mathbb{N}$. *Consider the Hamiltonian*

$$\mathbf{H} := \mathbf{H}_{X,\alpha,q} = -\frac{d^2}{dx^2} - \sum_{k=1}^\infty \frac{2}{d_k}\delta(x - x_k) \tag{2.10}$$

Define the function $f : \mathbb{R}_+ \to \mathbb{R}$ *as follows:* $f(x) = x$ *on* $[0, 1]$; $f(x) = x_{4k-2} - x$ *if* $x \in [x_{4k-3}, x_{4k-1}]$ *and* $f(x) = x - x_{4k}$ *if* $x \in [x_{4k-1}, x_{4k+1}]$, $k \in \mathbb{N}$. *Clearly,* $f''(x) = 0$ *for all* $x \in \mathbb{R}_+ \setminus X$ *and*

$$\int_0^\infty |f(x)|^2 dx = \sum_{k=1}^\infty \frac{d_k^3}{3} = \sum_{k=1}^\infty \frac{2}{3k^3} < \infty.$$

Moreover, it is straightforward to check that the function f *satisfies boundary conditions* (2.3) *for all* $k \in \mathbb{N}$. *Therefore,* $f \in \operatorname{dom}(\mathbf{H}^*)$. *However,* $f'(x) = 1$ *for all* $x \in \mathbb{R}_+ \setminus X$ *and hence* $f \notin W^{1,2}(\mathbb{R}_+ \setminus X)$.

Note that the operator $\mathbf{H}$ *is not lower semibounded (see Theorem* 2.15*(ii)). Moreover, it can be checked that it is symmetric with* $n_\pm(\mathbf{H}) = 1$.

Apparently, it is possible to construct examples of self-adjoint Hamiltonians $\mathbf{H}_{X,\alpha} = \mathbf{H}^*_{X,\alpha}$ *such that the embedding* (2.9) *fails.*

Further references: an extension of the Savchuk–Shkalikov approach to the case of more general Sturm–Liouville equations, as well as to operators with matrix-valued coefficients, can be found in [**28**], [**29**], [**30**], [**49**], [**50**], [**85**].

Sturm–Liouville operators on finite intervals with singular potentials have also been considered in the framework of the inverse spectral theory. In particular, the inverse spectral problems of reconstruction of the potential from the corresponding spectral data (from two spectra or one spectrum and the set of norming constants) have successfully been solved in the paper [**112**] for potentials that are signed measures and in [**54**] and [**100**] for potentials that are distributions in W_2^{-1}. Sturm–Liouville operators in impedance form, i.e., of the form

$$-a^{-2}(x)\frac{d}{dx}a^2(x)\frac{d}{dx}$$

with a positive impedance function a, were discussed in the papers [**12**], [**13**], [**97**], [**26**], [**4**]. For a regular enough function a, such an operator (under, say, the Dirichlet boundary conditions) is unitarily equivalent to a Sturm–Liouville operator in a potential form with the potential $q = a''/a$. The inverse spectral problem for impedance Sturm–Liouville operators with a of bounded variation was studied in [**12**]; note that then the corresponding q formally contains singularities of the form δ'. The case $a \in W_2^1$ was also completely analyzed in [**13**], [**97**], [**26**], while $a \in W_p^1$ with $p \geq 1$ in [**12**] (partially) and in [**4**]. In the papers [**102**] and [**53**] the global uniform stability in the inverse spectral problem of reconstruction of singular Sturm–Liouville operators from either two spectra or one spectrum and the norming constants is established; the potentials are from the Sobolev spaces W_2^θ with $\theta \geq -1$.

2.2. Self-adjointness. In the seminal paper [**39**], Gesztesy jointly with Kirsch proved the following very important result.

THEOREM 2.4 ([**39**]). *Let the Hamiltonian* $\mathbf{H}_{X,\alpha,q}$ *be defined by* (2.2)–(2.3). *Assume that the set* X *satisfies* (1.4) *and the potential* $q \in L^1_{\mathrm{loc}}(\mathbb{R})$ *is such that for any* $\varepsilon < d_*/2$ *the negative part of the potential*

$$(2.11)\qquad q_\varepsilon(x) := q(x)\chi_\varepsilon(x), \quad \chi_\varepsilon(x) := \begin{cases} 1, & x \in \cup_{k=0}^\infty (x_k + \varepsilon, x_{k+1} - \varepsilon) \\ 0, & x \notin \cup_{k=0}^\infty (x_k + \varepsilon, x_{k+1} - \varepsilon) \end{cases},$$

is form-bounded with respect to the free Hamiltonian $H_0 = -\frac{d^2}{dx^2}$ *with relative bound* $a_\varepsilon < 1$. *Then* $\mathbf{H}_{X,\alpha,q}$ *is self-adjoint.*

COROLLARY 2.5 ([**39**]). *If* q *is lower semibounded,* $q(x) \geq -c$ *a.e. on* $\mathbb{R}_+$, *and* (1.4) *holds true, then the operator* $\mathbf{H}_{X,\alpha,q}$ *is self-adjoint.*

REMARK 2.6. *If* X *is unbounded, previous investigations of Hamiltonians with* δ*-interactions either used the resolvent of* $\mathbf{H}_{X,\alpha,q}$ *(see* [**3**, §III.2] *and references therein) or the technique of local partitions* [**87**] *in order to define* $\mathbf{H}_{X,\alpha,q}$ *by the method of forms. In the one-dimensional case, Theorem* 2.4 *provides a powerful alternative to such methods which even applies if Hamiltonians* $\mathbf{H}_{X,\alpha,q}$ *unbounded from below are involved.*

It turned out that both assumptions on the negative part of the potential and on the set X are essential. If the potential q is "very negative", then one needs to make an additional assumption on interaction strengths α in order to ensure the self-adjointness of the Hamiltonian $\mathbf{H}_{X,\alpha,q}$.

THEOREM 2.7 ([**104**]). *Let the Hamiltonian* $\mathbf{H}_{X,\alpha,q}$ *be defined by* (2.2)–(2.3). *Assume that the set* X *satisfies* (1.4) *and there are positive constants* C_1, C_2, C_3, C_4 *such that*

$$(2.12)\qquad q(x) \geq -C_1 x^2 - C_2, \quad \alpha_k \geq -C_3|x_k| - C_4.$$

Then $\mathbf{H}_{X,\alpha,q}$ *is self-adjoint.*

If the set X does not satisfy (1.4), then, as it was first observed by Shubin and Stolz [**104**], the Hamiltonian $\mathbf{H}_{X,\alpha,q}$ might be symmetric with the nontrivial deficiency indices even in the case of zero potential q. Namely (see [**104**, p. 496]), they proved that the Hamiltonian

$$(2.13)\qquad \mathbf{H} = -\frac{d^2}{dx^2} - \sum_{k=1}^\infty (2k+1)\delta(x - x_k), \quad x_{k+1} - x_k = \frac{1}{k},$$

is symmetric with deficiency indices $n_\pm(\mathbf{H}) = 1$.

2.3. Connection with Jacobi matrices. The progress on the case $d_* = 0$ was made by the authors in the recent papers [**71**], [**72**], [**73**] in the framework of extension theory of symmetric operators. The main tool in [**71**] is the concept of boundary triplets and the corresponding Weyl functions (see [**46**], [**27**], [**22**]). The main ingredient of this approach is the following abstract version of the Green formula for the adjoint A^* of a symmetric operator $A \in \mathcal{C}(\mathfrak{H})$,

$$(A^*f, g)_{\mathfrak{H}} - (f, A^*g)_{\mathfrak{H}} = (\Gamma_1 f, \Gamma_0 g)_{\mathcal{H}} - (\Gamma_0 f, \Gamma_1 g)_{\mathcal{H}}, \qquad f, g \in \operatorname{dom}(A^*). \tag{2.14}$$

Here $\mathcal{H}$ is an auxiliary Hilbert space and the mapping $\Gamma := \begin{pmatrix}\Gamma_0\\ \Gamma_1\end{pmatrix} : \operatorname{dom}(A^*) \to \mathcal{H}\oplus\mathcal{H}$ is required to be surjective. The mapping Γ leads to a natural parametrization of self-adjoint (symmetric) extensions of A by means of self-adjoint (symmetric) linear relations in $\mathcal{H}$, see [**27, 46**]. For instance, every extension $\widetilde{A} = \widetilde{A}^*$, which is disjoint with $A_0 := A^* \restriction \ker(\Gamma_0)$, admits a representation

$$\widetilde{A} = A_B := A^* \restriction \ker\bigl(\Gamma_1 - B\Gamma_0\bigr) \tag{2.15}$$

where $B = B^* \in \mathcal{C}(\mathcal{H})$ is the "boundary" operator and its graph in $\mathcal{H}$ is given by $\Gamma \operatorname{dom}(\widetilde{A}) := \{\{\Gamma_0 f, \Gamma_1 f\} : f \in \operatorname{dom}(\widetilde{A})\}$. As distinguished from the J. von Neumann approach, (2.15) yields a natural parametrization of all self-adjoint (symmetric) extensions directly in terms of (abstract) boundary condition.

Assuming $q \in L^\infty(\mathbb{R}_+)$, we consider the Hamiltonian $\mathbf{H}_{X,\alpha,q}$ as an extension of the minimal symmetric operator $\mathbf{H}_{X,q} = \bigoplus_{k\in\mathbb{N}} \mathbf{H}_{q,k}$, where

$$\mathbf{H}_{q,k} f := \tau_q f = -f'' + qf, \quad \operatorname{dom}(\mathbf{H}_{q,k}) = W_0^{2,2}[x_{k-1}, x_k].$$

To construct an appropriate boundary triplet $\Pi = \{\mathcal{H}, \Gamma_0, \Gamma_1\}$ for the operator $\mathbf{H}^*_{X,q} = \bigoplus_{k\in\mathbb{N}} \mathbf{H}^*_{q,k}$ we apply the construction elaborated in [**79**] and [**71**] (note that a direct sum of boundary triplets is not necessarily a boundary triplet if either $d_* = 0$ or $q \notin L^\infty(\mathbb{R}_+)$, see [**67**] and [**71**]). Based on this construction, it is shown that the domain of $\mathbf{H}_{X,\alpha,q}$ admits the following representation

$$\operatorname{dom}(\mathbf{H}_{X,\alpha,q}) = \ker(\Gamma_1 - B_{X,\alpha}\Gamma_0),$$

$$B_{X,\alpha} = \begin{pmatrix} r_1^{-2}\bigl(\alpha_1 + \frac{1}{d_1} + \frac{1}{d_2}\bigr) & (r_1 r_2 d_2)^{-1} & 0 & \dots \\ (r_1 r_2 d_2)^{-1} & r_2^{-2}\bigl(\alpha_2 + \frac{1}{d_2} + \frac{1}{d_3}\bigr) & (r_2 r_3 d_3)^{-1} & \dots \\ 0 & (r_2 r_3 d_3)^{-1} & r_3^{-2}\bigl(\alpha_3 + \frac{1}{d_3} + \frac{1}{d_4}\bigr) & \dots \\ \dots & \dots & \dots & \dots \end{pmatrix}, \tag{2.16}$$

and $r_n = \sqrt{d_n + d_{n+1}}$, $n \in \mathbb{N}$. This parameterization implies that certain spectral properties of the operator $\mathbf{H}_{X,\alpha,q}$ correlate with the corresponding spectral properties of the Jacobi matrix $B_{X,\alpha}$. Namely, the following result was established in [**71**], [**72**].

THEOREM 2.8 ([**71, 72**]). *Let $q \in L^\infty(\mathbb{R}_+)$ and let $X = \{x_k\}_{k=1}^\infty \subset \mathbb{R}_+$ be strictly increasing and such that $d^* := \sup_k d_k < \infty$. Then:*

(i) *The deficiency indices of the operators $\mathbf{H}_{X,\alpha,q}$ and $B_{X,\alpha}$ coincide and*

$$n_\pm(\mathbf{H}_{X,\alpha,q}) = n_\pm(B_{X,\alpha}) \le 1. \tag{2.17}$$

In particular, the operator $\mathbf{H}_{X,\alpha,q}$ is self-adjoint if and only if so is the Jacobi matrix $B_{X,\alpha}$.

(ii) *The operator* $\mathbf{H}_{X,\alpha,q}$ *is lower semibounded precisely if so is the matrix* $B_{X,\alpha}$.
(iii) *The self-adjoint operator* $\mathbf{H}_{X,\alpha,q}$ *has purely discrete spectrum if and only if* $d_k \to 0$ *as* $k \to \infty$ *and the spectrum of* $B_{X,\alpha}$ *is purely discrete.*
(iv) *If the operator* $\mathbf{H}_{X,\alpha}$ *is self-adjoint, then*

$$\kappa_-(\mathbf{H}_{X,\alpha}) = \kappa_-(B_{X,\alpha}). \tag{2.18}$$

In particular, the operator $\mathbf{H}_{X,\alpha}$ *is nonnegative if and only if so is the matrix* $B_{X,\alpha}$.
(v) *If the operator* $\mathbf{H}_{X,\alpha}$ *is self-adjoint, then for any* $p \in [1,\infty]$

$$\mathbf{H}^-_{X,\alpha} \in \mathfrak{S}_p(L^2) \iff B^-_{X,\alpha} \in \mathfrak{S}_p(l^2). \tag{2.19}$$

In particular, $\sigma_{\text{ess}}(\mathbf{H}_{X,\alpha,q}) \subseteq \mathbb{R}_+$ *if and only if* $\sigma_{\text{ess}}(B_{X,\alpha}) \subseteq \mathbb{R}_+$.

REMARK 2.9. *(i) In the case* $d_* > 0$, *the boundary triplets approach was first applied for the study of spectral properties of Hamiltonians with local point interactions by Kochubei* [**68**] *(see also Mikhailets* [**80**]*).*

(ii) Let us mention that using a different approach the inequality $n_\pm(\mathbf{H}_{X,\alpha,q}) \leq 1$ *was first established by Minami* [**83**] *for arbitrary potentials* q, *not necessarily bounded (see also* [**23**] *and* [**104**]*).*

(iii) In the case $d_* > 0$, *the connection between Hamiltonians with point interactions and Jacobi matrices goes back at least to the papers by Phariseau* [**92**], [**93**] *and Bellissard et. al.* [**16**] *(for further details we refer to* [**3**, §III.2]*).*

Theorem 2.8 allows us to apply the well developed spectral theory of Jacobi operators (see, e.g., [**1**], [**17**], [**75**], [**76**], [**110**]) for the study of spectral properties of operators $\mathbf{H}_{X,\alpha,q}$. For instance, applying the Carleman test (see [**1**]) to the matrix $B_{X,\alpha}$, we immediately obtain the following improvement of Corollary 2.5 in the case $q \in L^\infty(\mathbb{R}_+)$.

COROLLARY 2.10. *Let* $q \in L^\infty(\mathbb{R}_+)$ *and let* X *be such that* $\sum_{k=1}^\infty d_k^2 = \infty$. *Then* $\mathbf{H}_{X,\alpha,q}$ *is self-adjoint.*

Let us mention that the condition $\{d_k\} \notin l^2$ is sharp (see [**71**, Proposition 5.9]). Let us conclude this subsection with the following example (see [**71**, Example 5.12] and also [**72**, Proposition 3]).

EXAMPLE 2.11. *Let* $\mathcal{I} = \mathbb{R}_+$, $x_0 = 0$, $x_k - x_{k-1} = d_k := 1/k$, $k \in \mathbb{N}$. *Consider the operator*

$$\mathbf{H}_A := -\frac{\mathrm{d}^2}{\mathrm{d}x^2} + \sum_{k=1}^\infty \alpha_k \delta(x - x_k). \tag{2.20}$$

Clearly, $\{d_k\}_{k=1}^\infty \in l_2$ *and we can not apply Corollary* 2.10. *However, the following statements are true:*

(i) *If* $\sum_{k=1}^\infty \frac{|\alpha_k|}{k^3} = \infty$, *then the operator* $\mathbf{H}_A$ *is self-adjoint.*
(ii) *If* $\alpha_k \leq -2(2k+1) + O(k^{-1})$, *then* $\mathbf{H}_A$ *is self-adjoint.*
(iii) *If* $\alpha_k \geq -\frac{C}{k}$, $k \in \mathbb{N}$, $C \equiv const > 0$, *then* $\mathbf{H}_A$ *is self-adjoint.*
(iv) *If* $\alpha_k = -(2k+1) + O(k^{-\varepsilon})$ *with some* $\varepsilon > 0$, *then* $\mathrm{n}_\pm(\mathbf{H}_A) = 1$.
(v) *If* $\alpha_k = -a(2k+1) + O(k^{-1})$ *with some* $a \in (0,2)$, *then* $\mathrm{n}_\pm(\mathbf{H}_A) = 1$.

REMARK 2.12. *Example 2.11 is inspired by the example of Shubin and Stolz (cf. (2.13)) and its proof is based on various self-adjointess tests for Jacobi matrices (cf. [1] and [17]). In particular, the proof of (iv) is based on the recent improvement by Kostyuchenko and Mirzoev* [**76**, Theorem 1] *of the well-known Berezanskii condition* [**17**, Theorem VII.1.5].

Further notes: A generalization and further developments of Example 2.11 can be found in [**65**]. Using the approach based on quasi-derivatives, it was noticed in [**59**] and [**85**], [**69**] that the analysis of [**57**], [**58**] and [**84**] extends to the case of Hamiltonians with δ-interactions. In particular, using this approach one can extend Corollary 2.10 to the case of semibounded potentials q, $q(.) \geq -c$ a.e. on $\mathbb{R}_+$.

2.4. Semiboundedness. As we already mentioned in Section 2.1, the Hamiltonian $\mathbf{H}_{X,\alpha,q}$ may be defined via the energy form

$$\mathfrak{t}^0_{X,\alpha,q}[f] := \int_{\mathbb{R}_+} (|f'(x)|^2 + q(x)|f(x)|^2)\,dx + \sum_{k=1}^{\infty} \alpha_k |f(x_k)|^2, \tag{2.21}$$

$$\operatorname{dom}(\mathfrak{t}^0_{X,\alpha,q}) = \{f \in W^{1,2}(\mathbb{R}_+) \cap L^2_{\text{comp}}(\mathbb{R}_+) : \mathfrak{t}^0_{X,\alpha,q}[f] < \infty\}. \tag{2.22}$$

Clearly, this form admits the representation

$$\mathfrak{t}^0_{X,\alpha,q}[f] = (\mathbf{H}^0_{X,\alpha,q} f, f)_{L^2}, \quad f \in \operatorname{dom}(\mathbf{H}^0_{X,\alpha,q}). \tag{2.23}$$

So, one is interested in conditions on X and α such that the form $\mathfrak{t}^0_{X,\alpha,q}$ is lower semibounded (and hence closable) and then to describe its closure.

THEOREM 2.13 ([**6**]). *If the Hamiltonian $\mathbf{H}_{X,\alpha,q}$ is lower semibounded, then it is self-adjoint. In particular, if the form $\mathfrak{t}^0_{X,\alpha,q}$ is lower semibounded, then it is closable and the self-adjoint operator associated with its closure $\mathfrak{t}_{X,\alpha,q} := \overline{\mathfrak{t}^0_{X,\alpha,q}}$ coincides with $\mathbf{H}_{X,\alpha,q}$.*

REMARK 2.14. *Theorem 2.13 is the analog of the celebrated Glazman–Povzner–Wienholtz Theorem* [**17**], [**41**], [**111**] *(see also the paper* [**24**] *by Clark and Gesztesy, where the case of matrix-valued Schrödinger operators was treated). An alternative proof of Theorem 2.13 has recently been proposed by Hryniv and Mykytyuk* [**56**]. *Let us also mention that a connection between lower-semiboundedness and self-adjointness for general Sturm–Liouville operators was first observed by Hartman* [**52**] *and Rellich* [**96**]. *Further details as well as a comprehensive list of references can be found in* [**24**].

The following result was obtained by Brasche in [**19**].

THEOREM 2.15 ([**19**]). *Assume that the negative parts of the potential q and the sequence α satisfy the following conditions*

$$\sup_{x>0} \int_x^{x+1} q_-(t)\,dt < \infty, \quad \sup_{x>0} \sum_{x_k \in [x,x+1]} \alpha_k^- < \infty, \tag{2.24}$$

where $q_- = (|q| - q)/2$ and $\alpha_k^- = (|\alpha_k| - \alpha_k)/2$. Then:

(i) *The form $\mathfrak{t}^0_{X,\alpha,q}$ is lower semibounded.*
(ii) *If both the potential q and the sequence α are negative, then the condition (2.24) is necessary and sufficient for the form $\mathfrak{t}^0_{X,\alpha,q}$ to be lower semibounded.*

REMARK 2.16. *(i) Theorem* 2.15 *immediately implies that the operator* $\mathbf{H}_{X,\alpha,q}$ *is self-adjoint and lower semibounded if* q *is lower semibounded,* $q(x) \geq -c$*, and* α *is a positive sequence.*

(ii) The condition (2.24) *is only sufficient if* q *and* α *take values of both signs. Examples of* α *and* q*, which do not satisfy* (2.24) *but such that the operator is lower semibounded can be found in* [**21**], [**19**, Example 2] *(see also Example* 2.17 *below).*

(iii) If conditions (2.24) *are satisfied, then* $\mathrm{dom}(\mathbf{H}_{X,\alpha,q})$ *is continuously embedded into* $W^{1,2}(\mathbb{R}_+)$ *(cf. Remark* 2.2*),*

$$\mathrm{dom}(\mathbf{H}_{X,\alpha,q}) \hookrightarrow \mathrm{dom}(\mathfrak{t}_{X,\alpha,q}) \hookrightarrow W^{1,2}(\mathbb{R}_+). \tag{2.25}$$

Apparently this embedding might be false even in the case of semibounded (hence self-adjoint) Hamiltonians $\mathbf{H}_{X,\alpha,q}$*. For further results and examples in the case of Hamiltonian* $\mathbf{H}_q$ *with locally integrable potentials we refer to* [**32**], [**33**], [**64**].

EXAMPLE 2.17 ([**19**]). *Choose any* $a > 1$ *and set* $x_{2k-1} = k$ *and* $x_{2k} = k + a^{-3k}$*. Let also* $\alpha_{2k-1} = a^k$ *and* $\alpha_{2k} = -a^k$*. Clearly,*

$$\sum_{x_k \in [n,n+1]} \alpha_k^- = a^n \to +\infty \quad as \quad n \to \infty,$$

and hence the second condition in (2.24) *is not fulfilled. However (see* [**19**, Example 2]*), the Hamiltonian*

$$\mathbf{H} := \mathbf{H}_{X,\alpha,} = -\frac{d^2}{dx^2} + \sum_{k=1}^{\infty} \alpha_k \delta(x - x_k) \tag{2.26}$$

is lower semibounded and hence self-adjoint in $L^2(\mathbb{R}_+)$*.*

Let us also mention that the function $f(x) = a^{-x/2}$ *is in the form domain, that is,* $\mathfrak{t}_{\mathbf{H}}[f] < \infty$*. Moreover,* $f \in W^{1,2}(\mathbb{R}_+)$*. However,*

$$\sum_{k \in \mathbb{N}} \alpha_k^- |f(x_k)|^2 = \sum_{k=1}^{\infty} a^k a^{-k-a^{-3k}} = \infty.$$

Note that in this example the embedding (2.25) *holds true and is continuous* [**19**].

Let us note that in the special case when there is a positive lower bound between interactions centers, i.e., $d_* > 0$, the criterion obtained in Theorem 2.8(ii) immediately implies the following statement.

COROLLARY 2.18. *Let* q *be bounded and* $d_* > 0$*. Then the operator* $\mathbf{H}_{X,\alpha,q}$ *is lower semibounded precisely if so is the sequence* α*.*

REMARK 2.19. *Note that Corollary* 2.18 *was first established by Brasche* [**19**] *by using the form approach. A different proof based on the boundary triplets approach was given in* [**80**] *(see also* [**71**]*).*

2.5. Spectral types. The literature on characterization of spectral types of Hamiltonians with δ-interactions is enormous and for a comprehensive treatment of operators with periodic potentials, short range perturbations etc. we refer the reader to the monograph [**3**]. In this subsection we shall review recent developments in the case $d_* = 0$.

Discreteness. We begin with the criteria for the operator $\mathbf{H}_{X,\alpha,q}$ to have a discrete spectrum. First of all, let us mention that the analog of the classical A.M. Molchanov discreteness criterion [**86**] (see also [**21**], [**41**]) holds true.

THEOREM 2.20 ([**6**]). *Assume that the negative parts of q and α satisfy conditions* (2.24). *Then the lower semibounded operator $\mathbf{H}_{X,\alpha,q}$ has purely discrete spectrum if and only if for every $\varepsilon > 0$*

$$\int_x^{x+\varepsilon} q(t)\,dt + \sum_{x_k\in[x,x+\varepsilon]} \alpha_k \to +\infty \quad \text{as} \quad x \to +\infty. \tag{2.27}$$

In the case $q \in L^\infty$, we immediately arrive at the following result.

COROLLARY 2.21. *If $q \in L^\infty(\mathbb{R}_+)$ and α satisfies the second condition in* (2.24), *then the operator $\mathbf{H}_{X,\alpha,q}$ has purely discrete spectrum if and only if for every $\varepsilon > 0$*

$$\sum_{x_k\in[x,x+\varepsilon]} \alpha_k \to +\infty \quad \text{as} \quad x \to +\infty. \tag{2.28}$$

In particular, the spectrum of $\mathbf{H}_{X,\alpha,q}$ is purely discrete whenever

$$d_k \to 0 \quad \text{and} \quad \frac{\alpha_k}{d_k} \to +\infty \quad \text{as} \quad k \to \infty. \tag{2.29}$$

Note that Theorem 2.20 applies only in the case of lower semibounded operators. Thus Theorem 2.8(iii) completes Theorem 2.20 in the non lower semibounded case. In particular, applying the Chihara condition to the matrix $B_{X,\alpha}$, we arrive at the following result.

COROLLARY 2.22 ([**71**]). *Let $q \in L^\infty(\mathbb{R}_+)$ and X be such that $d^* < \infty$. Let the matrix $B_{X,\alpha}$ be self-adjoint and let also $d_k \to 0$ and*

$$\lim_{k\to\infty} \frac{|\alpha_k|}{d_k} = \infty \quad \text{and} \quad \lim_{k\to\infty} \frac{1}{\alpha_k d_k} > -\frac{1}{4}. \tag{2.30}$$

Then the self-adjoint operator $\mathbf{H}_{X,\alpha,q}$ has a purely discrete spectrum.

Using the approach developed in [**58**] for smooth potentials, Ismagilov and Kostyuchenko [**59**] obtained the following result.

PROPOSITION 2.23 ([**59**]). *Assume that $q \in L^\infty(\mathbb{R}_+)$, $\alpha_k < 0$ for all $k \in \mathbb{N}$, $d_k \to 0$ and*

$$\frac{|\alpha_k|}{d_k + d_{k+1}} - \frac{2}{d_k d_{k+1}} \to +\infty, \quad k \to \infty. \tag{2.31}$$

Then the operator $\mathbf{H}_{X,\alpha,q}$ is non lower semibounded and its spectrum is purely discrete.

Simple examples show that the condition (2.30) does not imply (2.31) and visa versa. Thus Corollary 2.22 and Proposition 2.23 complete each other.

Let us mention that Corollary 2.22 and Proposition 2.23 enable us to construct examples of Hamiltonians $\mathbf{H}_{X,\alpha,q}$, which are non lower semibounded, self-adjoint and their spectra are purely discrete.

EXAMPLE 2.24 ([**72**]). *Set $x_k = 2\sqrt{k}$ and $\alpha_k = -C\sqrt{k}$ with $C \in \mathbb{R}_+ \setminus \{4\}$. Then $d_k = x_k - x_{k-1} \sim \frac{1}{\sqrt{k}}$, $k \to \infty$. The spectrum of the Hamiltonian*

$$\mathbf{H} = -\frac{d^2}{dx^2} - C\sum_{k\in\mathbb{N}} \sqrt{k}\delta(x - 2\sqrt{x})$$

is non lower semibounded. Moreover, the spectrum is discrete if and only if $C > 4$. The latter, in particular, implies that the second condition in (2.30) *is sharp.*

REMARK 2.25. *Combining Theorem* 2.20 *and Theorem* 2.8*(iii) one can obtain a discreteness criterion for Jacobi matrices (for further details see* [**6**, §7]*). This topic has attracted some attention recently, see* [**25**], [**60**], [**61**] *and references therein.*

Continuous spectrum. The next result is the extension of Birman's stability result [**18**] to the case of δ-potentials.

THEOREM 2.26 ([**6**]). *Assume that the negative part of the potential q satisfies* (2.24). *Then $\sigma_{\mathrm{ess}}(\mathbf{H}_{X,\alpha,q}) = \sigma_{\mathrm{ess}}(\mathbf{H}_q)$ provided that*

$$\lim_{x\to\infty} \sum_{x_k\in[x,x+1]} |\alpha_k| \to 0. \tag{2.32}$$

In particular, if in addition $q \to 0$ as $x \to \infty$, then $\sigma_{\mathrm{ess}}(\mathbf{H}_{X,\alpha,q}) = [0,+\infty)$.

COROLLARY 2.27. *If the negative part of the potential q satisfies* (2.24) *and*

$$\lim_{k\to\infty} \frac{\alpha_k}{d_k} = 0, \tag{2.33}$$

then $\sigma_{\mathrm{ess}}(\mathbf{H}_{X,\alpha,q}) = \sigma_{\mathrm{ess}}(\mathbf{H}_q)$.

REMARK 2.28. *In the case $d_* > 0$, the condition $\alpha_k \to 0$ as $k \to \infty$ is sufficient for the equality $\sigma_{\mathrm{ess}}(\mathbf{H}_{X,\alpha,q}) = \sigma_{\mathrm{ess}}(\mathbf{H}_q)$ to hold. However, if $d_* = 0$, then this conclusion is no longer true. It might even happen that $\sigma(\mathbf{H}_{X,\alpha})$ is purely discrete. For example, it suffices to set $x_k = \sqrt{k}$ and $\alpha_k = \frac{1}{k^\varepsilon}$ with $\varepsilon \in (0, \frac{1}{2})$, $k \in \mathbb{N}$ (cf. condition* (2.29) *and also* [**71**, Example 5.19]*).*

Absolutely continuous and singular spectra. Theorem 2.26 can be specified under additional assumptions on α and X.

THEOREM 2.29 ([**71**, **80**, **104**]). *Assume that $d^* < \infty$ and $q \in L^\infty(\mathbb{R}_+)$.*

(i) *Then $\sigma_{\mathrm{ac}}(\mathbf{H}_{X,\alpha,q}) = \sigma_{\mathrm{ac}}(\mathbf{H}_q)$ provided that*

$$\sum_{k=1}^{\infty} \frac{|\alpha_k|}{d_{k+1}} < \infty. \tag{2.34}$$

If in addition $q \in L^1(\mathbb{R}_+)$, then $\sigma_{\mathrm{ac}}(\mathbf{H}_{X,\alpha,q}) = [0,+\infty)$.

(ii) *If $q \equiv 0$, $d_* > 0$ and* (2.34) *is satisfied, then $\sigma(\mathbf{H}_{X,\alpha})$ is purely absolutely continuous in $(0,+\infty)$.*

REMARK 2.30. *The first statement of Theorem* 2.29 *is immediate by combining* [**71**, Corollary 5.15] *with the Kato–Rozenblum theorem* [**66**]. *Under an additional assumption $d_* > 0$ this statement was proved in* [**80**]. *The second part of Theorem* 2.29 *was established in* [**104**].

Let us also present one result on the absence of absolutely continuous spectrum.

THEOREM 2.31 ([**81**, **104**]). *Let X be such that $d_* > 0$. Then $\sigma_{\mathrm{ac}}(\mathbf{H}_{X,\alpha,q}) = \emptyset$ if at least one of the following conditions is satisfied:*

(i) *q is bounded from below, $\alpha_k \ge 0$ for all $k \in \mathbb{N}$ and*

$$\limsup_{k\to\infty} \alpha_k = +\infty,$$

(ii) *$q \in L^\infty(\mathbb{R}_+)$ and*

$$\limsup_{k\to\infty} |\alpha_k| = +\infty.$$

The first and the second parts of Theorem 2.31 were established in [**104**] and [**81**], respectively, by using a trace class technique similar to [**108**].

REMARK 2.32 ([**81**]). *Let ω be a Gaussian measure on the set of all real sequences. Then the subset of sequences which are semibounded (below or above) has a zero measure (see* [**106**, §3.5]*). Therefore, Theorem* 2.31 *implies that for any fixed X with $d_* > 0$ the set of Hamiltonians $\mathbf{H}_{X,\alpha,q}$ having nonempty absolutely continuous spectrum is of measure zero.*

The following result was obtained by Lotoreichik [**78**].

THEOREM 2.33 ([**78**]). *Assume that the set X is sparse, that is*

$$\lim_{k\to\infty} \frac{d_k}{d_{k-1}} = \infty. \tag{2.35}$$

Assume also that the intensities $\{\alpha_k\}_1^\infty$ are such that $\alpha_k \to \infty$ and

$$\liminf_{k\to\infty} \frac{d_k}{d_{k-1}\alpha_k^2} =: a \in (0,\infty) \cup \{\infty\}. \tag{2.36}$$

If $a \in (0,\infty)$, then:

(i) $\sigma_{\mathrm{ac}}(\mathbf{H}_{X,\alpha}) = \emptyset$,
(ii) $\sigma_{\mathrm{pp}}(\mathbf{H}_{X,\alpha}) \subseteq [0, a^{-1}]$,
(iii) $[a^{-1}, \infty] \subseteq \sigma_{\mathrm{sc}}(\mathbf{H}_{X,\alpha}) \subseteq [0, +\infty)$.

If $a = \infty$ and all $\alpha_k > 0$, then $\sigma(\mathbf{H}_{X,\alpha}) = \sigma_{\mathrm{sc}}(\mathbf{H}_{X,\alpha}) = [0, +\infty)$.

Further notes: In [**59**], Ismagilov and Kostyuchenko constructed a class of operators $\mathbf{H}_{X,\alpha}$ with purely point spectra having precisely two accumulation points 0 and $+\infty$. Note that Hamiltonians with δ-interactions form a good source of examples with exotic spectral properties. For example, Pearson in [**91**, §14.6] used Schrödinger operators with δ-interactions for constructing Hamiltonians with purely singular continuous spectrum. Let us also mention papers [**47**] and [**48**] for further examples of Schrödinger operators having exotic spectra.

2.6. Negative spectrum. During the last decade the problem on the number of negative eigenvalues for Schrödinger operators with δ-interactions attracted some attention. It is easy to observe from (2.21)–(2.23) that $\kappa_-(\mathbf{H}_{X,\alpha}) \le \kappa_-(\alpha)$, where $\kappa_-(\alpha)$ is the number of negative entries in the sequence α. However, the converse inequality is, in general, not true.

Albeverio and Nizhnik discovered in [**9**] the connection between this problem and certain continued fractions. The latter enabled them to construct the algorithm for computing the number of negative eigenvalues. Assuming that $d_* > 0$ and using the boundary triplets approach, their construction has been extended in [**42**] to the case of infinitely many δ-interactions (cf. Theorem 2.8(iv)). Note that a different matrix is used in [**7**] for the analysis of $\kappa_-(\mathbf{H}_{X,\alpha})$. For simplicity we restrict our considerations to the case of finitely many point interactions.

PROPOSITION 2.34 ([**7**]). *If $X = \{x_k\}_{k=1}^N$ and $\alpha = \{\alpha_k\}_{k=1}^N$, where $N \in \mathbb{N}$, then*

$$\kappa_-(\mathbf{H}_{X,\alpha}) = \kappa_+(M_{X,\alpha}) - \kappa_+(\alpha), \tag{2.37}$$

where

$$M_{X,\alpha} = \begin{pmatrix} \frac{1}{\alpha_1} + x_1 & x_1 & x_1 & \dots & x_1 \\ x_1 & \frac{1}{\alpha_2} + x_2 & x_2 & \dots & x_2 \\ x_1 & x_2 & \frac{1}{\alpha_3} + x_3 & \dots & x_3 \\ \dots & \dots & \dots & \dots & \dots \\ x_1 & x_2 & x_3 & \dots & \frac{1}{\alpha_N} + x_N \end{pmatrix}. \tag{2.38}$$

Next let us present the following extension of the celebrated Bargmann estimate (see, e.g., [**95**]).

THEOREM 2.35 ([**7**]). *Let q, X and α be such that the operator $\mathbf{H}_{X,\alpha,q}$ is self-adjoint. Then*

$$\kappa_-(\mathbf{H}_{X,\alpha,q}) < \int_{\mathbb{R}_+} |q_-(x)|\,dx + \sum_{k=1}^{\infty} |\alpha_k^-| x_k. \tag{2.39}$$

Finally, let us mention that combining Theorem 2.8(iv) with Theorem 2.35, we arrive at the following estimate for Jacobi matrices.

COROLLARY 2.36. *Let X and α be such that $d^* < \infty$ and let $B_{X,\alpha}$ given by* (2.16) *be self-adjoint. Then*

$$\kappa_-(B_{X,\alpha}) = \kappa_-(\mathbf{H}_{X,\alpha}) < \sum_{k=1}^{\infty} |\alpha_k^-| x_k. \tag{2.40}$$

Several different proofs of Theorem 2.35 can be found in [**7**]. Let us give a proof of Corollary 2.36 for the case of a finite number of δ-interactions based on Proposition 2.34.

PROOF. Firstly, assume that all α_k are negative, that is $\alpha = \alpha^-$. Then $\kappa_+(\alpha) = 0$ and hence, by (2.37), we get

$$\kappa_-(\mathbf{H}_{X,\alpha}) = \kappa_+(M_{X,\alpha}).$$

Set $\Lambda := \mathrm{diag}(|\alpha_1|, \dots, |\alpha_N|)$. Then we obtain from (2.37)

$$\kappa_+(M_{X,\alpha}) = \kappa_-(I_N - \Lambda^{1/2} M_X \Lambda^{1/2}), \quad M_X = \begin{pmatrix} x_1 & x_1 & \dots & x_1 \\ x_1 & x_2 & \dots & x_2 \\ \dots & \dots & \dots & \dots \\ x_1 & x_2 & \dots & x_N \end{pmatrix}.$$

Therefore, denoting $M_X^\Lambda := \Lambda^{1/2} M_X \Lambda^{1/2}$, we conclude that

$$\kappa_-(\mathbf{H}_{X,\alpha}) = \kappa_+(M_{X,\alpha}) \le \sum_{\lambda_j(M_X^\Lambda)>1} 1 < \sum_{\lambda_j(M_X^\Lambda)>1} \lambda_j(M_X^\Lambda) \le \operatorname{tr} M_X^\Lambda = \sum_{k=1}^{N} |\alpha_k| x_k.$$

To prove the statement in the case $\alpha \ne \alpha^-$, it suffices to note that $\kappa_-(\mathbf{H}_{X,\alpha}) \le \kappa_-(\mathbf{H}_{X,\alpha^-})$. □

REMARK 2.37. *(i) Let us mention that Theorem* 2.8*(iv) and Proposition* 2.34 *enables us to construct the operator $\mathbf{H}_{X,\alpha}$ having a given number of negative eigenvalues (for further details see* [**89**], [**90**], [**42**], [**7**]*).*

(ii) The above results demonstrate that Bargmann's bound is a one-sided estimate if the number of δ-interactions is greater than 1 *(see* [**7**, Example 4.10] *and also examples below).*

EXAMPLE 2.38. *Let $N \geq 2$. Assume that $\alpha_1 < 0$ and $\alpha_k > 0$ for all $k \geq 2$. Clearly, $\kappa_+(\alpha) = N-1$ and $\kappa_+(M_{X,\alpha}) \geq N-1$. If $\frac{1}{\alpha_1}+x_1 < 0$, then we immediately conclude that $\kappa_+(M_{X,\alpha}) = N-1$ and hence, by (2.37), $\kappa_-(\mathbf{H}_{X,\alpha}) = 0$, i.e., the operator is positive. Note that in this case the positivity also follows from the Bargmann estimate (2.40).*

Next, if $\frac{1}{\alpha_1}+x_1 > 0$ and

$$\Delta_2 := \det \begin{pmatrix} \frac{1}{\alpha_1}+x_1 & x_1 \\ x_1 & \frac{1}{\alpha_2}+x_2 \end{pmatrix} < 0, \tag{2.41}$$

then again we conclude $\kappa_+(M_{X,\alpha}) = N-1$ and hence, by (2.37), $\kappa_-(\mathbf{H}_{X,\alpha}) = 0$. Notice that in this case the Bargmann estimate (2.40) only gives the inequality $\kappa_-(\mathbf{H}_{X,\alpha}) \leq 1$. Let us also mention that under the additional assumption $N=2$, the positivity of the determinant in (2.41) implies that $\kappa_-(\mathbf{H}_{X,\alpha}) = 1$.

EXAMPLE 2.39. *Let $N \geq 3$. Assume that $\alpha_1 < 0$, $\alpha_2 < 0$ and $\alpha_k > 0$ for all $k \geq 3$. Clearly, $\kappa_+(\alpha) = N-2$ and $\kappa_+(M_{X,\alpha}) \geq N-2$. If $\frac{1}{\alpha_1}+x_1 < 0$ and the determinant in (2.41) is positive, then $\frac{1}{\alpha_2}+x_2 < 0$ and $\kappa_+(M_{X,\alpha}) = N-2$. Therefore, (2.37) yields the equality $\kappa_-(\mathbf{H}_{X,\alpha}) = 0$. On the other hand, the Bargmann estimate (2.40) only provides the inequality $\kappa_-(\mathbf{H}_{X,\alpha}) < 2$.*

3. Hamiltonians with δ'-interactions

3.1. Definition of δ'-interactions. The main object of this section is the Hamiltonian formally given by the differential expression (1.3). The existence of the model (1.3) was pointed out in 1980 by Grossmann, Hoegh–Krohn and Mebkhout [**51**]. However, the first rigorous treatment of (1.3) was made by Gesztesy and Holden in [**37**]. Namely, they defined the Hamiltonian $\mathbf{H}_{X,\beta,q}$ by using the method of boundary conditions. To be precise, let us consider (1.3) on the interval $[0,b)$, $0 < b \leq +\infty$, assuming that the sequence $X = \{x_k\}_1^\infty$ is strictly increasing and accumulates at b. Then define the operator

$$\mathbf{H}^0_{X,\beta,q} f := \tau_q f = -f'' + q(x)f, \quad f \in \operatorname{dom}(\mathbf{H}^0_{X,\beta,q}), \tag{3.1}$$

$$\operatorname{dom}(\mathbf{H}^0_{X,\beta,q}) := \Big\{ f \in W^{2,1}_{\text{comp}}([0,b)\setminus X) : f(0) = 0, \quad \begin{matrix} f'(x_k+) = f'(x_k-) \\ f(x_k+) - f(x_k-) = \beta_k f'(x_k) \end{matrix}, \ \tau_q f \in L^2(\mathbb{R}_+) \Big\}. \tag{3.2}$$

Clearly, $\mathbf{H}^0_{X,\beta,q}$ is symmetric. Let us denote its closure by $\mathbf{H}_{X,\beta,q}$:

$$\mathbf{H}_{X,\beta,q} := \overline{\mathbf{H}^0_{X,\beta,q}}. \tag{3.3}$$

For $q = \mathbf{0}$ we set $\mathbf{H}_{X,\beta} := \mathbf{H}_{X,\beta,\mathbf{0}}$. If $\beta_k = \infty$, then the boundary condition at x_k reads as $f'(x_k+) = f'(x_k-) = 0$. Therefore, the operator $\mathbf{H}_{X,\infty,q}$ becomes

$$\mathbf{H}_{X,\infty,q} := \mathbf{H}^N_{X,q} = \bigoplus_{k\in\mathbb{N}} \mathbf{H}^N_{q,k}, \quad \operatorname{dom}(\mathbf{H}^N_{X,q}) = \bigoplus_{k\in\mathbb{N}} \operatorname{dom}(\mathbf{H}^N_{q,k}), \tag{3.4}$$

where $\mathbf{H}^N_{q,k}$ is the Neumann realization of $\tau_q = -\frac{d^2}{dx^2} + q$ in $L^2(x_{k-1}, x_k)$.

Up to now it was not clear how to apply the form approach in order to rigorously define a δ'-interaction on X (cf. [**36**, Section 7.2]). Indeed, a very naive guess is to consider a single δ'-interaction at x_0 as the following form sum

$$\mathfrak{t}'[f] = \int_{\mathbb{R}} |f'(x)|^2\,dx + \beta_0|f'(x_0)|^2$$

defined on the domain

$$\mathrm{dom}(\mathfrak{t}') = \{f \in W^{1,2}(\mathbb{R}) : f'(x_0) \text{ exists and is finite}\}.$$

Clearly, the form $\mathfrak{t}'$ is not closable. However (see [**74**][2]), one needs to consider a δ'-interaction as a form sum of two forms $\mathfrak{t}_N$ and $\mathfrak{b}$, where

$$\mathfrak{t}_N[f] := \int_{\mathbb{R}} |f'(x)|^2\,dx, \quad \mathrm{dom}(\mathfrak{t}_N) := W^{1,2}(\mathbb{R}\setminus\{x_0\}), \tag{3.5}$$

and

$$\mathfrak{b}[f] := \frac{|f(x_0+) - f(x_0-)|^2}{\beta_0}, \quad \mathrm{dom}(\mathfrak{t}_N) := W^{1,2}(\mathbb{R}\setminus\{x_0\}). \tag{3.6}$$

Let us note that the operator

$$\mathbf{H}^N_{x_0} := -\frac{d^2}{dx^2}, \quad \mathrm{dom}(\mathbf{H}^N_{x_0}) = \{f \in W^{2,2}(\mathbb{R}\setminus\{x_0\}) : f'(x_0+) = f'(x_0-) = 0\}, \tag{3.7}$$

is associated with the form $\mathfrak{t}_N$. Clearly, $\mathbf{H}^N_{x_0}$ is the direct sum of Neumann realizations of $-\frac{d^2}{dx^2}$ in $L^2(-\infty, x_0)$ and $L^2(x_0, +\infty)$, respectively. Note that the form $\mathfrak{b}$ is infinitesimally form bounded with respect to the form $\mathfrak{t}_N$ and hence, by the KLMN theorem, the form

$$\mathfrak{t}'[f] := \mathfrak{t}_N[f] + \mathfrak{b}[f], \quad \mathrm{dom}(\mathfrak{t}') := W^{1,2}(\mathbb{R}\setminus\{x_0\}), \tag{3.8}$$

is closed and lower semibounded and gives rise to a self-adjoint operator

$$\begin{aligned} \mathbf{H}' &= -\frac{d^2}{dx^2}, \\ \mathrm{dom}(\mathbf{H}') &:= \left\{ f \in W^{2,2}(\mathbb{R}\setminus\{x_0\}) : \begin{array}{c} f'(x_0+) = f'(x_0-) \\ f(x_0+) - f(x_0-) = \beta_0 f'(x_0+) \end{array} \right\}. \end{aligned} \tag{3.9}$$

REMARK 3.1. *(i) Let us emphasize that the definition of a δ'-interaction via the form sum* (3.8) *allows to observe the key difference between δ and δ'-interactions. Namely, δ-interactions are considered as a perturbation of the free Hamiltonian. However, δ'-interactions can be viewed as a perturbation of the operator $\mathbf{H}^N_{X,q}$ defined by* (3.4). *In particular, in the case of infinitely many interaction centers, the free Hamiltonian has purely absolutely continuous spectrum though the spectrum of $\mathbf{H}^N_{X,q}$ is purely point. Let us also mention that the idea to consider Hamiltonians with δ'-interactions $\mathbf{H}_{X,\beta,q}$ as a perturbation of the Neumann realization $\mathbf{H}^N_{X,q}$ was used by Exner in* [**35**] *in order to prove that the spectra of δ' Wannier–Stark Hamiltonians have no absolutely continuous parts.*

(ii) Similar to δ-interactions, Hamiltonians with δ'-interactions can also be considered as quasi-differential operators. For example, set $p(x) = x + \beta_0\chi_{[x_0,+\infty)(x)}$ and consider in $L^2(\mathbb{R})$ the following differential expression $\tau_p := -\frac{d}{dx}\frac{d}{dp(x)}$. It can be shown (cf. [**28**, §3] *and* [**31**]*), that the corresponding self-adjoint operator coincides with $\mathbf{H}'$ given by* (3.9). *Note that this definition enables us to introduce*

[2]In the paper [**15**], which appeared during the preparation of [**74**], Hamiltonians with a δ'-interaction supported on a hypersurface are treated in a similar way.

δ'-interaction on an arbitrary set of Lebesgue measure zero and this will be done in the forthcoming paper [**31**]. *Let us also mention that using a different approach these operators have been studied recently by Albeverio and Nizhnik* [**11**] *and Brasche and Nizhnik* [**20**].

As in the case of δ-interactions, the domain of $\mathbf{H}_{X,\beta,q}$ can be further specified if $q \in L^\infty(\mathbb{R}_+)$. Let us equip $\mathrm{dom}(\mathbf{H}_{X,\beta,q})$ with the graph norm of $\mathbf{H}_{X,\beta,q}$.

PROPOSITION 3.2. *Let $q \in L^\infty(\mathbb{R}_+)$. Then:*

(i) *The operator $\mathbf{H}_{X,\beta,q}$ is self-adjoint and its domain is given by*

(3.10)
$$\mathrm{dom}\left(\mathbf{H}_{X,\beta,q}\right) := \left\{ f \in W^{2,2}(\mathbb{R}_+ \setminus X) : f(0) = 0, \ \begin{array}{c} f'(x_k+) = f'(x_k-) \\ f(x_k+) - f(x_k-) = \beta_k f(x_k) \end{array} \right\}.$$

(ii) *The embedding*

$$W^{2,2}(\mathbb{R}_+ \setminus X) \hookrightarrow W^{1,2}(\mathbb{R}_+ \setminus X),$$

holds and is continuous if and only if $d_* > 0$.

(iii) *If $d_* > 0$, then the embedding*

$$\mathrm{dom}(\mathbf{H}_{X,\beta,q}) \hookrightarrow W^{1,2}(\mathbb{R}_+ \setminus X) \tag{3.11}$$

holds and is continuous.

REMARK 3.3. *Self-adjointness of $\mathbf{H}_{X,\beta,q}$ was established in* [**23**] *(see also Section 3.2). The proof of Proposition 3.2 can be found in* [**74**].

Further remarks: There is one more approach to define δ'-interactions. Namely, a single δ'-interaction can be treated as an $\mathcal{H}_{-2}$-perturbation of the free Hamiltonian. For further details and results we refer to the monographs [**8**] and [**70**].

Let us also mention that there is a difference between δ'-interactions and δ'-potentials. During the last few years there was some activity in understanding the Hamiltonians with δ'-potentials. In this connection we refer to the recent papers [**43**], [**44**], [**45**] and [**20**] (see also the references therein).

3.2. Self-adjointness and connection with Jacobi matrices. The first results on the self-adjointness for Hamiltonians with δ'-interactions were obtained by Gesztesy and Holden [**37**] (see also [**3**, §III.3]). Using the approach introduced by Phariseau in [**92**] for δ-interactions, Gesztesy and Holden [**37**] established self-adjointness in the case $q \equiv 0$ and $d_* > 0$. Let us stress that the analysis becomes much more complicated if either $d_* = 0$ or $q \notin L^\infty$.

The next step was made by Buschmann, Stolz and Weidmann [**23**]. Namely, in contrast to Hamiltonians with δ-interactions, it was observed in [**23**] that the Hamiltonian $\mathbf{H}_{X,\beta,q}$ is always self-adjoint provided that $q \in L^\infty$ and $b = +\infty$. However, as in the case of δ-interactions, Buschmann, Stolz and Weidmann [**23**] proved that $n_\pm(\mathbf{H}_{X,\beta,q}) \leq 1$ and the deficiency indices can be characterized by using Weyl's limit point/limit circle criterion.

Using the boundary triplets approach, it was shown in [**71**] that in the case $q \in L^\infty(0,b)$ certain spectral properties of $\mathbf{H}_{X,\beta,q}$ are closely connected with those of the following Jacobi matrix

$$(3.12)\quad B_{X,\beta} := \begin{pmatrix} d_1^{-2} & d_1^{-2} & 0 & 0 & 0 & \dots \\ d_1^{-2} & \frac{d_1^{-1}}{\beta_1} + d_1^{-2} & \frac{d_1^{-1/2} d_2^{-1/2}}{\beta_1} & 0 & 0 & \dots \\ 0 & \frac{d_1^{-1/2} d_2^{-1/2}}{\beta_1} & \frac{d_2^{-1}}{\beta_1} + d_2^{-2} & d_2^{-2} & 0 & \dots \\ 0 & 0 & d_2^{-2} & \frac{d_2^{-1}}{\beta_2} + d_2^{-2} & \frac{d_2^{-1/2} d_3^{-1/2}}{\beta_2} & \dots \\ 0 & 0 & 0 & \frac{d_2^{-1/2} d_3^{-1/2}}{\beta_2} & \frac{d_3^{-1}}{\beta_2} + d_3^{-2} & \dots \\ \dots & \dots & \dots & \dots & \dots & \dots \end{pmatrix}.$$

More precisely, under a suitable choice of a boundary triplet $\Pi = \{l^2(\mathbb{N}), \Gamma_0, \Gamma_1\}$ for the operator $\mathbf{H}_{X,q}^*$ (see Section 2.3), the operator $\mathbf{H}_{X,\beta,q}$ admits the following representation

$$\operatorname{dom}(\mathbf{H}_{X,\beta,q}) = \{f \in \operatorname{dom}(\mathbf{H}_{X,q}^*) : \ \Gamma_1 = B_{X,\beta}\Gamma_0\}.$$

The next result was established in [**71**], [**73**].

THEOREM 3.4 ([**71**, **73**]). *Let $\mathbf{H}_{X,\beta,q}$ be given by* (3.1)–(3.3) *and let $B_{X,\beta}$ be the matrix* (3.12). *Let also $q \in L^\infty$ and $d^* < \infty$. Then:*

(i) *$n_\pm(\mathbf{H}_{X,\beta,q}) = n_\pm(B_{X,\beta})$. In particular, $\mathbf{H}_{X,\beta,q}$ is self-adjoint if and only if so is $B_{X,\beta}$.*
(ii) *The operator $\mathbf{H}_{X,\beta,q}$ is lower semibounded if and only if so is $B_{X,\beta}$.*
(iii) *If $\mathbf{H}_{X,\beta,q}$ is self-adjoint, then its spectrum is purely discrete if and only if $d_k \to 0$ and the spectrum of $B_{X,\beta}$ is purely discrete.*
(iv) *If $\mathbf{H}_{X,\beta}$ is self-adjoint, then*

$$(3.13)\qquad \kappa_-(\mathbf{H}_{X,\beta}) = \kappa_-(B_{X,\beta}).$$

(v) *If $\mathbf{H}_{X,\beta}$ is self-adjoint, then for any $p \in [1,\infty]$*

$$(3.14)\qquad \mathbf{H}_{X,\beta}^- \in \mathfrak{S}_p(L^2) \Longleftrightarrow B_{X,\beta}^- \in \mathfrak{S}_p(l^2).$$

In particular, $\sigma_{\rm ess}(\mathbf{H}_{X,\alpha}) \subseteq \mathbb{R}_+$ if and only if $\sigma_{\rm ess}(B_{X,\alpha}) \subseteq \mathbb{R}_+$.

It is interesting to note that the matrix $B_{X,\beta}$ admits the representation

$$(3.15)\qquad B_{X,\beta} = R_X^{-1}(I+U)D_{X,\beta}^{-1}(I+U^*)R_X^{-1},$$

where U is the unilateral shift on $l^2(\mathbb{N})$ and

$$(3.16)\qquad R_X = \bigoplus_{k=1}^{\infty} \begin{pmatrix} \sqrt{d_k} & 0 \\ 0 & \sqrt{d_k} \end{pmatrix}, \quad D_{X,\beta} = \bigoplus_{k=1}^{\infty} \begin{pmatrix} d_k & 0 \\ 0 & \beta_k \end{pmatrix}.$$

This observation immediately implies a connection of the Hamiltonian $\mathbf{H}_{X,\beta,q}$ with Krein–Stieltjes string operators [**1**, Appendix], [**63**] (see also [**71**, §6] and [**73**] for further details). Namely, if all β_k are positive, then setting $l_{2k-1} := d_k$, $l_{2k} := \beta_k$, $m_{2k-1} = m_{2k} := d_k$, $k \in \mathbb{N}$, the difference equation associated with the matrix $B_{X,\beta}$ describes the motion of an inhomogeneous string (Krein–Stieltjes string) with the mass distribution $\mathcal{M}(y) = \sum_{y_k<y} m_k$, where $y_k - y_{k-1} = l_k$ and $y_0 = 0$. This class of matrices is studied sufficiently well. In particular, applying Hamburger's Theorem [**1**, Theorem 0.5] to the matrix $B_{X,\beta}$, we arrive at the following self-adjointness criterion.

THEOREM 3.5 ([**71**]). *Deficiency indices of the operator* $\mathbf{H}_{X,\beta}$ *are equal and are not greater than one,* $n_+(\mathbf{H}_{X,\beta}) = n_-(\mathbf{H}_{X,\beta}) \le 1$. *Furthermore,* $\mathbf{H}_{X,\beta}$ *is self-adjoint if and only if at least one of the following conditions hold:*

(i) $\sum_{n=1}^{\infty} d_n = \infty$, *i.e.,* $b = +\infty$;

(ii) $\sum_{n=1}^{\infty} \left[d_{n+1}\big|\sum_{i=1}^{n}(\beta_i + d_i)\big|^2\right] = \infty$.

REMARK 3.6. *As distinguished from the case of* δ*-interactions, by Theorem 3.5(i), the operator* $\mathbf{H}_{X,\beta}$ *is self-adjoint in* $L^2(\mathbb{R}_+)$ *for any* $\beta \subset \mathbb{R}$ *(cf. Example 2.11). This fact was first observed in* [**23**]. *Let us also mention that statement (ii) provides the self-adjointness criterion in the case of a finite interval* $[0, b)$, $b < \infty$.

3.3. Semiboundedness. Let $b = +\infty$ and $X = \{x_k\}_{k=1}^{\infty}$ be a strictly increasing sequence accumulating at $+\infty$. Consider the following energy form

$$\mathfrak{t}^0_{X,\beta,q}[f] := \int_{\mathbb{R}_+} (|f'(x)|^2 + q(x)|f(x)|^2)\,dx + \sum_{k=1}^{\infty} \frac{|f(x_k+) - f(x_k-)|^2}{\beta_k}, \tag{3.17}$$

$$\operatorname{dom}(\mathfrak{t}^0_{X,\beta,q}) = \{f \in W^{1,2}(\mathbb{R}_+ \setminus X) \cap L^2_{\mathrm{comp}}(\mathbb{R}_+) : \mathfrak{t}^0_{X,\beta,q}[f] < \infty\}. \tag{3.18}$$

Integrating by parts, one gets that the form $\mathfrak{t}^0_{X,\beta,q}$ admits the representation

$$\mathfrak{t}^0_{X,\beta,q}[f] = (\mathbf{H}^0_{X,\beta,q} f, f)_{L^2}, \quad f \in \operatorname{dom}(\mathbf{H}^0_{X,\beta,q}), \tag{3.19}$$

and hence the form $\mathfrak{t}^0_{X,\beta,q}$ is closable whenever it is lower semibounded.

Firstly, let us mention that similar to the case of δ-interactions, the analog of the Glazman–Povzner–Wienholtz Theorem holds true in the case of δ'-interactions.

THEOREM 3.7 ([**74**]). *If the Hamiltonian* $\mathbf{H}_{X,\beta,q}$ *is lower semibounded, then it is self-adjoint.*

Combining this theorem with the representation (3.19), we immediately arrive at the following result.

COROLLARY 3.8. *If the form* $\mathfrak{t}^0_{X,\beta,q}$ *is lower semibounded, then it is closable and the self-adjoint operator associated with its closure* $\mathfrak{t}_{X,\beta,q} := \overline{\mathfrak{t}^0_{X,\beta,q}}$ *coincides with* $\mathbf{H}_{X,\beta,q}$.

Next we state the counterpart of Theorem 2.15.

THEOREM 3.9 ([**74**]). *Assume that* $d^* < \infty$ *and there exist positive constants* $C_0, C_1 > 0$ *such that*

$$\frac{1}{d_k}\int_{x_{k-1}}^{x_k} q_-(x)dx \le C_0, \quad \frac{1}{\beta_k^-} \le C_1 \min\{d_k, d_{k+1}\}, \quad k \in \mathbb{N}.\text{[3]} \tag{3.20}$$

Then:

(i) *The form* $\mathfrak{t}^0_{X,\beta,q}$ *is lower semibounded and the Hamiltonian* $\mathbf{H}_{X,\beta,q}$ *is lower semibounded and self-adjoint,*

(ii) *If both the potential* q *and the sequence* β *are negative, then the conditions* (3.20) *are also necessary for the form* $\mathfrak{t}^0_{X,\beta,q}$ *(and hence for the operator* $\mathbf{H}_{X,\beta,q}$*) to be lower semibounded.*

[3]Here we formally set $\frac{1}{\beta_k^-} := 0$ if $\beta_k^- = 0$, i.e., the corresponding inequality automatically holds true if β_k is positive.

REMARK 3.10. *(i) Theorem 3.9 immediately implies that the operator $\mathbf{H}_{X,\beta,q}$ is self-adjoint and lower semibounded if q is bounded from below, $q(x) \geq -c$, and β is a positive sequence.*

(ii) Note also that conditions (3.20) are only sufficient if q and β take values of both signs. Let us also mention that (3.20) imply the corresponding conditions (2.24) for q. However, the converse is not true.

Finally, let us present some simple conditions, which are necessary for the operator $\mathbf{H}_{X,\beta,q}$ to be lower semibounded (for further conditions see [**71**] and [**74**]).

LEMMA 3.11 ([**74**]). *Let $q = \mathbf{0}$. If the form $\mathfrak{t}^0_{X,\beta}$ is lower semibounded, that is $\mathfrak{t}^0_{X,\beta} \geq -C$ for some $C \geq 0$, then:*

(i) *for all $\beta_k^- \neq 0$*

$$\frac{1}{\beta_k^-} \leq 1 + \frac{C}{3}, \qquad k \in \mathbb{N}, \tag{3.21}$$

(ii)

$$\frac{1}{\beta_j^-} := \frac{1}{|\beta_{k_j}|} \leq C \min\{d_j^-, d_{j+1}^-\}, \qquad j \in \mathbb{N}, \tag{3.22}$$

where $X^- = \{x_j^-\}_{j=1}^\infty := \{x_{k_j}\}$ is the subsequence supporting negative intensities and $d_j^- := x_j^- - x_{j-1}^- = x_{k_j} - x_{k_{j-1}}$.

3.4. Spectral types. Hamiltonians with periodically arranged δ'-interactions were first discussed by Gesztesy and Holden in [**37**]. Namely, they investigated in great detail the spectral properties of $\mathbf{H}_{X,\beta}$ in the cases when $\alpha_k \equiv \alpha \in \mathbb{R}$ and $X = a\mathbb{Z}$ (crystal) or $X = a\mathbb{N}$ (half-crystal). Also, in [**37**], it was studied how the introduction of impurities affects spectral properties of crystals. The analysis of various types of ordered alloys, both deterministic and random, for this model was done in [**38**] where, e.g., the Saxon–Hunter conjecture [**103**], concerning gaps in the spectrum was proved. For a comprehensive treatment of these models we refer to the monograph [**3**]. The main aim of this subsection is to review recent developments in the case $d_* = 0$.

Discreteness. Using Theorem 3.4(iii) and the Kac–Krein discreteness criterion [**62**], one can prove the following result.

PROPOSITION 3.12 ([**71**]). *Let $\mathcal{I} = \mathbb{R}_+$ and $d_k \to 0$. The spectrum of the operator $\mathbf{H}_{X,\beta}$ is not discrete if at least one of the following conditions hold:*

(i) $\lim_{k\to\infty} x_k \sum_{j=k}^\infty d_j^3 > 0$;
(ii) $\beta_k \geq -C d_k^3, \quad k \in \mathbb{N}, \quad C > 0$;
(iii) $\beta_k^- \leq -C(d_k^{-1} + d_{k+1}^{-1}), \quad k \in \mathbb{N}, \quad C > 0.$

It follows from Proposition 3.12 that discreteness of the spectrum is a very rare property. For instance, the spectrum of the operator $\mathbf{H}_{X,\beta}$ is not discrete if either $\beta_n > 0$ for all $n \in \mathbb{N}$ or $\{d_n\}_{n=1}^\infty \notin l^3(\mathbb{N})$. However, it is possible to indicate certain conditions on X and β which guarantee the discreteness.

PROPOSITION 3.13 ([**71**]). *Assume that $\beta_k + d_k \geq 0$ for all $k \in \mathbb{N}$ and $\mathcal{I} = \mathbb{R}_+$. Then the spectrum of $\mathbf{H}_{X,\beta}$ is purely discrete if and only if*

$$\lim_{k\to\infty} x_k \sum_{j=k}^\infty d_j^3 = 0 \quad \text{and} \quad \lim_{k\to\infty} x_k \sum_{j=k}^\infty (\beta_j + d_j) = 0. \tag{3.23}$$

As it was already mentioned, the Hamiltonian $\mathbf{H}_{X,\beta,q}$ can be considered as a form sum perturbation of the operator

$$\mathbf{H}^N_{X,q} := \bigoplus_{k\in\mathbb{N}} \mathbf{H}^N_{q,k}, \tag{3.24}$$

where $\mathbf{H}^N_{q,k}$ is the Neumann realization of $-\frac{d^2}{dx^2} + q(x)$ in $L^2(x_{k-1}, x_k)$. The next result provides a discreteness criterion for the operator $\mathbf{H}^N_{X,q}$.

THEOREM 3.14 ([**74**]). *Assume that $d^* < \infty$, $q \in L^1_{\text{loc}}(\mathbb{R}_+)$, and q_- satisfies the first condition in* (3.20). *Then the spectrum of $\mathbf{H}^N_{X,q}$ is discrete if and only if the following conditions are satisfied:*

$$\text{for every } \varepsilon > 0 \quad \int_x^{x+\varepsilon} q(t)dt \to +\infty \quad \text{as} \quad x \to \infty. \tag{3.25}$$

$$\frac{1}{d_k}\int_{x_{k-1}}^{x_k} q(x)dx \to +\infty \quad \text{as} \quad k \to \infty \tag{3.26}$$

It is an immediate corollary of Theorem 3.14 that both conditions (3.25) and (3.26) are sufficient for the discreteness of the spectrum of $\mathbf{H}_{X,\beta,q}$. Moreover, (3.25) remains to be necessary although (3.26) is no longer necessary.

THEOREM 3.15 ([**74**]). *Assume that $q \in L^1_{\text{loc}}(\mathbb{R}_+)$, $d^* < \infty$ and conditions* (3.20) *are satisfied.*

(i) *If q satisfies* (3.25) *and* (3.26), *then the spectrum of $\mathbf{H}_{X,\beta,q}$ is discrete.*
(ii) *If the spectrum of the lower semibounded Hamiltonian $\mathbf{H}_{X,\beta,q}$ is purely discrete, then q satisfies* (3.25) *and*

$$\frac{1}{d_k}\Big(\int_{x_{k-1}}^{x_k} q(x)dx + \frac{1}{\beta_{k-1}} + \frac{1}{\beta_k}\Big) \to +\infty.$$

The next result complements Proposition 3.12.

PROPOSITION 3.16 ([**74**]). *Let $b = +\infty$ and $q \in L^\infty(\mathbb{R}_+)$. If the Hamiltonian $\mathbf{H}_{X,\beta,q}$ is lower semibounded, then its spectrum is not discrete. In particular, if the operator $\mathbf{H}_{X,\beta} := \mathbf{H}_{X,\beta,\mathbf{0}}$ is lower semibounded, then its spectrum is not discrete.*

There is a gap between necessary and sufficient conditions in Theorem 3.15. Indeed, the next result shows that condition (3.26) is only sufficient and in cases when (3.26) is not satisfied the discreteness of $\sigma(\mathbf{H}_{X,\beta,q})$ depends on q and β. In particular, the spectrum of the Hamiltonian $\mathbf{H}_{X,\beta,q}$ might be discrete although the spectrum of the corresponding Neumann realization $\mathbf{H}^N_{X,q}$ is not.

PROPOSITION 3.17 ([**74**]). *Let $X = \{x_k\}_1^\infty \subset \mathbb{R}_+$ be such that $d^* < \infty$ and*

$$\inf_{k\in\mathbb{N}} d_{2k-1} =: \varepsilon_0 > 0 \quad \text{and} \quad \lim_{k\to\infty} d_{2k} = 0. \tag{3.27}$$

Let q satisfy (3.20) *and Molchanov's condition* (3.25). *If β satisfies* (3.20) *and*

$$\lim_{k\to\infty} d_{2k}\beta_{2k-1} = 0, \tag{3.28}$$

then the spectrum $\sigma(\mathbf{H}_{X,\beta,q})$ of the Hamiltonian $\mathbf{H}_{X,\beta,q}$ is purely discrete.

Continuous spectrum.

THEOREM 3.18 ([**74**]). *Assume that* $q \in L^1_{\rm loc}(\mathbb{R}_+)$ *and the first condition in* (3.20) *is satisfied. Then* $\sigma_{\rm ess}(\mathbf{H}_{X,\beta,q}) = \sigma_{\rm ess}(\mathbf{H}^N_{X,q})$ *if*

$$\frac{|\beta_k|^{-1}}{\min\{d_k, d_{k+1}\}} \to 0 \quad as \quad k \to \infty. \tag{3.29}$$

If, in addition,

$$\lim_{k\to\infty} \frac{1}{d_k} \int_{x_{k-1}}^{x_k} |q(x)|dx = 0, \tag{3.30}$$

then

$$\sigma_{\rm ess}(\mathbf{H}_{X,\beta,q}) = \sigma_{\rm ess}(\mathbf{H}^N_{X,q}) = \sigma_{\rm ess}(\mathbf{H}^N_X). \tag{3.31}$$

Noting that the spectrum of $\mathbf{H}^N_{X,q}$ is pure point, we can construct various examples of operators $\mathbf{H}_{X,\beta,q}$ with exotic essential spectra. In particular, (3.31) implies that the structure of $\sigma_{\rm ess}(\mathbf{H}_{X,\beta,q})$ depends only on a "geometry" of X.

COROLLARY 3.19 ([**74**]). *Let the assumptions of Theorem* 3.18 *be satisfied. Assume also that* q *satisfies* (3.30) *and* $\lim_{k\to\infty} d_k = 0$. *Then*

$$\sigma_{\rm ess}(\mathbf{H}_{X,\beta,q}) = \{0\}, \tag{3.32}$$

i.e. the spectrum of $\mathbf{H}_{X,\beta,q}$ *is pure point and accumulates only at* 0 *and* ∞.

3.5. Negative spectrum. In contrast to the case of δ-interactions, the number of negative squares is determined by the number of negative intensities.

THEOREM 3.20 ([**42, 73**]). *If the operator* $\mathbf{H}_{X,\beta}$ *is self-adjoint, then*

$$\kappa_-(\mathbf{H}_{X,\beta}) = \kappa_-(\beta). \tag{3.33}$$

In particular, the operator $\mathbf{H}_{X,\beta}$ *is nonnegative if and only if* $\beta_k \geq 0$, $k \in \mathbb{N}$.

PROOF. By Theorem 3.4(iv), we get $\kappa_-(\mathbf{H}_{X,\beta}) = \kappa_-(B_{X,\beta})$. On the other hand, it follows from the factorization (3.15)–(3.16) that $\kappa_-(B_{X,\beta}) = \kappa_-(\beta)$. Combining both equalities we complete the proof. □

REMARK 3.21. *The equality* (3.33) *was observed in* [**10**] *in the special case when* $|X| = N < \infty$ *and all intensities are negative. In the case* $d_* > 0$, *Theorem* 3.20 *was established in* [**42**]. *The assumption* $d_* > 0$ *was removed in* [**73**] *by using a different method.*

Finally, let us mention that Theorem 3.20 enables us to give a different proof of Corollary 3.16.

PROOF OF COROLLARY 3.16. Clearly, it suffices to prove Corollary 3.16 in the case $q \equiv 0$. By Proposition 3.12(ii), if the spectrum of $\mathbf{H}_{X,\beta}$ is purely discrete, then $\kappa_-(\beta) = \infty$. Therefore, by Theorem 3.20, $\kappa_-(\mathbf{H}_{X,\beta}) = \infty$. However, if $\mathbf{H}_{X,\beta}$ is lower semibounded, then the negative spectrum of $\mathbf{H}_{X,\beta}$ has at least one finite accumulation point. This contradiction completes the proof. □

Further notes: In [**88**], [**11**], [**20**], Nizhnik with co-authors introduced δ'-interactions on sets of a Lebesgue measure zero, for example, on Cantor type sets. In these papers, the self-adjointness and basic spectral properties of these operators have been analyzed. A different approach to analyze the spectral properties of Hamiltonians with δ'-interactions on Cantor type sets is proposed in [**31**].

Acknowledgments

The authors are grateful to Rostyslav Hryniv for the careful reading of the manuscript and helpful hints with respect to the literature. We are also grateful to Gerald Teschl and the anonymous referee for useful remarks.

References

[1] N. I. Akhiezer, *The classical moment problem and some related questions in analysis*, Translated by N. Kemmer, Hafner Publishing Co., New York, 1965. MR0184042 (32 #1518)

[2] S. Albeverio, F. Gesztesy, R. Høegh-Krohn, and Helge Holden, *Solvable models in quantum mechanics*, Texts and Monographs in Physics, Springer-Verlag, New York, 1988. MR926273 (90a:81021)

[3] S. Albeverio, F. Gesztesy, R. Høegh-Krohn, and H. Holden, *Solvable models in quantum mechanics*, 2nd ed., AMS Chelsea Publishing, Providence, RI, 2005. With an appendix by Pavel Exner. MR2105735 (2005g:81001)

[4] S. Albeverio, R. Hryniv, and Ya. Mykytyuk, *Inverse spectral problems for Sturm-Liouville operators in impedance form*, J. Funct. Anal. **222** (2005), no. 1, 143–177, DOI 10.1016/j.jfa.2004.08.010. MR2129769 (2005j:34011)

[5] S. Albeverio, R. Hryniv, and Ya. Mykytyuk, *Inverse spectral problems for Bessel operators*, J. Differential Equations **241** (2007), no. 1, 130–159, DOI 10.1016/j.jde.2007.06.017. MR2356213 (2008k:34035)

[6] Sergio Albeverio, Aleksey Kostenko, and Mark Malamud, *Spectral theory of semibounded Sturm-Liouville operators with local interactions on a discrete set*, J. Math. Phys. **51** (2010), no. 10, 102102, 24, DOI 10.1063/1.3490672. MR2761288 (2012a:47105)

[7] S. Albeverio, A. Kostenko, M. Malamud, and H. Neidhardt, *Spherical Schrödinger operators with δ-interactions*, ArXiv:1211:4048, 24 pp (submitted).

[8] S. Albeverio and P. Kurasov, *Singular perturbations of differential operators*, London Mathematical Society Lecture Note Series, vol. 271, Cambridge University Press, Cambridge, 2000. Solvable Schrödinger type operators. MR1752110 (2001g:47084)

[9] Sergio Albeverio and Leonid Nizhnik, *On the number of negative eigenvalues of a one-dimensional Schrödinger operator with point interactions*, Lett. Math. Phys. **65** (2003), no. 1, 27–35, DOI 10.1023/A:1027396004785. MR2019389 (2004h:34178)

[10] Sergio Albeverio and Leonid Nizhnik, *Schrödinger operators with a number of negative eigenvalues equal to the number of point interactions*, Methods Funct. Anal. Topology **9** (2003), no. 4, 273–286. MR2040624 (2004k:81118)

[11] Sergio Albeverio and Leonid Nizhnik, *A Schrödinger operator with a δ'-interaction on a Cantor set and Krein-Feller operators*, Math. Nachr. **279** (2006), no. 5-6, 467–476, DOI 10.1002/mana.200310371. MR2213587 (2007f:47038)

[12] Lars-Erik Andersson, *Inverse eigenvalue problems with discontinuous coefficients*, Inverse Problems **4** (1988), no. 2, 353–397. MR954898 (89g:34015)

[13] Lars-Erik Andersson, *Inverse eigenvalue problems for a Sturm-Liouville equation in impedance form*, Inverse Problems **4** (1988), no. 4, 929–971. MR966766 (90g:34010)

[14] F. V. Atkinson, W. N. Everitt, and A. Zettl, *Regularization of a Sturm-Liouville problem with an interior singularity using quasiderivatives*, Differential Integral Equations **1** (1988), no. 2, 213–221. MR922562 (89b:34054)

[15] J. Behrndt, M. Langer, and V. Lotoreichik, *Schrödinger operators with δ and δ'–potentials supported on hypersurfaces*, Ann. Henri Poincaré **14** (2013), 385–423.

[16] J. Bellissard, A. Formoso, R. Lima, D. Testard, *Quasi–periodic interaction with a metal–insulator transition*, Phys. Rev. B **26** (1982), 3024–3030.

[17] Ju. M. Berezans′kiĭ, *Expansions in eigenfunctions of selfadjoint operators*, Translated from the Russian by R. Bolstein, J. M. Danskin, J. Rovnyak and L. Shulman. Translations of Mathematical Monographs, Vol. 17, American Mathematical Society, Providence, R.I., 1968. MR0222718 (36 #5768)

[18] M. Š. Birman, *On the spectrum of singular boundary-value problems*, Mat. Sb. (N.S.) **55 (97)** (1961), 125–174 (Russian). MR0142896 (26 #463)

[19] Johannes Brasche, *Perturbation of Schrödinger Hamiltonians by measures—selfadjointness and lower semiboundedness*, J. Math. Phys. **26** (1985), no. 4, 621–626, DOI 10.1063/1.526598. MR785672 (87d:81024)

[20] J. F. Brasche and L. P. Nizhnik, *One-dimensional Schrödinger operator with δ'-interactions on a set of Lebesgue measure zero*, ArXiv:1112.2545, (2011), 22 pp.

[21] Inge Brinck, *Self-adjointness and spectra of Sturm-Liouville operators*, Math. Scand. **7** (1959), 219–239. MR0112999 (22 #3844)

[22] Jochen Brüning, Vladimir Geyler, and Konstantin Pankrashkin, *Spectra of self-adjoint extensions and applications to solvable Schrödinger operators*, Rev. Math. Phys. **20** (2008), no. 1, 1–70, DOI 10.1142/S0129055X08003249. MR2379246 (2008m:47029)

[23] D. Buschmann, G. Stolz, and J. Weidmann, *One-dimensional Schrödinger operators with local point interactions*, J. Reine Angew. Math. **467** (1995), 169–186, DOI 10.1515/crll.1995.467.169. MR1355927 (97c:47055)

[24] Steve Clark and Fritz Gesztesy, *On Povzner-Wienholtz-type self-adjointness results for matrix-valued Sturm-Liouville operators*, Proc. Roy. Soc. Edinburgh Sect. A **133** (2003), no. 4, 747–758, DOI 10.1017/S0308210500002651. MR2006200 (2004h:47067)

[25] Petru A. Cojuhari and Jan Janas, *Discreteness of the spectrum for some unbounded Jacobi matrices*, Acta Sci. Math. (Szeged) **73** (2007), no. 3-4, 649–667. MR2380070 (2009g:47082)

[26] Carol Flynn Coleman and Joyce R. McLaughlin, *Solution of the inverse spectral problem for an impedance with integrable derivative. I, II*, Comm. Pure Appl. Math. **46** (1993), no. 2, 145–184, 185–212, DOI 10.1002/cpa.3160460203. MR1199197 (93m:34017)

[27] V. A. Derkach and M. M. Malamud, *Generalized resolvents and the boundary value problems for Hermitian operators with gaps*, J. Funct. Anal. **95** (1991), no. 1, 1–95, DOI 10.1016/0022-1236(91)90024-Y. MR1087947 (93d:47046)

[28] J. Eckhardt and G. Teschl, *Sturm–Liouville operators with measure-valued coefficients*, J. Anal. Math. (to appear).

[29] J. Eckhardt, F. Gesztesy, R. Nichols and G. Teschl, *Supersymmetry and Schrödinger–type operators with distributional matrix–valued potentials*, ArXiv:1206.4966, 35 pp.

[30] J. Eckhardt, F. Gesztesy, R. Nichols, and G. Teschl, *Weyl–Titchmarsh theory for Sturm–Liouville operators with distributional coefficients*, Opuscula Math. (to appear), (ArXiv: 1206.4677).

[31] J. Eckhardt, A. Kostenko, M. Malamud and G. Teschl, *Schrödinger operators with δ'-interactions on Cantor type sets*, in preparation.

[32] W. N. Everitt, *On a result of Brinck in the limit-point theory of second-order differential expressions*, Applicable Anal. **15** (1983), no. 1-4, 71–89, DOI 10.1080/00036818308839440. MR710184 (85a:34029)

[33] W. N. Everitt, M. Giertz, and J. Weidmann, *Some remarks on a separation and limit-point criterion of second-order, ordinary differential expressions*, Math. Ann. **200** (1973), 335–346. MR0326047 (48 #4393)

[34] S. Albeverio, F. Gesztesy, R. Høegh-Krohn, and H. Holden, *Solvable models in quantum mechanics*, 2nd ed., AMS Chelsea Publishing, Providence, RI, 2005. With an appendix by Pavel Exner. MR2105735 (2005g:81001)

[35] P. Exner, *The absence of the absolutely continuous spectrum for δ' Wannier-Stark ladders*, J. Math. Phys. **36** (1995), no. 9, 4561–4570, DOI 10.1063/1.530908. MR1347099 (96g:81050)

[36] Pavel Exner, *Leaky quantum graphs: a review*, Analysis on graphs and its applications, Proc. Sympos. Pure Math., vol. 77, Amer. Math. Soc., Providence, RI, 2008, pp. 523–564. MR2459890 (2010d:81110)

[37] F. Gesztesy and H. Holden, *A new class of solvable models in quantum mechanics describing point interactions on the line*, J. Phys. A **20** (1987), no. 15, 5157–5177. MR914699 (89a:81016)

[38] F. Gesztesy, H. Holden, and W. Kirsch, *On energy gaps in a new type of analytically solvable model in quantum mechanics*, J. Math. Anal. Appl. **134** (1988), no. 1, 9–29, DOI 10.1016/0022-247X(88)90003-0. MR958850 (90c:81032)

[39] F. Gesztesy and W. Kirsch, *One-dimensional Schrödinger operators with interactions singular on a discrete set*, J. Reine Angew. Math. **362** (1985), 28–50. MR809964 (87e:34034)

[40] F. Gesztesy and P. Šeba, *New analytically solvable models of relativistic point interactions*, Lett. Math. Phys. **13** (1987), no. 4, 345–358, DOI 10.1007/BF00401163. MR895297 (89b:81034)

[41] I. M. Glazman, *Pryamye metody kachestvennogo spektralnogo analiza*, Gosudarstv. Izdat. Fiz.-Mat. Lit., Moscow, 1963 (Russian). MR0185471 (32 #2938)

[42] N. Goloschapova and L. Oridoroga, *On the negative spectrum of one-dimensional Schrödinger operators with point interactions*, Integral Equations Operator Theory **67** (2010), no. 1, 1–14, DOI 10.1007/s00020-010-1759-x. MR2629973 (2011h:47083)

[43] Yu. D. Golovaty and S. S. Man'ko, *Solvable models for the Schrödinger operators with δ'-like potentials*, Ukr. Math. Bull. **6** (2009), no.2, 179–212.

[44] Yu. D. Golovaty and R. O. Hryniv, *On norm resolvent convergence of Schrödinger operators with δ'-like potentials*, J. Phys. A **43** (2010), no. 15, 155204, 14, DOI 10.1088/1751-8113/43/15/155204. MR2608272 (2011d:81100)

[45] Yu. D. Golovaty and R. O. Hryniv, *Norm resolvent convergence of singularly scaled Schrödinger operators and δ'-potentials*, ArXiv:1108.5345, 30 pp.

[46] V. I. Gorbachuk and M. L. Gorbachuk, *Boundary Value Problems for Operator Differential Equations*, Translated and revised from the 1984 Russian original. Mathematics and its Applications (Soviet Series), vol. 48, 1991. 347pp. MR1154792 (92m:34133)

[47] A. Gordon, J. Holt, and S. Molchanov, *Spectral theory of 1-D Schrödinger operators with unbounded potentials*, Probability and mathematical physics, CRM Proc. Lecture Notes, vol. 42, Amer. Math. Soc., Providence, RI, 2007, pp. 181–198. MR2352269 (2008j:34133)

[48] A. Ya. Gordon, S. A. Molchanov, and B. Tsagani, *Spectral theory of one-dimensional Schrödinger operators with strongly fluctuating potentials*, Funktsional. Anal. i Prilozhen. **25** (1991), no. 3, 89–92, DOI 10.1007/BF01085500 (Russian); English transl., Funct. Anal. Appl. **25** (1991), no. 3, 236–238 (1992). MR1139884 (93a:34097)

[49] Andrii Goriunov and Vladimir Mikhailets, *Regularization of singular Sturm-Liouville equations*, Methods Funct. Anal. Topology **16** (2010), no. 2, 120–130. MR2667807 (2011e:34060)

[50] A. S. Goryunov and V. A. Mikhaĭlets, *Resolvent convergence of Sturm-Liouville operators with singular potentials*, Mat. Zametki **87** (2010), no. 2, 311–315, DOI 10.1134/S0001434610010372 (Russian); English transl., Math. Notes **87** (2010), no. 1-2, 287–292. MR2731483 (2011h:34050)

[51] A. Grossmann, R. Høegh-Krohn, and M. Mebkhout, *A class of explicitly soluble, local, many-center Hamiltonians for one-particle quantum mechanics in two and three dimensions. I*, J. Math. Phys. **21** (1980), no. 9, 2376–2385, DOI 10.1063/1.524694. MR585589 (83c:81013)

[52] P. Hartman, *Differential equations with non-oscillatory eigenfunctions*, Duke Math. J. **15** (1948), 697–709.

[53] Rostyslav O. Hryniv, *Analyticity and uniform stability in the inverse singular Sturm-Liouville spectral problem*, Inverse Problems **27** (2011), no. 6, 065011, 25, DOI 10.1088/0266-5611/27/6/065011. MR2805768 (2012k:34035)

[54] R. O. Hryniv and Ya. V. Mykytyuk, *Inverse spectral problems for Sturm-Liouville operators with singular potentials*, Inverse Problems **19** (2003), no. 3, 665–684, DOI 10.1088/0266-5611/19/3/312. MR1984883 (2004e:34016)

[55] R. O. Hryniv and Y. V. Mykytyuk, *Eigenvalue asymptotics for Sturm-Liouville operators with singular potentials*, J. Funct. Anal. **238** (2006), no. 1, 27–57, DOI 10.1016/j.jfa.2006.04.015. MR2234122 (2007h:34155)

[56] Rostyslav O. Hryniv and Yaroslav V. Mykytyuk, *Self-adjointness of Schrödinger operators with singular potentials*, Methods Funct. Anal. Topology **18** (2012), no. 2, 152–159. MR2978191

[57] R. S. Ismagilov, *On the self-adjointness of the Sturm-Liouville operator*, Uspehi Mat. Nauk **18** (1963), no. 5 (113), 161–166 (Russian). MR0155037 (27 #4979)

[58] R. S. Ismagilov, *The spectrum of the Sturm-Liouville equation with oscillating potential*, Mat. Zametki **37** (1985), no. 6, 869–879, 942 (Russian). MR802430 (86i:34034)

[59] R. S. Ismagilov and A. G. Kostyuchenko, *Spectral asymptotics of the Sturm-Liouville operator with point interaction*, Funktsional. Anal. i Prilozhen. **44** (2010), no. 4, 14–20, DOI 10.1007/s10688-010-0036-8 (Russian, with Russian summary); English transl., Funct. Anal. Appl. **44** (2010), no. 4, 253–258. MR2768561 (2011k:34052)

[60] J. Janas and S. Naboko, *Multithreshold spectral phase transitions for a class of Jacobi matrices*, Recent advances in operator theory (Groningen, 1998), Oper. Theory Adv. Appl., vol. 124, Birkhäuser, Basel, 2001, pp. 267–285. MR1839840 (2002d:47044)

[61] S. N. Naboko and Ya. Yanas, *Criteria for semiboundedness in a class of unbounded Jacobi operators*, Algebra i Analiz **14** (2002), no. 3, 158–168 (Russian, with Russian summary); English transl., St. Petersburg Math. J. **14** (2003), no. 3, 479–485. MR1921992 (2003m:47057)

[62] I. S. Kac and M. G. Kreĭn, *Criteria for the discreteness of the spectrum of a singular string*, Izv. Vysš. Učebn. Zaved. Matematika **1958** (1958), no. 2 (3), 136–153 (Russian). MR0139804 (25 #3232)

[63] I. S. Kac and M. G. Krein, *On the spectral functions of the string*, Transl. AMS, ser. (2), **103** (1974), 19–102.

[64] Hubert Kalf, *Remarks on some Dirichlet type results for semibounded Sturm-Liouville operators*, Math. Ann. **210** (1974), 197–205. MR0355177 (50 #7653)

[65] I. I. Karpenko and D. L. Tyshkevich, *On self–adjointness of 1–D Schrödinger operators with δ–interactions*, Meth. Funct. Anal. Topology **4** (2012), 13 pp.

[66] Tosio Kato, *Perturbation theory for linear operators*, Die Grundlehren der mathematischen Wissenschaften, Band 132, Springer-Verlag New York, Inc., New York, 1966. MR0203473 (34 #3324)

[67] A. N. Kočubeĭ, *Symmetric operators and nonclassical spectral problems*, Mat. Zametki **25** (1979), no. 3, 425–434, 477 (Russian). MR534098 (80j:47040)

[68] A. N. Kochubeĭ, *One-dimensional point interactions*, Ukrain. Mat. Zh. **41** (1989), no. 10, 1391–1395, 1439, DOI 10.1007/BF01057262 (Russian); English transl., Ukrainian Math. J. **41** (1989), no. 10, 1198–1201 (1990). MR1034685 (91e:81020)

[69] N. N. Konechnaya, *On the asymptotic integration of second-order symmetric quasidifferential equations*, Mat. Zametki **90** (2011), no. 6, 875–884, DOI 10.1134/S0001434611110241 (Russian, with Russian summary); English transl., Math. Notes **90** (2011), no. 5-6, 850–858. MR2962962

[70] V. D. Koshmanenko, *Singulyarnye bilineinye formy v teorii vozmushchenii samosopryazhennykh operatorov*, "Naukova Dumka", Kiev, 1993 (Russian, with Russian and Ukrainian summaries). MR1817308 (2001m:47042)

[71] Aleksey S. Kostenko and Mark M. Malamud, *1-D Schrödinger operators with local point interactions on a discrete set*, J. Differential Equations **249** (2010), no. 2, 253–304, DOI 10.1016/j.jde.2010.02.011. MR2644117 (2011g:34206)

[72] A. S. Kostenko and M. M. Malamud, *On the one-dimensional Schrödinger operator with δ-interactions*, Funktsional. Anal. i Prilozhen. **44** (2010), no. 2, 87–91, DOI 10.1007/s10688-010-0019-9 (Russian); English transl., Funct. Anal. Appl. **44** (2010), no. 2, 151–155. MR2681961 (2011j:47135)

[73] A. S. Kostenko and M. M. Malamud, *The Schrödinger operator with δ′-interactions and the Kreĭn-Stieltjes string*, Dokl. Akad. Nauk **432** (2010), no. 1, 12–17, DOI 10.1134/S1064562410030026 (Russian); English transl., Dokl. Math. **81** (2010), no. 3, 342–347. MR2723573 (2012b:47118)

[74] A. Kostenko and M. Malamud, *Spectral theory of semibounded Sturm–Liouville operators with δ′-interactions on a discrete set*, Ann. Henri Poincaré (to appear), (ArXiv:1212.1691).

[75] A. G. Kostyuchenko and K. A. Mirzoev, *Generalized Jacobi matrices and deficiency indices of ordinary differential operators with polynomial coefficients*, Funktsional. Anal. i Prilozhen. **33** (1999), no. 1, 30–45, 96, DOI 10.1007/BF02465140 (Russian, with Russian summary); English transl., Funct. Anal. Appl. **33** (1999), no. 1, 25–37. MR1711874 (2000i:47056)

[76] A. G. Kostyuchenko and K. A. Mirzoev, *Conditions for the complete indefiniteness of Jacobi matrices with matrix elements*, Funktsional. Anal. i Prilozhen. **35** (2001), no. 4, 32–37, 95, DOI 10.1023/A:1013174423473 (Russian, with Russian summary); English transl., Funct. Anal. Appl. **35** (2001), no. 4, 265–269. MR1879117 (2002h:47044)

[77] R. de L. Kronig and W. G. Penney, *Quantum mechanics of electrons in crystal lattices*, Proc. Roy. Soc. (London) **130A** (1931), 499–513.

[78] Vladimir Lotoreichik, *Singular continuous spectrum of half-line Schrödinger operators with point interactions on a sparse set*, Opuscula Math. **31** (2011), no. 4, 615–628, DOI 10.7494/OpMath.2011.31.4.615. MR2823483 (2012j:34201)

[79] Mark Malamud and Hagen Neidhardt, *Sturm-Liouville boundary value problems with operator potentials and unitary equivalence*, J. Differential Equations **252** (2012), no. 11, 5875–5922, DOI 10.1016/j.jde.2012.02.018. MR2911417

[80] V. A. Mikhaĭlets, *A one-dimensional Schrödinger operator with point interactions*, Dokl. Akad. Nauk **335** (1994), no. 4, 421–423 (Russian); English transl., Russian Acad. Sci. Dokl. Math. **49** (1994), no. 2, 345–349. MR1281688 (95d:47062)

[81] V. A. Mikhaĭlets, *The structure of the continuous spectrum of the one-dimensional Schrödinger operator with point interactions*, Funktsional. Anal. i Prilozhen. **30** (1996), no. 2, 90–93, DOI 10.1007/BF02509459 (Russian); English transl., Funct. Anal. Appl. **30** (1996), no. 2, 144–146. MR1402089 (97h:81046)

[82] Vladimir Mikhailets and Volodymyr Molyboga, *One-dimensional Schrödinger operators with singular periodic potentials*, Methods Funct. Anal. Topology **14** (2008), no. 2, 184–200. MR2432767 (2009f:47066)

[83] Nariyuki Minami, *Schrödinger operator with potential which is the derivative of a temporally homogeneous Lévy process*, Probability theory and mathematical statistics (Kyoto, 1986), Lecture Notes in Math., vol. 1299, Springer, Berlin, 1988, pp. 298–304, DOI 10.1007/BFb0078486. MR936002 (89m:60146)

[84] K. A. Mirzoev, *Cauchy function and $\mathcal{L}^p_w$-properties of solutions of quasidifferential equations*, Uspekhi Mat. Nauk **46** (1991), no. 4(280), 161–162, DOI 10.1070/RM1991v046n04ABEH002835 (Russian); English transl., Russian Math. Surveys **46** (1991), no. 4, 190–191. MR1138966 (93d:34012)

[85] K. A. Mirzoev and T. A. Safonova, *Singular Sturm-Liouville operators with distribution potential in the space of vector functions*, Dokl. Akad. Nauk **441** (2011), no. 2, 165–168, DOI 10.1134/S106456241107012X (Russian); English transl., Dokl. Math. **84** (2011), no. 3, 791–794. MR2953790

[86] A. M. Molčanov, *On conditions for discreteness of the spectrum of self-adjoint differential equations of the second order*, Trudy Moskov. Mat. Obšč. **2** (1953), 169–199 (Russian). MR0057422 (15,224g)

[87] John D. Morgan III, *Schrödinger operators whose potentials have separated singularities*, J. Operator Theory **1** (1979), no. 1, 109–115. MR526292 (81i:35044)

[88] L. P. Nizhnik, *A Schrödinger operator with δ'-interaction*, Funktsional. Anal. i Prilozhen. **37** (2003), no. 1, 85–88, DOI 10.1023/A:1022932229094 (Russian); English transl., Funct. Anal. Appl. **37** (2003), no. 1, 72–74. MR1988013 (2004j:81048)

[89] Osamu Ogurisu, *On the number of negative eigenvalues of a Schrödinger operator with point interactions*, Lett. Math. Phys. **85** (2008), no. 2-3, 129–133, DOI 10.1007/s11005-008-0258-3. MR2443934 (2009g:34219)

[90] Osamu Ogurisu, *On the number of negative eigenvalues of a Schrödinger operator with δ interactions*, Methods Funct. Anal. Topology **16** (2010), no. 1, 42–50. MR2656130 (2011d:34171)

[91] D. B. Pearson, *Quantum scattering and spectral theory*, Techniques of Physics, vol. 9, Academic Press Inc. [Harcourt Brace Jovanovich Publishers], London, 1988. MR1099604 (91k:81198)

[92] P. Phariseau, *Surface states in a one-dimensional perfect semi-infinite crystal*, Physica **26** (1960), 737–743. MR0135920 (24 #B1962)

[93] P. Phariseau, *The energy spectrum of an amorphous substance*, Physica **26** (1960), 1185–1191. MR0127344 (23 #B390)

[94] Michael Reed and Barry Simon, *Methods of modern mathematical physics. II. Fourier analysis, self-adjointness*, Academic Press [Harcourt Brace Jovanovich Publishers], New York, 1975. MR0493420 (58 #12429b)

[95] Michael Reed and Barry Simon, *Methods of modern mathematical physics. IV. Analysis of operators*, Academic Press [Harcourt Brace Jovanovich Publishers], New York, 1978. MR0493421 (58 #12429c)

[96] F. Rellich, *Halbbeschränkte gewöhnliche Differentialoperatoren zweiter Ordnung*, Math. Ann. **122** (1951), 343–368.

[97] William Rundell and Paul E. Sacks, *The reconstruction of Sturm-Liouville operators*, Inverse Problems **8** (1992), no. 3, 457–482. MR1166492 (93c:34038)

[98] A. M. Savchuk and A. A. Shkalikov, *Sturm-Liouville operators with singular potentials*, Mat. Zametki **66** (1999), no. 6, 897–912, DOI 10.1007/BF02674332 (Russian, with Russian summary); English transl., Math. Notes **66** (1999), no. 5-6, 741–753 (2000). MR1756602 (2000m:34061)

[99] A. M. Savchuk and A. A. Shkalikov, *Sturm-Liouville operators with distribution potentials*, Tr. Mosk. Mat. Obs. **64** (2003), 159–212 (Russian, with Russian summary); English transl., Trans. Moscow Math. Soc. (2003), 143–192. MR2030189 (2004j:34198)

[100] A. M. Savchuk and A. A. Shkalikov, *Inverse problem for Sturm-Liouville operators with distribution potentials: reconstruction from two spectra*, Russ. J. Math. Phys. **12** (2005), no. 4, 507–514. MR2201315 (2006i:34024)

[101] A. M. Savchuk and A. A. Shkalikov, *On the properties of mappings associated with inverse Sturm-Liouville problems*, Tr. Mat. Inst. Steklova **260** (2008), no. Teor. Funkts. i Nelinein. Uravn. v Chastn. Proizvodn., 227–247, DOI 10.1134/S0081543808010161 (Russian, with Russian summary); English transl., Proc. Steklov Inst. Math. **260** (2008), no. 1, 218–237. MR2489515 (2009j:34064)

[102] A. M. Savchuk and A. A. Shkalikov, *Inverse problems for the Sturm-Liouville operator with potentials in Sobolev spaces: uniform stability*, Funktsional. Anal. i Prilozhen. **44** (2010), no. 4, 34–53, DOI 10.1007/s10688-010-0038-6 (Russian, with Russian summary); English transl., Funct. Anal. Appl. **44** (2010), no. 4, 270–285. MR2768563 (2012b:34033)

[103] D. S. Saxon and R. A. Hunter, *Some electronic properties of a one-dimensional crystal model*, Philips Res. Rep. **4** (1949), 81–122.

[104] Carol Shubin Christ and Günter Stolz, *Spectral theory of one-dimensional Schrödinger operators with point interactions*, J. Math. Anal. Appl. **184** (1994), no. 3, 491–516, DOI 10.1006/jmaa.1994.1218. MR1281525 (95k:47072)

[105] Petr Šeba, *Some remarks on the δ'-interaction in one dimension*, Rep. Math. Phys. **24** (1986), no. 1, 111–120, DOI 10.1016/0034-4877(86)90045-5. MR932938 (89e:81024)

[106] G. E. Šilov and Fan Dyk Tin′, *Integral, mera i proizvodnaya na lineinykh prostranstvakh*, With a supplement by M. G. Sonis, Izdat. "Nauka", Moscow, 1967 (Russian). MR0225964 (37 #1554)

[107] B. Simon, *On the number of bound states of two body Schrödinger operators — a review*, in: Studies in Math. Phys.: Essays in honor of V. Bargmann, by Lieb E. et. al. eds., Princeton, New Jersey, 1976; pp. 305–326.

[108] Barry Simon and Thomas Spencer, *Trace class perturbations and the absence of absolutely continuous spectra*, Comm. Math. Phys. **125** (1989), no. 1, 113–125. MR1017742 (91g:81018)

[109] Barry Simon, *Trace ideals and their applications*, 2nd ed., Mathematical Surveys and Monographs, vol. 120, American Mathematical Society, Providence, RI, 2005. MR2154153 (2006f:47086)

[110] Gerald Teschl, *Jacobi operators and completely integrable nonlinear lattices*, Mathematical Surveys and Monographs, vol. 72, American Mathematical Society, Providence, RI, 2000. MR1711536 (2001b:39019)

[111] Ernst Wienholtz, *Halbbeschränkte partielle Differentialoperatoren zweiter Ordnung vom elliptischen Typus*, Math. Ann. **135** (1958), 50–80 (German). MR0094576 (20 #1090)

[112] V. V. Zhikov, *On inverse Sturm–Liouville problems on a finite segment*, Izv. Akad. Nauk SSSR, **35** (1967), no. 5, 965–976 (in Russian).

Fakultät für Mathematik, Universität Wien, Nordbergstr. 15, 1090 Wien, Austria
E-mail address: `Oleksiy.Kostenko@univie.ac.at`
E-mail address: `duzer80@gmail.com`

Institute of Applied Mathematics and Mechanics, NAS of Ukraine, R. Luxemburg str. 74, Donetsk 83114, Ukraine
E-mail address: `mmm@telenet.dn.ua`

Proceedings of Symposia in Pure Mathematics
Volume **87**, 2013
http://dx.doi.org/10.1090/pspum/087/01441

Inverse problem for small oscillations

Yu. I. Lyubarskii and V. A. Marchenko

Dedicated to Fritz Gesztesy on the occasion of his 60th birthday

ABSTRACT. We study small oscillations near the stable equilibrium of a mechanical system which consists of a finite number of interacting particles. For a given subset of particles we consider the problem of reconstruction of the characteristics of the whole system from observation motion of the particles which belong to this subset. In the last section we apply these results to study the inverse problem of multichannel scattering.

1. Introduction

We consider small oscillations of a mechanical system near its stable equilibrium. The system consists of n particles $\alpha, \beta, \dots$ with masses $m(\alpha)$, $m(\beta)\dots$ interacting with each other and with an external field. Let $\mathfrak{A}$ be the set of all particles and

$$\vec{r}(\alpha) = \sum_{i\in\mathfrak{I}(3)} \xi^i(\alpha)\vec{e}_i,$$

be the deviation of the particle $\alpha \in \mathfrak{A}$ off its equilibrium position, here $\mathfrak{I}(3) = \{1,2,3\}$ and $\{\vec{e}_i\}$ is the standard orthonormal basis for the Euclidean space. The deviation of the whole system from the equilibrium is defined by its generalized coordinates $\xi^i(\alpha)$, $i \in \mathfrak{I}(3)$, $\alpha \in \mathfrak{A}$; the total number of degrees of freedom is $N = 3n$. We identify the pairs (α, i) with the points in the product set $A = \mathfrak{A}\times\mathfrak{I}(3)$, these points we denote by $x, y, \dots$. It is convenient to consider the generalized coordinates as functions $\xi(x)$ on A:

$$\xi(x) = \xi^i(\alpha) \quad \text{for} \quad x = (\alpha, i).$$

The potential energy of the system is a positive quadratic form of its generalized coordinates

$$U = \frac{1}{2}\sum_{x\in A}\sum_{y\in A} U(x,y)\xi(y)\xi(x), \tag{1}$$

2010 *Mathematics Subject Classification.* Primary 34L25; Secondary 34A55, 39A21, 15A29.

Key words and phrases. Inverse spectral problem, small oscillations, divided difference operator, inverse scattering problem, Sturm–Liouville operator.

The first author is partly supported by the Research Council of Norway grant 213638 and by the Center of Advanced Study, Oslo.

here $U(x,y)=U(y,x)$. Therefore the Lagrange function of the system is

$$\mathcal{L}=\frac{1}{2}\sum_{x\in A}\Big\{m(x)\dot{\xi}(x)^2-\sum_{y\in A}U(x,y)\xi(x)\xi(y)\Big\},$$

and the equations of motion take the form

$$m(x)\ddot{\xi}(x)+\sum_{y\in A}U(x,y)\xi(y)=0, \tag{2}$$

where $m(x)=m(\alpha)$ for $x=(\alpha,i)$, $i\in\mathfrak{I}(3)$.

The substitution

$$\eta(x)=\sqrt{m(x)}\,\xi(x)$$

reduces these equations to the standard form

$$\ddot{\eta}(x)+\sum_{y\in A}L(x,y)\eta(y)=0, \tag{3}$$

where

$$L(x,y)=\frac{U(x,y)}{\sqrt{m(x)m(y)}}=L(y,x).$$

Let $L^2(A)$ be the Hilbert space of functions on A endowed with the standard inner product

$$(\eta,\xi)=\sum_{x\in A}\eta(x)\overline{\xi(x)}.$$

Let $e_y(x)=\delta_y^x$, $y\in A$ be the canonical basis in $L^2(A)$. We denote the linear operators in $L^2(A)$ by $\widehat{P}$, $\widehat{Q},\dots$ while $P(x,y)$, $Q(x,y),\dots$ stay for their matrices in the canonical basis. The equation of motion (3) takes the form

$$\ddot{\eta}+\widehat{L}\eta=0,$$

here $\widehat{L}$ is a positive operator in $L^2(A)$, $L(x,y)$ is the matrix of $\widehat{L}$, and $\eta=\eta(x,t)\in L^2(A)$ for each t.

Let $\psi(j,x)$ and $\lambda(j)$, $j=1,2,\dots,N$ be the orthonormal system of eigenfunctions and the corresponding set of eigenvalues of $\widehat{L}$. We assume

$$\begin{aligned}0<\lambda(1)=\lambda(2)=\dots=\lambda(p_1)=\mu(1)<\lambda(p_1+1)=\dots=\lambda(p_2)=\mu(2)<\cdots\\ =\lambda(p_{K-1})=\mu(K-1)<\lambda(p_{K-1}+1)=\dots=\lambda(p_K)=\lambda(N)=\mu(K),\end{aligned} \tag{4}$$

where $\mu(1)<\mu(2)<\dots<\mu(K)$ is the spectrum of $\widehat{L}$.

Let $H(\mu(k))$ be the corresponding eigenspaces of $\widehat{L}$, i.e., the kernels of $\widehat{L}-\mu(k)\widehat{I}$; $\widehat{E}_k$ be the orthogonal projectors on $H(\mu(k))$ and

$$E_k(x,y)=\sum_{p_{k-1}<j\le p_k}\psi(j,x)\overline{\psi(j,y)}, \tag{5}$$

be their matrix elements. We also denote $n_k=\dim H(\mu(k))=p_k-p_{k-1}$, here we assume $p_0=0$.

The solution of (3) which satisfies the initial conditions

$$\eta(x,0)=\eta_0(x),\quad \dot{\eta}(x,0)=\eta_1(x)$$

can be expressed via $E_k(x,y)$ and $\mu(k)$ as

$$\eta(x,t) = \sum_{y\in A}\sum_{k=1}^{K} E_k(x,y)\left\{\eta_0(y)\cos\omega_k t + \eta_1(y)\frac{\sin\omega_k t}{\omega_k}\right\},$$

where $\omega_k = \sqrt{\mu(k)}$. Since $\eta(x,t) = \sqrt{m(x)}\,\xi(x,t)$, the solution of the original equation (2) satisfying the initial conditions has the form $\xi(x,0) = \xi_0(x)$, $\dot{\xi}(x,0) = \xi_1(x)$ is

$$\xi(x,t) = \sum_{y\in A}\sum_{k=1}^{K} Q_k(x,y)\left\{\xi_0(y)\cos\omega_k t + \xi_1(y)\frac{\sin\omega_k t}{\omega_k}\right\},$$

here

(6) $$Q_k(x,y) = m(x)^{-\frac{1}{2}}E_k(x,y)m(y)^{\frac{1}{2}}.$$

If for example the system is perturbed by applying an impulse to the particle β such that at the initial moment the particle gains unit velocity in the direction $\vec{e}_i$ i.e.,

$$\xi(x,0) = 0, \quad \dot{\xi}(x,0) = \delta_y^x,\ y = (\beta, i),$$

then the corresponding solution $\xi_y(x,t)$ of (2) has the form

(7) $$\xi_y(x,t) = \sum_{k=1}^{K} Q_k(x,y)\frac{\sin\omega_k t}{\omega_k}.$$

The inverse problem in the theory of small oscillations may be formulated as follows.

Let a subset $\mathfrak{A}_0 \subset \mathfrak{A}$ of all particles be available for observation. This means that the only oscillations one can study are those generated by the initial perturbations of the particles in $\mathfrak{A}_0$, and for such oscillations only the motion of particles in $\mathfrak{A}_0$ can be observed. In other words, the set of observable data consists of the functions $\xi_y(x,t)$, $0 < t < \infty$, for which $x, y \in A_0 = \mathfrak{A}_0 \times \mathfrak{I}(3)$ only.

Given this set of data when can one reconstruct the whole matrix $L(x,y)$ (x, $y \in A$) and how can this be done?

In what follows, we assume that none of eigenfunctions of $\widehat{L}$ vanish on the whole set A_0. This is of course a necessary condition of solvability since such functions cannot be detected from observations of the particles in $\mathfrak{A}_0$ only. We remark that the hypothesis of Theorem 2 below ensures that there is no eigenfunctions of $\widehat{L}$ which vanish identically on A_0.

Similar problems related to scattering on the quantum graphs have been intencively studied in the last decade see, e.g., [**2**–**11**, **14**], and references therein.

2. Reduction to the inverse spectral problem for Hermitian matrices

We begin with a description of those spectral data of $\widehat{L}$ which can be determined in terms of the observable functions $\xi_y(x,t)$, $0 < t < \infty$, x, $y \in A_0$ defined in (7). The coefficients $Q_k(x,y)$, x, $y \in A_0$ can be found by using the standard machinery of almost periodic functions. It follows from (5), (6) that

$$\sum_{x\in A_0} Q_k(x,x) = \sum_{x\in A_0} E_k(x,x) = \sum_{p_{k-1}<j\le p_k}\ \sum_{x\in A_0} |\psi(j,x)|^2,$$

and we have

$$\sum_{x\in A_0} Q_k(x,x) > 0, \quad k = 1,2,\dots,K,$$

since by our assumption the eigenfunctions of $\widehat{L}$ cannot vanish identically on A_0. Thus the spectrum of the (observable) function

$$\sum_{x\in A_0} \xi_x(x,t) = \sum_{k=1}^{K} \Big(\omega_k^{-1} \sum_{x\in A_0} Q_k(x,x)\Big) \sin \omega_k t$$

includes all frequencies $\{\omega_k = \sqrt{\mu(k)},\ k = 1,2,\dots,K\}$ of our system. We thus get the spectrum of $\widehat{L}$, $\{\mu(k) = \omega_k^2,\ k = 1,2,\dots,K\}$ together with the matrix elements $Q_k(x,y)$, $x,\ y\in A_0$, $k = 1,2,\dots,K$. It follows from (6) that

$$Q_k(x,y)\sqrt{\frac{m(x)}{m(y)}} = E_k(x,y) = \overline{E_k(y,x)} = \overline{Q_k(y,x)}\sqrt{\frac{m(y)}{m(x)}},$$

and

$$\frac{m(x)}{m(y)} = \frac{\overline{Q_k(y,x)}}{Q_k(x,y)}, \quad \text{if} \quad Q_k(x,y) \neq 0.$$

Finally

$$E_k(x,y) = \begin{cases} Q_k(x,y)\sqrt{\frac{\overline{Q_k(y,x)}}{Q_k(x,y)}}, & \text{if} \quad Q_k(x,y) \neq 0, \\ 0, & \text{if} \quad Q_k(x,y) = 0. \end{cases}$$

Thus given the set of the observable functions $\xi_y(x,t)$, $0 < t < \infty$, $x,\ y\in A_0$ we can determine the spectrum of $\widehat{L}$: $\{\mu(k),\ k = 1,2,\dots,K\}$ and the subset $E_k(x,y)$, $x,\ y \in A_0$ of all matrix elements of the orthogonal projections on the eigenspaces $H(\mu(k))$, $k = 1,2,\dots,K$.

Remark. One needs infinite observation time in order to find the precise values of the spectral data of the almost periodic functions $\xi_y(x,t)$, $0 < t < \infty$, $x,\ y \in A_0$. From the observations during finite time T one can find just the approximate values, the error decays as $T^{-1/2}$, $T \to \infty$.

In what follows we treat as *known* all quantities which can be obtained via the observable functions $\xi_y(x,t)$, $0 < t < \infty$, $x,\ y \in A_0$, by applying explicitly described procedures. For example the spectrum of $\widehat{L}$ as well as the matrix elements $E_k(x,y)$, $x,\ y\in A_0$ are known.

Let $\chi_0(x)$ be the characteristic function of the set A_0 and $\widehat{\chi}_0$ be the operator of multiplication by this function. Consider the non-negative operators $\widehat{\chi}_0\widehat{E}_k\widehat{\chi}_0$. Their matrices are known:

$$\chi_0(x)E_k(x,y)\chi_0(y) = \chi_0(x)\left(\sum_{p_{k-1}<j\le p_k} \psi(j,x)\overline{\psi(j,y)}\right)\chi_0(y). \tag{8}$$

Let

$$0 < \varkappa_k(1)^2 \le \varkappa_k(2)^2 \le \cdots \le \varkappa_k(m_k)^2$$

be the positive eigenvalues (with account of multiplicities) of $\widehat{\chi}_0\widehat{E}_k\widehat{\chi}_0$ and $q_k(i,x)$ be the corresponding system of orthonormal eigenvectors. The standard techniques allow one to reconstruct the eigenvalues $\varkappa_k(i)$ and the eigenvectors $q_k(i,x)$ from the matrix elements (8) thus they are also known.

In the lemma below we construct special orthonormal bases for the spaces $H(\mu(k))$.

LEMMA 1. *Suppose that none of the eigenfunctions of $\widehat{L}$ vanish identically on A_0. Then*

(1) *The numbers $n_k = \dim H(\mu(k))$ equal to the numbers m_k of positive eigenvalues of $\widehat{\chi}_0\widehat{E}_k\widehat{\chi}_0$, thus n_k are known.*

(2) *The operator $\widehat{L}$ has a complete orthonormal system of eigenfunctions $\varphi(j,x)$, $j = 1,2,\ldots,N$, such that the restrictions of $\varphi(j,x)$ to A_0 are known:*

$$(\widehat{L} - \lambda(j)\widehat{I})\varphi(j) = 0, \quad \chi_0(x)\varphi(j,x) = \varkappa_k(i)q_k(i,x),$$

here $j = p_{k-1} + i$, $i = 1,2,\ldots,n_k$, $p_0 = 0$, $p_{k+1} = p_k + n_k$, $k = 0,1,\ldots,K-1$.

PROOF. It follows from (8) that the range of $\widehat{\chi}_0\widehat{E}_k\widehat{\chi}_0$ is contained in the linear span of the functions $\chi_0(x)\psi(j,x)$, $p_{k-1} < j \le p_k$. These functions are linearly independent, otherwise there exists a linear combination $\sum\limits_{p_{k-1}<j\le p_k} c(j)\psi(j,x)$, vanishing on A_0. Therefore there exists a system of functions $\{g(j,x),\, p_{k-1} < j \le p_k\}$, which is biorthogonal to $\{\chi_0(x)\psi(j,x),\, p_{k-1} < j \le p_k\}$ and such that $g(j,x) = 0$ for $x \in A \setminus A_0$. By applying (8) once again we obtain

$$\begin{aligned}
\widehat{\chi}_0\widehat{E}_k\widehat{\chi}_0 g(s) &= \sum_{p_{k-1}<j\le p_k} \chi_0(x)\psi(j,x) \sum_{y\in A_0} g(s,y)\chi_0(y)\overline{\psi(j,y)} \\
&= \sum_{p_{k-1}<j\le p_k} \chi_0(x)\psi(j,x)\delta_j^s = \chi_0(x)\psi(s,x),
\end{aligned}$$

i.e., the functions $\chi_0(x)\psi(s,x)$, $p_{k-1} < s \le p_k$ are in the range of $\widehat{\chi}_0\widehat{E}_k\widehat{\chi}_0$. Since they are linearly independent they form a basis for the range of $\widehat{\chi}_0\widehat{E}_k\widehat{\chi}_0$, so the dimension of this space is n_k.

On the other hand the eigenfunctions $q_k(i,x)$, $i = 1,2,\ldots,m_k$ of $\widehat{\chi}_0\widehat{E}_k\widehat{\chi}_0$ corresponding to the positive eigenvalues also form a basis for the range of $\widehat{\chi}_0\widehat{E}_k\widehat{\chi}_0$. Therefore $n_k = m_k$.

In order to prove the second part of the lemma we note that

$$\varkappa_k(i)^2 q_k(i,x) = \chi_0(x) \sum_{p_{k-1}<j\le p_k} \psi(j,x)c_i^j(k), \tag{9}$$

where

$$c_i^j(k) = \sum_{y\in A} \overline{\psi(j,y)}\chi_0(y)q_k(i,y).$$

Since $q_k(i,x)$, $i = 1,2,\ldots,m_k$ form an orthonormal system we obtain

$$\varkappa_k(i)^2\delta_{i'}^i = \sum_{p_{k-1}<j\le p_k} \overline{c_i^j(k)}c_{i'}^j(k). \tag{10}$$

Now define the functions $\varphi(j,x)$, $p_{k-1} < j \le p_k$, $k = 1,2,\ldots,K$ as

$$\varphi(s,x) = \varkappa_k(i)^{-1} \sum_{p_{k-1}<j\le p_k} \psi(j,x)c_i^j(k) \quad p_{k-1} < s \le p_k,$$

where $i = s - p_{k-1}$. These functions belong to $H(\mu(k))$ and, by (9), satisfy the relations

$$\chi_0(x)\varphi(s,x) = \varkappa_k(i)q_k(i,x), \quad i = s - p_{k-1},\ k = 1,2,\dots\ k,\ p_{k-1} < s \leq p_k.$$

The right-hand sides of these relations are known.

It remains to prove that $\varphi(s,x)$, $p_{k-1} < s \leq p_k$, $k = 1,2,\dots,K$ are orthogonal. Indeed the subspaces $H(\mu(k))$ are mutually orthogonal so the functions $\varphi(s,x)$ which correspond to different k are orthogonal. If $\varphi(s,x)$ and $\varphi(s',x)$ belong to the same $H(\mu(k))$, then

$$\big(\varphi(s),\varphi(s')\big) = \big(\varkappa_k(i)\varkappa_k(i')\big)^{-1} \sum_{p_{k-1}<j\leq p_k} \sum_{p_{k-1}<j'\leq p_k} c_i^j(k)\overline{c_{i'}^{j'}(k)} \sum_{x\in A} \psi(j,x)\overline{\psi(j',x)}$$

and, since $\psi(j,x)$ are orthogonal, relation (10) yields

$$\begin{aligned}\big(\varphi(s),\varphi(s')\big) &= \big(\varkappa_k(i)\varkappa_k(i')\big)^{-1} \sum_{p_{k-1}<j\leq p_k} \sum_{p_{k-1}<j'\leq p_k} c_i^j(k)\overline{c_{i'}^{j'}(k)}\delta_{j'}^{j} \\ &= \big(\varkappa_k(i)\varkappa_k(i')\big)^{-1} \sum_{p_{k-1}<j\leq p_k} c_i^j(k)\overline{c_{i'}^{j}(k)} = \frac{\varkappa_k(i')}{\varkappa_k(i)}\delta_i^{i'} = \delta_s^{s'}.\end{aligned}$$

Thus $\{\varphi(s,x)\}$ is a complete set of eigenfunctions of $\widehat{L}$. These functions take the known values $\varkappa_k(i)q_k(i,x)$, here $i = s - p_{k-1}$, $p_{k-1} < s \leq p_k$ on A_0. □

By the spectral data of Hermitian matrix we mean the set of its eigenvalues $\lambda(j)$ (with account of their multiplicities) and the corresponding eigenfunctions $\varphi(j,x)$. Clearly the whole matrix can be reconstructed form its spectral data

$$L(x,y) = \sum_j \lambda(j)\varphi(j,x)\overline{\varphi(j,y)}. \tag{11}$$

One obtains *restriction of the spectral data to the set* A_0 by taking the restrictions of the eigenfunctions to this set.

Inverse problem of spectral analysis for Hermitian matrices can be formulated as follows: **reconstruct the whole matrix from the restriction of its spectral data to some set.** *In other words, given the spectrum, one has to reconstruct the eigenfunctions from their restriction to a given set.*

It follows from Lemma 1 that, given the observable data (i.e., the functions $\xi_y(x,t)$, $0 < t < \infty$, x, $y \in A_0$) one can find the eigenvalues $\lambda(j)$ (and their multiplicities) of the matrix $L(x,y)$ as well as the restrictions of its eigenfunctions $\varphi(j,x)$ to the set A_0, that is the restriction of the spectral data of L to the set A_0. So inverse problems in the theory of small oscillations can be reduced to inverse problems of spectral analysis of Hermitian matrices.

3. Inverse problems of spectral analysis of Hermitian matrices (preliminary results)

Let $\{\varphi(j,x),\ j = 1,2,\dots,N\}$ be an orthonormal system of eigenfunctions of the Hermitian matrix $L(x,y)$. Consider the isometry mapping

$$f(x) \mapsto \widetilde{f}(j) = \sum_{x\in A} f(x)\overline{\varphi(j,x)}$$

of the space $L^2(A)$ onto N-dimensional Hilbert space $L^2(\mathfrak{I}(N))$ of functions $\widetilde{f}(j)$, $j \in \mathfrak{I}(N) = \{1, 2, \ldots, N\}$. The standard orthonormal basis in $L^2(A)$ has the form $\{e_y\}_{y\in A}$, where $e_y(x) = \delta^x_y$. Its image

$$\widetilde{e}_y(j) = \sum_{x\in A} e_y(x)\overline{\varphi(j,x)} = \overline{\varphi(j,y)}, \quad j = 1, 2, \ldots, N$$

is an orthonormal basis in $L^2(\mathfrak{I}(N))$, the elements of this basis are indexed by the points $y \in A$.

Assume that the values of the eigenfunctions $\varphi(j,x)$ are known for $x \in A_0 \subset A$. This means that the only basis elements $\widetilde{e}_y(j) = \overline{\varphi(j,y)}$ which are known are those indexed by the points $y \in A_0$. Let $H = \mathrm{Span}(\{\widetilde{e}_y,\ y \in A_0\})$. The unknown basis elements $\widetilde{e}_y(j) = \overline{\varphi(j,y)}$, $\ y \notin A_0$ form an orthonormal basis for $H^\perp$. One can construct some orthonormal basis $Q(j,z)$, $z \in A \setminus A_0$, $j = 1, 2, \ldots, N$ for $H^\perp$, then the general solution of the inverse spectral problem has the form

$$\varphi(j,y) = \sum_{z\in\overline{A_0}} V(y,z)Q(j,z), \quad y \in A \setminus A_0,$$

here the functions $Q(j,z)$ are known while $V(y,z)$, $y,\ z \in A \setminus A_0$ is an arbitrary unitary matrix.

So the solution of the inverse spectral problem is not unique; in order to make it unique one needs additional information on the matrix $L(x,y)$.

We consider the systems where each particle interacts with it closest neighbors only and with the external field. For such systems the potential energy matrix $U(x,y)$ and, hence, $L(x,y) = (m(x)m(y))^{-1/2}U(x,y)$ has a lot of zero entries, their positions are known. This additional information provides unique solvability of the corresponding inverse problems.

Consider the situation in more details. Let $[\alpha] = \{(\alpha,1), (\alpha,2), (\alpha,3)\}$ be the set of degrees of freedom of the particle α. Then expression (1) for the potential energy takes the form

$$U = \frac{1}{2}\sum_{\alpha\in\mathfrak{A}}\sum_{\beta\in\mathfrak{A}}\sum_{x\in[\alpha]}\sum_{y\in[\beta]} U(x,y)\xi(x)\xi(y) = \frac{1}{2}\sum_{\alpha\in\mathfrak{A}}\sum_{\beta\in\mathfrak{A}}\Big(\widehat{U}^{\alpha\beta}\vec{r}(\alpha), \vec{r}(\beta)\Big), \tag{12}$$

here $\vec{r}(\alpha)$, $\vec{r}(\beta)$ denote the deviations of the particles α, β of the equilibrium and $\widehat{U}^{\alpha\beta}$ denote the operators with the matrices

$$U^{\alpha\beta} = U^{\alpha\beta}(i,j) = U^{\beta\alpha}(j,i)^* = U(x,y), \quad x \in [\alpha], y \in [\beta],$$

these operators act in the three dimensional Hilbert space $L^2(\mathfrak{I}(3))$.

The matrix $U(x,y)$, $x, y \in A$ is represented as a block matrix with the entries $U^{\alpha\beta}$, α, $\beta \in \mathfrak{A}$. We consider the similar representation for $L(x,y)$. The corresponding entries have the form

$$L^{\alpha\beta} = L^{\alpha\beta}(i,j) = L(x,y) = \big(m(\alpha)m(\beta)\big)^{-\frac{1}{2}}U(x,y), \quad x \in [\alpha], y \in [\beta].$$

We say that *the particles α, β ($\alpha \neq \beta$) strongly interact if the operator $\widehat{U}^{\alpha\beta}$ is invertible (i.e.,* $\mathrm{Det}(U^{\alpha\beta}) \neq 0$*)*, we say that *the particles α, β ($\alpha \neq \beta$) interact if* $\mathrm{Det}(U^{\alpha\beta}) = 0$ *but* $\widehat{U}^{\alpha\beta} \neq 0$, finally we say that *the particles α, β ($\alpha \neq \beta$) do not interact if $\widehat{U}^{\alpha\beta} = 0$.*

For a wide class of the mechanical systems the total potential energy is the sum of energies of interaction with the external field and the energies of pairwise interactions. We denote by E^{α} the energy of interaction of the particle α with external field and by $E^{\alpha\beta}$ the energy of interaction between the particles α and β ($\alpha \neq \beta$). The value E^{α} depends on the position of the particle α only while $E^{\alpha\beta}$ depends only on mutual location of the particles α and β. One can write the expansions of E^{α} and $E^{\alpha\beta}$ near the equilibrium as

$$E^{\alpha} = E_0^{\alpha} + \big(\vec{E}_1^{\alpha}, \vec{r}(\alpha)\big) + \frac{1}{2}\big(\widehat{E}_2^{\alpha}\vec{r}(\alpha), \vec{r}(\alpha)\big) + \dots ;$$

$$E^{\alpha\beta} = E_0^{\alpha\beta} + \big(\vec{E}_1^{\alpha\beta}, \vec{r}(\alpha) - \vec{r}(\beta)\big) + \frac{1}{2}\big(\widehat{E}_2^{\alpha\beta}(\vec{r}(\alpha) - \vec{r}(\beta)), \vec{r}(\alpha) - \vec{r}(\beta)\big) + \dots .$$

Here $\vec{E}_1^{\alpha}$, $\vec{E}_1^{\alpha\beta} = -\vec{E}_1^{\beta\alpha}$ are vectors in $L^2(\mathfrak{I}(3))$ and $\widehat{E}_2^{\alpha}$, $\widehat{E}_2^{\alpha\beta} = \widehat{E}_2^{\beta\alpha}$ are non-negative operators in this space. We have $\widehat{E}_2^{\alpha} = 0$ (respectively $\widehat{E}_2^{\alpha\beta} = 0$) if the particle α does not interact with the external field (respectively α and β do not interact) and $\widehat{E}_2^{\alpha} > 0$ if the particle α strongly interacts with the external field (respectively $\widehat{E}_2^{\alpha\beta} > 0$ in the case when α and β strongly interact).

The sum of the linear terms vanishes since we consider expansion near the equilibrium. Besides the sum of E_0^{α} and $E_0^{\alpha\beta}$ is zero since we assume that the system has zero energy at the equilibrium. We also ignore the higher order terms. The expression for the potential energy takes the form

$$U = \frac{1}{2}\sum_{\alpha\in\mathfrak{A}}\big(\widehat{E}_2^{\alpha}\vec{r}(\alpha), \vec{r}(\alpha)\big) + \frac{1}{4}\sum_{\alpha\in\mathfrak{A}}\sum_{\beta\in\mathfrak{A}}\big(\widehat{E}_2^{\alpha\beta}(\vec{r}(\alpha) - \vec{r}(\beta)), \vec{r}(\alpha) - \vec{r}(\beta)\big),$$

the operators $\widehat{E}_2^{\alpha\alpha}$ can be chosen arbitrary for they do not contribute to the total energy. Taking into account that $\widehat{E}_2^{\alpha\beta} = \widehat{E}_2^{\beta\alpha}$, we obtain

$$U = \frac{1}{2}\sum_{\alpha\in\mathfrak{A}}\Big(\Big\{\widehat{E}_2^{\alpha} + \sum_{\beta\neq\alpha}\widehat{E}_2^{\alpha\beta} + \widehat{E}_2^{\alpha\alpha}\Big\}\vec{r}(\alpha), \vec{r}(\alpha)\Big) - \frac{1}{2}\sum_{\alpha\in\mathfrak{A}}\sum_{\beta\in\mathfrak{A}}\big(\widehat{E}_2^{\alpha\beta}\vec{r}(\alpha), \vec{r}(\beta)\big).$$

Since the operators $\widehat{E}_2^{\alpha\alpha}$ can be chosen arbitrary we set

$$\widehat{E}_2^{\alpha\alpha} = -\Big(\widehat{E}_2^{\alpha} + \sum_{\alpha\neq\beta}\widehat{E}_2^{\alpha\beta}\Big)$$

in order to obtain the final expression:

$$U = \frac{1}{2}\sum_{\alpha\in\mathfrak{A}}\sum_{\beta\in\mathfrak{A}}\big(-\widehat{E}_2^{\alpha\beta}\vec{r}(\alpha), \vec{r}(\beta)\big),$$

This is a special case of equation (12) with additional important porperty:

the non-diagonal blocks $U^{\alpha\beta} = U^{\beta\alpha}$ *are non-negative matrices of order three.* Clearly the matrix $L(x, y)$ also possesses this property.

Above we assumed that each particle has three degrees of freedom. The same arguments are valid if one replaces particles by other objects having any fixed number $m \geq 1$ degrees of freedom. That is why we consider *the inverse problem of spectral analysis in the class* $\mathfrak{M}_0$ *consisting of positively defined matricies* $L(x, y)$ *of order* mn *and such that their non-diagonal blocks*

$$L^{\alpha\beta}(i, j) = L^{\beta\alpha}(i, j) = L(x, y)\Big|_{\substack{x = (\alpha, i)\\ y = (\beta, j)}}, \quad \alpha \neq \beta, \quad i, j \in \mathfrak{I}(m) = \{1, 2, \dots, m\}$$

are non-negative matrices of order m.

We denote the set of all Hermitian matricies of order mn by $\mathfrak{M}$. The set of all observable particles we still denote by $\mathfrak{A}_0 \subset \mathfrak{A}$, the set of degrees of freedom of a particle α we denote by $[\alpha] = \{(\alpha, 1), (\alpha, 2), \dots, (\alpha, m)\}$. The set of all degrees of freedom is $A = \mathfrak{A} \times \mathfrak{I}(m)$, where $\mathfrak{I}(m) = \{1, 2, \dots, m\}$. We assume that the restrictions of the spectral data to the set $A_0 = \mathfrak{A}_0 \times \mathfrak{I}(m)$ are known. As before we denote the block entries of the matrix $M = M(x, y)$, x, $y \in A$ by

$$M^{\alpha\beta} = M^{\alpha\beta}(i, j) = M(x, y)\Big|_{\substack{x = (\alpha, i) \\ y = (\beta, j)}}, \quad i, j \in \mathfrak{I}(m); \quad \alpha, \beta \in \mathfrak{A}.$$

Let a matrix $M = M(x, y) \in \mathfrak{M}$ be fixed.

DEFINITION 1. *We say that a set* $\mathfrak{B} \subset \mathfrak{A}$ *admits a* M*-prolongation if there exists a particle* $\alpha \in \mathfrak{B}$ *which interacts strongly with one particle* $\overline{\alpha} \in \overline{\mathfrak{B}} = \mathfrak{A} \backslash \mathfrak{B}$ *(i.e.,* $\mathrm{Det}(M^{\alpha\overline{\alpha}}) \neq 0$*) and does not interact with the rest of the particles in* $\overline{\mathfrak{B}}$*, i.e.* $M^{\alpha\overline{\beta}} = 0$ *for all* $\overline{\beta} \in \overline{\mathfrak{B}}$, $\overline{\beta} \neq \overline{\alpha}$*).*

We say that the set $\mathfrak{B}^{(1)} = \mathfrak{B} \cup \{\overline{\alpha}\}$ is a prolongation of $\mathfrak{B}$. In the case $\mathfrak{B}^{(1)}$ admits M-prolongation as well, we denote its prolongation $\left(\mathfrak{B}^{(1)}\right)^{(1)}$ by $\mathfrak{B}^{(2)}$ and further by induction $\mathfrak{B}^{(k+1)} = \left(\mathfrak{B}^{(k)}\right)^{(1)}$.

We say that the chain of such prolongations

$$\mathfrak{B} = \mathfrak{B}^{(0)} \subset \mathfrak{B}^{(1)} \subset \dots \subset \mathfrak{B}^{(p)}, \tag{13}$$

which begins at a set $\mathfrak{B} \subset \mathfrak{B}$ is *maximal* if the last set $\mathfrak{B}^{(p)}$ does not admit prolongation. There could be several maximal chains which begin at the given set $\mathfrak{B}$, generally speaking.

LEMMA 2. *Given a subset* $\mathfrak{B} \subset \mathfrak{A}$*, all maximal chains that begin at* $\mathfrak{B}$ *end with the same set* $[\mathfrak{B}] \subset \mathfrak{A}$*.*

DEFINITION 2. *The set* $[\mathfrak{B}]$ *is called the maximal* M*-prolongation of the set* $\mathfrak{B}$*.*

PROOF. It suffices to prove that given two maximal chains

$$\mathfrak{B} = \mathfrak{B}_1^{(0)} \subset \mathfrak{B}_1^{(1)} \subset \dots \subset \mathfrak{B}_1^{(p_1)}, \quad \mathfrak{B} = \mathfrak{B}_2^{(0)} \subset \mathfrak{B}_2^{(1)} \subset \dots \subset \mathfrak{B}_2^{(p_2)}, \tag{14}$$

we have $\mathfrak{B}_1^{(p_1)} \subset \mathfrak{B}_2^{(p_2)}$.

Assume the opposite: $\mathfrak{B}_1^{(p_1)} \not\subset \mathfrak{B}_2^{(p_2)}$. Take m such that

$$\mathfrak{B}_1^{(m)} \not\subset \mathfrak{B}_2^{(p_2)}, \quad \mathfrak{B}_1^{(m-1)} \subset \mathfrak{B}_2^{(p_2)}.$$

Since $\mathfrak{B}_1^{(0)} = \mathfrak{B} = \mathfrak{B}_2^{(0)}$ we have $m \geq 1$.

Let $\mathfrak{B}_1^{(m)} = \mathfrak{B}_1^{(m-1)} \cup \{\bar{\alpha}\}$. There exists $\alpha \in \mathfrak{B}_1^{(m-1)}$ which strongly interacts with $\bar{\alpha}$ and does not interact with other particles in $\mathfrak{A} \setminus \mathfrak{B}_1^{(m-1)}$. In particular α does not interact with the particles in $\left(\mathfrak{A} \setminus \mathfrak{B}_2^{(p_2)}\right) \backslash \{\bar{\alpha}\}$. Since $\alpha \in \mathfrak{B}_1^{(m-1)} \subset \mathfrak{B}_2^{(p_2)}$ the second chain in (14) cannot be maximal.

□

DEFINITION 3. *We say that a subset* $\mathfrak{B} \subset \mathfrak{A}$ *admits a complete* M*-prolongation, if there exists a chain* (13)*, which ends with the whole set* $\mathfrak{A}$*, i.e.,* $\mathfrak{A} = [\mathfrak{B}]$*.*

It follows from Lemma 2 that any maximal chain can be taken in order to determine whether the set $\mathfrak{B}$ admits a complete M-prolongation.

LEMMA 3. *Let a set $\mathfrak{B} \subset \mathfrak{A}$ admit a complete M-prolongation. Then the eigenfunctions of the matrix $M = M(x,y) \in \mathfrak{M}$ cannot vanish identically on $\mathfrak{B} \times \mathfrak{I}(m)$.*

PROOF. Denote $\overline{\mathfrak{B}} = \mathfrak{A}\backslash\mathfrak{B}$. Let a function $\varphi(x)$ satisfy

$$\lambda\varphi(x) = \sum_{y\in A} M(x,y)\varphi(y)$$

and vanish on $\mathfrak{B} \times \mathfrak{I}(m)$. Then for all $x = (\alpha, i) \in \mathfrak{B} \times \mathfrak{I}(m)$ we have

$$0 = \sum_{y\in\overline{\mathfrak{B}}\times\mathfrak{I}(m)} M(x,y)\varphi(y) = \sum_{\overline{\beta}\in\overline{\mathfrak{B}}}\sum_{j=1}^{m} M^{\alpha\overline{\beta}}(i,j)\varphi((\overline{\beta},j)).$$

Since the set $\mathfrak{B}$ admits M-prolongation, there exists a couple of particles $\alpha \in \mathfrak{B}$, $\overline{\alpha} \in \overline{\mathfrak{B}}$, such that $\mathrm{Det}(M^{\alpha\overline{\alpha}}) \neq 0$ and $M^{\alpha\overline{\beta}} = 0$ for all $\overline{\beta} \in \overline{\mathfrak{B}} \setminus \{\overline{\alpha}\}$. The previous relation yields

$$0 = \sum_{j=1}^{m} M^{\alpha\overline{\alpha}}(i,j)\varphi((\overline{\alpha},j)),$$

and, since $\mathrm{Det}(M^{\alpha\overline{\alpha}}) \neq 0$, we obtain $\varphi((\overline{\alpha},j)) = 0$ for all $j \in \mathfrak{I}(m)$.

Thus $\varphi(x) = 0$ for all $x \in \mathfrak{B}^{(1)} \times \mathfrak{I}(m)$, where $\mathfrak{B}^{(1)} = \mathfrak{B} \cup \{\overline{\alpha}\}$ is an M-prolongation of $\mathfrak{B}$. By repeating this arguments and using the fact that $\mathfrak{B}$ admits complete M-prolongation we obtain that $\varphi(x) = 0$ for all $x \in A = \mathfrak{A} \times \mathfrak{I}(m)$. So the functions which vanish on $x \in \mathfrak{B} \times \mathfrak{I}(m)$ cannot be eigenfunctions of M. □

In the previous section we assumed that the eigenfunctions of the matrix $L(x,y)$ do not vanish identically on $A_0 = \mathfrak{A}_0 \times \mathfrak{I}(m)$. It follows from Lemma 3 that this assumption holds automatically in the case when the set $\mathfrak{A}_0$ of observable particles admits complete L-prolongation.

4. Solution of inverse problem of spectral analysis for Hermitian matricies

Let a matrix $M = M(x,y)$, $x,y \in A$ belong to $\mathfrak{M}$ and $\lambda(s)$, $\varphi(s,x)$, $s = 1,2,\dots,N = mn$, $x \in A$ be its spectral data. Let also B be an arbitrary subset of A and $\overline{B} = A \setminus B$ be its complement.

LEMMA 4. *Denote*

$$F(s,x) = \sum_{s'=1}^{N} P(s,s')\lambda(s')\varphi(s',x),$$

$$D(x,x') = \sum_{s=1}^{N}\sum_{s'=1}^{N} P(s',s)\lambda(s)\lambda(s')\varphi(s,x)\overline{\varphi(s',x')},$$

and

$$P(s,s') = \delta_s^{s'} - \sum_{y\in B}\varphi(s,y)\overline{\varphi(s',y)}.$$

Then the relations

$$\sum_{y\in\overline{B}} M(x,y)\varphi(s,y) = F(s,x), \quad \sum_{y\in\overline{B}} M(x,y)\overline{M(x',y)} = D(x,x'), \tag{15}$$

hold for all $x, x' \in B$ *and all* $s = 1, 2, \ldots, N$.

The right-hand sides in (15) can be expressed in terms of the restrictions of the spectral data of the matrix M to the set B.

PROOF. We use the relation

$$M(x,y) = \sum_s \lambda(s)\varphi(s,x)\overline{\varphi(s,y)}.$$

Then

$$F(s,x) = \sum_{y\in\overline{B}} M(x,y)\varphi(s,y) = \sum_{y\in\overline{B}}\left(\sum_{s'=1}^{N}\lambda(s')\varphi(s',x)\overline{\varphi(s',y)}\right)\varphi(s,y)$$
$$= \sum_{s'=1}^{N}\left(\sum_{y\in\overline{B}}\varphi(s,y)\overline{\varphi(s',y)}\right)\lambda(s')\varphi(s',x) = \sum_{s'=1}^{N} P(s,s')\lambda(s')\varphi(s',x'),$$

where

$$P(s,s') = \sum_{y\in\overline{B}}\varphi(s,y)\overline{\varphi(s',y)} = \sum_{y\in A}\varphi(s,y)\overline{\varphi(s',y)} - \sum_{y\in B}\varphi(s,y)\overline{\varphi(s',y)} \tag{16}$$
$$= \delta_s^{s'} - \sum_{y\in B}\varphi(s,y)\overline{\varphi(s',y)}.$$

Similarly

$$D(x,x') = \sum_{y\in\overline{B}} M(x,y)\overline{M(x',y)}$$
$$= \sum_{y\in\overline{B}}\left(\sum_{s=1}^{N}\lambda(s)\varphi(s,x)\overline{\varphi(s,y)}\right)\left(\sum_{s'=1}^{N}\lambda(s')\overline{\varphi(s',x')}\varphi(s',y)\right)$$
$$= \sum_{s'=1}^{N}\sum_{s=1}^{N}\left(\sum_{y\in\overline{B}}\varphi(s',y)\overline{\varphi(s,y)}\right)\lambda(s)\lambda(s')\varphi(s,x)\overline{\varphi(s',x)}$$

and now relation (16) yields

$$D(x,x') = \sum_{s'=1}^{N}\sum_{s=1}^{N} P(s',s)\lambda(s')\lambda(s)\overline{\varphi(s',x')}\varphi(s,x).$$

□

In what follows, we consider functions as column vectors of length N, their restrictions to the set $[\alpha] = \{(\alpha,1),(\alpha,2),\ldots,(\alpha,m)\}$ are column vectors of length m. We denote by $S(M)$ the spectral data of M and by $S_B(M)$ we denote the restriction of these data to the subset $B \subset A$.

Given a matrix V of the block structure we say that V is a block-diagonal matrix if all non-diagonal block entries in V are 0.

LEMMA 5. *Let a subset $\mathfrak{B} \subset \mathfrak{A}$ admit a M-prolongation $\mathfrak{B}^{(1)}$. Then there exists a unitary block-diagonal matrix V_1 such that the restriction of the spectral data of the matrix $M_1 = V_1 M V_1^*$ to the set $B^1 = \mathfrak{B}^{(1)} \times \mathfrak{I}(m)$ can be expressed in terms of the restriction of the spectral data of the matrix M to the set $B = \mathfrak{B} \times \mathfrak{I}(m)$.*

If in addition M belongs to $\mathfrak{M}_0$, then V_1 is the unit matrix and $M_1 = M$, in this case the restriction of the spectral data of the matrix $M \in \mathfrak{M}_0$ to B^1 can be expressed via their restriction to $B = \mathfrak{B} \times \mathfrak{I}(m)$.

PROOF. Take the particles $\alpha \in \mathfrak{B}$ and $\overline{\alpha} \in \overline{\mathfrak{B}} = \mathfrak{A} \backslash \mathfrak{B}$, such that $\mathfrak{B}^{(1)} = \mathfrak{B} \cup \{\overline{\alpha}\}$, and also $\operatorname{Det}(M^{\alpha\overline{\alpha}}) \neq 0$ and $M^{\alpha\overline{\beta}} = 0$ for all $\overline{\beta} \in \overline{\mathfrak{B}} \setminus \{\overline{\alpha}\}$. Lemma 4 with $B = \mathfrak{B} \times \mathfrak{I}(m)$, $x = (\alpha, i)$, $x' = (\alpha, i')$, yields

$$\sum_{y \in [\overline{\alpha}]} M(x,y)\varphi(s,y) = F(s,x); \quad \sum_{y \in [\overline{\alpha}]} M(x,y)\overline{M(x',y)} = D(x,x'),$$

The right-hand side of these relations can be expressed in terms of $S_B(M)$. We have

$$M^{\alpha\overline{\alpha}} \varphi_{\overline{\alpha}}(s) = F_\alpha(s); \quad M^{\alpha\overline{\alpha}} \left(M^{\alpha\overline{\alpha}}\right)^* = D_\alpha, \tag{17}$$

where the $m \times m$ matrix $D_\alpha = D_\alpha(p, p')$ is defined as

$$D_\alpha(p,p') = D_\alpha(x,x') \Big|_{\substack{x = (\alpha, p) \\ x' = (\alpha, p')}} .$$

The matrix D_α can be expressed in terms of $S_B(M)$. Therefore the matrices $D_\alpha^{\frac{1}{2}} = \sqrt{M^{\alpha\overline{\alpha}} \left(M^{\alpha\overline{\alpha}}\right)^*}$ and $D_\alpha^{-\frac{1}{2}}$ also can be found. Then

$$M^{\alpha\overline{\alpha}} = D_\alpha^{\frac{1}{2}} V_{\overline{\alpha}} \tag{18}$$

where $V_{\overline{\alpha}}$ is some unitary matrix and

$$F_\alpha(s) = M^{\alpha\overline{\alpha}} \varphi_{\overline{\alpha}}(s) = D_\alpha^{\frac{1}{2}} V_{\overline{\alpha}} \varphi_{\overline{\alpha}}(s).$$

Finally

$$V_{\overline{\alpha}} \varphi_{\overline{\alpha}}(s) = D_\alpha^{-\frac{1}{2}} F_\alpha(s). \tag{19}$$

The matrix $D_\alpha^{-\frac{1}{2}} F_\alpha(s)$ can be expressed through $S_B(M)$.

If in addition the matrix $M^{\alpha\overline{\alpha}}$ is negative, in particular if $M \in \mathfrak{M}_0$ realation (18) yields $V_{\overline{\alpha}} = -I(m)$, where $I(m)$ is the unit matrix of order m.

Denote by V_1 the unitary block-diagonal matrix with diagonal entries of the form

$$V_1^{\beta\beta} = \begin{cases} I(m), & \text{if } \beta \neq \overline{\alpha}, \\ -V_{\overline{\alpha}}, & \text{if } \beta = \overline{\alpha}. \end{cases} \tag{20}$$

The eigenvalues of the matrices M and $M_1 = V_1 M V_1^*$ coincide (counting their multiplicities) while the eigenfunctions $\varphi^{(1)}(s) = \varphi^{(1)}(s,x)$ of the matrix M_1 can be expressed via the eigenfunctions $\varphi(s) = \varphi(s,x)$ of the matrix M as

$$\varphi^{(1)}(s) = V_1 \varphi(s).$$

It follows now from (19) and (20), that

$$\varphi^{(1)}(s,x) = \begin{cases} \varphi(s,x), & \text{if } x \notin [\overline{\alpha}], \\ -V_{\overline{\alpha}} \varphi_{\overline{\alpha}}(s,x) = -D_\alpha^{-\frac{1}{2}} F_\alpha(x,s), & \text{if } x \in [\overline{\alpha}]. \end{cases}$$

We have $S_B(M_1) = S_B(M)$ and, since $-D_\alpha^{-\frac{1}{2}} F_\alpha(s)$ can be expressed in terms of $S_B(M)$, we see that $S_{B^1}(M_1)$ also can be expressed through $S_B(M)$. If in addition $M \in \mathfrak{M}_0$, then $V_{\overline{\alpha}} = -I(m)$, $V_1 = I$ and $M_1 = M$, i.e., in this case $S_{B^1}(M)$ can be expressed in terms of $S_B(M)$.

Below we list the operations to be done in order to find the set $S_{B^1}(M_1)$ in terms of $S_B(M)$:

1) Evaluate

$$P(s,s') = \delta_s^{s'} - \sum_{y \in B} \varphi(s,y)\overline{\varphi(s',y)}.$$

2) Evaluate

$$F_\alpha(s,p) = \sum_{s'=1}^{N} P(s,s')\lambda(s')\varphi_\alpha(s',p).$$

3) Evaluate the elements $D_\alpha(p,p')$ of the matrix D_α

$$D_\alpha(p,p') = \sum_{s=1}^{N}\sum_{s'=1}^{N} P(s,s')\lambda(s)\lambda(s')\varphi_\alpha(s,p)\overline{\varphi_\alpha(s',p')}.$$

4) Find the elements $D_\alpha^{-\frac{1}{2}}(p,p')$ of the matrix $(\sqrt{D_\alpha})^{-1}$ using the relation

$$D_\alpha^{-\frac{1}{2}}(p,p') = \sum_{s=1}^{m} \frac{\nu(s,p)\overline{\nu(s,p')}}{\sqrt{\mu_s}},$$

here μ_s and $\nu(s,p)$ $(s,\, p \in \mathfrak{I}(m))$ are the eigenvalues and normalized eigenfunctions of D_α

5) Evaluate the restriction of the eigenfunctions $\varphi^{(1)}(s,x)$ of the matrix M_1 to B^1 by using the relation

$$\varphi^{(1)}(s,x) = \begin{cases} \varphi(s,x), & x \in B, \\ -\sum\limits_{p'=1}^{m} D_\alpha^{-\frac{1}{2}}(p,p')F_\alpha(s,p'), & x = (\overline{\alpha},p). \end{cases}$$

□

One can represent transition from $S_B(M)$ to $S_{B^1}(M_1)$ as:

$$S_B(M) \to S_{B^1}(M_1), \quad M_1 = V_1 M V_1^*,$$

here the arrow corresponds to the above described operations over the set $S_B(M)$ and V_1 is a block-diagonal unitary matrix. This matrix is unknown in the general case. If in addition $M \in \mathfrak{M}_0$, then V_1 is the unit matrix.

Theorem 1. *Let $\mathfrak{B} \subset \mathfrak{A}$ and $[\mathfrak{B}]$ be its maximal M-prolongation. Then there exists a block-diagonal unitary matrix Usuch that the restriction to the set $[B] = [\mathfrak{B}] \times \mathfrak{I}(m)$ of the spectral data of the matrix UMU^* can be writen in terms of restriction of the spectral data of the matrix M to $B = \mathfrak{B} \times \mathfrak{I}(m)$.*

If in addition $M \in \mathfrak{M}_0$ then $U = I$ and the restriction to $[B] = [\mathfrak{B}] \times \mathfrak{I}(m)$ of the spectral data of the matrix M can be expressed in terms of the restriction of its spectral data to the set $B = \mathfrak{B} \times \mathfrak{I}(m)$.

Proof. Let V be an arbitrary unitary block-diagonal matrix. Then the block entries of the matrices M and $M_1 = VMV^*$ are related by the equation

$$M_1^{\alpha\beta} = V^{\alpha\alpha} M^{\alpha\beta} (V^{\beta\beta})^*.$$

Therefore a subset $\mathfrak{B} \subset \mathfrak{A}$ admits M_1 prolongation if and only if it admits M prolongation.

Let M

$$\mathfrak{B} = \mathfrak{B}^{(0)} \subset \mathfrak{B}^{(1)} \subset \cdots \subset \mathfrak{B}^{(p)} = [\mathfrak{B}],$$

be a maximal chain of M-prolongations of the set $\mathfrak{B}$. We apply Lemma 5 in order to proceed

$$S_B(M) \to S_{B^1}(M_1), \quad M_1 = V_1 M V_1^*,$$

here V_1 is a block-diagonal matrix and $B^1 = \mathfrak{B}^{(1)} \times \mathfrak{I}(m)$. It follows from the remark above that $\mathfrak{B}^{(1)}$ admits M_1-prolongation and we can apply Lemma 5, once again in order to make the next transition

$$S_{B^1}(M_1) \to S_{B^2}(M_2), \quad M_2 = V_2 M_1 V_2^* = V_2 V_1 M (V_2 V_1)^*,$$

and so on. The process results in a sequence of consequent transitions

$$S_B(M) \to S_{B^1}(M_1) \to \cdots \to S_{B^p}(M_p),$$

$$M_k = (V_k V_{k-1} \dots V_1) M (V_k V_{k-1} \dots V_1)^*, \quad B^k = \mathfrak{B}^{(k)} \times \mathfrak{I}(m),$$

each follows the procedure described in Lemma 5. We have $M_p = UMU^*$, where $U = V_p V_{p-1} \cdots V_1$ is a block-diagonal unitary matrix and $B^p = \mathfrak{B}^{(p)} \times \mathfrak{I}(m) = [\mathfrak{B}] \times \mathfrak{I}(m) = [B]$, and finally

$$S_{B^p}(M_p) = S_{[B]}(UMU^*),$$

i.e., the restriction to the set $[B]$ of the spectral data of the matrix UMU^* can be expressed in terms of the restriction to $B = \mathfrak{B} \times \mathfrak{I}(m)$ of the spectral data of M by applying the sequence of operations explictely described in Lemma 3.

If in addition $M \in \mathfrak{M}_0$, then $V_k = I$ and $M_p = M$, i.e., the restriction of the spectral data of a matrix $M \in \mathfrak{M}_0$ to the set $[B] = [\mathfrak{B}] \times \mathfrak{I}(m)$ can be explicitely described in terms of their restriction to the set $B = \mathfrak{B} \times \mathfrak{I}(m)$. □

Let L be the matrix of the mechanical system considered in §2. If we assume that the eigenfunctions of L do not vanish identically on the set $\mathfrak{A}_0$ of observable particles, then the inverse problem for small oscillations reduces to the inverse problem of spectral analysis for L. Lemma 3 and Theorem 1 now yield the following result.

Theorem 2. *If the matrix L belongs to the set $\mathfrak{M}_0$ and the set $\mathfrak{A}_0 \subset \mathfrak{A}$ of observable particles admits complete L-prolongation then the inverse problem for small oscillations admits unique solution. This solution can be constructed by the algorithms in §3 and Theorem 1.*

5. Examples

This section contains examples of the systems for which one can find the maximal L-prolongations of sets $\mathfrak{B} \subset \mathfrak{A}$ and also examples of the sets which contain a given set $\mathfrak{B}$ and admit a complete L-prolongation. At the end of the section we describe the general form of block matrices of order N which correspond to a set which admits a complete L-prolongation and consists of $q < N$ particles.

In order to study L-prolongation it suffices to know the set Π of pairs of interacting particles (α, β) and its subset Π' which consists of pairs of strongly interacting particles.

We say that the pair $\Gamma = (\mathfrak{A}, \Pi)$ is the *graph of interactions* the particles $\alpha \in \mathfrak{A}$ are the *vertices*, and the pairs $(\alpha, \beta) \in \Pi$ are *edges* og Γ. The pairs $(\alpha', \beta') \in \Pi'$ are called the *marked edges* of Γ.

The number of edges which contain a given vertex $\alpha \in \mathfrak{A}$ (i.e. the number of particles which interact with α) is called the *multiplicity* of α, we denote it by $\mathrm{Mult}(\alpha)$. We denote the *degree* of the graph as $\deg(\Gamma) = \max\{\mathrm{Mult}(\alpha);\ \alpha \in \mathfrak{A}\}$ and say that $\alpha \in \mathfrak{A}$ is an *inner* vertex if $\mathrm{Mult}(\alpha)=\deg(\Gamma)$, otherwise we say that α is an *external* vertex in Γ.

Given a set $\mathfrak{B} \subset \mathfrak{A}$ and a particle $\alpha \in \mathfrak{B}$, we define the *external multiplicity* of α (with respect to $\mathfrak{B}$) as the number of particles $\beta \in \mathfrak{A} \setminus \mathfrak{B}$ which interact with α. We denote it as $\mathrm{Mult}_{\mathfrak{B}}(\alpha)$.

We will employ the graphical representation, where vertices are represented by points on the plane or in the space (it may be convenient to locate them in the equilibrium position) while edges are represented by disjoint smooth curves connecting the interacting particles.

In order to find the maximal L-prolongation $[\mathfrak{B}]$ of the set $\mathfrak{B} \subset \mathfrak{A}$, we need to find a particle $\beta_0 \in \mathfrak{B}$ with $\mathrm{Mult}_{\mathfrak{B}}(\beta_0) = 1$, and which strongly interact with a particle $\alpha_1 \in \mathfrak{A} \setminus \mathfrak{B}$. If there are no such particles then $\mathfrak{B}$ does not admit L-prolongation, i.e., $[\mathfrak{B}] = \mathfrak{B}$. Otherwise one considers the $\mathfrak{B}^{(1)} = \mathfrak{B} \cup \{\alpha_1\}$ and then repeat the procedure obtaining a chain of L-prolongations

$$\mathfrak{B} = \mathfrak{B}^{(0)} \subset \mathfrak{B}^{(1)} \subset \dots \subset \mathfrak{B}^{(p)}.$$

This procedure ends at a set $\mathfrak{B}^{(p)}$, which does not admit L-prolongation. We then have $[\mathfrak{B}] = \mathfrak{B}^{(p)}$.

On each step of this procedure we have $\mathfrak{B}^{(i)} = \mathfrak{B}^{(i-1)} \cup \alpha_i$, where α_i is a particle in $\mathfrak{A} \setminus \mathfrak{B}^{(i-1)}$ which strongly interact with $\beta_{i-1} \in \mathfrak{B}^{(i-1)}$. The edge which connects β_{i-1} and α_i belongs to Π', we provide this edge with an arrow directed from β_{i-1} to α_i, also we mark by $\bullet$ the particles in the original set $\mathfrak{B}$ and by $\circ$ the added particles, they belong to $[\mathfrak{B}] \setminus \mathfrak{B}$. We just denote $\mathfrak{B}(\bullet) = \mathfrak{B}$ and $\mathfrak{B}(\circ) = [\mathfrak{B}] \setminus \mathfrak{B}$.

Let $[\mathfrak{B}(\bullet)] \neq \mathfrak{A}$. In order to find a set which contains $\mathfrak{B}$ and also admits L-prolongation to the whole $\mathfrak{A}$, one has to add some subset of $\mathfrak{A} \setminus [\mathfrak{B}(\bullet)]$. We mark the points form this set by $\blacksquare$, the set of all points added we denote by $\mathfrak{B}(\blacksquare)$. Then we consider the set $[\mathfrak{B}(\bullet, \blacksquare)]$ which is the maximal L prolongatin of $[\mathfrak{B}(\bullet)] \cup \mathfrak{B}(\blacksquare)$ and consider the set of added points $\mathfrak{B}(\square) = [\mathfrak{B}(\bullet, \blacksquare)] \setminus ([\mathfrak{B}(\bullet)] \cup \mathfrak{B}(\blacksquare)$. In case $[\mathfrak{B}(\bullet, \blacksquare)] \neq \mathfrak{A}$ we repeat the procedure by adding $\blacktriangle$, say, and so on. Below the realization of this procedure is illustrated for some concrete graphs.

We say that the system is irreducible if it cannot be decomposed into a union of two systems which do not interact. Graphs of such systems are always connected.

The structure of irreducible systems of degree two is very simple: they contain either two (Fig. 1 a-c) or none (Fig. 1 d) external particles. For such systems one can uniquely reconstruct matrices $L \in \mathfrak{M}_0$ in terms of restriction of their spectral data to $\mathfrak{B}(\bullet) \times \mathfrak{I}(m)$, yet this problem is overdetermined when $\mathfrak{B}(\bullet)$ consists of two particles.

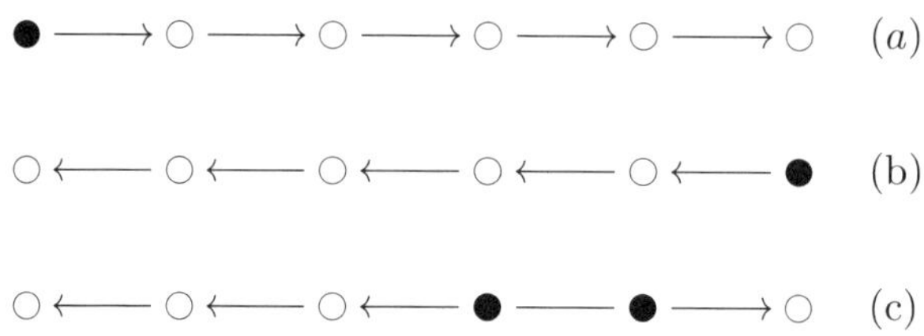

Fig. 1

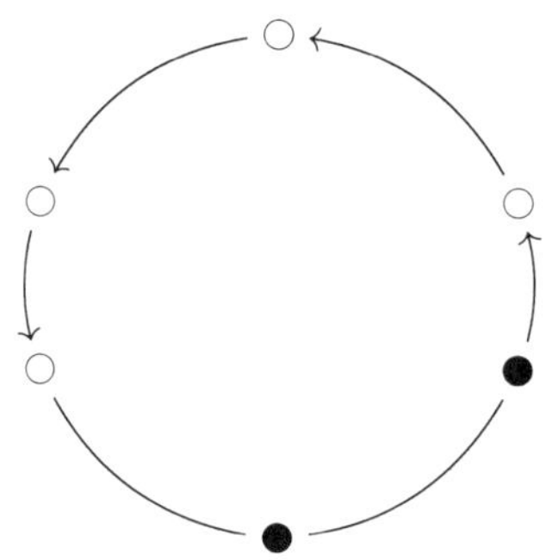

Fig. 1 d

One can easily find sets which admits L prolongation for the systems whose graph is a tree: assuming that all interactions are strong, the set of all particles of multiplicity 1 admits complete prolongation.

On Fig. 2 we presented an example of a graph of the third degree, each external vertex is of multiplicity 2, therefore all sets which admits complete prolongation have to contain two particles at least. The maximal prolongation of the two points set $\mathfrak{B}(\bullet)$ consists of five particles, and $[\mathfrak{B}(\bullet)] \neq \mathfrak{A}$. By adding the set $\mathfrak{B}(\blacksquare)$ one obtains the set $\mathfrak{B}(\bullet, \blacksquare)$, its maximal prolongation consists of 14 particles, still $[\mathfrak{B}(\bullet, \blacksquare)] \neq \mathfrak{A}$. Finally adding the set $\mathfrak{B}(\blacktriangle)$ we obtain the set $\mathfrak{B}(\bullet, \blacksquare, \blacktriangle)$ which admits a complete L-prolongation. This set is minimal in the sense that none of its proper parts admits L-prolongation to the whole $\mathfrak{A}$.

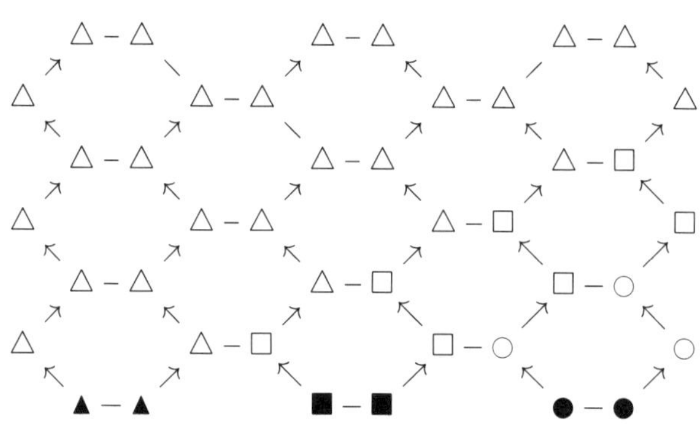

Fig. 2

A simplest graph of degree 4 is presented on Fig. 3.

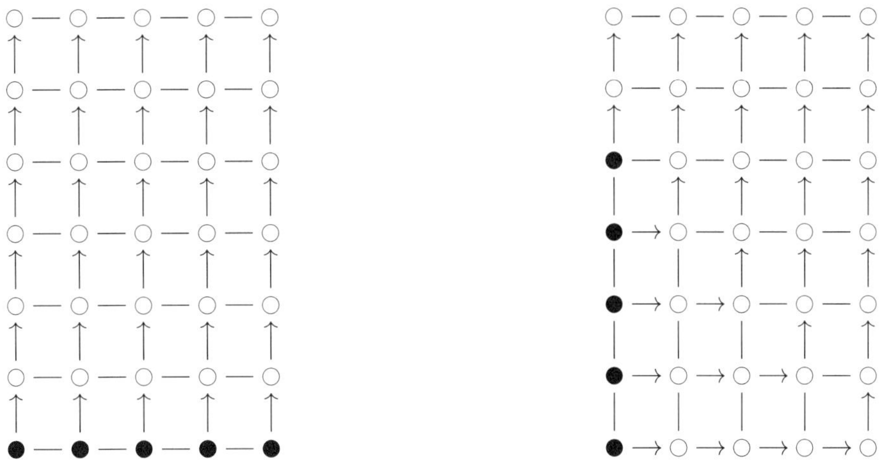

(a) (b)

Fig. 3
The sets of all vertices which belong to one of the sides admit complete L-prolongation. One can see that the set of points which belong to the shorter side (see Fig 3a) is minimal, while the set of points which belong to the longer side contains smaller set which already admits a complete L-prolongation set (see Fig. 3b).

In Figure 4 we present examples of the graphs of degree five (Fig. 4a) and six (Fig. 4b) As before, the set $\mathfrak{B}(\bullet)$ which admits complete L prolongation consists of external points. The prolongation procedure starts from the points with external multiplicity one, such points are marked with larger bullets.

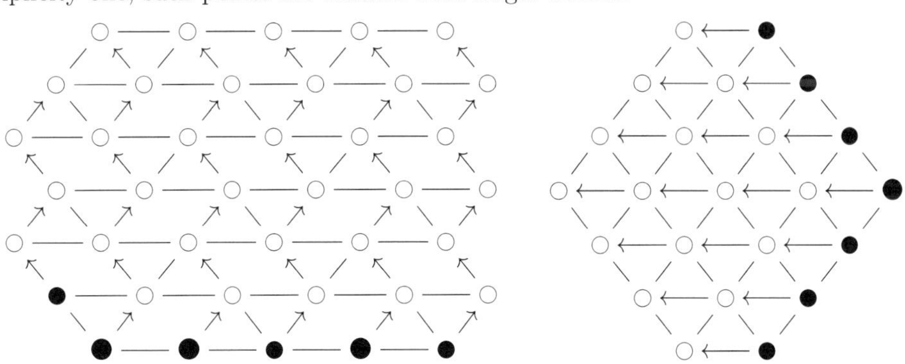

(a) (b)

Fig. 4

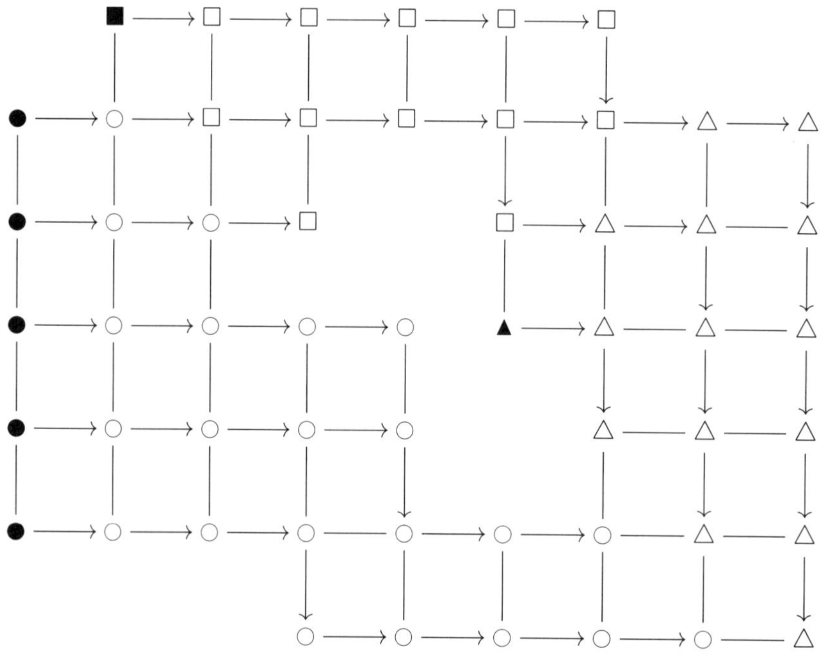

Fig. 5

Yet anothr example is given in Figure 5. It shows the influence of "external" vertices of the graph. The set $\mathfrak{B}(\bullet, \blacksquare, \blacktriangle) = \mathfrak{B}(\bullet) \cup \mathfrak{B}(\blacksquare) \cup \mathfrak{B}(\blacktriangle)$ admits a complete L-prolongation.

In the examples above we considered planar graphs, where each particle interacts with the nearest neighbor only.

Now we consider two graphs which can be obtained from that on Fig. 3 by adding extra edges. We add the edges $(\alpha(1), \beta(1))$, $(\alpha(2), \beta(2)), \dots, (\alpha(7), \beta(7))$ in order to obtain the graph in Figure 6a, and the edges $(\alpha(1), \beta(7))$, $(\alpha(2), \beta(5)), \dots,$ $(\alpha(7), \beta(1))$ in order to obtain the graph in Figure 6b. In the equilibrium position the corresponding systems are located on a cylindrical surface (6a) or on a Mobius band (6b).

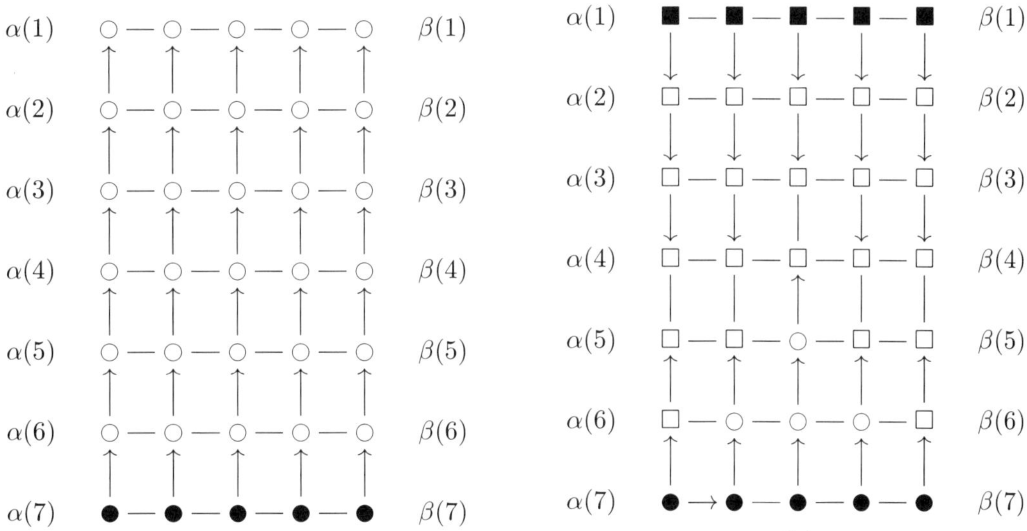

(a) we added the edges $(\alpha(k), \beta(k))$, $k = 1, 2, \ldots 7$

(b) we added the edges $(\alpha(k), \beta(8-k))$, $k = 1, 2, \ldots 7$

Fig. 6

In the case of "three dimensional" graphs the set of external vertices is of "dimension two". It can be chosen as a completely prolongated set. (Fig. 7)

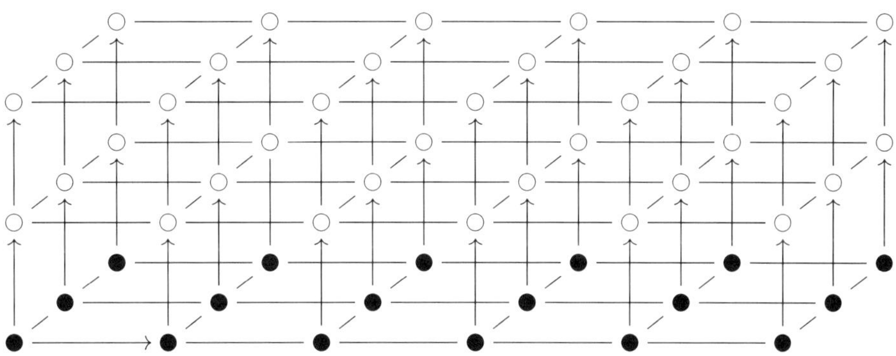

Fig. 7

In the more complicated cases one has to take into account the inner structure of the graph as it was shown on Fig. 5 for the two-dimensional case.

Finally we discuss the structure of the block matrix $M = (M(\alpha, \beta))_{\alpha,\beta \in \mathfrak{A}}$, here $M(\alpha, \beta) = M(\beta, \alpha)^*$ are blocks of order m, such that there exists a subset $\mathfrak{B} \subset \mathfrak{A}$ which admits complete M prolongation and consists of q particles. In other words there exists a chain of subsets

$$\mathfrak{B} = \mathfrak{B}^{(0)} \subset \mathfrak{B}^{(1)} \subset \ldots \subset \mathfrak{B}^{(n-q)} = \mathfrak{A},$$

here $q = \#\mathfrak{B}$ and $n = \#\mathfrak{A}$, which meets the following conditions:

A) $\mathfrak{B}^{(j)} = \mathfrak{B}^{(j-1)} \cup \{\alpha^{(j)}\}$, for some $\alpha^{(j)} \in \mathfrak{A} \setminus \mathfrak{B}^{(j-1)}$. Each set $\mathfrak{B}^{(j-1)}$ contains a particle $\beta^{(j-1)} \subset \mathfrak{B}^{(j-1)}$ which strongly interact with $\alpha^{(j)} \in \mathfrak{B}^{(j)}$ and does not interact with the particles in $\mathfrak{A} \setminus \mathfrak{B}^{(j)}$.

We observe that $\mathfrak{A} \setminus \mathfrak{B}^{(j'-1)} \subset \mathfrak{A} \setminus \mathfrak{B}^{(j)}$ for $j' > j$, therefore $\beta^{(j'-1)} \neq \beta^{(j-1)}$ for $j \neq j'$.

We enumerate all particles in $\mathfrak{A}$ as follows. The particles in $\mathfrak{B} = \mathfrak{B}^{(0)}$ are arbitrarily numerated by $1, 2, \dots, q$, thus $\mathfrak{B}^{(0)} = \cup_{l=1}^{q} \{\alpha(l)\}$. The particle $\alpha^{(j)} \in \mathfrak{B}^{(j)} \setminus \mathfrak{B}^{(j-1)}$ is provided with the number $q + j$, thus $\alpha^{(j)} = \alpha(q + j)$ $\mathfrak{B}^{(j)} = \cup_{k=1}^{j+q} \{\alpha(k)\}$, $j = 0, 1, \dots, n - q$. The matrix M can be then rewritten in the form $M = \big(M_{l,p}\big)$, $l, p = 1, 2, \dots, n$, $M_{l,p} = M(\alpha(l), \alpha(p))$.

Let $k(j)$ be the indices of the particles $\beta^{(j-1)}$, i.e. let $\beta^{(j-1)} = \alpha(k(j))$, $j = 1, 2, \dots n - q$. Then condition A) can be rewritten as

A′) The relation

$$\operatorname{Det}(M_{k(j),q+j}) \neq 0, \;\; M_{k(j),q+j'} = 0,$$

holds for all $j' > j = 1, 2, \dots, n - q$.

Since $\beta^{(j-1)} \in \mathfrak{B}^{(j-1)} = \cup_{s=1}^{q+j-1} \{\alpha(s)\}$ we have $k(j) \leq q + j - 1$ and since $\beta^{(j'-1)} \neq \beta^{(j-1)}$ we have $k(j) \neq k(j')$ for $j \neq j'$. Thus we have an embedding $k(j) : \{1, 2, \dots, n - q\} \to \{1, 2, \dots, n - 1\}$ satisfying $k(j) < q + j$. It is clear that the inverse is also true, so we come to the following statement:

Theorem 3. *In order that there exists a completely M-prolongating set consisting of q particles it is necessary and sufficient that there exists an enumeration $\alpha(1), \alpha(2), \dots \alpha(n)$ of the set $\mathfrak{A}$ and an embedding $k(j) : \{1, 2, \dots, n - q\} \to \{1, 2, \dots, n - 1\}$ satisfying $k(j) < q + j$ and such that the block entries M satisfy condition A′).*

Remark. Conditions A′) are related to those blocks of the matrix M which are located over the main diagonal of the matrix and belong to $n - q$ rows with the numbers $k(j)$, $j = 1, 2, \dots, n - q$. The blocks located in the rest q rows over the main diagonal are arbitrary. The blocks located below the man diagonal are defined by the symmetry. So necessary and sufficient conditions for existence of completely M-prolongating set consisting of q particles can be formulated as follows:

For some enumeration of the particles there exist $n - q$ rows, each containing at least one non-degenerated block entry and such that all block entries on the right of it are zero.

An example of such matrix is presented in Figure 8. In this example $n = 17$, $q = 5$. The symbols $\times$ and ■, denote the non-degenerated and zero blocks respectively. The block entries which can be chosen arbitrarily are left blank. In particular the over diagonal blocks in rows 1,5,11, and 13 are arbitrary

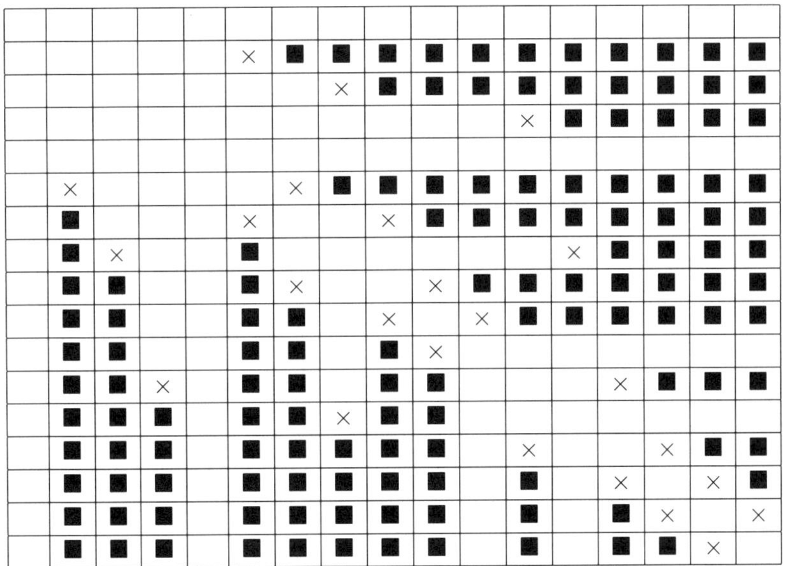

Fig. 8

6. Inverse problem of multichannel scattering

In this section we consider an example of inverse spectral problem for infinite matrices.

Let $M(A)$ be the space of complex-valued functions $\xi(x)$, $\eta(x), \dots$ on a countable set A. The elements $x, y, z, \dots \in A$ are called particles[1].

We consider the matrix

$$L = \big(L(x, x')\big),\ L(x, x') = \overline{L(x', x)},\ x, x' \in A$$

such that each row contains only a finite number of nonzero entries, this number may vary from row to row. Define the operator $\hat{L} : \xi(x) \mapsto \eta(x)$ in $M(A)$ as

$$\eta(x) = \sum_{x' \in A} L(x, x')\xi(x').$$

As always we say that the number λ and the function $\xi(x, \lambda) \not\equiv 0$, are eigenvalue and eigenfunction of the matrix L (or operator $\hat{L}$) if

$$\sum_{x' \in A} L(x, x')\xi(x', \lambda) = \lambda\xi(x, \lambda). \tag{21}$$

The set of all eigenvalues is the spectrum of the matrix L (or operator $\hat{L}$). Since L contains infinitely many elements, its spectrum may contain segments of the real axis.

We say that a particle x interacts with a particle x' (or a subset $B \subset A$) if $L(x, x') \neq 0$ (respectively there is $x' \in B$ such that $L(x, x') \neq 0$).

As before we say that $B \subset A$ admits a L-prolongation if there exists $x \in B$ which interacts with some $\bar{x} \in A \setminus B$ and does not interact with the rest of the particles in $A \setminus B$. The set $B \cup \{\bar{x}\}$ is a L-prolongation of B.

[1]For simplicity we assume that each particle has one degree of freedom.

We say that the set $B \subset A$ admits a complete L-prolongation if there is a chain of prolongations

$$B = B^{(0)} \subset B^{(1)} \subset \cdots \subset B^{(k)} \subset \cdots$$

such that $A = \cup_{k \geq 0} B^{(k)}$, now this chain may be of infinite length. We observe that Lemma 3 is still valid in this case.

Let our system of particles consist of two parts $A = A_0 \cup A_1$, $A_0 \cap A_1 = \emptyset$ and and in addition $\# A_1 = N_1 < \infty$. By $\Gamma_i \subset A_i$, $i = 0, 1$ we denote the set of particles which interact with $A \setminus A_i$.

We will always assume the following

I) *The set* $\Gamma_0 \cup \Gamma_1$ *admits a complete* L*-prolongation;*
$N_0 := \# \Gamma_0 < \infty$;
each particle $y \in \Gamma_0$ *interacts with precisely one particle in* $A \setminus A_0$, *we denote this particle by* $[y]$.

II) *The spectrum of* L *contains a segment* $[a, b]$ *in the real axis, and, for each* $\lambda \in [a, b]$ *equation* (21) *has* N_0 *solutions*

$$\xi(x, \lambda, 1),\ \xi(x, \lambda, 2),\ \ldots\ , \xi(x, \lambda, N_0), \tag{22}$$

such that their restrictions to $\Gamma_0 \cup \Gamma_1$ *are continuous functions of* $\lambda \in [a, b]$, *and their restrictions to* Γ_0 *are linearly independent.*

We use the following notation:

$\xi_i = \xi_i(x)$, $i = 0, 1$ denote restrictions of a function $\xi(x) \in M(A)$ to A_i; $\hat{L}_i$ denotes an operator on $M(A_i)$ with the matrix $L_i = \Big(L(x, x')\Big)$, $x, x' \in A_i$;

$l^2(A_1)$ denote the N_1-dimensional space $M(A_1)$ provided with the standard inner product $(\xi, \eta) = \sum_{x \in A_1} \xi(x) \overline{\eta(x)}$.

The operator $\hat{L}_1$ becomes selfadjoint in this space. By $\{\psi(j, x)\}_{j=1,2,\ldots,N_1} \subset l^2(A_1)$ and $\{\lambda(j)\}_{j=1,2,\ldots,N_1}$ we denote respectively the (orthonormalized) eigenfunctions and eigenvalues of $\hat{L}_1$. The numeration is chosen so that

$$\begin{aligned} \lambda(1) = \lambda(2) = \cdots = \lambda(p_1) = \mu(1) < \lambda(p_1 + 1) = \cdots = \lambda(p_2) = \mu(2) < \cdots \\ = \mu(K - 1) < \lambda(p_{K-1} + 1) = \cdots = \lambda(p_2) = \mu(K); \end{aligned}$$

here $S(L_1) = \{\mu(1), \mu(2), \ldots , \mu(K)\}$ is the spectrum of L_1.

We also consider the orthogonal projections onto the corresponding eigenspaces $\mathrm{Ker}(\hat{L}_1 - \mu(k)\hat{I})$

$$E_k = \Big(E_k(x, x')\Big),\ \ E_k(x, x') = \sum_{p_{k-1} < j \leq p_k} \psi(j, x)\overline{\psi(j, x')} \tag{23}$$

and the resolvent of $\hat{L}_1$

$$(\hat{L}_1 - \lambda \hat{I})^{-1} = R(\lambda) = \Big(R(x, x', \lambda)\Big),\ R(x, x', \lambda) = \sum_{k=1}^{K} \frac{E_k(x, x')}{\mu(k) - \lambda}. \tag{24}$$

The eigenfunction equations for L can be written as

$$\sum_{x' \in A_0} L(x, x')\xi(x', \lambda) + \sum_{x' \in A_1} L(x, x')\xi(x', \lambda) - \lambda\xi(x) = 0,\ x \in A,$$

this is a system of equations with respect to the functions $\xi_0 = \xi(y,\lambda)$, $y \in A_0$ and $\xi_1 = \xi(x,\lambda)$, $x \in A_1$:

$$\left(\hat{L}_1 - \lambda\hat{I}\right)(\xi_1)(x) = -\sum_{y'\in A_0} L(x,y')\xi_0(y',\lambda),\ x \in A_1,$$

$$\left(\hat{L}_0 - \lambda\hat{I}\right)(\xi_0)(y) = -\sum_{x'\in A_1} L(y,x')\xi_1(x',\lambda),\ y \in A_0.$$

For $\lambda \notin S(L_1)$ this system can be written as

$$\xi_1(x,\lambda) = -\sum_{x'\in A_1}\sum_{y'\in A_0} R(x,x',\lambda)L(x',y')\xi_0(y',\lambda),\ x \in A_1;$$

(25)
$$\sum_{x'\in A_1} L(y,x')\xi_1(x',\lambda) = -\sum_{x''\in A_1}\sum_{x'\in A_1}\sum_{y'\in A_0} L(y,x'')R(x'',x',\lambda)L(x',y')\xi_0(y',\lambda);$$

$$\sum_{x'\in A_1} L(y,x')\xi_1(x',\lambda) = -\left(\hat{L}_0 - \lambda\hat{I}\right)(\xi_0)(y),\ y \in A_0.$$

It follows from condition I) that the only pairs $(x,y) \in A_1 \times A_0$ such that $L(x,y) \neq 0$ are of the form $([y],y)$, $y \in \Gamma_0$. Denote the characteristic function of Γ_0 by $\gamma_0(y)$ and let

(26)
$$b(y) = L([y],y),\ y \in \Gamma_0.$$

Relation (25) can be written as

$$\xi_1(x,\lambda) = -\sum_{y'\in\Gamma_0} R(x,[y'],\lambda)b(y')\xi_0(y',\lambda),\ x \in A_1$$

$$\gamma_0(y)\overline{b(y)}\xi_1([y],\lambda) = \gamma_0(y)\sum_{y'\in\Gamma_0}\overline{b(y)}R([y],[y'],\lambda)b(y')\xi_0(y',\lambda),$$

(27)
$$\gamma_0(y)\overline{b(y)}\xi_1([y],\lambda) = -\left(\hat{L}_0 - \lambda\hat{I}\right)(\xi_0(\lambda))(y),\ y \in A_0.$$

Let $y(1), y(2), \ldots, y(N_0)$ be some numeration of particles in Γ_0. Consider the matrices of order N_0:

$$U_0(\lambda) = \Big(\xi([y(n)],\lambda,m)\Big),\ U_1 = \Big(\xi(y(n),\lambda,m)\Big),$$

$$R_1(\lambda) = \Big(R([y(n)],[y(m)],\lambda)\Big),\ B = \Big(b(y(n))\delta_m^n\Big),$$

here $\xi(x,\lambda,m)$ are the solutions of (21), described in condition II). We also assume that $\lambda \in [a,b]$ and λ does not belong to the spectrum of L_1. We denote the spectrum of L_1 by $S(L_1)$.

It follows from (6) that

(28)
$$BU_0(\lambda) = B^* R_1(\lambda) B U_1(\lambda).$$

The columns of the matrix $U_1(\lambda)$ are linearly independent since, by condition II) the restrictions of the eigenfunctions $\xi(x,\lambda,m)$ to Γ_0 are linearly independent. Therefore the matrix $U_1(\lambda)$ is invertible for all $\lambda \in [a,b]$.

Assuming conditions I) and II) to be fulfilled we obtain

$$B^* U_0(\lambda)\big(U_1(\lambda)\big)^{-1} = B^* R_1(\lambda) B,\ \lambda \in [a,b] \setminus S(L_1),$$

for all $\lambda \in [a, b,] \setminus S(L_1)$. In adition the left-hand side of this relation is continuous on $\lambda \in [a, b]$ as follows from condition II). We use relation (24) in order to express the right-hand side of this relation in terms of spectral data of L_1:

$$B^* R_1(\lambda) B = \sum_{k=1}^{M_0} \frac{\mathcal{E}_k}{\mu(k) - \lambda},$$

here

$$\mathcal{E}_k = \Big(\mathcal{E}_k(n, m)\Big), \ \mathcal{E}_k(n, m) = \overline{b(y(n))} E_k([y(n)][y(m)]) b(y(m)) = \sum_{p_{k-1} < j \le p_k} \overline{b(y(n))} \psi(j, [y(n)]) \overline{\psi(j, [y(m)])} b(y(m)). \tag{29}$$

Let for simplicity

$$B^* U_0(\lambda) \big(U_1(\lambda)\big)^{-1} = D(\lambda) = \Big(D(n, m; \lambda)\Big), \tag{30}$$

then

$$D(\lambda) = \sum_{k=1}^{K} \frac{\mathcal{E}_k}{\mu(k) - \lambda}, \quad \lambda \in [a, b] \setminus S(L_1). \tag{31}$$

Relation (30) establishes connection between the eigenfunctions $\psi(j, x)$ of the matrix L_1 and the eigenfunctions $\xi(x, \lambda, m)$ of L.

Theorem 4. *Let conditions I) and II) be fulfilled. Let in addition all non-diagonal elements of L_1 be non-positive and $b(y) = L([y], y) < 0$, $y \in \Gamma_0$. Then the spectral data of the matrix L_1, (and hence the matrix L_1 itself) can be expressed in terms of the matrix $D(\lambda) = \big(BU_0(\lambda)\big)\big(U_1(\lambda)\big)^{-1}$, $\lambda \in [a, b]$.*

Proof. The set $\Gamma_1 = \cup_{n=1}^{N_0} \{[y(n)]\}$ admits a complete L-prolongation. Therefore the eigenfunctions $\psi(j, x)$ of L_1 cannot vanish on the whole Γ_1 and we have

$$\mathrm{Tr}\mathcal{E}_k = \sum_{n=1}^{N_1} \sum_{p_{k-1} < j \le p_k} |b(y(n))|^2 |\psi(j, [y(n)])|^2 > 0. \tag{32}$$

Now relation (31) yields

$$\mathrm{Tr} D(\lambda) = \sum_{k=1}^{K} \frac{\mathrm{Tr}\mathcal{E}_k}{\mu(k) - \lambda}, \quad \lambda \in [a, b], \tag{33}$$

The left-hand side of (33) is bounded on $[a, b]$, while the right-hand side is unbounded near $S(L_1)$. Therefore $S(L_1) \cap [a, b] = \emptyset$.

The Zhukovskii mapping

$$\tilde{\lambda}(\theta) = \frac{a+b}{2} + \frac{b-a}{4}(\theta + \theta^{-1})$$

maps the unit circle $\mathbb{T} = \{\theta : |\theta| = 1\}$ onto $[a, b]$. Let θ_k, θ_k^{-1} be the solutions of the equation $\tilde{\lambda}(\theta) = \mu(k)$, we have $\theta_k, \theta_k^{-1} \in \mathbb{R} \setminus \{-1, 1\}$. We may assume $\theta_k \in (-1, 0) \cup (0, 1)$ and $\theta_k^{-1} \in (-\infty, 1) \cup (1, \infty)$. The residus theorem yields

$$\frac{1}{2\pi i} \int_{\mathbb{T}} \theta^l \mathrm{Tr} D(\tilde{\lambda}(\theta)) d\theta = \sum_{k=1}^{K} \frac{4\mathrm{Tr}\mathcal{E}_k}{(b-a)(\theta_k^{-2} - 1)} \theta_k^l =: s_l, \ l = 1, 2, \dots \ .$$

The sequence $\{s_l\}$ is the moment sequence for a measure located at the finite set of real points $\theta_1, \theta_2, \dots, \theta_K$, the load at each point θ_k is

$$\rho(\{\theta_k\}) = \frac{4\mathrm{Tr}\mathcal{E}_k}{(b-a)(\theta_k^{-2}-1)} > 0.$$

It is known (see, e.g., [**1**]) that the determinants

$$\Delta_l = \mathrm{Det}\begin{pmatrix} s_0 & s_1 & \dots & s_l \\ s_1 & s_2 & \dots & s_{l+1} \\ \dots & \dots & \dots & \dots \\ s_l & s_{l+1} & \dots & s_{2l} \end{pmatrix}$$

are strictly positive as $l < K$ and $\Delta_K = 0$. In addition the set $\{\theta_1, \theta_2, \dots, \theta_K\}$ is the zero set of the polynomial

$$\Pi(\theta) = \begin{pmatrix} s_0 & s_1 & \dots & s_K \\ s_1 & s_2 & \dots & s_{K+1} \\ \dots & \dots & \dots & \dots \\ s_{K-1} & s_K & \dots & s_{2K-1} \\ 1 & \theta & \dots & \theta^K \end{pmatrix}.$$

Therefore given the matrix $D(\lambda) = \big(B^* U_0(\lambda)\big)\big(U_1(\lambda)\big)^{-1}$, one can find s_l's and then the first vanishing determinant Δ_K, then the polynomial $\Pi(\theta)$, its zero set $\theta_1, \theta_2, \dots \theta_K$ and finally the the eigenvalues $\mu(k) = \tilde{\lambda}(\theta_k)$, $k = 1, 2, \dots, , K$ of L_1.

Now relation (31) yields expression for $\mathcal{E}_k$, $k = 1, 2, \dots, K$:

$$\frac{1}{2\pi i}\int_{\mathbb{T}} D(\tilde{\lambda}(\theta))\frac{\Pi(\theta)}{\theta-\theta_k}d\theta = \frac{1}{2\pi i}\int_{\mathbb{T}}\sum_{q=1}^{K}\frac{\mathcal{E}_q\Pi(\theta)}{(\mu(q)-\tilde{\lambda}(\theta))(\theta-\theta_k)}d\theta = \frac{\mathcal{E}_k\Pi'(\theta_k)}{(b-a)(\theta_k^{-2}-1)},$$

therefore

$$\mathcal{E}_k = \frac{(b-a)(\theta_k^{-2}-1)}{2\pi i\Pi'(\theta_k)}\int_{\mathbb{T}}\frac{D(\tilde{\lambda}(\theta))\Pi(\theta)}{\theta-\theta_k}d\theta,$$

and

$$\mathcal{E}_k(n,m) = \frac{(b-a)(\theta_k^{-2}-1)}{2\pi i\Pi'(\theta_k)}\int_{\mathbb{T}}\frac{D(n,m,\tilde{\lambda}(\theta))\Pi(\theta)}{\theta-\theta_k}d\theta,$$

We now use relation (29). Since $\{\psi(j,x)\}$ is an orthonormal system we obtain

$$\sum_{k=1}^{K}\mathcal{E}_k(n,m) = \overline{b(y(n))}\left(\sum_{j=1}^{N_1}\psi(j,[y(n)])\overline{\psi(j,[y(m)])}\right),$$

$$b(y(m)) = \overline{b(y(n))}\delta_{[y(m)]}^{[y(n)]}b(y(m)).$$

In particular,

$$\sum_{k=1}^{M_1}\mathcal{E}_k(n,n) = |b(y(n))|^2.$$

Since all coefficients $b(y)$, $y \in \Gamma_0$ are negative we obtain

$$b(y(n)) = -\sqrt{\sum_{k=1}^{K}\mathcal{E}_k(n,n)}$$

and

$$E_k([y(n)],[y(m)]) = \sum_{j=1}^{N_0} \psi(j,[y(n)])\overline{\psi(j,[y(m)])} = \frac{\mathcal{E}_k(n,m)}{\sqrt{\sum_{k=1}^K \mathcal{E}_k(n,n)\sum_{k=1}^K \mathcal{E}_k(m,m)}}.$$

Given $x \in \Gamma_1$ we denote by $\Gamma_1(x)$ the set of particles in Γ_0, which interact with x. Then

$$E_k(x,x')|_{x,x'\in\Gamma_1} = E_k([y(n)],[y(m)])|_{y(n)\in\Gamma_1(x),y(m)\in\Gamma_1(x')}.$$

Thus given the matrix $D(\lambda)$, $\lambda \in [a,b]$, one can determine the eigenvalues $\mu(k)$ of the matrix L_1 as well as the matrix entries $E_k(x,x')|_{x,x'\in\Gamma}$ of the orthogonal projector operators $\hat{E}_k$, $k = 1,2,\dots K$ on the corresponding eigenspaces. In other words one can find the restriction of the spectral data of L_1 on the set Γ_1. Now the theorem follows from Theorem 2. □

We apply Theorem 4 to study the inverse problem of multichannel scattering. In this setting the set A_0 consists of finite number of channels. By channel σ we mean a sequence of particles $\sigma(1),\sigma(2),\dots,\sigma(k),\dots$ such that each particle $\sigma(k)$ interacts with its two nearest neighbors $\sigma(k-1)$ and $\sigma(k+1)$ and with external field. The particle $\sigma(0) \in A_1$ is the endpoint of the channel, it does not belong to σ. It may happen that different channels have the same endpoint. The set of all channels is denoted by $\mathfrak{S}$. We thus have

$$A_0 = \cup_{\sigma\in\mathfrak{S}} \cup_{k=1}^{\infty} \sigma(k),\ \Gamma_0 = \cup_{\sigma\in\mathfrak{S}}\sigma(1),\ \Gamma_1 = \cup_{\sigma\in\mathfrak{S}}\sigma(0),\ \#\Gamma_0 = \#\mathfrak{S} = N_0$$

and $A_1 \supset \Gamma_1$ is a finite set of particles.

It follows from the definition of the channel that $L(\sigma(k),y) \neq 0$ for $y = \sigma(k-1),\sigma(k),\sigma(k+1)$ only. We use the notation

$$L(\sigma(k),\sigma(k-1)) = L(\sigma(k-1),\sigma(k)) = -b_\sigma(k-1),\quad (b_\sigma(k-1) > 0)$$
$$L(\sigma(k),\sigma(k)) = a_\sigma(k).$$

We assume that the channels are homogeneous at infinity, i.e., there exist the limits

$$\lim_{k\to\infty} b_\sigma(k) = b,\ \lim_{k\to\infty} a_\sigma(k) = a,\quad a,b \neq 0,$$

and

$$\sum_{k=1}^{\infty} k\left\{|b_\sigma(k) - b| + |a_\sigma(k) - a|\right\} < \infty,$$

a and b are independent of $\sigma \in \mathfrak{S}$. Without loss of generality one may assume $a = 2$, $b = 1$. For $x = \sigma(k)$, $(\sigma \in \mathfrak{S},\ k = 1,2,\dots)$ equations (21) take the form

$$-b_\sigma(k-1)\xi(\sigma(k-1),\lambda) + a_\sigma(k)\xi(\sigma(k),\lambda) - b_\sigma(k)\xi(\sigma(k+1),\lambda) = \lambda\xi(\sigma(k),\lambda).$$

Let $|\theta| \le 1$ and $\lambda = \lambda(\theta) = 2-\theta-\theta^{-1}$. By $e_\sigma(k,\theta)$ we denote the Jost solutions, which are defined by their asymptotic behavior at infinity $e_\sigma(k,\theta) = \theta^k(1+o(1))$, $k \to \infty$.

According [**12**], for each $\lambda = \lambda(\theta) \in (0,4)\backslash S(L)$, equation (21) admits $N_0 = \#\mathfrak{S}$ solutions $\xi(x,\lambda,\nu)$, $\nu \in \mathfrak{S}$ and such that the restrictions $\xi(\sigma(k),\lambda(\theta),\nu)$, $\sigma \in \mathfrak{S}$, $k = 0,1,2,\dots$ of these solutions to $A_0 \cup \Gamma_1$ are the columns of the matrix

$$U_k(\lambda(\theta)) = E(k,\theta^{-1}) + E(k,\theta)S(\theta).$$

Here

$$E(k,\theta) = \mathrm{diag}\{e_\sigma(k,\theta)\}_{\sigma\in\mathfrak{S}},\ S(\theta) = -T(\theta)^{-1}T(\theta^{-1}),$$

$$T(\theta) = E(0,\theta) - R_1(\theta)B(0)E(1,\theta), R_1(\theta) = (R([\sigma(1)], [\nu(1)], \lambda(\theta)))_{\sigma,\nu\in\mathfrak{S}}$$

and

$$B(0) = \operatorname{diag}\{b_\sigma(0)\}_{\sigma\in\mathfrak{S}}.$$

It is also proved in [**12**] (see relation (41)) that, for $|\theta| = 1$, $\lambda(\theta) \in (0,4)\backslash S(L_1)$,

$$U_1(\lambda(\theta)) = (\theta^{-1} - \theta)T(\theta)^{-1}B(0)^{-1}E(1,\theta)^{-1},$$

and hence $U_1(\lambda(\theta))$ is invertible for all $\lambda(\theta) \in (0,4) \setminus S(L_1)$.

Let in addition the set $\Gamma_1 = \cup_{\sigma\in\mathfrak{S}}\{\sigma(0)\}$ (i.e. the set of endpoints of the channels) admit complete L_1-prolongation. Then conditions of Theorem 4 are fulfilled and the matrix L_1 can be found in terms of the matrix functions $B(0)U_0(\lambda(\theta))$ and $U_1(\lambda(\theta))$, $\lambda(\theta) \in [a,b]$, here $[a,b]$ is any segment in $(0,4) \setminus \{S(L_1) \cup \{\pm 1\}\}$.

It is proved in [**12**] that the matrices L_0, $U_1(\lambda(\theta))$ and $B(0)U_0(\lambda(\theta))$ can be found in terms of the scattering data at infinity i.e. from observations at the (practically) homogeneous parts of the channels in case when the system is in generic position. Therefore the following statement holds:

Theorem 5. *Let the set of the channel endpoints Γ_1 admit complete L_1 prolongation. Then in case of generic position the inverse multichannel scattering problem admits unique solution, i.e., the whole matrix L can be reconstructed in terms of the scattering data.*

References

[1] N. I. Akhiezer, *The classical moment problem and some related questions in analysis*, Translated by N. Kemmer, Hafner Publishing Co., New York, 1965. MR0184042 (32 #1518)

[2] Sergei Avdonin, Pavel Kurasov, and Marlena Nowaczyk, *Inverse problems for quantum trees II: recovering matching conditions for star graphs*, Inverse Probl. Imaging **4** (2010), no. 4, 579–598, DOI 10.3934/ipi.2010.4.579. MR2726415 (2011j:81112)

[3] M. I. Belishev and A. F. Vakulenko, *Inverse problems on graphs: recovering the tree of strings by the BC-method*, J. Inverse Ill-Posed Probl. **14** (2006), no. 1, 29–46, DOI 10.1163/156939406776237474. MR2218385

[4] Robert Carlson, *Inverse eigenvalue problems on directed graphs*, Trans. Amer. Math. Soc. **351** (1999), no. 10, 4069–4088, DOI 10.1090/S0002-9947-99-02175-3. MR1473434 (99m:34189)

[5] N. I. Gerasimenko and B. S. Pavlov, *A scattering problem on noncompact graphs*, Teoret. Mat. Fiz. **74** (1988), no. 3, 345–359, DOI 10.1007/BF01016616 (Russian, with English summary); English transl., Theoret. and Math. Phys. **74** (1988), no. 3, 230–240. MR953298 (90f:47010)

[6] Boris Gutkin and Uzy Smilansky, *Can one hear the shape of a graph?*, J. Phys. A **34** (2001), no. 31, 6061–6068, DOI 10.1088/0305-4470/34/31/301. MR1862642 (2002k:05205)

[7] V. Kostrykin, R. Schrader, The inverse scattering problem for metric graphs and the traveling salesman problem, *ArXiv, math-ph/0603010*, pp.1 – 68, 2006.

[8] Peter Kuchment, *Quantum graphs. II. Some spectral properties of quantum and combinatorial graphs*, J. Phys. A **38** (2005), no. 22, 4887–4900, DOI 10.1088/0305-4470/38/22/013. MR2148631 (2006a:81035)

[9] Peter Kuchment, *Quantum graphs: an introduction and a brief survey*, Analysis on graphs and its applications, Proc. Sympos. Pure Math., vol. 77, Amer. Math. Soc., Providence, RI, 2008, pp. 291–312. MR2459876 (2010b:81058)

[10] V. Pivovarchik, *Inverse problem for the Sturm-Liouville equation on a star-shaped graph*, Math. Nachr. **280** (2007), no. 13-14, 1595–1619, DOI 10.1002/mana.200410567. MR2354982 (2009d:34019)

[11] Pavel Kurasov and Marlena Nowaczyk, *Inverse spectral problem for quantum graphs*, J. Phys. A **38** (2005), no. 22, 4901–4915, DOI 10.1088/0305-4470/38/22/014. MR2148632 (2006c:81218)

[12] Yu. I. Lyubarskiĭ and V. A. Marchenko, *Direct and inverse problems of multichannel scattering*, Funktsional. Anal. i Prilozhen. **41** (2007), no. 2, 58–77, 112, DOI 10.1007/s10688-007-0012-0 (Russian, with Russian summary); English transl., Funct. Anal. Appl. **41** (2007), no. 2, 126–142. MR2345041 (2008k:81341)

[13] V.A. Marchenko, *Introduction to the theory of inverse problems of spectral analysis*, (Russian), Kharkov, 2005, pp.1–135.

[14] V. A. Yurko, *An inverse problem for Sturm-Liouville operators on arbitrary compact spatial networks*, Dokl. Akad. Nauk **432** (2010), no. 3, 318–321, DOI 10.1134/S1064562410030208 (Russian); English transl., Dokl. Math. **81** (2010), no. 3, 410–413. MR2723599 (2011g:34028)

Department of Mathematical Sciences, Norwegian University of Science and Technology, NO–7491, Trondheim, Norway

E-mail address: yura@math.ntnu.no

Mathematical Division, Institute for Low Temperature Physics and Engineering, National Academy of Sciences of Ukraine, 47 Lenin Ave., Kharkov 61103 , Ukraine

E-mail address: marchenko@ilt.kharkov.ua

Proceedings of Symposia in Pure Mathematics
Volume **87**, 2013
http://dx.doi.org/10.1090/pspum/087/01452

On the Weyl-Titchmarsh and Livšic functions

K. A. Makarov and E. Tsekanovskiĭ

Dedicated with great pleasure to our friend and colleague Fritz Gesztesy on the occasion of his 60th birthday anniversary

ABSTRACT. We establish a mutual relationship between main analytic objects for the dissipative extension theory of a symmetric operator $\dot{A}$ with deficiency indices $(1,1)$. In particular, we introduce the Weyl-Titchmarsh function $\mathcal{M}$ of a maximal dissipative extension $\widehat{A}$ of the symmetric operator $\dot{A}$. Given a reference self-adjoint extension A of $\dot{A}$, we introduce a von Neumann parameter κ, $|\kappa|<1$, characterizing the domain of the dissipative extension $\widehat{A}$ against $\mathrm{Dom}(A)$ and show that the pair $(\kappa, \mathcal{M})$ is a complete unitary invariant of the triple $(\dot{A}, A, \widehat{A})$, unless $\kappa = 0$. As a by-product of our considerations we obtain a relevant functional model for a dissipative operator and get an analog of the formula of Krein for its resolvent.

1. Introduction

In 1946 M. Livšic obtained the following remarkable result [**19**, a part of Theorem 13] (for a textbook exposition see [**3**]).

THEOREM 1.1 ([**19**]). *Suppose that $\dot{A}$ and $\dot{B}$ are closed, prime*[1]*, densely defined symmetric operators with deficiency indices $(1,1)$. Then $\dot{A}$ and $\dot{B}$ are unitarily equivalent if, and only if, for an appropriate choice of normalized deficiency elements*

$$g_{\pm} \in \mathrm{Ker}((\dot{A})^* \mp iI) \quad \textit{and} \quad f_{\pm} \in \mathrm{Ker}((\dot{B})^* \mp iI)$$

the following equality

$$\frac{(g_z, g_-)}{(g_z, g_+)} = \frac{(f_z, f_-)}{(f_z, f_+)}, \qquad z \in \mathbb{C}_+,$$

holds, where $g_z \neq 0$ and $f_z \neq 0$ are arbitrary deficiency elements of the symmetric operators $\dot{A}$ and $\dot{B}$,

$$g_z \in \mathrm{Ker}((\dot{A})^* - zI) \quad \textit{and} \quad f_z \in \mathrm{Ker}((\dot{B})^* - zI).$$

Livšic suggested to call the function

$$s(z) = \frac{z-i}{z+i} \cdot \frac{(g_z, g_-)}{(g_z, g_+)}, \qquad z \in \mathbb{C}_+, \tag{1.1}$$

2010 *Mathematics Subject Classification.* Primary 81Q10, Secondary 35P20, 47N50.

Key words and phrases. Deficiency indices, quasi-self-adjoint extensions, Weyl–Titchmarsh functions.

[1]A symmetric operator $\dot{A}$ is prime if there does not exist a subspace invariant under $\dot{A}$ such that the restriction of $\dot{A}$ to this subspace is self-adjoint.

the characteristic function of the symmetric operator $\dot{A}$.

Theorem 1.1 identifies the function $s(z)$ (modulo z-independent unimodular factor) with a complete unitary invariant of a prime symmetric operator with deficiency indices $(1,1)$ that determines the operator uniquely up to unitary equivalence.

Livšic also gave a criterion [**19**, Theorem 15] (also see [**3**]) for a contractive analytic mapping from the upper half-plane $\mathbb{C}_+$ to the unit disk $\mathbb{D}$ to be the characteristic function of a densely defined symmetric operator with deficiency indices $(1,1)$.

THEOREM 1.2 ([**19**]). *For an analytic mapping s from the upper half-plane to the unit disk to be the characteristic function of a densely defined symmetric operator with deficiency indices $(1,1)$ it is necessary and sufficient that*

$$s(i)=0 \quad \text{and} \quad \lim_{z\to\infty} z(s(z)-e^{2i\alpha})=\infty \quad \text{for all} \quad \alpha\in[0,\pi), \tag{1.2}$$

$$0<\varepsilon\le arg(z)\le\pi-\varepsilon.$$

In the same article, Livšic put forward a concept of a characteristic function of a quasi-self-adjoint dissipative extension of a symmetric operator with deficiency indices $(1,1)$.

Let us recall Livšic's construction.

Suppose that $\dot{A}$ is a symmetric operator with deficiency indices $(1,1)$ and that $g_\pm$ are its normalized deficiency elements,

$$g_\pm\in \operatorname{Ker}((\dot{A})^*\mp iI), \quad \|g_\pm\|=1.$$

Suppose that $\widehat{A}\neq(\widehat{A})^*$ is a maximal dissipative extension of $\dot{A}$,

$$\operatorname{Im}(\widehat{A}f,f)\ge 0, \quad f\in \operatorname{Dom}(\widehat{A}).$$

Since $\dot{A}$ is symmetric, its dissipative extension $\widehat{A}$ is automatically quasi-self-adjoint [**24**], [**26**], that is,

$$\dot{A}\subset\widehat{A}\subset(\dot{A})^*,$$

and hence

$$g_+-\kappa g_-\in \operatorname{Dom}(\widehat{A}) \quad \text{for some } |\kappa|<1. \tag{1.3}$$

Based on the parameterization (1.3) of the domain of the extension $\widehat{A}$, which is an analog of the von Neumann formulae[2], Livšic suggested to call the Möbius transformation

$$S(z)=\frac{s(z)-\kappa}{\overline{\kappa}\,s(z)-1}, \quad z\in\mathbb{C}_+, \tag{1.4}$$

where s is given by (1.1), the characteristic function of the dissipative extension $\widehat{A}$ [**19**].

A culminating point of Livšic's considerations was the discovery of the following result [**19**, the remaining part of Theorem 13].

THEOREM 1.3 ([**19**]). *Suppose that $\dot{A}$ and $\dot{B}$ are closed prime densely defined symmetric operators with deficiency indices $(1,1)$. Assume, in addition, that $\widehat{A}$ and $\widehat{B}$ are their maximal dissipative extensions, respectively ($\widehat{A}\neq(\widehat{A})^*$, $\widehat{B}\neq(\widehat{B})^*$).*

[2]Throughout this paper κ will be called the von Neumann extension parameter.

Then, $\widehat{A}$ and $\widehat{B}$ are unitarily equivalent if, and only if, the corresponding characteristic functions coincide up to a unimodular constant factor.

In 1965 Donoghue [**8**] introduced a concept of the Weyl-Titchmarsh function $M(\dot{A}, A)$ associated with a pair $(\dot{A}, A)$ by

$$M(\dot{A}, A)(z) = ((Az + I)(A - zI)^{-1} g_+, g_+), \quad z \in \mathbb{C}_+,$$

$$g_+ \in \operatorname{Ker}((\dot{A})^* - iI), \quad \|g_+\| = 1,$$

where $\dot{A}$ is a symmetric operator with deficiency indices $(1,1)$, $\operatorname{def}(\dot{A}) = (1,1)$, and A is its self-adjoint extension[3].

Furthermore, Donoghue showed that the Weyl-Titchmarsh function admits the following Herglotz-Nevanlinna representation

$$M(\dot{A}, A)(z) = \int_{\mathbb{R}} \left(\frac{1}{\lambda - z} - \frac{\lambda}{1 + \lambda^2} \right) d\mu \tag{1.5}$$

for some infinite Borel measure μ such that

$$\int_{\mathbb{R}} \frac{d\mu(\lambda)}{1 + \lambda^2} = 1,$$

and discovered the following result (for a full presentation see [**10**]).

THEOREM 1.4 ([**8**], [**10**]). *Suppose that $\dot{A}$ and $\dot{B}$ are closed, prime, symmetric operators with deficiency indices $(1,1)$. Assume, in addition, that A and B are some self-adjoint extensions of $\dot{A}$ and $\dot{B}$, respectively.*

Then

(i) *the pairs $(\dot{A}, A)$ and $(\dot{B}, B)$ are unitarily equivalent[4] if, and only if, the Weyl-Titchmarsh functions $M(\dot{A}, A)$ and $M(\dot{B}, B)$ coincide;*

(ii) *the pair $(\dot{A}, A)$ is unitarily equivalent to a model pair $(\dot{\mathcal{B}}, \mathcal{B})$ in the Hilbert space $L^2(\mathbb{R}; d\mu)$, where $\mathcal{B}$ is the multiplication (self-adjoint) operator by the independent variable, $\dot{\mathcal{B}}$ is its restriction on*

$$\operatorname{Dom}(\dot{\mathcal{B}}) = \left\{ f \in \operatorname{Dom}(\mathcal{B}) \,\middle|\, \int_{\mathbb{R}} f(\lambda) d\mu(\lambda) = 0 \right\}, \tag{1.6}$$

and μ is the Borel measure from the Herglotz-Nevanlinna representation (1.5) for the Weyl-Titchmarsh function $M = M(\dot{A}, A)$.

Theorem 1.4, on the one hand, recognizes the Weyl-Titchmarsh function M as a (complete) unitary invariant of the pair of a symmetric operator with deficiency indices $(1,1)$ and its self-adjoint extension that determines the operators uniquely up to unitary equivalence. On the other hand, this result provides a general model for a symmetric operator with deficiency indices $(1,1)$ and its family of self-adjoint extensions.

The main goal of this paper is

---

[3]The concept of the Weyl–Titchmarsh function in the general case where $\operatorname{def}(\dot{A}) = (n,n)$, $n \in \mathbb{N} \cup \{\infty\}$ is due to Saakjan [**25**]. Different approaches to the concept can be found in [**1**], [**7**], [**9**], [**10**], [**13**] and the bibliography therein.

[4]We say that pairs of operators $(\dot{A}, A)$ and $(\dot{B}, B)$ in Hilbert spaces $\mathcal{H}_A$ and $\mathcal{H}_B$ are unitarily equivalent if there is a unitary map $\mathcal{U}$ from $\mathcal{H}_A$ onto $\mathcal{H}_B$ such that $\dot{B} = \mathcal{U}\dot{A}\mathcal{U}^{-1}$ and $B = \mathcal{U}A\mathcal{U}^{-1}$.

(i) to establish a relationship between the classical analytic objects of the extension theory such as the characteristic function $s(z)$ of a symmetric operator $\dot{A}$, the characteristic function $S(z)$ of its dissipative extensions $\widehat{A}$, and the Weyl-Titchmarsh function $M(z)$ associated with the pair $(\dot{A}, A)$;
(ii) to introduce the Weyl-Titchmarsh function for dissipative extensions $\widehat{A}$ of $\dot{A}$, and
(iii) to provide a functional model for the triple $(\dot{A}, \widehat{A}, A)$ and obtain an analog of the formula of Krein [**12**] for the resolvents of $\widehat{A}$ and A.

The paper is organized as follows.

In Sec. 2 we propose to consider the characteristic function of a symmetric operator $\dot{A}$ to be the one associated with a pair $(\dot{A}, A)$ where A is a special reference self-adjoint extension of $\dot{A}$ uniquely determined by the choice of the basis $\{g_\pm\}$ in the deficiency subspace (see, eq. (2.1)). We call this function *the Livšic function* associated with the pair $(\dot{A}, A)$. For different definitions of the characteristic functions in the unbounded case we refer to [**11**], [**15**], [**22**], [**26**], and [**29**].

We also show that the Weyl-Titchmarsh and the Livšic functions associated with the pair $(\dot{A}, A)$ are related by the Cayley transformation (see Theorem 2.2 (ii)). Based on this transformation law, we show that Livšic's Theorem 1.3 and Donoghue's Theorem 1.4 can be deduced from one another.

In Sec. 3 we introduce the Weyl-Titchmarsh function of a dissipative quasi-self-adjoint extension $\widehat{A}$ of $\dot{A}$ and extend Theorem 2.2 to the dissipative case (see Theorem 3.2).

In Sec. 4 we introduce the characteristic function associated with the triple of operators $(\dot{A}, \widehat{A}, A)$ and provide a functional model for such triples, continuing a list of various functional models for non-self-adjoint operators discovered by M. Livšic [**19**], [**20**], B. Sz.-Nagy and C. Foias [**21**], L. de Branges and J. Rovnyak [**4**], B. S. Pavlov [**22**], and others (also see [**1**], [**15**], [**17**], [**18**], [**23**], [**26**], [**27**], and references therein).

Based on the functional model, we obtain a refinement of Livšic's uniqueness result of Theorem 1.3 (see Theorem 4.1).

In Sec. 5 we perform initial spectral analysis of a dissipative triple and obtain an analog of Krein's resolvent formula in the rank-one dissipative extension theory (cf., [**9**], [**10**], [**28**]).

In Appendix A some spectral properties of the model symmetric operator are summarized.

Our exposition is based on the classical von Neumann extension theory. The proofs are straightforward and no knowledge of the rigged Hilbert space approach [**1**], [**29**] and/or the boundary triplets theory [**7**], [**16**], [**26**] is required.

2. The Livšic and the Weyl-Titchmarsh functions

Livšic's definition of a characteristic function of a symmetric operator (see eq. (1.1)) has some ambiguity related to the choice of the deficiency elements $g_\pm$. To avoid this ambiguity we proceed as follows.

Suppose that A is a self-adjoint extension of a symmetric operator $\dot{A}$ with deficiency indices $(1,1)$. Let $g_\pm$ be deficiency elements $g_\pm \in \operatorname{Ker}((\dot{A})^* \mp iI)$, $\|g_+\| = 1$. Assume, in addition, that

$$g_+ - g_- \in \operatorname{Dom}(A). \tag{2.1}$$

Introduce the Livšic function $s(\dot{A}, A)$ associated with the pair $(\dot{A}, A)$ by

$$s(z) = \frac{z-i}{z+i} \cdot \frac{(g_z, g_-)}{(g_z, g_+)}, \quad z \in \mathbb{C}_+, \tag{2.2}$$

where $0 \neq g_z \in \operatorname{Ker}((\dot{A})^* - zI)$ is an arbitrary (deficiency) element.

The following result establishes a standard relationship between the Weyl-Titchmarsh and the Livšic functions associated with the pair $(\dot{A}, A)$ [**1**], [**5**], [**6**], [**29**] (also see [**7**], where a linear–fractional transformation between the Livšic characteristic function of a symmetric operator and the Krein–Langer Q-function in the framework of boundary triplets theory was established).

THEOREM 2.1. *Denote by $M = M(\dot{A}, A)$ and by $s = s(\dot{A}, A)$ the Weyl-Titchmarsh function and the Livšic function associated with the pair $(\dot{A}, A)$, respectively.*

Then

$$s(z) = \frac{M(z) - i}{M(z) + i}, \quad z \in \mathbb{C}_+. \tag{2.3}$$

PROOF. Let μ be the measure from the Herglotz-Nevanlinna representation (1.5) of the Weyl-Titchmarsh function M associated with the pair $(\dot{A}, A)$.

By splitting off the self-adjoint part of $\dot{A}$, we may assume that $\dot{A}$ is a prime symmetric operator and then, in accordance with Theorem 1.4, one may also assume without loss that $\dot{A}$ and A are already chosen in their model representation in the Hilbert space $L^2(\mathbb{R}; d\mu)$, so that the deficiency elements are given by

$$g_z(\lambda) = \frac{1}{\lambda - z}, \quad \mu\text{-a.e.}\,, \quad \operatorname{Im}(z) \neq 0,$$

and

$$g_\pm(\lambda) = g_{\pm i}(\lambda) = \frac{1}{\lambda \mp i}, \quad \mu\text{-a.e.}\ .$$

Notice that $g_+ - g_- \in \operatorname{Dom}(A)$ and hence

$$s(z) = \frac{z-i}{z+i} \cdot \frac{(g_z, g_-)}{(g_z, g_+)} = \frac{(z-i)\int_{\mathbb{R}} \frac{d\mu(\lambda)}{(\lambda - z)(\lambda - i)}}{(z+i)\int_{\mathbb{R}} \frac{d\mu(\lambda)}{(\lambda - z)(\lambda + i)}} = \frac{M(z) - i}{M(z) + i},$$

completing the proof. □

As a corollary we obtain the following analog of Theorem 1.4.

THEOREM 2.2. *Suppose that $\dot{A}$ and $\dot{B}$ are closed, prime, densely defined symmetric operators with deficiency indices $(1,1)$. Assume, in addition, that A and B are some self-adjoint extensions of $\dot{A}$ and $\dot{B}$ respectively.*

Then

(i) *the pairs $(\dot{A}, A)$ and $(\dot{B}, B)$ are unitarily equivalent if, and only if, the Livšic functions $s(\dot{A}, A)$ and $s(\dot{B}, B)$ coincide;*

(ii) *the pair $(\dot{A}, A)$ is unitarily equivalent to the model pair $(\dot{\mathcal{B}}, \mathcal{B})$ in the Hilbert space $L^2(\mathbb{R}; d\mu)$, where μ is the representing measure for the Weyl-Titchmarsh function $M = M(\dot{A}, A)$.*

In this case,

$$M(z) = \frac{1}{i} \cdot \frac{s(z) + 1}{s(z) - 1}, \quad z \in \mathbb{C}_+, \quad \text{with} \quad s = s(\dot{A}, A).$$

3. The Weyl-Titchmarsh function in the dissipative case

The Weyl-Titchmarsh function can also be introduced in a more general context of dissipative extensions of a symmetric operator. We refer to [**2**] (also see [**1**]) where the concept of the Weyl-Titchmarsh function for bounded non-self-adjoint operators has been discussed.

To be more specific, suppose that $\dot{A}$ is a densely defined, closed, symmetric operator with deficiency indices $(1,1)$ and $\widehat{A}$ is its maximal dissipative extension, that is,

$$\dot{A} \subset \widehat{A} \subset (\dot{A})^*$$

and

$$\operatorname{Im}(\widehat{A}f, f) \geq 0, \quad f \in \operatorname{Dom}(\widehat{A}).$$

Suppose that g_+ is a deficiency element $g_+ \in \operatorname{Ker}((\dot{A})^* - iI)$ satisfying the normalization condition $\|g_+\| = 1$.

We introduce the Weyl-Titchmarsh function $\mathcal{M} = \mathcal{M}(\dot{A}, \widehat{A})$ associated with the pair $(\dot{A}, \widehat{A})$ by

$$\mathcal{M}(z) = \Big(((\widehat{A})^* z + I)((\widehat{A})^* - zI)^{-1} g_+, g_+\Big), \quad z \in \mathbb{C}_+.$$

We remark that in contradistinction to the self-adjoint case, the dependence of $\mathcal{M}$ on the first argument, the symmetric operator $\dot{A}$, can be suppressed for, under our assumptions,

$$\dot{A} = \widehat{A}|_{\operatorname{Dom}(A) \cap \operatorname{Dom}((\dot{A})^*)}.$$

In other words, the symmetric operator $\dot{A}$ is uniquely determined by the extension $\widehat{A}$, provided that $\widehat{A}$ is not self-adjoint.

In order to establish a relationship between the Livšic function associated with the pair $(\dot{A}, A)$ and the Weyl-Titchmarsh function of the dissipative extension $\widehat{A}$ of $\dot{A}$ it is convenient to use an analog of the von Neumann parameterization of $\operatorname{Dom}(\widehat{A})$.

To set up the notation, introduce the following hypothesis.

HYPOTHESIS 3.1. Suppose that $\widehat{A} \neq (\widehat{A})^*$ is a maximal dissipative extension of a symmetric operator $\dot{A}$ with deficiency indices $(1,1)$. Assume, in addition, that A is a self-adjoint extension of $\dot{A}$. Suppose, that the deficiency elements $g_\pm \in \operatorname{Ker}((\dot{A})^* \mp iI)$ are normalized, $\|g_\pm\| = 1$, and chosen in such a way that

$$g_+ - g_- \in \operatorname{Dom}(A) \text{ and } g_+ - \kappa g_- \in \operatorname{Dom}(\widehat{A}) \text{ for some } |\kappa| < 1. \tag{3.1}$$

The following result is an extension of Theorem 2.1 to the dissipative case.

THEOREM 3.2. *Assume Hypothesis* 3.1. *Denote by* $s = s(\dot{A}, A)$ *the Livšic function associated with the pair* $(\dot{A}, A)$ *and by* $\mathcal{M} = \mathcal{M}(\widehat{A})$ *the Weyl–Titchmarsh function of the dissipative operator* $\widehat{A}$.

Then

$$\overline{\kappa} s(z) = \frac{\mathcal{M}(z) - i}{\mathcal{M}(z) + i}, \quad z \in \mathbb{C}_+. \tag{3.2}$$

PROOF. Since the extension $\widehat{A}$ of $\dot{A}$ is a restriction of $(\dot{A})^*$ on $\operatorname{Dom}(\widehat{A})$, one gets that

$$(\widehat{A} - iI)(\widehat{A} - zI)^{-1} g_+ \in \operatorname{Ker}((\dot{A})^* - zI), \quad z \in \rho(\widehat{A}), \tag{3.3}$$

where $\rho(\widehat{A})$ denotes the resolvent set of $\widehat{A}$.

Indeed,

$$(\widehat{A}-iI)(\widehat{A}-zI)^{-1}g_+ = g_+ + (z-i)(\widehat{A}-zI)^{-1}g_+$$

and hence

$$\begin{aligned}((\dot{A})^*-zI)(\widehat{A}-iI)(\widehat{A}-zI)^{-1}g_+ &= ((\dot{A})^*-zI)g_+ \\ &\quad + (z-i)((\dot{A})^*-zI)(\widehat{A}-zI)^{-1}g_+ \\ &= (i-z)g_+ + (z-i)(\widehat{A}-zI)(\widehat{A}-zI)^{-1}g_+ = 0\end{aligned}$$

which proves (3.3).

From (3.3) one obtains that $(\widehat{A}-iI)(\widehat{A}+iI)^{-1}g_+ \in \operatorname{Ker}((\dot{A})^*+iI)$ and hence $(\widehat{A}-iI)(\widehat{A}+iI)^{-1}g_+ = \alpha g_-$ for some $\alpha \in \mathbb{C}$.

On the other hand,

$$\alpha g_- = (\widehat{A}-iI)(\widehat{A}+iI)^{-1}g_+ = g_+ - 2i(\widehat{A}+iI)^{-1}g_+ \tag{3.4}$$

and therefore

$$\alpha g_- - g_+ = -2i(\widehat{A}+iI)^{-1}g_+ \in \operatorname{Dom}(\widehat{A}).$$

Taking into account the characterization (3.1) of $\operatorname{Dom}(\widehat{A})$, one obtains that $\alpha = \kappa$ and therefore

$$\kappa g_- = (\widehat{A}-iI)(\widehat{A}+iI)^{-1}g_+, \tag{3.5}$$

as it follows from (3.4).

Introducing the elements

$$g_z = (\widehat{A}^*-iI)(\widehat{A}^*-zI)^{-1}g_+, \quad z \in \rho(\widehat{A}^*), \tag{3.6}$$

and taking into account that the adjoint operator $(\widehat{A})^*$ is also a quasi-self-adjoint extension of $\dot{A}$, one concludes that

$$g_z \in \operatorname{Ker}((\dot{A})^*-zI), \quad z \in \rho(\widehat{A}),$$

which shows that the Livšic function $s = s(\dot{A}, A)$ admits the representation (see definition (2.2))

$$s(z) = \frac{z-i}{z+i}\cdot\frac{(g_z, g_-)}{(g_z, g_+)}, \quad z \in \mathbb{C}_+.$$

Therefore, in view of (3.5) and (3.6), one computes

$$\begin{aligned}\overline{\kappa}\, s(z) &= \frac{z-i}{z+i}\cdot\frac{(g_z,\kappa g_-)}{(g_z,g_+)}\\ &= \frac{z-i}{z+i}\cdot\frac{((\widehat{A}^*-iI)(\widehat{A}^*-zI)^{-1}g_+,(\widehat{A}-iI)(\widehat{A}+iI)^{-1}g_+)}{((\widehat{A}^*-iI)(\widehat{A}^*-zI)^{-1}g_+,g_+)}\\ &= \frac{z-i}{z+i}\cdot\frac{((\widehat{A}^*+iI)(\widehat{A}^*-zI)^{-1}g_+,g_+)}{((\widehat{A}^*-iI)(\widehat{A}^*-zI)^{-1}g_+,g_+)}\\ &= \frac{z-i}{z+i}\cdot\frac{(1+(z+i)((\widehat{A}^*-zI)^{-1}g_+,g_+))}{(1+(z-i)((\widehat{A}^*-zI)^{-1}g_+,g_+))}\\ &= \frac{z-i+(z^2+1)((\widehat{A}^*-zI)^{-1}g_+,g_+)}{z+i+(z^2+1)((\widehat{A}^*-zI)^{-1}g_+,g_+)}\\ &= \frac{\mathcal{M}(z)-i}{\mathcal{M}(z)+i},\quad z\in\rho(\widehat{A}^*),\end{aligned}$$

proving the claim.

□

REMARK 3.3. Combining Theorems 2.1 and 3.2, it is easy to see that under Hypothesis 3.1 the Weyl–Titchmarsh functions $M=M(\dot{A},A)$ and $\mathcal{M}(\widehat{A})$ are related as follows

$$\frac{\mathcal{M}-i}{\mathcal{M}+i}=\overline{\kappa}\cdot\frac{M-i}{M+i}.$$

Our next result shows that the Weyl–Titchmarsh function of a dissipative operator admits the Herglotz-Nevanlinna representation with a (locally) absolutely continuous representing measure.

COROLLARY 3.4. Assume Hypothesis 3.1. Then the Weyl–Titchmarsh function $\mathcal{M}=\mathcal{M}(\widehat{A})$ of the dissipative operator $\widehat{A}$ is an analytic function mapping the upper half-plane into the disk of radius $\frac{2|\kappa|}{1-|\kappa|^2}$ centered at the point $\left(0,\frac{1+|\kappa|^2}{1-|\kappa|^2}\right)$ of the xy-plane.

Moreover, the function $\mathcal{M}$ admits the Herglotz-Nevanlinna representation

$$\mathcal{M}(z)=\int_{\mathbb{R}}\left(\frac{1}{\lambda-z}-\frac{\lambda}{1+\lambda^2}\right)f(\lambda)d\lambda,\quad z\in\mathbb{C}_+, \tag{3.7}$$

where

$$\frac{1}{\pi}\cdot\frac{1-|\kappa|}{1+|\kappa|}\le f(\lambda)\le\frac{1}{\pi}\cdot\frac{1+|\kappa|}{1-|\kappa|}\quad \text{a.e.} \tag{3.8}$$

and

$$\int_{\mathbb{R}}\frac{f(\lambda)}{1+\lambda^2}d\lambda=1. \tag{3.9}$$

PROOF. Since the Cayley transform

$$z\mapsto\frac{z-i}{z+i}$$

maps the disk $\frac{2|\kappa|}{1-|\kappa|^2}\mathbb{D}+i\frac{1+|\kappa|^2}{1-|\kappa|^2}$ onto the disk $|\kappa|\mathbb{D}$ and the Livšic function $s(\dot{A},A)$ is contractive in $\mathbb{C}_+$, from (3.2) follows that

$$\text{Range }(\mathcal{M})\subset\frac{2|\kappa|}{1-|\kappa|^2}\mathbb{D}+i\frac{1+|\kappa|^2}{1-|\kappa|^2}.$$

In particular,

$$\frac{1-|\kappa|}{1+|\kappa|}\le\operatorname{Im}\mathcal{M}(z)\le\frac{1+|\kappa|}{1-|\kappa|},\quad z\in\mathbb{C}_+. \tag{3.10}$$

Since $\mathcal{M}(i)=i$ and $\mathcal{M}$ has a bounded imaginary part in the upper half-plane, using Fatou's Lemma, one proves the Herglotz-Nevanlinna representation

$$\mathcal{M}(z)=\int_{\mathbb{R}}\left(\frac{1}{\lambda-z}-\frac{\lambda}{1+\lambda^2}\right)d\mu(\lambda),\quad z\in\mathbb{C}_+,$$

where μ is a (locally) absolutely continuous measure with the Radon-Nikodym density

$$f(\lambda)=\frac{1}{\pi}\lim_{\varepsilon\downarrow 0}\operatorname{Im}\mathcal{M}(\lambda+i\varepsilon),\quad \text{a.e.},$$

which proves (3.8), using (3.10). Now (3.9) follows from the observation that $\mathcal{M}(i)=i$. □

4. The characteristic function and the functional model of a dissipative operator

Under Hypothesis 3.1, we introduce the characteristic function $S=S(\dot{A},\widehat{A},A)$ associated with the triple of operators $(\dot{A},\widehat{A},A)$ as the Möbius transformation

$$S(z)=\frac{s(z)-\kappa}{\overline{\kappa}\,s(z)-1},\quad z\in\mathbb{C}_+, \tag{4.1}$$

of the Livšic function $s=s(\dot{A},A)$ associated with the pair $(\dot{A},A)$.

We remark that given a triple $(\dot{A},\widehat{A},A)$, one can always find a basis $g_\pm$ in the deficiency subspace $\operatorname{Ker}((\dot{A})^*-iI)\dot{+}\operatorname{Ker}((\dot{A})^*+iI)$,

$$\|g_\pm\|=1,\quad g_\pm\in((\dot{A})^*\mp iI),$$

such that

$$g_+-g_-\in\operatorname{Dom}(A)\quad\text{and}\quad g_+-\kappa g_-\in\operatorname{Dom}(\widehat{A}),$$

and then, in this case,

$$\kappa=S(\dot{A},\widehat{A},A)(i). \tag{4.2}$$

Our next goal is to introduce a *functional model* of a prime dissipative triple[5] parameterized by the characteristic function.

Given a contractive analytic map S,

$$S(z)=\frac{s(z)-\kappa}{\overline{\kappa}\,s(z)-1},\quad z\in\mathbb{C}_+, \tag{4.3}$$

where $|\kappa|<1$ and s is an analytic, contractive function in $\mathbb{C}_+$ satisfying the Livšic criterion (1.2), introduce the function

$$M(z)=\frac{1}{i}\cdot\frac{s(z)+1}{s(z)-1},\quad z\in\mathbb{C}_+,$$

[5]We call a triple $(\dot{A},\widehat{A},A)$ a prime triple if $\dot{A}$ is a prime symmetric operator.

so that

$$M(z)=\int_{\mathbb{R}}\left(\frac{1}{\lambda-z}-\frac{\lambda}{1+\lambda^2}\right)d\mu(\lambda),\quad z\in\mathbb{C}_+,$$

for some infinite Borel measure with

$$\int_{\mathbb{R}}\frac{d\mu(\lambda)}{1+\lambda^2}=1.$$

In the Hilbert space $L^2(\mathbb{R};d\mu)$ introduce the multiplication (self-adjoint) operator by the independent variable $\mathcal{B}$ on

$$\text{Dom}(\mathcal{B})=\left\{f\in L^2(\mathbb{R};d\mu)\,\middle|\,\int_{\mathbb{R}}\lambda^2|f(\lambda)|^2d\mu(\lambda)<\infty\right\},\tag{4.4}$$

denote by $\dot{\mathcal{B}}$ its restriction on

$$\text{Dom}(\dot{\mathcal{B}})=\left\{f\in\text{Dom}(\mathcal{B})\,\middle|\,\int_{\mathbb{R}}f(\lambda)d\mu(\lambda)=0\right\},\tag{4.5}$$

and let $\widehat{\mathcal{B}}$ be the dissipative restriction of the operator $(\dot{\mathcal{B}})^*$ on

$$\text{Dom}(\widehat{\mathcal{B}})=\text{Dom}(\dot{\mathcal{B}})\dot{+}\text{lin span}\left\{\frac{1}{\cdot-i}-S(i)\frac{1}{\cdot+i}\right\}.\tag{4.6}$$

We will refer to the triple $(\dot{\mathcal{B}},\widehat{\mathcal{B}},\mathcal{B})$ as *the model triple* in the Hilbert space $L^2(\mathbb{R};d\mu)$.

Our next result shows that a triple $(\dot{A},A,\widehat{A})$ with the characteristic function S is unitarily equivalent to the model triple $(\dot{\mathcal{B}},\widehat{\mathcal{B}},\mathcal{B})$ in the Hilbert space $L^2(\mathbb{R};d\mu)$ whenever the underlying symmetric operator $\dot{A}$ is prime.

The triple $(\dot{\mathcal{B}},\widehat{\mathcal{B}},\mathcal{B})$ will therefore be called *the functional model* for $(\dot{A},A,\widehat{A})$.

THEOREM 4.1. *Suppose that $\dot{A}$ and $\dot{B}$ are prime, closed, densely defined symmetric operators with deficiency indices $(1,1)$. Assume, in addition, that A and B are some self-adjoint extensions of $\dot{A}$ and $\dot{B}$ and that $\widehat{A}$ and $\widehat{B}$ are maximal dissipative extensions of $\dot{A}$ and $\dot{B}$, respectively ($\widehat{A}\neq(\widehat{A})^*$, $\widehat{B}\neq(\widehat{B})^*$).*

Then

(i) *the triples $(\dot{A},\widehat{A},A)$ and $(\dot{B},\widehat{B},B)$ are unitarily equivalent*[6] *if, and only if, the characteristic functions $S_A=S(\dot{A},\widehat{A},A)$ and $S_B=S(\dot{B},\widehat{B},B)$ of the triples coincide;*

(ii) *the triple $(\dot{A},\widehat{A},A)$ is unitarily equivalent to the model triple $(\dot{\mathcal{B}},\widehat{\mathcal{B}},\mathcal{B})$ in the Hilbert space $L^2(\mathbb{R};d\mu)$, where μ is the representing measure for the Weyl–Titchmarsh function $M=M(\dot{A},A)$ associated with the pair $(\dot{A},A)$.*

PROOF. (i). Since $S_A=S_B$ and

$$s(\dot{A},A)=\frac{S_A-S_A(i)}{\overline{S_A(i)}S_A-1}=\frac{S_B-S_B(i)}{\overline{S_B(i)}S_B-1}=s(\dot{B},B),$$

one concludes that the Livšic functions $s(\dot{A},A)$ and $s(\dot{B},B)$ coincide,

By Theorem 1.4 (i), there exists a unitary map $\mathcal{U}$ such that

$$\mathcal{U}\dot{A}\mathcal{U}^*=\dot{B}\quad\text{and}\quad\mathcal{U}A\mathcal{U}^*=B.\tag{4.7}$$

[6]We say that triples of operators $(\dot{A},\widehat{A},A)$ and $(\dot{B},\widehat{B},B)$ in Hilbert spaces $\mathcal{H}_A$ and $\mathcal{H}_B$ are unitarily equivalent if there is a unitary map $\mathcal{U}$ from $\mathcal{H}_A$ onto $\mathcal{H}_B$ such that $\dot{B}=\mathcal{U}\dot{A}\mathcal{U}^{-1}$, $\widehat{B}=\mathcal{U}\widehat{A}\mathcal{U}^{-1}$, and $B=\mathcal{U}A\mathcal{U}^{-1}$.

Let $g_\pm \in \operatorname{Ker}((\dot{A})^* \mp iI)$, $\|g_\pm\| = 1$, such that

$$g_+ - g_- \in \operatorname{Dom}(A).$$

Set

$$f_\pm = \mathcal{U} g_\pm .$$

Equalities (4.7) yield $f_\pm \in \operatorname{Ker}((\dot{B})^* \mp iI)$, $\|f_\pm\| = 1$, and

$$f_+ - f_- \in \operatorname{Dom}(B).$$

In this case, since $\kappa = S_A(i) = S_B(i)$, the membership

$$g_+ - \kappa g_- \in \operatorname{Dom}(\widehat{A})$$

means that $f_+ - \kappa f_- \in \operatorname{Dom}(\widehat{B})$. Thus,

$$\mathcal{U}\operatorname{Dom}(\widehat{A}) = \operatorname{Dom}(\widehat{B}),$$

and hence

$$\mathcal{U}\widehat{A}\mathcal{U}^* = \widehat{B},$$

proving that the triples $(\dot{A}, \widehat{A}, A)$ and $(\dot{B}, \widehat{B}, B)$ are unitarily equivalent.

(ii). By Theorem 1.4 (ii), the Weyl–Titchmarsh function $M(\dot{\mathcal{B}}, \mathcal{B})$ coincides with $M(\dot{A}, A)$ and hence the corresponding Livšic functions coincide, that is, $s(\dot{A}, A) = s(\dot{\mathcal{B}}, \mathcal{B})$.

Since the deficiency elements

$$f_\pm(\lambda) = \frac{1}{\lambda \mp i}$$

are normalized by one, $f_+ - f_- \in \operatorname{Dom}(\mathcal{B})$, and, by hypothesis,

$$f_+ - S(i) f_- \in \operatorname{Dom}(\widehat{\mathcal{B}}),$$

one computes that

$$S_B = \frac{s(\dot{\mathcal{B}}, \mathcal{B}) - S_A(i)}{\overline{S_A(i)}\, s(\dot{\mathcal{B}}, \mathcal{B}) - 1} = S_A.$$

Therefore, the triples $(\dot{A}, \widehat{A}, A)$ and $(\dot{\mathcal{B}}, \widehat{\mathcal{B}}, \mathcal{B})$ are unitarily equivalent by the first part of the proof.

The proof is complete. □

REMARK 4.2. We remark that the Weyl-Titchmarsh function $M = M(\dot{A}, A)$ can be recovered from the characteristic function S_A of the triple $(\dot{A}, \widehat{A}, A)$ by the equation

$$M(z) = \frac{1}{i} \cdot \frac{s(z) + 1}{s(z) - 1} \quad \text{with} \quad s(z) = \frac{S_A(z) - S_A(i)}{\overline{S_A(i)} S_A(z) - 1}, \quad z \in \mathbb{C}_+.$$

We conclude this section by showing that the characteristic function S associated with the triple $(\dot{A}, \widehat{A}, A)$ and the Weyl-Titchmarsh function $\mathcal{M}(\widehat{A})$ of the dissipative operator $\widehat{A}$ are related by a linear transformation.

THEOREM 4.3. *Let $\mathcal{M}$ be the Weyl-Titchmarsh function $\mathcal{M}(\widehat{A})$ of a dissipative operator $\widehat{A}$ and $S = S(\dot{A}, \widehat{A}, A)$ the characteristic function associated with the triple $(\dot{A}, \widehat{A}, A)$.*

Then

$$\overline{\kappa} S(z) = \frac{|\kappa|^2 - 1}{2i} \mathcal{M}(z) + \frac{|\kappa|^2 + 1}{2}, \quad \text{with } \kappa = S(i), \quad z \in \mathbb{C}_+. \tag{4.8}$$

PROOF. By definition, the characteristic function S and the Livšic function s are related by the Möbius transformation

$$S = \frac{s-\kappa}{\overline{\kappa}s-1}.$$

By Theorem 3.2, one obtains that

$$\overline{\kappa}s = \frac{\mathcal{M}-i}{\mathcal{M}+i}$$

and therefore

$$\begin{aligned}\overline{\kappa}S &= \overline{\kappa}\cdot\frac{s-\kappa}{\overline{\kappa}s-1} = \frac{\overline{\kappa}s-|\kappa|^2}{\overline{\kappa}s-1} = \frac{\frac{\mathcal{M}-i}{\mathcal{M}+i}-|\kappa|^2}{\frac{\mathcal{M}-i}{\mathcal{M}+i}-1}\\ &= \frac{i}{2}\left((1-|\kappa|^2)\mathcal{M}-i(1+|\kappa|^2)\right)\end{aligned}$$

which proves (4.8). □

REMARK 4.4. Theorem 4.3 shows that in case when the von Neumann parameter κ does not vanish, the characteristic function S of a (prime) triple is uniquely determined by the pair $(\kappa, \mathcal{M})$. Therefore, along with the characteristic function S, the pair $(\kappa, \mathcal{M})$, with $\kappa \neq 0$, can also be considered to be a complete unitary invariant of a prime dissipative tripple $(\dot{A}, \widehat{A}, A)$.

However, if $\kappa = 0$, and therefore by (3.2), $\mathcal{M}(z) = i$ for all $z \in \mathbb{C}_+$, the quasi-self-adjoint extension $\widehat{A}$ coincides with the restriction of the adjoint operator $(\dot{A})^*$ on

$$\mathrm{Dom}(\widehat{A}) = \mathrm{Dom}(\dot{A}) \dot{+} \mathrm{Ker}((\dot{A})^* - iI).$$

Hence, the prime triples $(\dot{A}, \widehat{A}, A)$ with $\kappa = 0$ are in one-to-one correspondence with the set of prime symmetric operators.

In this case, the characteristic function S and the Livšic function s coincide (up to a sign),

$$S(z) = -s(z), \quad z \in \mathbb{C}_+.$$

Therefore, for S to be the characteristic function of a triple $(\dot{A}, \widehat{A}, A)$ with $\kappa = 0$ and $\mathcal{M}(z) = i$ for all z in the upper half-plane it is necessary and sufficient that S satisfy the Livšic criterion (1.2).

5. The spectral analysis of the model dissipative operator

In the suggested functional model in the Hilbert space $L^2(\mathbb{R}; d\mu)$, the eigenfunctions of the model dissipative operator $\widehat{\mathcal{B}}$ from the triple $(\dot{\mathcal{B}}, \widehat{\mathcal{B}}, \mathcal{B})$ look exceptionally simple.

LEMMA 5.1. *Suppose that $(\dot{\mathcal{B}}, \widehat{\mathcal{B}}, \mathcal{B})$ is the model triple in $L^2(\mathbb{R}; d\mu)$. Then a point $z_0 \in \mathbb{C}_+$ is an eigenvalue of the dissipative operator $\widehat{\mathcal{B}}$ if, and only if, $S(\dot{\mathcal{B}}, \widehat{\mathcal{B}}, \mathcal{B})(z_0) = 0$.*

In this case, the corresponding eigenfunction f is of the form

$$f(\lambda) = \frac{1}{\lambda - z_0}, \qquad \mu\text{-a.e.}$$

PROOF. Suppose that $z_0 \in \mathbb{C}_+$ is an eigenvalue of $\widehat{\mathcal{B}}$ and that f, $f \in L^2(\mathbb{R}; d\mu)$, is the corresponding eigenvector, that is,

$$\widehat{\mathcal{B}}f = z_0 f, \quad f \in \mathrm{Dom}(\widehat{\mathcal{B}}).$$

Since $f \in \mathrm{Dom}(\widehat{\mathcal{B}})$, the element f admits the representation

$$f(\lambda) = f_0(\lambda) + K\left(\frac{1}{\lambda - i} - \kappa\frac{1}{\lambda + i}\right),$$

where $f_0 \in \mathrm{Dom}(\dot{\mathcal{B}})$ and K is some constant. Then

$$0 = ((\widehat{\mathcal{B}} - z_0 I)f)(\lambda) = (\lambda - z_0)f_0(\lambda) + K\left(\frac{i - z_0}{\lambda - i} + \kappa\frac{i + z_0}{\lambda + i}\right)$$

and hence

$$f_0(\lambda) = -\frac{K}{\lambda - z_0}\left(\frac{i - z_0}{\lambda - i} + \kappa\frac{i + z_0}{\lambda + i}\right).$$

Since $f_0 \in \mathrm{Dom}(\dot{\mathcal{B}})$,

$$\int_{\mathbb{R}} f_0(\lambda)d\mu(\lambda) = 0$$

and hence

$$\begin{aligned}
0 &= \int_{\mathbb{R}} \frac{1}{\lambda - z_0}\left(\frac{i - z_0}{\lambda - i} + \kappa\frac{i + z_0}{\lambda + i}\right) d\mu(\lambda) \\
&= -\int_{\mathbb{R}}\left(\frac{1}{\lambda - z_0} - \frac{1}{\lambda - i}\right) d\mu(\lambda) + \kappa\int_{\mathbb{R}}\left(\frac{1}{\lambda - z_0} - \frac{1}{\lambda + i}\right) d\mu(\lambda) \\
&= -M(z_0) + M(i) + \kappa(M(z_0) - M(-i)) \\
&= -M(z_0) + i + \kappa(M(z_0) + i).
\end{aligned}$$

Therefore,

$$\kappa = \frac{M(z_0) - i}{M(z_0) + i} = s(\dot{\mathcal{B}}, \widehat{\mathcal{B}})(z_0)$$

and hence the characteristic function $S(\dot{\mathcal{B}}, \widehat{\mathcal{B}}, \mathcal{B})$ vanishes at the point z_0,

$$S(\dot{\mathcal{B}}, \widehat{\mathcal{B}}, \mathcal{B})(z_0) = \frac{s(\dot{\mathcal{B}}, \widehat{\mathcal{B}})(z_0) - \kappa}{\overline{\kappa}s(\dot{\mathcal{B}}, \widehat{\mathcal{B}})(z_0) - 1} = 0.$$

In this case,

$$\begin{aligned}
f(\lambda) &= K\left[\left(\frac{1}{\lambda - i} - \kappa\frac{1}{\lambda + i}\right) - \frac{1}{\lambda - z_0}\left(\frac{i - z_0}{\lambda - i} + \kappa\frac{i + z_0}{\lambda + i}\right)\right] \\
&= K\left[\frac{1}{\lambda - i}\left(1 - \frac{i - z_0}{\lambda - z_0}\right) - \kappa\frac{1}{\lambda + i}\left(1 + \frac{i + z_0}{\lambda - z_0}\right)\right] \\
&= K\frac{1 - \kappa}{\lambda - z_0}.
\end{aligned}$$

So, we have shown that if z_0 is an eigenvalue of of $\widehat{\mathcal{B}}$, then

$$S(\dot{\mathcal{B}}, \widehat{\mathcal{B}}, \mathcal{B})(z_0) = 0$$

and that the corresponding eigenelement f is of the form

$$f(\lambda) = \frac{1}{\lambda - z_0}. \tag{5.1}$$

Repeating the same reasoning in the reverse order, one easily shows that if $S(\dot{\mathcal{B}}, \widehat{\mathcal{B}}, \mathcal{B})(z_0) = 0$, then the function f given by (5.1) belongs to $\mathrm{Dom}(\widehat{\mathcal{B}})$ and $\widehat{\mathcal{B}} f = z_0 f$. □

For the resolvents of the model dissipative operator $\widehat{\mathcal{B}}$ and the self-adjoint (reference) operator $\mathcal{B}$ from the model triple $(\dot{\mathcal{B}}, \widehat{\mathcal{B}}, \mathcal{B})$ one gets the following resolvent formula.

THEOREM 5.2. *Suppose that $(\dot{\mathcal{B}}, \widehat{\mathcal{B}}, \mathcal{B})$ is the model triple in the Hilbert space $L^2(\mathbb{R}; d\mu)$.*

Then the resolvent of the model dissipative operator $\widehat{\mathcal{B}}$ in $L^2(\mathbb{R}; d\mu)$ has the form

$$(\widehat{\mathcal{B}} - zI)^{-1} = (\mathcal{B} - zI)^{-1} - p(z)(\cdot\,, g_{\overline{z}})g_z,$$

with

$$p(z) = \left(M(\dot{\mathcal{B}}, \mathcal{B})(z) + i\frac{\kappa+1}{\kappa-1} \right)^{-1},$$

$$z \in \rho(\widehat{\mathcal{B}}) \cap \rho(\mathcal{B}).$$

Here $M(\dot{\mathcal{B}}, \mathcal{B})$ is the Weyl-Titchmarsh function associated with the pair $(\dot{\mathcal{B}}, \mathcal{B})$ continued to the lower half-plane by the Schwarz reflection principle, and the deficiency elements g_z are given by

$$g_z(\lambda) = \frac{1}{\lambda - z}, \qquad \mu\text{-a.e.}\ .$$

PROOF. Given $h \in L^2(\mathbb{R}; d\mu)$ and $z \in \rho(\widehat{\mathcal{B}})$, suppose that

$$(\widehat{\mathcal{B}} - zI)f = h \quad \text{for some } f \in \mathrm{Dom}(\widehat{\mathcal{B}}). \tag{5.2}$$

Since $f \in \mathrm{Dom}(\widehat{\mathcal{B}})$, one gets the representation

$$f(\lambda) = f_0(\lambda) + K\left(\frac{1}{\lambda - i} - \kappa \frac{1}{\lambda + i} \right) \tag{5.3}$$

for some $f_0 \in \mathrm{Dom}(\dot{\mathcal{B}})$ and $K \in \mathbb{C}$. Eq. (5.2) yields

$$(\lambda - z) f_0(\lambda) + K\left(\frac{i - z}{\lambda - i} + \kappa \frac{i + z}{\lambda + i} \right) = h(\lambda)$$

and hence

$$f_0(\lambda) = \frac{h(\lambda)}{\lambda - z} - \frac{K}{\lambda - z}\left(\frac{i - z}{\lambda - i} + \kappa \frac{i + z}{\lambda + i} \right). \tag{5.4}$$

Since $f_0 \in \mathrm{Dom}(\dot{A})$, and therefore

$$\int_{\mathbb{R}} f_0(\lambda) d\mu(\lambda) = 0,$$

integrating (5.4) against μ, one obtains that

$$K \int_{\mathbb{R}} \frac{1}{\lambda - z}\left(\frac{i - z}{\lambda - i} + \kappa \frac{i + z}{\lambda + i} \right) d\mu(\lambda) = \int_{\mathbb{R}} \frac{h(\lambda)}{\lambda - z} d\mu(\lambda). \tag{5.5}$$

Observing that

$$\int_{\mathbb{R}} \frac{1}{\lambda - z}\left(\frac{i - z}{\lambda - i} + \kappa \frac{i + z}{\lambda + i} \right) d\mu(\lambda) = i - M(z) + \kappa(M(z) + i),$$

with $M(z) = M(\dot{\mathcal{B}}, \widehat{\mathcal{B}})(z)$, and solving (5.5) for K, one obtains

$$K = \frac{\int_{\mathbb{R}} \frac{h(\lambda)}{\lambda - z} d\mu(\lambda)}{(\kappa - 1)M(z) + i(1+\kappa)}.$$

Combining (5.3) and (5.4), for the element f we have the representation

$$\begin{aligned} f(\lambda) &= \frac{h(\lambda)}{\lambda - z} + \frac{K}{\lambda - z}\left(\frac{\lambda - z}{\lambda - i} - \kappa\frac{\lambda - z}{\lambda + i} - \left[\frac{i - z}{\lambda - i} + \kappa\frac{i + z}{\lambda + i}\right)\right] \\ &= \frac{h(\lambda)}{\lambda - z} - K\frac{\kappa - 1}{\lambda - z} \\ &= \frac{h(\lambda)}{\lambda - z} - \left(M(z) + i\frac{\kappa + 1}{\kappa - 1}\right)^{-1} \frac{1}{\lambda - z}\int_{\mathbb{R}} \frac{h(\lambda)}{\lambda - z} d\mu(\lambda), \\ &\qquad z \in \rho(\widehat{\mathcal{B}}) \cap \rho(\mathcal{B}), \end{aligned}$$

which proves the claim. □

REMARK 5.3. It is easy to see, using (2.3), that the poles of the function p in the upper half-plane coincide with the roots of the equation

$$s(\dot{\mathcal{B}}, \mathcal{B})(z) = \kappa, \quad z \in \mathbb{C}_+,$$

provided that $\kappa \neq 0$ and $M(z) \neq i$ identically in the upper half-plane. Therefore, the zeros of the characteristic function $S(\dot{\mathcal{B}}, \widehat{\mathcal{B}}, \mathcal{B})$ in the upper half-pane determine the poles of the resolvent of the dissipative operator $\widehat{\mathcal{B}}$ (cf., Lemma 5.1).

We also remark that if $\kappa = 0$ and $M(z) = i$ for all $z \in \mathbb{C}_+$, then the point spectrum of the dissipative operator $\widehat{\mathcal{B}}$ fills in the whole open upper half-plane $\mathbb{C}_+$.

Given a triple $(\dot{A}, \widehat{A}, A)$ satisfying Hypothesis 3.1, the following corollary provides an analog of the Krein formula for resolvents for all quasi-self-adjoint dissipative extensions of the symmetric operator $\dot{A}$ with deficiency indices $(1,1)$ (cf., [**16**], where in the framework of the boundary triplets theory the general resolvent formula for dual pairs of linear relations was obtained).

COROLLARY 5.4. Under Hypothesis 3.1, the following resolvent formula

$$(\widehat{A} - zI)^{-1} = (A - zI)^{-1} - p(z)(\cdot\,, g_{\overline{z}})g_z, \tag{5.6}$$

$$z \in \rho(\widehat{A}) \cap \rho(A),$$

holds.

Here

(a)

$$p(z) = \left(M(\dot{A}, A)(z) + i\frac{\kappa + 1}{\kappa - 1}\right)^{-1} \tag{5.7}$$

$$= i\left(\frac{s(\dot{A}, A)(z) + 1}{s(\dot{A}, A)(z) - 1} - \frac{\kappa + 1}{\kappa - 1}\right)^{-1}; \tag{5.8}$$

(b) $M(\dot{A}, A)$ and $s(\dot{A}, A)$ are the Weyl-Titchmarsh and the Livšic function of the pair $(\dot{A}, A)$, respectively;

(c) g_z are deficiency elements of $\dot{A}$,

$$g_z \in \operatorname{Ker}((\dot{A})^* - zI),$$

satisfying the normalization condition

$$\|g_z\| = \left(\int_{\mathbb{R}} \frac{d\mu(\lambda)}{|\lambda - z|^2} \right)^{1/2} \tag{5.9}$$

(the deficiency elements g_z can be chosen to be analytic in $z \in \rho(\widehat{A}) \cap \rho(A)$);

(d) μ is the measure from the Herglotz-Nevanlinna representation

$$M(\dot{A}, A)(z) = \int_{\mathbb{R}} \left(\frac{1}{\lambda - z} - \frac{\lambda}{1 + \lambda^2} \right) d\mu;$$

and

(e) κ is the von Neumann parameter characterizing the domain of the dissipative extension $\widehat{A}$,

$$g_i - g_{-i} \in \operatorname{Dom}(A) \quad \text{and} \quad g_i - \kappa g_{-i} \in \operatorname{Dom}(\widehat{A}). \tag{5.10}$$

REMARK 5.5. We would like to stress that

i) the von Neumann parameter κ from (5.10),
ii) the Livšic function $s(\dot{A}, A)$,
and
iii) the Weyl-Titchmarsh function $M(\dot{A}, A)$,

can easily be recovered from the functional parameter of the (prime) triple, the characteristic function $S = S(\dot{A}, \widehat{A}, A)$, which is a complete unitary invariant of $(\dot{A}, \widehat{A}, A)$.

Indeed,

$$\kappa = S(i),$$
$$s(\dot{A}, A)(z) = \frac{S(z) - \kappa}{\overline{\kappa} S(z) - 1},$$
$$M(\dot{A}, A)(z) = \frac{1}{i} \cdot \frac{s(\dot{A}, A)(z) + 1}{s(\dot{A}, A)(z) - 1},$$
$$z \in \mathbb{C}_+,$$

with $M(\dot{A}, A)$ continued to the lower half-plane by the Schwarz reflection principle

$$M(\dot{A}, A)(z) = \overline{M(\dot{A}, A)(\overline{z})}, \quad z \in \mathbb{C}_-.$$

REMARK 5.6. The resolvent formula (5.6)–(5.8) also holds if $|\kappa| = 1$ and hence $\widehat{A}$ is self-adjoint. In this case, it provides the standard Krein resolvent formula for self-adjoint extensions of $\dot{A}$ via the von Neumann extension parameter κ and the Weyl-Titchmarsh (Livšic) function.

We also notice that although it appears unlikely that the explicit normalization condition (5.9) has been missed in the literature, we were not able to locate an appropriate reference.

REMARK 5.7. We remark that if two triples $(\dot{A}, A, \widehat{A}_1)$ and $(\dot{A}, A, \widehat{A}_2)$ satisfy Hypothesis 3.1 with the von Neumann parameters κ_1 and κ_2, respectively, then one gets the following resolvent formula for the dissipative extensions $\widehat{A}_1$ and $\widehat{A}_2$ refining, in the rank one setting, a result in [**14**]:

$$(\widehat{A}_2 - zI)^{-1} = (\widehat{A}_1 - zI)^{-1} - q(z)(\cdot, g_{\overline{z}})g_z,$$

where $q(z) = p_2(z) - p_1(z)$ with

$$\begin{aligned} p_k(z) &= \left(M(\dot{A}, A)(z) + i\frac{\kappa_k + 1}{\kappa_k - 1}\right)^{-1}, \\ &= i\left(\frac{s(\dot{A}, A)(z) + 1}{s(\dot{A}, A)(z) - 1} - \frac{\kappa_k + 1}{\kappa_k - 1}\right)^{-1}, \quad k = 1, 2, \end{aligned}$$

$$z \in \rho(A) \cap \rho(\widehat{A}_1) \cap \rho(\widehat{A}_2).$$

We recall that if S_1 and S_2 are the characteristic functions of the triples $(\dot{A}, A, \widehat{A}_1)$ and $(\dot{A}, A, \widehat{A}_2)$, respectively, then

$$s(\dot{A}, A) = \frac{S_k - \kappa_k}{\overline{\kappa_k} S - 1}, \quad k = 1, 2.$$

Appendix A. The spectral properties of the model symmetric operator

For completeness, we provide a summary of spectral properties of the model symmetric operator $\dot{\mathcal{B}}$ in $L^2(\mathbb{R}; d\mu)$ (we refer to a relevant discussion in [**10**]) and obtain the characterization of the core of its spectrum in terms of the measure μ.

Denote by $\mathfrak{m}$ the class of infinite Lebesgue-Stieltjes measures μ on the real axis, $\mu(\mathbb{R}) = \infty$, such that

$$\int_{\mathbb{R}} \frac{d\mu(\lambda)}{1 + \lambda^2} = 1. \tag{A.1}$$

Our main goal is to show that every measure μ from the measure class $\mathfrak{m}$ gives rise to a prime symmetric operator in the Hilbert space $L^2(\mathbb{R}; d\mu)$ with deficiency indices $(1, 1)$ via the following construction.

Given $\mu \in \mathfrak{m}$, in the Hilbert space $L^2(\mathbb{R}; d\mu)$ introduce the multiplication (self-adjoint) operator $\mathcal{B}$ by the independent variable on

$$\mathrm{Dom}(\mathcal{B}) = \left\{ f \in L^2(\mathbb{R}; d\mu) \,\middle|\, \int_{\mathbb{R}} \lambda^2 |f(\lambda)|^2 d\mu(\lambda) < \infty \right\} \tag{A.2}$$

and denote by $\dot{\mathcal{B}}$ its restriction on

$$\mathrm{Dom}(\dot{\mathcal{B}}) = \left\{ f \in \mathrm{Dom}(\mathcal{B}) \,\middle|\, \int_{\mathbb{R}} f(\lambda) d\mu(\lambda) = 0 \right\}. \tag{A.3}$$

We remark that the membership $f \in \mathrm{Dom}(\mathcal{B})$ means that

$$\int_{\mathbb{R}} (1 + \lambda^2) |f(\lambda)|^2 d\mu(\lambda) < \infty.$$

Therefore, using Cauchy-Schwarz,

$$\int_{\mathbb{R}} |f(\lambda)| d\mu(\lambda) \le \left(\int_{\mathbb{R}} (1+\lambda^2)|f(\lambda)|^2 d\mu(\lambda)\right)^{1/2} \left(\int_{\mathbb{R}} \frac{d\mu(\lambda)}{1+\lambda^2}\right)^{1/2}$$

$$(\text{A.4}) \qquad = \left(\int_{\mathbb{R}} (1+\lambda^2)|f(\lambda)|^2 d\mu(\lambda)\right)^{1/2} < \infty.$$

Inequality (A.4) shows that the following unbounded functional ℓ on $\mathrm{Dom}(\ell) = \mathrm{Dom}(\mathcal{B})$ given by

$$\ell(f) = \int_{\mathbb{R}} f(\lambda) d\mu(\lambda), \quad f \in \mathrm{Dom}(\ell),$$

is well defined and hence the restriction $\dot{\mathcal{B}}$ on $\mathrm{Dom}(\dot{\mathcal{B}})$ given by (A.3) is well defined as well.

THEOREM A.1. *Suppose that* $(\dot{\mathcal{B}}, \mathcal{B})$ *is the model pair in the Hilbert space* $L^2(\mathbb{R}; d\mu)$ *as defined by* (A.2) *and* (A.3). *Then* $\dot{\mathcal{B}}$ *is a prime, densely defined, closed, symmetric operator with deficiency indices* $(1,1)$.

PROOF. Since $\dot{\mathcal{B}}$ is a restriction of a self-adjoint operator, the operator $\dot{\mathcal{B}}$ is symmetric.

To show that $\dot{\mathcal{B}}$ is closed, assume that $\{f_n\}_{n=1}^{\infty}$ is a sequence such that $f_n \in \mathrm{Dom}(\dot{\mathcal{B}})$ for all n and

$$\lim_{n\to\infty} f_n = f \quad \text{and} \quad \lim_{n\to\infty} \dot{\mathcal{B}} f_n = g$$

for some $f, g \in L^2(\mathbb{R}; d\mu)$. Since $\dot{\mathcal{B}}$ is a restriction of the self-adjoint operator $\mathcal{B}$, one gets that

$$\lim_{n\to\infty} f_n = f \quad \text{and} \quad \lim_{n\to\infty} \mathcal{B} f_n = g$$

and hence $f \in \mathrm{Dom}(\mathcal{B})$ and $g = \mathcal{B} f$ for $\mathcal{B}$ is a closed operator.

Since $f_n \in \mathrm{Dom}(\dot{\mathcal{B}})$, one gets that

$$\int_{\mathbb{R}} f_n(\lambda) d\mu(\lambda) = \int_{\mathbb{R}} (\lambda - i) f_n(\lambda) \overline{\frac{1}{\lambda + i}} d\mu(\lambda) = 0$$

which means that $(\mathcal{B} - iI) f_n$ is orthogonal to the element $h \in L^2(\mathbb{R}; d\mu)$ given by

$$h(\lambda) = \frac{1}{\lambda + i}.$$

Therefore,

$$\lim_{n\to\infty} ((\mathcal{B} - iI) f_n, h) = ((\mathcal{B} - iI) f, h) = 0$$

which means that $\int_{\mathbb{R}} f(\lambda) d\mu(\lambda) = 0$ and hence $f \in \mathrm{Dom}(\dot{\mathcal{B}})$.

To prove that $\dot{\mathcal{B}}$ is densely defined, we observe first, that the quotient space $\mathrm{Dom}(\mathcal{B})/\mathrm{Dom}(\dot{\mathcal{B}})$ is one-dimensional. More specifically,

$$(\text{A.5}) \qquad \mathrm{Dom}(\mathcal{B}) = \mathrm{Dom}(\dot{\mathcal{B}}) \dot{+} \mathrm{span}\,\{g\},$$

where the function g is given by

$$g(\lambda) = \frac{1}{1+\lambda^2}, \quad \mu\text{-a.e.}.$$

Indeed, for any $f \in \mathrm{Dom}(\mathcal{B})$, the function

$$g(\lambda) = f(\lambda) - \frac{\int_{\mathbb{R}} f(s) d\mu(s)}{1+\lambda^2}$$

belongs to $\mathrm{Dom}(\dot{\mathcal{B}})$, since

$$\int_{\mathbb{R}} g(\lambda)d\mu(\lambda) = \int_{\mathbb{R}} f(\lambda)d\mu(\lambda) - \int_{\mathbb{R}} f(s)d\mu(s) \int_{\mathbb{R}} \frac{d\mu(\lambda)}{1+\lambda^2} = 0,$$

due to the normalization condition

$$\int_{\mathbb{R}} \frac{d\mu(\lambda)}{1+\lambda^2} = 1.$$

To show that $\mathrm{Dom}(\dot{\mathcal{B}})$ is dense in $L^2(\mathbb{R}; d\mu)$, assume that $h \perp \mathrm{Dom}(\dot{\mathcal{B}})$. Given $n \in \mathbb{N}$, introduce the function

$$h_n(\lambda) = h(\lambda)\chi_{(-n,n)}(\lambda),$$

where $\chi_{(-n,n)}(\cdot)$ is the characteristic function of the interval $(-n, n)$.

The function

$$g_n(\lambda, z) = \frac{h_n(\lambda)}{\lambda - z} - \frac{1}{1+\lambda^2} \int_{\mathbb{R}} \frac{h_n(s)}{s-z} d\mu(s)$$

obviously belongs to $\mathrm{Dom}(\dot{\mathcal{B}})$. In particular,

$$g_n(\cdot, z) \perp h, \quad n \in \mathbb{N},$$

which means that

$$\begin{aligned} \int_{\mathbb{R}} \frac{h_n(\lambda)\overline{h(\lambda)}}{\lambda - z} d\mu(\lambda) &= \int_{\mathbb{R}} \frac{\overline{h(\lambda)}}{1+\lambda^2} d\mu(\lambda) \cdot \int_{\mathbb{R}} \frac{h_n(s)}{s-z} d\mu(s) \\ &= \int_{\mathbb{R}} \frac{h_n(\lambda)\overline{m}}{\lambda - z} d\mu(\lambda). \end{aligned}$$

Here we used the notation

$$m = \int_{\mathbb{R}} \frac{h(\lambda)}{1+\lambda^2} d\mu(\lambda).$$

By the uniqueness theorem for Cauchy integrals, one gets that

$$h_n(\lambda)\overline{h(\lambda)} = |h(\lambda)|^2 = h(\lambda)\overline{m} \quad \text{for } \mu\text{-a.e. } \lambda \in (-n, n).$$

Therefore,

$$h(\lambda) = m \quad \text{for } \mu\text{-a.e. } \lambda \in (-n, n).$$

Since n is arbitrary, one concludes that

$$h(\lambda) = m \quad \mu\text{-a.e. }.$$

According to the hypothesis, the measure μ is infinite. Therefore, $h \in L^2(\mathbb{R}; d\mu)$ only if the constant m is zero. That is, $h = 0$ which proves that $\mathrm{Dom}(\dot{\mathcal{B}})$ is dense in $L^2(\mathbb{R}; d\mu)$, and hence, the operator $\dot{\mathcal{B}}$ is densely defined.

Next, we prove that $\dot{\mathcal{B}}$ is a prime operator.

Assume that $\mathcal{H}_0$ reduces $\dot{\mathcal{B}}$ and that the part $\dot{\mathcal{B}}|_{\mathcal{H}_0}$ is a self-adjoint operator. Since the self-adjoint multiplication operator $\mathcal{B}$ in $\mathcal{H} = L^2(\mathbb{R}; d\mu)$ is an extension of the symmetric operator $\dot{\mathcal{B}}$, the subspace $\mathcal{H}_0$ also reduces the self-adjoint operator $\mathcal{B}$ and hence

$$\mathcal{H}_0 = \mathrm{Ran}\, E_{\mathcal{B}}(\delta) \quad \text{for some Borel set } \delta \subset \mathbb{R}$$

for $\mathcal{B}$ has a simple spectrum. Here $E_{\mathcal{B}}(\cdot)$ denotes the projection-valued spectral measure of the self-adjoint operator $\mathcal{B}$.

One observes that the function $f(\cdot) = \chi_{\delta\cap(-n,n)}(\cdot)$ belongs to $\mathrm{Dom}(\mathcal{B}|_{\mathcal{H}_0}) \subset \mathrm{Dom}(\dot{\mathcal{B}})$. In particular,

$$\int_{(-n,n)\cap\delta} d\mu(\lambda) = \mu\left((-n,n)\cap\delta\right) = 0 \quad \text{for all } n \in \mathbb{N}.$$

Hence, $\mu(\delta) = 0$ and therefore, $\mathcal{H}_0 = 0$, proving that $\dot{\mathcal{B}}$ is a prime symmetric operator.

Finally, we prove that $\dot{\mathcal{B}}$ has deficiency indices $(1,1)$.

We claim that

$$\mathrm{Ker}((\dot{\mathcal{B}})^* \mp iI) = \mathrm{span}\{g_\pm\},$$

where

$$g_\pm(\lambda) = \frac{1}{\lambda \mp i}, \quad \mu\text{-a.e. }.$$

First, we prove the inclusion

$$\mathrm{Ker}((\dot{\mathcal{B}})^* \mp iI) \subset \mathrm{span}\{g_\pm\}.$$

Indeed, it suffices to show that $((\dot{\mathcal{B}} \pm iI)f, g_\pm) = 0$ for all $f \in \mathrm{Dom}(\dot{\mathcal{B}})$. One computes

$$(\dot{\mathcal{B}} \pm iI)f, g_\pm) = \int_{\mathbb{R}} (\lambda \pm i) f(\lambda) \overline{\frac{1}{\lambda \mp i}} d\mu(\lambda) = \int_{\mathbb{R}} f(\lambda) d\mu(\lambda) = 0,$$

which shows that $\dot{\mathcal{B}}$ has deficiency indices at least $(1,1)$. Now the claim follows from (A.5) and the observation that

$$\frac{1}{2i}(g_+(\lambda) - g_-(\lambda)) = \frac{1}{\lambda^2+1}, \quad \mu\text{-a.e. }.$$

□

The following result characterizes the core of the spectrum of the symmetric operator $\dot{\mathcal{B}}$ as the set of non-isolated points of the support of the measure μ.

Recall that a point $\lambda \in \mathbb{C}$ is said to be quasi-regular for an operator T if $T - \lambda I$ has a continuous inverse on $\mathrm{Ran}(T - \lambda I)$ and the core of the spectrum of T is defined to be the complement to the set of its quasi-regular points.

THEOREM A.2. *Assume that $\dot{\mathcal{B}}$ is a symmetric operator from the model pair in the Hilbert space $L^2(\mathbb{R}, d\mu)$.*

Then a point λ_0, $\lambda_0 \in \mathbb{R}$, is a quasi-regular point for the symmetric operator $\dot{\mathcal{B}}$ if and only if there exists an $\varepsilon > 0$ such that

$$\mu\left((-\varepsilon + \lambda_0, \lambda_0) \cup (\lambda_0, \lambda_0 + \varepsilon)\right) = 0. \tag{A.6}$$

PROOF. *"Only If"* Part. Suppose that (A.6) holds true. If $\mu(\{\lambda_0\}) = 0$, then the point λ_0 belongs to the resolvent set of the self-adjoint operator $\mathcal{B}$ and hence λ_0 is automatically a quasi-regular point for the symmetric restriction $\dot{\mathcal{B}}$ of $\mathcal{B}$.

If $\mu(\{\lambda_0\}) > 0$, one proceeds as follows.

Suppose that

$$u \in \mathrm{Dom}(\dot{\mathcal{B}}). \tag{A.7}$$

Then

$$\|u\|^2 = \int_{\mathbb{R}} |u(\lambda)|^2 d\mu(\lambda) = \int_{|\lambda-\lambda_0|\geq\varepsilon} |u(\lambda)|^2 d\mu(\lambda) + |u(\lambda_0)|^2\mu(\{\lambda_0\})$$

$$\leq \left(\int_{|\lambda-\lambda_0|\geq\varepsilon} \frac{|u(\lambda)|^2}{(\lambda-\lambda_0)^2} d\mu(\lambda)\right)^{1/2} \left(\int_{|\lambda-\lambda_0|\geq\varepsilon} (\lambda-\lambda_0)^2|u(\lambda)|^2 d\mu(\lambda)\right)^{1/2}$$

$$+ |u(\lambda_0)|^2\mu(\{\lambda_0\}) \leq \varepsilon^{-1}\|u\| \, \|(\dot{\mathcal{B}} - \lambda_0 I)u\| + |u(\lambda_0)|^2\mu(\{\lambda_0\}).$$

Since (A.7) implies

$$\int_{|\lambda-\lambda_0|\geq\varepsilon} u(\lambda)d\mu(\lambda) + u(\lambda_0)\mu(\{\lambda_0\}) = 0,$$

and hence

$$u(\lambda_0) = -(\mu(\{\lambda_0\}))^{-1} \int_{|\lambda-\lambda_0|\geq\varepsilon} u(\lambda)d\mu(\lambda),$$

one gets the inequality

$$\|u\|^2 \leq \varepsilon^{-1}\|u\| \, \|(\dot{\mathcal{B}} - \lambda_0 I)u\| + (\mu(\{\lambda_0\})^{-1} \left(\int_{|\lambda-\lambda_0|\geq\varepsilon} u(\lambda)d\mu(\lambda)\right)^2$$

$$\leq \varepsilon^{-1}\|u\| \, \|(\dot{\mathcal{B}} - \lambda_0 I)u\| + (\mu(\{\lambda_0\})^{-1} \int_{|\lambda-\lambda_0|\geq\varepsilon} \frac{d\mu(\lambda)}{(\lambda-\lambda_0)^2} \|(\dot{\mathcal{B}} - \lambda_0 I)u\|^2.$$

Thus, the ratio

$$r = \frac{\|u\|}{\|(\dot{\mathcal{B}} - \lambda_0 I)u\|}$$

satisfies the inequality

$$r^2 \leq \varepsilon^{-1} r + (\mu(\{\lambda_0\}))^{-1} \int_{|\lambda-\lambda_0|\geq\varepsilon} \frac{d\mu(\lambda)}{(\lambda-\lambda_0)^2},$$

and therefore r is uniformly bounded with respect to ε which proves that the point λ_0 is a quasi-regular point.

"If" Part. If (A.6) does not hold, and therefore λ_0 is not an isolated point from the support of the measure μ, we proceed as follows.

Take an $\varepsilon > 0$ and choose measurable sets $\delta_\pm$ such that

$$\delta_+ \cup \delta_- = (\lambda_0 - \varepsilon, \lambda_0 + \varepsilon) \setminus \{\lambda_0\},$$

with $\mu(\delta_\pm) > 0$. Set

$$u_\varepsilon(\lambda) = (\mu(\delta_+))^{-1}\chi_{\delta_+}(\lambda) - (\mu(\delta_-))^{-1}\chi_{\delta_-}(\lambda).$$

Clearly, $u_\varepsilon \in L^2(\mathbb{R}; d\mu)$ and

$$\int_{\mathbb{R}} u_\varepsilon(\lambda)d\lambda = 0.$$

Thus, $u_\varepsilon \in \mathrm{Dom}(\dot{\mathcal{B}})$. One computes that

$$\|u_\varepsilon\|^2 = (\mu(\delta_+))^{-1} + (\mu(\delta_-))^{-1}.$$

However,

$$\|(\dot{\mathcal{B}} - \lambda_0 I)u_\varepsilon\|^2 \leq \varepsilon^2 \left((\mu(\delta_+))^{-1} + (\mu(\delta_-))^{-1}\right) = \varepsilon^2 \|u_\varepsilon\|^2$$

which proves that λ_0 belongs to the core of the spectrum of $\dot{A}$ since ε can be chosen arbitrarily small.

The proof is complete. □

References

[1] Yuri Arlinskii, Sergey Belyi, and Eduard Tsekanovskii, *Conservative realizations of Herglotz-Nevanlinna functions*, Operator Theory: Advances and Applications, vol. 217, Birkhäuser/Springer Basel AG, Basel, 2011. MR2828331 (2012f:47049)

[2] Yury Arlinskiĭ, Seppo Hassi, and Henk de Snoo, *Q-functions of quasi-selfadjoint contractions*, Operator theory and indefinite inner product spaces, Oper. Theory Adv. Appl., vol. 163, Birkhäuser, Basel, 2006, pp. 23–54, DOI 10.1007/3-7643-7516-7_2. MR2215857 (2007b:47008)

[3] N. I. Akhiezer and I. M. Glazman, *Theory of linear operators in Hilbert space*, Dover Publications Inc., New York, 1993. Translated from the Russian and with a preface by Merlynd Nestell; Reprint of the 1961 and 1963 translations; Two volumes bound as one. MR1255973 (94i:47001)

[4] Louis de Branges and James Rovnyak, *Canonical models in quantum scattering theory*, Perturbation Theory and its Applications in Quantum Mechanics (Proc. Adv. Sem. Math. Res. Center, U.S. Army, Theoret. Chem. Inst., Univ. of Wisconsin, Madison, Wis., 1965), Wiley, New York, 1966, pp. 295–392. MR0244795 (39 #6109)

[5] M. S. Brodskiĭ, *Triangular and Jordan representations of linear operators*, American Mathematical Society, Providence, R.I., 1971. Translated from the Russian by J. M. Danskin; Translations of Mathematical Monographs, Vol. 32. MR0322542 (48 #904)

[6] M. S. Brodskiĭ and M. S. Livšic, *Spectral analysis of non-self-adjoint operators and intermediate systems*, Uspehi Mat. Nauk (N.S.) **13** (1958), no. 1(79), 3–85 (Russian). MR0100793 (20 #7221)

[7] V. A. Derkach and M. M. Malamud, *Generalized resolvents and the boundary value problems for Hermitian operators with gaps*, J. Funct. Anal. **95** (1991), no. 1, 1–95, DOI 10.1016/0022-1236(91)90024-Y. MR1087947 (93d:47046)

[8] William F. Donoghue Jr., *On the perturbation of spectra*, Comm. Pure Appl. Math. **18** (1965), 559–579. MR0190761 (32 #8171)

[9] Fritz Gesztesy, Konstantin A. Makarov, and Eduard Tsekanovskii, *An addendum to Krein's formula*, J. Math. Anal. Appl. **222** (1998), no. 2, 594–606, DOI 10.1006/jmaa.1998.5948. MR1628437 (99g:47047)

[10] Fritz Gesztesy and Eduard Tsekanovskii, *On matrix-valued Herglotz functions*, Math. Nachr. **218** (2000), 61–138, DOI 10.1002/1522-2616(200010)218:1¡61::AID-MANA61¿3.3.CO;2-4. MR1784638 (2001j:47018)

[11] A. N. Kočubeĭ, *Characteristic functions of symmetric operators and their extensions*, Izv. Akad. Nauk Armyan. SSR Ser. Mat. **15** (1980), no. 3, 219–232, 247 (Russian, with English and Armenian summaries). MR592863 (81k:47017)

[12] M. G. Krein, *On resolvent of Hermitian operator with defect numbers* (m, m), Dokl. Akad. Mauk SSSR **52**, no. 8, (1946), 657–660 (Russian).

[13] M. G. Kreĭn and H. Langer, *Über die Q-Funktion eines π-hermiteschen Operators im Raume* Π_κ, Acta Sci. Math. (Szeged) **34** (1973), 191–230 (German). MR0318958 (47 #7504)

[14] A. V. Kuzhel′, *An analogue of M. G. Kreĭn's formula for resolvents of nonselfadjoint extensions of a Hermitian operator*, Teor. Funktsiĭ Funktsional. Anal. i Prilozhen. **36** (1981), 49–55, 127 (Russian). MR645309 (83e:47018)

[15] A. Kuzhel, *Characteristic functions and models of nonselfadjoint operators*, Mathematics and its Applications, vol. 349, Kluwer Academic Publishers Group, Dordrecht, 1996. MR1383821 (97e:47014)

[16] M. M. Malamud and V. I. Mogilevskii, *Kreĭn type formula for canonical resolvents of dual pairs of linear relations*, Methods Funct. Anal. Topology **8** (2002), no. 4, 72–100. MR1942823 (2003j:47002)

[17] Nikolaĭ K. Nikolskiĭ and Vasily I. Vasyunin, *A unified approach to function models, and the transcription problem*, The Gohberg anniversary collection, Vol. II (Calgary, AB, 1988), Oper. Theory Adv. Appl., vol. 41, Birkhäuser, Basel, 1989, pp. 405–434. MR1038349 (91c:47017)

[18] N. K. Nikol′skiĭ and S. V. Khrushchëv, *A functional model and some problems of the spectral theory of functions*, Trudy Mat. Inst. Steklov. **176** (1987), 97–210, 327 (Russian). Translated in Proc. Steklov Inst. Math. **1988**, no. 3, 101–214; Mathematical physics and complex analysis (Russian). MR902373 (88j:47012)

[19] M. S. Livšic, *On a class of linear operators in Hilbert space*, Amer. Math. Soc. Transl. (2) **13** (1960), 61–83. MR0113142 (22 #3981a)

[20] M. S. Livšic, *On the spectral resolution of linear non-selfadjoint operators*, Amer. Math. Soc. Transl. (2) **5** (1957), 67–114. MR0083703 (18,748f)

[21] Béla Sz.-Nagy and Ciprian Foiaş, *Harmonic analysis of operators on Hilbert space*, Translated from the French and revised, North-Holland Publishing Co., Amsterdam, 1970. MR0275190 (43 #947)

[22] B. S. Pavlov, *Functional model and spectral singularities*, Scattering theory. Theory of oscillations (Russian), Probl. Mat. Fiz., vol. 9, Leningrad. Univ., Leningrad, 1979, pp. 113–121, 186 (Russian). MR566391 (83m:47011)

[23] B. S. Pavlov, *Dilation theory and spectral analysis of nonselfadjoint differential operators*, Mathematical programming and related questions (Proc. Seventh Winter School, Drogobych, 1974), Theory of operators in linear spaces (Russian), Central. Èkonom. Mat. Inst. Akad. Nauk SSSR, Moscow, 1976, pp. 3–69 (Russian). MR0634807 (58 #30375)

[24] R. S. Phillips, *On dissipative operators*, in "Lecture Series in Differential Equations," Vol. II (A. K. Aziz, ed.) von Nostrand, (1969), 65–113.

[25] Š. N. Saakjan, *Theory of resolvents of a symmetric operator with infinite defect numbers*, Akad. Nauk Armjan. SSR Dokl. **41** (1965), 193–198 (Russian, with Armenian summary). MR0196497 (33 #4684)

[26] A. V. Štraus, *Extensions and characteristic function of a symmetric operator*, Izv. Akad. Nauk SSSR Ser. Mat. **32** (1968), 186–207 (Russian). MR0225193 (37 #788)

[27] A. V. Shtraus, *Functional models and generalized spectral functions of symmetric operators*, Algebra i Analiz **10** (1998), no. 5, 1–76 (Russian, with Russian summary); English transl., St. Petersburg Math. J. **10** (1999), no. 5, 733–784. MR1660000 (99m:47009)

[28] Barry Simon, *Spectral analysis of rank one perturbations and applications*, Mathematical quantum theory. II. Schrödinger operators (Vancouver, BC, 1993), CRM Proc. Lecture Notes, vol. 8, Amer. Math. Soc., Providence, RI, 1995, pp. 109–149. MR1332038 (97c:47008)

[29] E. R. Tsekanovskii and Yu. L. Smuljan, *The theory of biextensions of operators in rigged Hilbert spaces. Unbounded operator colligations and characteristic functions*, Russian Math. Surv. **31** (1977), 73–131.

Department of Mathematics, University of Missouri, Columbia, Missouri 63211
E-mail address: makarovk@missouri.edu

Department of Mathematics, Niagara University, P.O. Box 2044, New York 14109
E-mail address: tsekanov@niagara.edu

Proceedings of Symposia in Pure Mathematics
Volume **87**, 2013
http://dx.doi.org/10.1090/pspum/087/01430

Stability for the inverse resonance problem for the CMV operator

Roman Shterenberg, Rudi Weikard, and Maxim Zinchenko

Dedicated with great pleasure to Fritz Gesztesy on the occasion of his 60th birthday.

Abstract. For the class of unitary CMV operators with super-exponentially decaying Verblunsky coefficients we give a new proof of the inverse resonance problem of reconstructing the operator from its resonances - the zeros of the Jost function. We establish a stability result for the inverse resonance problem that shows continuous dependence of the operator coefficients on the location of the resonances.

1. Introduction

In recent years there has been a substantial interest in inverse resonance problems for operators that arise in mathematical physics. An inverse resonance problems asks whether an operator is determined by its resonances. One is typically interested in reconstructing coefficients of an operator from a given set of resonances and in stability of the solution of the inverse resonance problem under small perturbation of the resonances. The case of Schrödinger and Jacobi operators have received most of the attention, see for instance [**2**, **3**, **5**–**11**] and the references therein.

The inverse resonance problem for a class of unitary operators known as CMV operators has been considered in [**14**]. These operators are represented by five-diagonal infinite unitary matrices giving rise to unitary operators on $\ell^2(\mathbb{N}_0)$. They are closely connected with trigonometric moment problems and with orthogonal polynomials and finite measures on the unit circle. The entries in a CMV matrix are determined by the Verblunsky coefficients, a sequence of complex numbers in the unit disk. For background information on CMV matrices we refer the reader to Simon's two-volume monograph [**12**, **13**].

In this paper we continue our study of the inverse resonance problem for unitary CMV operators associated with sequences of exponentially decaying Verblunsky coefficients. For these operators we review the concept of a resonance and give a simplified proof of the uniqueness result in [**14**] that the location of all resonances determines the Verblunsky coefficients uniquely. The main result of the present

2010 *Mathematics Subject Classification.* 47B36, 34L40, 39A70.

Key words and phrases. Inverse problem, eigenvalues and resonances, Jost function, CMV.

The authors were supported in part by NSF grants DMS-0901015, DMS-0800906, and DMS-0965411, respectively.

study is a stability result: Suppose two CMV operators in our class have resonances z_n and $\breve{z}_n$, respectively. Further suppose that these resonances are respectively close to each other as long as they are not too large. Then the Verblunsky coefficients are respectively close to each other, too. We will now make these statements precise. Throughout this paper $\gamma > 1$, $\eta > 0$, and $Q > 1$ denote fixed constants. We define the class $B_0(\gamma, \eta, Q)$ as the set of those sequences of Verblunsky coefficients $\alpha : \mathbb{N} \to \mathbb{D}$ satisfying the following two conditions:

(1) $|\alpha_k| \le \eta \exp(-k^\gamma)$ for all $k \in \mathbb{N}$ and
(2) $\prod_{j=1}^\infty (1 - |\alpha_j|) \ge 1/Q$.

We note that the above two conditions are essentially independent as the first one enforces the decay of $|\alpha_k|$ for large k and the second one bounds $|\alpha_k|$ away from 1 for small k. Also, two sets $\{z_1, ..., z_N\}$ and $\{\breve{z}_1, ..., \breve{z}_N\}$ of complex numbers will be called respectively ε-close if $|z_n - \breve{z}_n| < \varepsilon$ for all $n \in \{1, ..., N\}$.

The following technical result required for our proof of stability for the inverse resonance problem is of independent interest. We will prove it in Section 4.2.

THEOREM 1.1. *Suppose α is a sequence of Verblunsky coefficients in $B_0(\gamma, \eta, Q)$ and U is the associated CMV operator. Then there is a positive number δ such that U has no resonances in the disk $\{z : |z| < 1 + \delta\}$.*

The main purpose of this paper is to prove the following theorem, which we will do in Section 4.

THEOREM 1.2. *Suppose α and $\breve{\alpha}$ are two sequences of Verblunsky coefficients in $B_0(\gamma, \eta, Q)$ and U and $\breve{U}$ are the associated CMV operators. Let δ be the number introduced in the previous theorem. Further suppose that, for two numbers $R > 1$ and $\varepsilon \in (0, \delta/2)$, the resonances of U and $\breve{U}$ in the circle $|z| < R$, if there are any, are respectively ε-close. Then there is a constant A_0, depending only on γ, η, and Q, such that*

$$|\alpha_n - \breve{\alpha}_n| \le A_0 (6Q^2)^n \left(\varepsilon + \frac{(\log R)^{\gamma/(\gamma-1)}}{R} \right)$$

for all $n \in \mathbb{N}$.

Theorem 1.2 extends earlier results on stability of the inverse resonance problems for Schrödinger and Jacobi operators [**9**–**11**] to the case of unitary CMV operators. In this note we present a new approach to the stability of the inverse resonance problems. Unlike the earlier work our approach does not rely on a heavy machinery of the transformation operators but instead uses the Schur algorithm - a simple recursion relation that arises naturally in the context of CMV operators. We point out that this approach is not specific to CMV operators only. There is also a similar simple recursive approach to the stability result of the inverse resonance problem for Jacobi operators.

The paper is organized as follows. In Section 2 we introduce the basics of CMV operators, define the main objects, and state some known facts that are central to our study. In Section 3 we set the stage for the stability result and give a new proof of the inverse resonance problem for CMV operators that first appeared in [**14**]. Section 4 is devoted to the stability of the inverse resonance problem and contains the proof of our main Theorem 1.2.

Notation: In the following, we denote the set of all complex-valued sequences defined on $\mathbb{N}_0$ by $\mathbb{C}^{\mathbb{N}_0}$. The Hilbert space of all square summable complex-valued sequences is $\ell^2(\mathbb{N}_0)$ and its scalar product $\langle\cdot,\cdot\rangle$ is linear in the second argument. Recall that the vectors $\delta_k \in \ell^2(\mathbb{N}_0)$, $k \in \mathbb{N}_0$, defined by the requirement that $\delta_k(n)$ equals Kronecker's $\delta_{k,n}$, form the standard basis in $\ell^2(\mathbb{N}_0)$. The open unit disk in the complex plane is denoted by $\mathbb{D}$.

2. Preliminaries

2.1. The CMV equations and the CMV operator. The CMV equations are defined through a sequence of coefficients $\alpha : \mathbb{N} \to \mathbb{D}$; these are called Verblunsky coefficients. For $z \in \mathbb{C}\backslash\{0\}$ the CMV equations are the recursive equations

$$\begin{pmatrix} u \\ v \end{pmatrix}(z,k) = T(z,k)\begin{pmatrix} u \\ v \end{pmatrix}(z,k-1), \quad k \in \mathbb{N} \tag{2.1}$$

where, using the abbreviation $\rho_k = \sqrt{1-|\alpha_k|^2}$,

$$T(z,k) = \begin{cases} \frac{1}{\rho_k}\begin{pmatrix} \alpha_k & z \\ 1/z & \overline{\alpha}_k \end{pmatrix}, & k \text{ odd}, \\ \frac{1}{\rho_k}\begin{pmatrix} \overline{\alpha}_k & 1 \\ 1 & \alpha_k \end{pmatrix}, & k \text{ even}. \end{cases}$$

It is clear that the space of solutions of these equations is 2-dimensional. A basis of solutions is given by the sequences $\vartheta(z,\cdot)$ and $\varphi(z,\cdot)$ defined by the initial conditions

$$\vartheta(z,0) = (-1,1)^\top \text{ and } \varphi(z,0) = (1,1)^\top. \tag{2.2}$$

Clearly $\vartheta(\cdot,k)$ and $\varphi(\cdot,k)$ are analytic in $\mathbb{C}\backslash\{0\}$ for any $k \in \mathbb{N}_0$. Since

$$\overline{T(1/\overline{z},k)} = \begin{pmatrix} 0 & 1 \\ 1 & 0 \end{pmatrix} T(z,k) \begin{pmatrix} 0 & 1 \\ 1 & 0 \end{pmatrix}$$

$(\overline{v},\overline{u})^\top(1/\overline{z},\cdot)$ satisfies the CMV equations if $(u,v)^\top(z,\cdot)$ does. This implies, taking the initial conditions into account, that

$$\varphi(z,k) = \begin{pmatrix} 0 & 1 \\ 1 & 0 \end{pmatrix} \overline{\varphi(1/\overline{z},k)} \text{ and } \vartheta(z,k) = -\begin{pmatrix} 0 & 1 \\ 1 & 0 \end{pmatrix} \overline{\vartheta(1/\overline{z},k)} \tag{2.3}$$

whenever $z \in \mathbb{C}\backslash\{0\}$.

To define the CMV operator set first

$$\Theta_k = \begin{pmatrix} -\alpha_k & \rho_k \\ \rho_k & \overline{\alpha}_k \end{pmatrix}.$$

These blocks are then used to define $W = \bigoplus_{k=1}^\infty \Theta_{2k-1}$ and $V = 1 \oplus (\bigoplus_{k=1}^\infty \Theta_{2k})$ where 1 is interpreted as a 1×1 block. Finally the product VW is denoted by U and is called the CMV operator. U, V, and W are defined on $\mathbb{C}^{\mathbb{N}_0}$ (and map to that space). Their restrictions to $\ell^2(\mathbb{N}_0)$ are unitary operators which we denote using the same letters as the precise meaning will always be clear from the context.

The following lemma was established in [**4**]:

LEMMA 2.1. *Suppose $z \in \mathbb{C}\backslash\{0\}$. $(u,v)^\top(z,\cdot)$ is a solution of the CMV equations* (2.1) *if and only if*

$$\begin{pmatrix} U & 0 \\ 0 & V \end{pmatrix}\begin{pmatrix} u \\ v \end{pmatrix}(z,\cdot) = [u(z,\cdot) + (v(z,0) - u(z,0))\delta_0]\begin{pmatrix} z \\ 1 \end{pmatrix}.$$

Hence if $u(z,\cdot)$ is the first component of $\varphi(z,\cdot)$ we have $Uu = zu$.

Now suppose that $|z| \neq 1$ so that z is in the resolvent set of the unitary operator U and set $u(z,\cdot) = 2z(U-z)^{-1}\delta_0$ which is in $\ell^2(\mathbb{N}_0)$. Define

$$m(z) = 1 + u(z,0) = \langle \delta_0, (U+z)(U-z)^{-1}\delta_0\rangle \tag{2.4}$$

which is analytic in $\mathbb{D}$, and, assuming also $z \neq 0$,

$$\omega(z,\cdot) = \vartheta(z,\cdot) + m(z)\varphi(z,\cdot).$$

Both components of $\omega(z,\cdot)$ are square summable (the first component is u and the second is $V^{-1}(u + 2\delta_0)$). Moreover, $\omega(z,\cdot)$ and its constant multiples are the only square summable solutions of the CMV equations (2.1) since the unitary operator U can not have eigenvalues away from the unit circle. Thus, employing (2.3), we find that

$$-\begin{pmatrix} 0 & 1 \\ 1 & 0 \end{pmatrix} \overline{\omega(1/\overline{z},\cdot)} = \vartheta(z,\cdot) - \overline{m(1/\overline{z})}\varphi(z,\cdot)$$

is equal to $\omega(z,\cdot)$ which implies that $m(z) = -\overline{m(1/\overline{z})}$.

The function m is called the Weyl-Titchmarsh m-function while the sequence $\omega(z,\cdot)$ is called the Weyl-Titchmarsh solution of the CMV equations (2.1). We also note that, for $z = 0$ we get $u(0,\cdot) = 0$, $m(0) = 1$, and $v(0,\cdot) = 2\delta_0$ so that, for every $k \in \mathbb{N}_0$, the singularity of $\omega(\cdot,k)$ at 0 is removable.

It follows from (2.4) via the spectral theorem that

$$m(z) = \oint_{\partial\mathbb{D}} \frac{\zeta+z}{\zeta-z} d\mu(\zeta) = \frac{1}{2\pi}\int_{-\pi}^{\pi} \frac{\mathrm{e}^{it}+z}{\mathrm{e}^{it}-z} d\tilde{\mu}(t),$$

where $d\mu$ denotes the spectral measure associated with the unitary operator U and the cyclic vector δ_0 and $\tilde{\mu}(t) = 2\pi\mu(\mathrm{e}^{it})$ gives rise to the corresponding measure on $[-\pi,\pi]$. The case $z = 0$ shows that $d\mu$ is a probability measure. Since $(\mathrm{e}^{it} + z)/(\mathrm{e}^{it} - z)$ has positive real part for all $z \in \mathbb{D}$, it follows that m is a Caratheodory function, that is, m is analytic on $\mathbb{D}$, $m(0) = 1$, and $\operatorname{Re} m(z) > 0$ for $|z| < 1$.

Employing the Neumann series for $(U-z)^{-1}$ in (2.4) shows

$$m(z) = 1 + 2\sum_{n=1}^{\infty} z^n \langle \delta_0, U^{-n}\delta_0\rangle.$$

This implies $m^{(n)}(0) = 2n!\langle \delta_0, U^{-n}\delta_0\rangle$ so that

$$m'(0) = -2\bar{\alpha}_1 \text{ and } m''(0) = 4\bar{\alpha}_1^2 - 4\rho_1^2\bar{\alpha}_2. \tag{2.5}$$

2.2. Jost solutions, the Jost function, and resonances. Assuming super-exponential decay of the Verblunsky coefficients, i.e.,

$$|\alpha_k| \leq \eta \exp(-k^\gamma)$$

Jost solutions of the CMV equations were defined and investigated in [**14**]. We repeat here briefly the most important results.

Defining

$$\zeta_k = \begin{cases} z, & k \text{ odd} \\ 1, & k \text{ even} \end{cases}$$

it was proved in **[14]** that the Volterra-type equations

$$F(z,k) = \begin{pmatrix}1\\0\end{pmatrix} - \sum_{n=k+1}^{\infty} \begin{pmatrix}0 & \alpha_n\zeta_n\\ z^{n-k-1}\overline{\alpha}_n\zeta_{k+1} & 0\end{pmatrix} F(z,n), \quad k\in\mathbb{N}_0, \tag{2.6}$$

have a unique solution for any complex number z. Either component of $F(\cdot,k)$ is an entire function of growth order zero and, if $|z|\geq 1$,

$$\|F(z,0)\| \leq \exp(\eta + 2K(z)\log|z|)\prod_{n=1}^{\infty}(1+|\alpha_n|) \tag{2.7}$$

where $\|\cdot\|$ denotes the 2-norm in $\mathbb{C}^2$ and

$$K(z) = \lfloor (\log 2|z|^2)^{\frac{1}{\gamma-1}} \rfloor.$$

We also recall that

$$\|F(z,k) - \begin{pmatrix}1\\0\end{pmatrix}\| \leq \beta(z,k)\exp(\beta(z,k)) \tag{2.8}$$

where $\beta(z,k) = \sum_{n=k+1}^{\infty}|\alpha_n|\max\{1,|z|^{2n-1}\}$.

Setting $C_k = \prod_{j=k+1}^{\infty}\rho_j^{-1}$, it is straightforward to show that the sequence $\nu(z,\cdot)$ defined by

$$\nu(z,k) = 2z^{\lceil k/2\rceil}C_k\begin{pmatrix}0 & 1\\ 1 & 0\end{pmatrix}^{k+1} F(z,k) \tag{2.9}$$

satisfies the CMV equations (2.1) as does the sequence

$$\tilde{\nu}(z,k) = \begin{pmatrix}0 & 1\\ 1 & 0\end{pmatrix}\overline{\nu(1/\overline{z},k)}.$$

If $|z|<1$ both components of $\nu(z,\cdot)$ are in $\ell^2(\mathbb{N}_0)$ so that $\nu(z,\cdot)$ must be a multiple of the Weyl-Titchmarsh solution $\omega(z,\cdot)$, i.e.,

$$\nu(z,\cdot) = \psi_0(z)\omega(z,\cdot) = \psi_0(z)(\vartheta(z,\cdot)+m(z)\varphi(z,\cdot)), \quad |z|<1$$

for some appropriate function ψ_0. Evaluating at $k=0$ using (2.2) and (2.9) yields

$$\psi_0(z) = \frac{(-1,1)\nu(z,0)}{2} = C_0(1,-1)F(z,0), \tag{2.10}$$

$$\psi_0(z)m(z) = \frac{(1,1)\nu(z,0)}{2} = C_0(1,1)F(z,0), \tag{2.11}$$

hence ψ_0 and $\psi_0 m$ extend to entire functions. Consequently m extends to a meromorphic function on $\mathbb{C}$ which we will denote by M (we emphasize that $M(z)\neq m(z)$ for $|z|>1$). Examining also $\tilde{\nu}$ we obtain the relationships

$$\nu(z,\cdot) = \psi_0(z)(\vartheta(z,\cdot)+M(z)\varphi(z,\cdot))$$

and

$$\tilde{\nu}(z,\cdot) = \overline{\psi_0(1/\overline{z})}(-\vartheta(z,\cdot)+\overline{M(1/\overline{z})}\varphi(z,\cdot))$$

which are valid for any $z\in\mathbb{C}\backslash\{0\}$. The solutions $\nu(z,\cdot)$ and $\tilde{\nu}(z,\cdot)$ are called respectively the Jost solutions of the CMV equations if $|z|<1$ or $|z|>1$. The function ψ_0 is called the Jost function. It is an entire function of growth order zero. Its zeros are called resonances.

We end this section with the following observation. Since $\det T(z,k) = -1$ we find

$$\det(\nu(z,k),\tilde{\nu}(z,k)) = (-1)^k\det(\nu(z,0),\tilde{\nu}(z,0)).$$

The asymptotic behavior of $\nu(z,k)$ and $\tilde{\nu}(z,k)$ as k tends to infinity shows that the left hand side is equal to $4(-1)^{k+1}$. If we now pick z on the unit circle so that $z = 1/\overline{z}$ we obtain from this

$$1 = |\psi_0(z)|^2 \operatorname{Re}(M(z)). \tag{2.12}$$

2.3. The Schur algorithm. Closely related to the concept of a Caratheodory function is that of a Schur function, that is, a function defined and analytic on $\mathbb{D}$ whose modulus never exceeds one. Indeed if f is a Caratheodory function then $(f-1)/(f+1)$ is a Schur function, while $(1+g)/(1-g)$ is a Caratheodory function if g is a Schur function. Note also that, by Schwarz's lemma, $z \mapsto g(z)/z$ is a Schur functions if g is a Schur function and $g(0) = 0$.

For $k \in \mathbb{N}_0$ and $z \in \mathbb{D}$ we define now the functions

$$\Phi_{2k}(z) = \frac{1}{z}\frac{(1,0)\omega(z,2k)}{(0,1)\omega(z,2k)} \text{ and } \Phi_{2k+1}(z) = \frac{(0,1)\omega(z,2k+1)}{(1,0)\omega(z,2k+1)} \tag{2.13}$$

and we note that in place of ω we may as well put ν since these are multiples of each other.

The initial conditions satisfied by the Weyl-Titchmarsh solution ω show that

$$\Phi_0(z) = \frac{1}{z}\frac{m(z)-1}{m(z)+1} \tag{2.14}$$

which is a Schur function.

Using the CMV equations (2.1) one may check that

$$\Phi_k(z) = \frac{1}{z} S(\alpha_k, \Phi_{k-1}(z))$$

where $S(w,\cdot)$ is the Möbius transform

$$z \mapsto S(w,z) = \frac{z+\overline{w}}{1+wz}$$

which maps $\mathbb{D}$ to $\mathbb{D}$ provided $w \in \mathbb{D}$.

Next, taking (2.5) into account, one sees that $\Phi_0(0) = \frac{1}{2}m'(0) = -\bar{\alpha}_1$. This shows that Φ_1 is a Schur function and we find, again by (2.5), that $\Phi_1(0) = \Phi_0'(0)/(1-|\alpha_1|^2) = -\bar{\alpha}_2$.

Consider now the truncated sequence of Verblunsky coefficients $n \mapsto \alpha_{2N+n}$. The Jost solution for this problem is given by $k \mapsto z^{-N}\nu(z, 2N+k)$. Consequently, the function Φ_{2N} plays the same role for the truncated sequence as Φ_0 plays for the full sequence. Therefore we have $\Phi_{2N}(0) = -\bar{\alpha}_{2N+1}$ and $\Phi_{2N+1}(0) = -\bar{\alpha}_{2N+2}$. Thus any of the functions Φ_k is a Schur function and

$$\Phi_k(0) = -\bar{\alpha}_{k+1}. \tag{2.15}$$

The hyperbolic distance on $\mathbb{D}$, given by

$$d[w_1,w_2] = 2\tanh^{-1}\left|\frac{w_1-w_2}{1-\overline{w}_1 w_2}\right| = \log\frac{1+|w_1-w_2|/|1-\overline{w}_1 w_2|}{1-|w_1-w_2|/|1-\overline{w}_1 w_2|},$$

is invariant under Möbius transforms which map $\mathbb{D}$ onto itself. Hence, employing the triangle inequality and the fact that, for $|z|<1$, we have $d[z\Phi_k(z),0] \leq d[\Phi_k(z),0]$,

$$d[\Phi_{k-1}(z),0] = d[z\Phi_k(z),\overline{\alpha}_k] \leq d[\Phi_k(z),0] + d[0,\overline{\alpha}_k].$$

Inequality (2.8) combined with (2.9) and (2.13) implies that $\Phi_k(z)$ tends uniformly to zero as k tends to infinity. Hence we may sum up the telescoping series resulting from the previous inequality to get

$$d[\Phi_0(z), 0] \leq \sum_{k=1}^{\infty} d[0, \overline{\alpha}_k] = \sum_{k=1}^{\infty} \log \frac{1+|\alpha_k|}{1-|\alpha_k|} \leq \log Q^2.$$

This, in turn, gives us the estimate

$$\frac{1+|\Phi_0(z)|}{1-|\Phi_0(z)|} \leq \prod_{k=1}^{\infty} \frac{1+|\alpha_k|}{1-|\alpha_k|} \leq Q^2 \tag{2.16}$$

which holds for all $z \in \mathbb{D}$ and hence also in the closure of $\mathbb{D}$.

3. The inverse problem

The main purpose of [**14**] was to prove the following theorem.

THEOREM 3.1. *The locations (and multiplicities) of the zeros of the Jost function, that is, the resonances, associated with the CMV equations* (2.1) *determine uniquely the Verblunsky coefficients, provided these satisfy* $|\alpha_n| \leq \eta \exp(-n^\gamma)$.

The strategy in [**14**] was to show that the zeros of ψ_0 determine the Weyl-Titchmarsh m-function. The m-function, in turn, determines the Verblunsky coefficients as was shown in [**4**]. In the following we prove this fact directly, because the proof of our stability result relies on the details of the relationship between the resonances and the Verblunsky coefficients.

PROOF OF THEOREM 3.1. As we know from equation (2.15), the Verblunsky coefficients are determined by the Schur functions Φ_k. Of these the first one is determined by m while the subsequent ones are found recursively via the Möbius transform S.

To find m we call on Schwarz's integral formula which says that

$$m(z) = \frac{1}{2\pi} \int_{-\pi}^{\pi} \frac{re^{it}+z}{re^{it}-z} \operatorname{Re}(m(re^{it}))dt \tag{3.1}$$

as long as $|z| < r < 1$. According to the first inequality in (2.16) $m = (1 + z\Phi_0(z))/(1 - z\Phi_0(z))$ is uniformly bounded in $\mathbb{D}$ and hence has no poles in its closure. Therefore we may take the limit $r \to 1$ in (3.1) under the integral. This fact and equation (2.12) show that

$$d\tilde{\mu}(t) = \operatorname{Re}(M(e^{it}))dt = \frac{dt}{|\psi_0(e^{it})|^2}$$

so that

$$m(z) = \frac{1}{2\pi} \int_{-\pi}^{\pi} \frac{e^{it}+z}{e^{it}-z} \frac{dt}{|\psi_0(e^{it})|^2} \tag{3.2}$$

for $|z| < 1$.

Finally we have to show that ψ_0 is determined from its zeros (the resonances). Since it is an entire function of growth order zero Hadamard's factorization theorem gives $\psi_0(z) = \psi_0(0)\Pi(z)$ where

$$\Pi(z) = \prod_{n=1}^{\infty} (1 - z/z_n)$$

and where the z_n are the zeros of ψ_0 repeated according to their multiplicities. We claim that the value $\psi_0(0)$ is also determined by the z_n. Indeed, evaluating (3.2) at $z = 0$, using that $m(0) = 1$, gives

$$|\psi_0(0)|^2 = \frac{1}{2\pi} \int_{-\pi}^{\pi} \frac{dt}{|\Pi(e^{it})|^2}. \tag{3.3}$$

This will prove our claim if we can show that $\psi_0(0)$ is positive. To this end we note that the unique solution of the Volterra-type equations (2.6) satisfy $F(0, 2k) = (1, 0)^\top$ for $z = 0$. Thus

$$\psi_0(0) = C_0 = \prod_{j=1}^{\infty} \rho_j^{-1} \geq 1. \tag{3.4}$$

□

4. Stability

In this section we prove Theorem 1.2. Throughout the section we assume that any sequence of Verblunsky coefficients is from the class $B_0(\gamma, \eta, Q)$. Subsequently we will be using repeatedly the following elementary facts:

(1) If $|x| \leq 1/2$, then $|\log(1 - x)| \leq 2|x|$.
(2) $|e^u - 1| \leq |u|e^{|u|}$ for all $u \in \mathbb{C}$. Moreover, if $|u| \leq 1/2$, then $|e^u - 1| \leq 2|u|$.

4.1. Upper and lower bound for Π. Upper bounds on Π follow from (2.7) since $\Pi(z) = (1, -1)F(z, 0)$ by (2.10) and (3.4). Indeed, using that $\prod_{j=1}^{\infty}(1+|\alpha_j|) \leq Q$ for $\alpha \in B_0(\gamma, \eta, Q)$, we find

$$\log|\Pi(z)| \leq \log(\sqrt{2}Q) + \eta + (\log 2|z|^2)^{\gamma/(\gamma-1)} \tag{4.1}$$

as long as $|z| \geq 1$.

We need lower bounds on Π only on the unit circle. Recall that by (2.12) we have $\mathrm{Re}(M(z)) = |\psi_0(z)|^{-2} = C_0^{-2}|\Pi(z)|^{-2}$ if $|z| = 1$. Note also that

$$\mathrm{Re}(M(z)) = \frac{1 - |z\Phi_0(z)|^2}{|1 - z\Phi_0(z)|^2} \leq \frac{1 + |\Phi_0(z)|}{1 - |\Phi_0(z)|}.$$

Combining these facts with (3.4) and (2.16) gives

$$|\Pi(z)|^{-2} \leq C_0^2 \prod_{n=1}^{\infty} \frac{1 + |\alpha_n|}{1 - |\alpha_n|} = \prod_{n=1}^{\infty} \frac{1}{(1 - |\alpha_n|)^2} \leq Q^2. \tag{4.2}$$

4.2. Distribution of resonances. Now let $N(r)$ denote the number of zeros of Π in the open disk of radius r centered at zero. We know that $N(r) = 0$ for $r \leq 1$. To deal with $r \geq 1$ we use Jensen's formula, the estimate (4.1), and the inequality $2(a + b)^p \leq (2a)^p + (2b)^p$, which holds for $a, b \geq 0$ and $p \geq 1$, to find

$$N(r) \leq \int_r^{er} \frac{N(t)}{t} dt \leq \int_0^{er} \frac{N(t)}{t} dt = \frac{1}{2\pi} \int_0^{2\pi} \log|\Pi(ere^{it})| dt \leq A_1 + \frac{(\log r^4)^p}{2}$$

where $p = \gamma/(\gamma - 1)$ and A_1 is a suitable constant which depends only on Q, η, and γ. From this we obtain

$$\sum_{|z_n| \geq R} \frac{1}{|z_n|} = \int_R^{\infty} \frac{dN(t)}{t} = -\frac{N(R)}{R} + \int_R^{\infty} \frac{N(t)}{t^2} dt \leq \frac{A_1}{R} + \frac{4^p}{2}\Gamma(p + 1, \log R)$$

where Γ denotes the incomplete Gamma function [**1**, Sect. 6.5]. In particular, we get

$$\sum_{n=1}^{\infty} \frac{1}{|z_n|} \leq A_1 + \frac{4^p}{2}\Gamma(p+1)$$

by setting $R = 1$. The asymptotic behavior of the incomplete Gamma function [**1**, Eq. 6.5.32] shows now that

$$\sum_{|z_n|\geq R} \frac{1}{|z_n|} \leq A_2 \frac{(\log R)^p}{R} \tag{4.3}$$

if $R \geq 2$ and A_2 is a suitable constant (depending on Q, η, and γ).

Next we estimate how close resonances can be to the unit circle. From (4.1) (making use of the maximum principle) we know that there is a constant L depending only on Q, η, and γ such that $|\Pi(z)| \leq L$ whenever $|z| \leq \mathrm{e}$. Cauchy's estimate gives $|\Pi'(a)| \leq L/(\mathrm{e}-|a|)$ for any point a with $|a| < \mathrm{e}$. Let $1+\delta = (QL+\mathrm{e})/(QL+1)$ and z_0 a point on the unit circle. Then

$$|\Pi(tz_0)| \geq |\Pi(z_0)| - \int_1^t |\Pi'(z_0 s)|ds \geq \frac{1}{Q} - (t-1)\frac{L}{\mathrm{e}-t}.$$

Since this is positive as long as $1 \leq t < 1+\delta$ we have established that there are no zeros of ψ_0, i.e., no resonances, in the disk $|z| < 1+\delta$ for any operator U from the class $B_0(\gamma, \eta, Q)$.

4.3. Comparing Π and $\breve{\Pi}$. We assume now that we have two CMV operators U and $\breve{U}$ with Verblunsky coefficients α_n and $\breve{\alpha}_n$, respectively. More generally, any quantity associated with $\breve{U}$ will have a $\breve{}$ accent to distinguish it from the corresponding quantity associated with U. Both α and $\breve{\alpha}$ are in $B_0(\gamma, \eta, Q)$.

Our goal is to show that the differences $|\alpha_n - \breve{\alpha}_n|$ are arbitrarily small provided that the resonances of the associated operators U and $\breve{U}$ in a sufficiently large disk (of radius R) are respectively ε-close for a sufficiently small ε , i.e., to prove Theorem 1.2.

We will henceforth always assume $R \geq 2$ and $\varepsilon \leq \delta/2$. We begin by looking at those factors in Π and $\breve{\Pi}$ associated with resonances which are respectively close to each other, i.e., the resonances in a disk of radius R. Let N be their number and assume $|z| = 1$. Thus $|\breve{z}_n - z_n| \leq \varepsilon$ and $|\breve{z}_n - z|, |z_n - z| \geq \delta > 0$ for $1 \leq n \leq N$. Since $\varepsilon/\delta \leq 1/2$ we have the following estimate

$$\left|\log\left(\prod_{n=1}^{N} \frac{1-z/\breve{z}_n}{1-z/z_n}\right)\right| \leq \sum_{n=1}^{N}\left|\log\left(1 - z\frac{z_n - \breve{z}_n}{(z_n - z)\breve{z}_n}\right)\right| \leq \frac{2\varepsilon}{\delta}\sum_{n=1}^{\infty}|\breve{z}_n|^{-1}. \tag{4.4}$$

We showed above that the sum on the right is bounded by $A_1 + \Gamma(p+1)/2$.

Next we turn to the terms associated with large resonances and show that these are negligible. Indeed, we get for $|z_n|, |\breve{z}_n| \geq R \geq 2$ and $|z| = 1$

$$\left|\log \prod_{n=N+1}^{\infty} \frac{1-z/\breve{z}_n}{1-z/z_n}\right| \leq 2 \sum_{n=N+1}^{\infty}\left(\frac{1}{|\breve{z}_n|} + \frac{1}{|z_n|}\right)$$

so that, with the aid of (4.3), we arrive at the estimate

$$\left|\log \prod_{n=N+1}^{\infty} \frac{1-z/\breve{z}_n}{1-z/z_n}\right| \leq 4A_2\frac{(\log R)^p}{R}.$$

Combining this estimate with (4.4) and denoting by A_3 a suitable constant depending only on Q, η, and γ, we obtain

$$\left|\frac{\breve{\Pi}(z)}{\Pi(z)} - 1\right| \leq A_3\left(\varepsilon + \frac{(\log R)^p}{R}\right) \tag{4.5}$$

provided that $|z| \leq 1$.

4.4. Comparing ψ_0 and $\breve{\psi}_0$. Since, by (4.5) and (4.2),

$$\left|\frac{1}{|\Pi(e^{it})|^2} - \frac{1}{|\breve{\Pi}(e^{it})|^2}\right| \leq 2Q^2 A_3\left(\varepsilon + \frac{(\log R)^p}{R}\right)$$

and since $C_0, \breve{C}_0 \geq 1$ we get from (3.3) that

$$|C_0^{-2} - \breve{C}_0^{-2}| \leq |C_0^2 - \breve{C}_0^2| \leq 2Q^2 A_3(\varepsilon + (\log R)^p/R).$$

Thus, whenever $|z| \leq 1$,

$$\left|\frac{1}{|\psi_0(z)|^2} - \frac{1}{|\breve{\psi}_0(z)|^2}\right| \leq A_4\left(\varepsilon + \frac{(\log R)^p}{R}\right) \tag{4.6}$$

where A_4 depends only on Q, η, and γ.

4.5. Comparing Verblunsky coefficients. Suppose $1 - |w|, 1 - |\breve{w}| \geq 1/Q$ and $z, \breve{z}$ are in the closed unit disk. Then

$$|S(w, z) - S(\breve{w}, \breve{z})| \leq Q^2(4|w - \breve{w}| + 2|z - \breve{z}|).$$

Since by assumption $1 - |\alpha_n| \geq \prod_{j=1}^{\infty}(1 - |\alpha_j|) \geq 1/Q$, it follows that for all z on the unit circle,

$$|\Phi_k(z) - \breve{\Phi}_k(z)| \leq Q^2(4|\alpha_k - \breve{\alpha}_k| + 2|\Phi_{k-1}(z) - \breve{\Phi}_{k-1}(z)|). \tag{4.7}$$

Let $\|\cdot\|_p$ denote the L^p-norm on the unit circle with respect to the normalized Lebesgue measure. By Gauss's mean value theorem

$$|\alpha_k - \breve{\alpha}_k| \leq \|\Phi_{k-1} - \breve{\Phi}_{k-1}\|_1, \quad k \in \mathbb{N},$$

and hence (4.7) yields,

$$\|\Phi_k - \breve{\Phi}_k\|_1 \leq 6Q^2\|\Phi_{k-1} - \breve{\Phi}_{k-1}\|_1, \quad k \in \mathbb{N}.$$

Thus we have

$$|\alpha_k - \breve{\alpha}_k| \leq (6Q^2)^{k-1}\|\Phi_0 - \breve{\Phi}_0\|_1, \quad k \in \mathbb{N},$$

by induction. Since for all z on the unit circle $\operatorname{Re} M(z) = 1/|\psi_0(z)|^2$ and $\operatorname{Re} \breve{M}(z) = 1/|\breve{\psi}_0(z)|^2$ are nonnegative, it follows from (2.14) that $\|\Phi_0 - \breve{\Phi}_0\|_1 \leq 2\|M - \breve{M}\|_1 \leq 2\|M - \breve{M}\|_2$. The imaginary parts of M and $\breve{M}$ can be obtained from the Hilbert transform of the respective real parts. Since the Hilbert transform is unitary on the space of square integrable functions we have $\|\operatorname{Im} M - \operatorname{Im} \breve{M}\|_2 = \|\operatorname{Re} M - \operatorname{Re} \breve{M}\|_2$ and hence $\|M - \breve{M}\|_2 \leq \sqrt{2}\|\operatorname{Re} M - \operatorname{Re} \breve{M}\|_2$. Thus, we get from (4.6)

$$\begin{aligned} |\alpha_k - \breve{\alpha}_k| &\leq 2\sqrt{2}(6Q^2)^{k-1}\left\|\frac{1}{|\psi_0(z)|^2} - \frac{1}{|\breve{\psi}_0(z)|^2}\right\|_2 \\ &\leq 2\sqrt{2}(6Q^2)^{k-1}A_4\left(\varepsilon + \frac{(\log R)^p}{R}\right). \end{aligned} \tag{4.8}$$

Setting $A_0 = 2\sqrt{2}A_4/(6Q^2)$ completes the proof of Theorem 1.2.

Estimate (4.8) becomes worse with increasing k. Eventually, of course we will have $|\alpha_k - \breve{\alpha}_k| \leq 2\eta e^{-k^\gamma}$ just by using our hypothesis on super-exponential decay of the Verblunsky coefficients. Using the worst possible case and introducing yet another approriate constant A_5 gives us the uniform estimate

$$|\alpha_k - \breve{\alpha}_k| \leq A_5 \left(\varepsilon + \frac{(\log R)^p}{R} \right)^{1/\log(6eQ^2)}, \quad k \in \mathbb{N}.$$

References

[1] M. Abramowitz and I. A. Stegun, *Handbook of mathematical functions with formulas, graphs, and mathematical tables*, National Bureau of Standards Applied Mathematics Series, vol. 55, For sale by the Superintendent of Documents, U.S. Government Printing Office, Washington, D.C., 1964. MR0167642 (29 #4914)

[2] B. M. Brown, S. Naboko, and R. Weikard, *The inverse resonance problem for Jacobi operators*, Bull. London Math. Soc. **37** (2005), no. 5, 727–737, DOI 10.1112/S0024609305004674. MR2164835 (2006e:39032)

[3] B. M. Brown, S. Naboko, and R. Weikard, *The inverse resonance problem for Hermite operators*, Constr. Approx. **30** (2009), no. 2, 155–174, DOI 10.1007/s00365-008-9037-8. MR2519659 (2011b:47065)

[4] F. Gesztesy and M. Zinchenko, *Weyl-Titchmarsh theory for CMV operators associated with orthogonal polynomials on the unit circle*, J. Approx. Theory **139** (2006), no. 1-2, 172–213, DOI 10.1016/j.jat.2005.08.002. MR2220038 (2007f:47027)

[5] A. Iantchenko and E. Korotyaev, *Periodic Jacobi operator with finitely supported perturbation on the half-lattice*, Inverse Problems **27** (2011), no. 11, 115003, 26, DOI 10.1088/0266-5611/27/11/115003. MR2851909 (2012k:47010)

[6] A. Iantchenko and E. Korotyaev, *Periodic Jacobi operator with finitely supported perturbations: the inverse resonance problem*, J. Differential Equations **252** (2012), no. 3, 2823–2844, DOI 10.1016/j.jde.2011.09.034. MR2860642 (2012k:35598)

[7] A. Iantchenko and E. Korotyaev, *Resonances for periodic Jacobi operators with finitely supported perturbations*, J. Math. Anal. Appl. **388** (2012), no. 2, 1239–1253, DOI 10.1016/j.jmaa.2011.11.016. MR2869822 (2012k:47011)

[8] E. Korotyaev, *Stability for inverse resonance problem*, Int. Math. Res. Not. **73** (2004), 3927–3936, DOI 10.1155/S1073792804140609. MR2104289 (2005i:81186)

[9] M. Marletta, S. Naboko, R. Shterenberg, and R. Weikard, *On the inverse resonance problem for Jacobi operators—uniqueness and stability*, J. Anal. Math. **117** (2012), 221–247, DOI 10.1007/s11854-012-0020-8. MR2944096

[10] M. Marletta, R. Shterenberg, and R. Weikard, *On the inverse resonance problem for Schrödinger operators*, Comm. Math. Phys. **295** (2010), no. 2, 465–484, DOI 10.1007/s00220-009-0928-8. MR2594334 (2011a:34203)

[11] M. Marletta and R. Weikard, *Stability for the inverse resonance problem for a Jacobi operator with complex potential*, Inverse Problems **23** (2007), no. 4, 1677–1688, DOI 10.1088/0266-5611/23/4/018. MR2348728 (2008i:47065)

[12] B. Simon, *Orthogonal polynomials on the unit circle. Part 1*, American Mathematical Society Colloquium Publications, vol. 54, American Mathematical Society, Providence, RI, 2005. Classical theory. MR2105088 (2006a:42002a)

[13] B. Simon, *Orthogonal polynomials on the unit circle. Part 2*, American Mathematical Society Colloquium Publications, vol. 54, American Mathematical Society, Providence, RI, 2005. Spectral theory. MR2105089 (2006a:42002b)

[14] R. Weikard and M. Zinchenko, *The inverse resonance problem for CMV operators*, Inverse Problems **26** (2010), no. 5, 055012, 10, DOI 10.1088/0266-5611/26/5/055012. MR2647154 (2011j:47098)

Department of Mathematics, University of Alabama at Birmingham, Birmingham, Alabama 35226-1170
E-mail address: shterenb@math.uab.edu

Department of Mathematics, University of Alabama at Birmingham, Birmingham, Alabama 35226-1170
E-mail address: rudi@math.uab.edu

Department of Mathematics and Statistics, University of New Mexico, Albuquerque, New Mexico 87131-0001
E-mail address: maxim@math.unm.edu

Proceedings of Symposia in Pure Mathematics
Volume **87**, 2013
http://dx.doi.org/10.1090/pspum/087/01428

On a conjecture of A. Bikchentaev

F. A. Sukochev

ABSTRACT. In *Majorization for products of measurable operators*, A. M. Bikchentaev conjectured that for positive $\tau-$measurable operators a and b affiliated with an arbitrary semifinite von Neumann algebra $\mathcal{M}$, the operator $b^{1/2}ab^{1/2}$ is submajorized by the operator ab in the sense of Hardy-Littlewood. We prove this conjecture in full generality and present a number of applications to fully symmetric operator ideals, Golden-Thompson inequality and (singular) traces.

1. Introduction and preliminaries

In this paper we answer a question due to A. M. Bikchentaev (see [**1**, p. 573, Conjecture A]) in the affirmative (see also [**2**, **3**]). To formulate his conjecture, we need some notions from the theory of noncommutative integration. For details on von Neumann algebra theory, the reader is referred to e.g. [**7**], [**17**, **18**] or [**32**]. General facts concerning measurable operators may be found in [**21**], [**27**] (see also [**33**, Chapter IX] and the forthcoming book [**10**]). For the convenience of the reader, some of the basic definitions are recalled.

In what follows, H is a Hilbert space and $B(H)$ is the $*$-algebra of all bounded linear operators on H, and $\mathbf{1}$ is the identity operator on H. Let $\mathcal{M}$ be a von Neumann algebra on H.

A linear operator $x : \mathfrak{D}(x) \to H$, where the domain $\mathfrak{D}(x)$ of x is a linear subspace of H, is said to be *affiliated* with $\mathcal{M}$ if $yx \subseteq xy$ for all $y \in \mathcal{M}'$. A linear operator $x : \mathfrak{D}(x) \to H$ is termed *measurable* with respect to $\mathcal{M}$ if x is closed, densely defined, affiliated with $\mathcal{M}$ and there exists a sequence $\{p_n\}_{n=1}^{\infty}$ in $P(\mathcal{M})$ such that $p_n \uparrow 1$, $p_n(H) \subseteq \mathfrak{D}(x)$ and $1 - p_n$ is a finite projection (with respect to $\mathcal{M}$) for all n. It should be noted that the condition $p_n(H) \subseteq \mathfrak{D}(x)$ implies that $xp_n \in \mathcal{M}$. The collection of all measurable operators with respect to $\mathcal{M}$ is denoted by $S(\mathcal{M})$, which is a unital $*$-algebra with respect to strong sums and products (denoted simply by $x + y$ and xy for all $x, y \in S(\mathcal{M})$).

Let a be a self-adjoint operator affiliated with $\mathcal{M}$. We denote its spectral measure by $\{e^a\}$. It is known that if x is a closed operator affiliated with $\mathcal{M}$ with the polar decomposition $x = u|x|$, then $u \in \mathcal{M}$ and $e \in \mathcal{M}$ for all projections $e \in \{e^{|x|}\}$. Moreover, $x \in S(\mathcal{M})$ if and only if x is closed, densely defined, affiliated with $\mathcal{M}$ and $e^{|x|}(\lambda, \infty)$ is a finite projection for some $\lambda > 0$. It follows immediately

2010 *Mathematics Subject Classification.* Primary 47B10, 46A22.

Key words and phrases. Hardy-Littlewood submajorization, Golden-Thompson inequalities.
The author acknowledges support from the Australian Research Council.

that in the case when $\mathcal{M}$ is a von Neumann algebra of type III or a type I factor, we have $S(\mathcal{M}) = \mathcal{M}$. For type II von Neumann algebras, this is no longer true. From now on, let $\mathcal{M}$ be a semifinite von Neumann algebra equipped with a faithful normal semifinite trace τ (the reader unfamiliar with von Neumann algebra theory can assume that $\mathcal{M} = B(H)$ and τ is the standard trace on $B(H)$: the A. Bikchentaev's question retains its interest also in this special case).

An operator $x \in S(\mathcal{M})$ is called $\tau-$measurable if there exists a sequence $\{p_n\}_{n=1}^{\infty}$ in $P(\mathcal{M})$ such that $p_n \uparrow 1$, $p_n(H) \subseteq \mathfrak{D}(x)$ and $\tau(1-p_n) < \infty$ for all n. The collection $S(\tau)$ of all τ-measurable operators is a unital $*$-subalgebra of $S(\mathcal{M})$ denoted by $S(\mathcal{M}, \tau)$. It is well known that a linear operator x belongs to $S(\mathcal{M}, \tau)$ if and only if $x \in S(\mathcal{M})$ and there exists $\lambda > 0$ such that $\tau(e^{|x|}(\lambda, \infty)) < \infty$. Alternatively, an unbounded operator x affiliated with $\mathcal{M}$ is $\tau-$measurable (see [**11**]) if and only if

$$\tau\left(e^{|x|}(\frac{1}{n}, \infty)\right) = o(1), \quad n \to \infty.$$

Let L_0 be a space of Lebesgue measurable functions either on $(0,1)$ or on $(0,\infty)$, or on $\mathbb{N}$ finite almost everywhere (with identification $m-$a.e.). Here m is Lebesgue measure or else counting measure on $\mathbb{N}$. Define S as the subset of L_0 which consists of all functions x such that $m(\{|x| > s\})$ is finite for some s.

The notation $\mu(x)$ stands for the non-increasing right-continuous rearrangement of $x \in S$ given by

$$\mu(t;x) = \inf\{s \geq 0 : \ m(\{|x| \geq s\}) \leq t\}.$$

In the case when x is a sequence we denote by $\mu(x)$ the usual decreasing rearrangement of the sequence $|x|$.

Let a semifinite von Neumann algebra $\mathcal{M}$ be equipped with a faithful normal semi-finite trace τ. Let $x \in S(\mathcal{M}, \tau)$. The generalized singular value function $\mu(x) : t \to \mu(t;x)$ of the operator x is defined by setting

$$\mu(s;x) = \inf\{\|xp\| : \ p = p^* \in \mathcal{M} \text{ is a projection, } \tau(1-p) \leq s\}.$$

There exists an equivalent definition which involves the distribution function of the operator x. For every self-adjoint operator $x \in S(\mathcal{M}, \tau)$, setting

$$d_x(t) = \tau(e^x(t, \infty)), \quad t > 0,$$

we have (see e.g. [**11**])

$$\mu(t;x) = \inf\{s \geq 0 : \ d_{|x|}(s) \leq t\}.$$

Consider the algebra $\mathcal{M} = L^\infty(0,\infty)$ of all Lebesgue measurable essentially bounded functions on $(0,\infty)$. Algebra $\mathcal{M}$ can be seen as an abelian von Neumann algebra acting via multiplication on the Hilbert space $\mathcal{H} = L^2(0,\infty)$, with the trace given by integration with respect to Lebesgue measure m. It is easy to see that the set of all measurable (respectively, τ-measurable) operators affiliated with $\mathcal{M}$ can be identified with S (respectively, with L_0). It should also be pointed out that the generalized singular value function $\mu(f)$ is precisely the decreasing rearrangement $\mu(f)$ of the function f defined above.

If $\mathcal{M} = B(H)$ (respectively, l_∞) and τ is the standard trace Tr (respectively, the counting measure on $\mathbb{N}$), then it is not difficult to see that $S(\mathcal{M}) = S(\mathcal{M}, \tau) = \mathcal{M}$. In this case, for $x \in S(\mathcal{M}, \tau)$ we have

$$\mu(n;x) = \mu(t;x), \quad t \in [n, n+1), \quad n \geq 0.$$

The sequence $\{\mu(n;x)\}_{n\geq 0}$ is just the sequence of singular values of the operator x.

Let $a, b \in S(\mathcal{M}, \tau)$. We say that b is submajorized by a in the sense of Hardy-Littlewood-Polya if and only if

$$\int_0^t \mu(s;b)ds \leq \int_0^t \mu(s;a)ds, \quad \forall t > 0.$$

In this case, we write $b \prec\prec a$. Observe that $b \prec\prec a$ if and only if $\mu(b) \prec\prec \mu(a)$. Sometimes, we also write $a \prec\prec f$ instead of $\mu(a) \prec\prec \mu(f)$.

In the special case when a and b are *positive* self-adjoint operators from $S(\mathcal{M}, \tau)$ the following question was asked in [**1**].

QUESTION 1. *Let a and $b \geq 0$ be self-adjoint $\tau-$measurable operators affiliated with $\mathcal{M}$. Is it necessarily true that*

$$b^{1/2}ab^{1/2} \prec\prec ab?$$

If $\mathcal{M}$ is a matrix algebra, then the positive answer to Question 1 may be inferred from [**19**]. The main objective of the present article is to provide a different (stronger and more general) approach to Question 1. Our method allows us to produce a number of applications.

Observe that the inequality $\mu(b^{1/2}ab^{1/2}) \leq \mu(ab)$ fails even for the case $\mathcal{M} = M_2(\mathbb{C})$ (see Remark 2, p. 575 of [**2**]).

The author thanks A. Bikchentaev for drawing his attention to this problem and additional references and D. Zanin and B. de Pagter for a number of insightful comments which improved the article. Some results of this article have been announced in [**30**].

2. The main result

The gist of our approach to Question 1 is contained in Lemmas 4 and 8 below. In the proofs we use two properties of singular value functions (see (1) and (2) below (see also [**11**, Lemma 4.1], [**6**] and [**9**, Proposition 3.10]). For simplicity of exposition, we shall assume that $\mathcal{M}$ is an atomless von Neumann algebra. Indeed, this is done by a standard trick consisting in considering a von Neumann tensor product $\mathcal{M} \otimes L_\infty(0,1)$ (see details in e.g. [**1**, pp. 574-575]).

Let L_1 and L_∞ be Lebesgue spaces on $(0, \tau(\mathbf{1}))$.

Let $a \in S(\mathcal{M}, \tau)$. We say that $a \in L_1(\mathcal{M}, \tau)$ if and only if $\mu(a) \in L_1(0, \infty)$. It is well-known that

$$\|a\|_1 := \|\mu(a)\|_{L_1}$$

is a Banach norm on $L_1(\mathcal{M}, \tau)$. Similarly, we say that $a \in (L_1 + L_\infty)(\mathcal{M}, \tau)$ if and only if $\mu(a) \in (L_1 + L_\infty)(0, \infty)$. Here, we identify $\mathcal{M}$ with $L_\infty(\mathcal{M}, \tau)$. The space $(L_1 + L_\infty)(\mathcal{M}, \tau)$ can be also viewed as a sum of Banach spaces $L_1(\mathcal{M}, \tau)$ and $L_\infty(\mathcal{M}, \tau)$ (the latter space is equipped with the uniform norm, which we denote simply by $\|\cdot\|$).

For all $a, c \in (L_1 + L_\infty)(\mathcal{M})$, we have (see [**6, 11**])

$$\mu(ac) \prec\prec \mu(a)\mu(c). \tag{1}$$

LEMMA 2. *If $a \in (L_1 + L_\infty)(\mathcal{M}, \tau)$, then*

$$\int_0^t \mu(s;a)ds = \sup\{|\tau(ac)| : \ |\tau(s(c))| \leq t, \ \|c\| \leq 1\}, \tag{2}$$

where $s(c)$ denotes the support projection of the operator c.

PROOF. Recall that if $x \in S((\mathcal{M},\tau)$, then

$$\tau(s(x)) = \inf\{s \geq 0 : \mu(s;x) = 0\}.$$

Therefore, if $c \in \mathcal{M}$ satisfies $\|c\| \leq 1$ and $|\tau(s(c))| \leq t$, then $s(ac) \leq s(c)$ and so

$$|\tau(ac)| \leq \tau(|ac|) = \int_0^\infty \mu(s;ac)ds = \int_0^t \mu(s;ac)ds.$$

For the converse inequality, first recall the following fact (see [**10, 11**]): if $x \in S((\mathcal{M},\tau)$, then (under the assumption that there are no atoms in $\mathcal{M}$) we have

$$\int_0^t \mu(s;x)ds = \sup\{\tau(p|x|p) : \ p = p^* = p^2 \in \mathcal{M}, \ \tau(p) \leq t\}.$$

If $a \in (L_1 + L_\infty)(\mathcal{M},\tau)$, and $\tau(p) \leq t$ then pap, $ap \in L_1(\mathcal{M},\tau)$ and so, $\tau(p|a|p) = \tau(|a|p)$. Hence,

$$\int_0^t \mu(s;a)ds = \sup\{\tau(|a|p) : \ p = p^* = p^2 \in \mathcal{M}, \ \tau(p) \leq t\}$$

for all $a \in (L_1 + L_\infty)(\mathcal{M},\tau)$. If $p = p^* = p^2 \in \mathcal{M}$ with $\tau(p) \leq t$, then $|a|p = v^*ap$ (where $a = v|a|$ is the polar decomposition of a) and so, $\tau(|a|p) = \tau(v^*ap) = \tau(apv^*)$. Setting $c = pv^*$, it follows that $\mu(c) \leq \mu(p)$ and so, $\mu(s;c) = 0$ for all $s \geq \tau(p)$, that is, $\tau(s(c)) \leq \tau(p) \leq$t. The result of the lemma follows. □

The following remark is well-known (and trivial) and stated here for convenience of the reader.

REMARK 3. *Let $b \in \mathcal{M}$. The mapping $z \to e^{zb}$ takes its values in $\mathcal{M}$ and is holomorphic on $\mathbb{C}$.*

LEMMA 4. *For any self-adjoint operators $a, b \in \mathcal{M}$ and every $\theta \in (0,1)$, we have*

$$e^{\theta b}ae^{(1-\theta)b} \prec\prec ae^b.$$

PROOF. Appealing to Remark 3, we see that the mapping $z \to e^{zb}ae^{(1-z)b}$ takes values in $\mathcal{M}$ and is holomorphic on $\mathbb{C}$.

Fix an operator $c \in \mathcal{M}$ such that $\|c\| \leq 1$ and $\tau(s(c)) \leq t$. The $\mathbb{C}-$valued function $F : z \to \tau(e^{bz}ae^{b(1-z)}c)$ is also holomorphic on $\mathbb{C}$. For every $0 \leq \Re z \leq 1$, it follows from (1) that

$$|F(z)| \leq t\|e^{zb}ae^{(1-z)b}\| \leq t\|e^{zb}\| \cdot \|a\| \cdot \|e^{(1-z)b}\| \leq t\|a\|e^{2\|b\|}.$$

Hence, F is a bounded function in the strip $0 \leq \Re z \leq 1$. Since F is holomorphic on $\mathbb{C}$, it follows that F is continuous on the boundary of the strip $0 \leq \Re z \leq 1$. By Hadamard three-lines theorem (see e.g. [**25**, p. 33-34]), we have

$$|F(z)| \leq \sup_{y \in \mathbb{R}} \max\{|F(iy)|, |F(1+iy)|\}.$$

To estimate $|F(iy)|$, we argue as follows:

$$|F(iy)| = |\tau(e^{iyb}ae^be^{-iyb}c)| \leq \tau(|ae^be^{-iyb}ce^{iyb}|) = \tau(|ae^bd|),$$

where $d = e^{-iyb}ce^{iyb}$. Since $\mu(ae^bd) \le \|ae^b\|\mu(d) = \|ae^b\|\mu(c)$ it is clear that $\mu(ae^bd;s) = 0$ for $s \ge t$. Hence,

$$|\tau(ae^bd)| = \int_0^\infty \mu(ae^bd;s)ds \le \int_0^t \mu(ae^bd;s)ds \le \int_0^t \mu(ae^b;s)ds.$$

Similarly,

$$|F(1+iy)| = |\tau(e^ba)| \le \int_0^t \mu(s;e^ba)ds$$

and, appealing to the assumption $a = a^*$, we may conclude that

$$|F(z)| = |\tau(e^{bz}ae^{b(1-z)}c)| \le \int_0^t \mu(s;ae^b)ds$$

for all $z \in \mathbb{C}$ with $0 \le \Re z \le 1$. Since this holds for all operators $c \in \mathcal{M}$ such that $|\tau(s(c))| \le t$ and $\|c\| \le 1$, we obtain from Lemma 2 that the estimate

$$\int_0^t \mu(s;e^{zb}ae^{(1-z)b})ds \le \int_0^t \mu(s;ae^b)ds$$

holds for all $z \in \mathbb{C}$ with $0 \le \Re z \le 1$. Setting $z = \theta \in (0,1)$, we conclude the proof. □

Observe that the assumption that $a \in \mathcal{M}$ is a self-adjoint operator was crucially used in the proof above, where we concluded that $\mu(e^ba) = \mu(ae^b)$. In fact, the assertion of the above lemma does not hold for a non-self-adjoint operator a.

EXAMPLE 5. *Let $M_2(\mathbb{C})$ be the von Neumann algebra of all 2×2 matrices. There exist $a, b \in M_2(\mathbb{C})$ such $b = b^*$ and such that the inequality*

$$e^{b/2}ae^{b/2} \prec\prec ae^b$$

fails.

PROOF. Let $\lambda, \mu \in \mathbb{R}$. We set

$$a = \begin{pmatrix} 0 & 1 \\ 0 & 0 \end{pmatrix}, b = \begin{pmatrix} \lambda & 0 \\ 0 & \mu \end{pmatrix}.$$

A direct computation yields $ae^b = e^\mu a$ and $e^{b/2}ae^{b/2} = e^{(\lambda+\mu)/2}a$. Setting $\lambda > \mu$, we obtain the assertion. □

However, a quick analysis of the proof of Lemma 4 yields a following strengthening.

LEMMA 6. *Let $a, b \in \mathcal{M}$. If $b = b^*$, then*

$$e^{\theta b}ae^{(1-\theta)b} \prec\prec \max\{\mu(ae^b),\ \mu(e^ba)\}$$

for every $\theta \in (0,1)$.

For every $\varepsilon, \delta > 0$, we define the set

$$V(\varepsilon,\delta) = \{x \in S(\mathcal{M},\tau) :\ \exists p = p^2 = p^* \in \mathcal{M} \text{ such that } \|x(1-p)\| \le \varepsilon,\ \tau(p) \le \delta\}.$$

The topology generated by the sets $V(\varepsilon,\delta)$, $\varepsilon, \delta > 0$, is called a measure topology. The following assertion is well-known. We incorporate the proof for convenience of the reader.

LEMMA 7. *Let $x_n, x \in S(\mathcal{M},\tau)$ be such that $x_n \to x$ in measure topology. It follows that $\mu(x_n) \to \mu(x)$ almost everywhere.*

PROOF. Let t be the continuity point of $\mu(x)$. Fix $\varepsilon > 0$ and select $\delta > 0$ such that $|\mu(t;x) - \mu(t \pm \delta;x)| \le \varepsilon$. Since $x_n - x \to 0$ in measure, it follows that $x_n - x \in V(\varepsilon, \delta)$ for every $n \ge N$. Thus, $\mu(\delta; x_n - x) \le \varepsilon$.

We have

$$\mu(t;x) \le \mu(t+\delta;x) + \varepsilon \le \varepsilon + \mu(t;x_n) + \mu(\delta;x_n - x) \le 2\varepsilon + \mu(t;x_n)$$

and

$$\mu(t;x_n) \le \mu(t-\delta;x) + \mu(\delta;x - x_n) \le 2\varepsilon + \mu(t;x).$$

Thus, $\mu(t;x_n) \to \mu(t;x)$. The assertion follows from the fact that $\mu(x)$ is almost everywhere continuous. □

LEMMA 8. *Let $a, b \in \mathcal{M}$*

(i) If a is self adjoint and $b \ge 0$, then

$$b^\theta a b^{1-\theta} \prec\prec ab$$

for every $\theta \in (0,1)$.

(ii) If a is an arbitrary operator and $b \ge 0$, then we have

$$b^\theta a b^{1-\theta} \prec\prec \max\{\mu(ab),\ \mu(ba)\}$$

for every $\theta \in (0,1)$.

PROOF. We shall prove only the first assertion (the proof of the second is the same via Lemma 6).

Suppose first that b is a positive invertible operator from $\mathcal{M}$. Then $\log(b)$ is a self-adjoint operator from $\mathcal{M}$ and applying Lemma 4 to the bounded operators a and $\log(b)$, we obtain the assertion. In general case, fix $n \in \mathbb{N}$, and consider the operator $b_n := b + \frac{1}{n}$ which is obviously invertible. It follows from the first part of the proof that

$$\int_0^t \mu(s; b_n^\theta a b_n^{1-\theta})ds \le \int_0^t \mu(s; ab_n)ds.$$

By Lemma 7

$$\mu(b_n^\theta a b_n^{1-\theta}) \to \mu(b^\theta a b^{1-\theta}), \quad \mu(ab_n) \to \mu(ab)$$

almost everywhere. Since the functions $\mu(b^\theta a b^{1-\theta})$ and $\mu(ab)$ are uniformly bounded, we easily infer from here that for every $t \ge 0$, we have

$$\int_0^t \mu(s; b^\theta a b^{1-\theta})ds \le \int_0^t \mu(s; ab)ds.$$

□

The resolution of Question 1 is contained in the first part of the theorem below. Observe that only the case $ab, ba \in (L_1 + L_\infty)(\mathcal{M}, \tau)$ should be treated. Indeed, if the latter assumption does not hold then the answer to Question 1 is trivially affirmative.

THEOREM 9. *Let $\mathcal{M}$ be a von Neumann algebra and let $a, b \in S(\mathcal{M}, \tau)$ be such operators that $b \ge 0$ and $ab \in (L_1 + L_\infty)(\mathcal{M}, \tau)$.*

(i) If a is self-adjoint, then

$$b^\theta a b^{1-\theta} \prec\prec ab$$

for every $\theta \in (0,1)$.

(ii) If a is an arbitrary operator, then we have

$$b^{\theta}ab^{1-\theta} \prec\prec \max\{\mu(ab),\ \mu(ba)\}$$

for every $\theta \in (0,1)$.

PROOF. We prove the second assertion. Let $p_n = ae^{|a|}[0,n)$ and let $q_n = e^b[0,n)$. The operators ap_n and $q_n bq_n = bq_n$ are bounded and evidently, $ap_n \to a$ and $bq_n \to b$ in measure as $n \to \infty$. Hence, $(bq_n)^{\theta} \to b^{\theta}$ in measure (see e.g. [**34**]) and, therefore,

$$(bq_n)^{\theta}(ap_n)(bq_n)^{1-\theta} \to b^{\theta}ab^{1-\theta}$$

in measure. By Lemma 7, we have

$$\mu((bq_n)^{\theta}(ap_n)(bq_n)^{1-\theta}) \to \mu(b^{\theta}ab^{1-\theta}) \tag{3}$$

almost everywhere. It follows now from Fatou lemma that

$$\int_0^t \mu(s; b^{\theta}ab^{1-\theta})ds \le \liminf_{n\to\infty} \int_0^t \mu(s; (bq_n)^{\theta}(ap_n)(bq_n)^{1-\theta})ds.$$

By Lemma 8, we have

$$\int_0^t \mu(s; b^{\theta}ab^{1-\theta})ds \le \liminf_{n\to\infty} \int_0^t \max\{\mu(s; (bq_n)(ap_n)), \mu(s; (ap_n)(bq_n))\}ds.$$

Since $|ad| = ||a|d|$ for all operators $a, d \in S(\mathcal{M}, \tau)$, it follows that

$$\mu((ap_n)(bq_n)) = \mu(|a|p_n(bq_n)) = \mu(p_n(|a|b)q_n) \le \mu(|a|b) = \mu(ab).$$

Also, we have

$$\mu((bq_n)(ap_n)) = \mu(q_n(ba)p_n) \le \mu(ba).$$

The assertion follows immediately. □

The result of Theorem 9 above extends and complements [**1**, Theorems 1 and 2], [**2**, Theorem 3 and Corollary 4], [**9**, Proposition 3.4]. More details are given in the next section.

We end this section with one more extension of Lemma 4.

PROPOSITION 10. *For any self-adjoint operators $a, b \in S(\mathcal{M}, \tau)$ and every $\theta \in (0,1)$, we have*

$$e^{\theta b}e^{a}e^{(1-\theta)b} \prec\prec e^{a}e^{b}.$$

PROOF. By [**23**, Lemma 3.1], we have $e^a, e^b \in S(\mathcal{M}, \tau)$. It is sufficient to prove the assertion only for the case $e^a e^b \in (L_1 + L_\infty)(\mathcal{M}, \tau)$. Introducing projections $p_n := e^{|a|}[0,n)$, $q_n := e^{|b|}[0,n)$, and operators $a_n := ap_n$, $b_n := bq_n$ we obtain from Lemma 4 that

$$e^{\theta b_n}e^{a_n}e^{(1-\theta)b_n} \prec\prec e^{a_n}e^{b_n},\ n \ge 1.$$

The same argument as in the proof of Theorem 9 completes the proof. □

3. Applications to ideals in $S(\mathcal{M},\tau)$

The best known examples of normed $\mathcal{M}$-bimodules of $S(\mathcal{M},\tau)$ are given by the so-called symmetric operator spaces (see e.g. [**8**, **10**, **16**, **29**]). We briefly recall relevant definitions (for more detailed information we refer to [**16**] and references therein).

Let E be a Banach space of real-valued Lebesgue measurable functions either on $(0,1)$ or $(0,\infty)$ (with identification $m-$a.e.) or on $\mathbb{N}$. The space E is said to be *absolutely solid* if $x\in E$ and $|y|\le|x|$, $y\in L_0$ implies that $y\in E$ and $||y||_E\le||x||_E$.

The absolutely solid space $E\subseteq S$ is said to be *symmetric* if for every $x\in E$ and every y the assumption $\mu(y)=\mu(x)$ implies that $y\in E$ and $||y||_E=||x||_E$ (see e.g. [**20**]).

If $E=E(0,1)$ is a symmetric space on $(0,1)$, then

$$L_\infty\subseteq E\subseteq L_1.$$

If $E=E(0,\infty)$ is a symmetric space on $(0,\infty)$, then

$$L_1\cap L_\infty\subseteq E\subseteq L_1+L_\infty.$$

If $E=E(\mathbb{N})$ is a symmetric space on $\mathbb{N}$, then

$$l_1\subseteq E\subseteq l_\infty,$$

where l_1 and l_∞ are classical spaces of all absolutely summable and bounded sequences respectively.

DEFINITION 11. *Let $\mathcal{E}$ be a linear subset in $S(\mathcal{M},\tau)$ equipped with a norm $\|\cdot\|_{\mathcal{E}}$. We say that $\mathcal{E}$ is a symmetric operator space (on $\mathcal{M}$, or in $S(\mathcal{M},\tau)$) if $x\in\mathcal{E}$ and every $y\in S(\mathcal{M},\tau)$ the assumption $\mu(y)\le\mu(x)$ implies that $y\in\mathcal{E}$ and $\|y\|_{\mathcal{E}}\le\|x\|_{\mathcal{E}}$.*

The fact that every symmetric operator space $\mathcal{E}$ is (an absolutely solid) $\mathcal{M}$-bimodule of $S(\mathcal{M},\tau)$ is well known (see e.g. [**16**, **29**] and references therein). In the special case, when $\mathcal{M}=B(H)$ and τ is a standard trace Tr, the notion of symmetric operator space introduced in Definition 11 coincides with the notion of symmetric operator ideal [**12**, **13**, **26**, **28**].

DEFINITION 12. *A linear subspace $\mathcal{I}$ in the von Neumann algebra $\mathcal{M}$ equipped with a norm $\|\cdot\|_{\mathcal{I}}$ is said to be a symmetric operator ideal if*

(1) $\|S\|_{\mathcal{I}}\ge\|S\|$ for all $S\in\mathcal{I}$.
(2) $\|S^*\|_{\mathcal{I}}=\|S\|_{\mathcal{I}}$ for all $S\in\mathcal{I}$.
(3) $\|ASB\|_{\mathcal{I}}\le\|A\|\,\|S\|_{\mathcal{I}}\|B\|$ for all $S\in\mathcal{I}$, $A,B\in\mathcal{M}$.

There exists a strong connection between symmetric function and operator spaces recently exposed in [**16**] (see earlier results in [**12**, **13**, **26**, **28**]).

Let E be a symmetric function space on the interval $(0,1)$ (respectively, on the semi-axis) and let $\mathcal{M}$ be a type II_1 (respectively, II_∞) von Neumann algebra. Define

$$E(\mathcal{M},\tau):=\{x\in S(\mathcal{M},\tau):\ \mu(x)\in E\},\ \|x\|_{E(\mathcal{M},\tau)}:=\|\mu(x)\|_E.$$

Main results of [**16**] assert that $(E(\mathcal{M},\tau),\|\cdot\|_{E(\mathcal{M},\tau)})$ is a symmetric operator space. Similarly, if $E=E(\mathbb{N})$ is a symmetric sequence space on $\mathbb{N}$, and the algebra $\mathcal{M}$ is a type I factor with standard trace, then (see [**16**]) setting

$$\mathcal{E}:=\{x\in\mathcal{M}:\ (\mu(n;x))_{n\ge0}\in E\},\ \|x\|_{\mathcal{E}}:=\|(\mu(n;x))_{n\ge0}\|_E$$

yields a symmetric operator ideal. Conversely, every symmetric operator ideal $\mathcal{E}$ in $\mathcal{M}$ defines a unique symmetric sequence space $E = E(\mathbb{N})$ by setting

$$E := \{a = (a_n)_{n\geq 0} \in l_\infty : \ \mu(a) = (\mu(n;x))_{n\geq 0} \text{ for some } x \in \mathcal{E}\}, \ \|a\|_E := \|x\|_{\mathcal{E}}$$

Finally, a symmetric space $E(\mathcal{M},\tau)$ is called fully symmetric if for every $a \in E(\mathcal{M},\tau)$ and every $b \in (L_1 + L_\infty)(\mathcal{M})$ with $b \prec\prec a$, we have $b \in E(\mathcal{M},\tau)$ and $\|b\|_E \leq \|a\|_E$. The following result now follows immediately from Theorem 9.

COROLLARY 13. *Let E be a fully symmetric function space on $(0,\tau(\mathbf{1}))$. If $a, b \in S(\mathcal{M},\tau)$ are such operators that $b \geq 0$ and $ab, ba \in E(\mathcal{M},\tau)$, then $b^\theta ab^{1-\theta} \in E(\mathcal{M},\tau)$ for every $\theta \in (0,1)$. In particular, if a is self-adjoint, then*

$$ab \in E(\mathcal{M},\tau) \Longrightarrow b^\theta ab^{1-\theta} \in E(\mathcal{M},\tau), \ \forall \theta \in (0,1)$$

and $\|b^\theta ab^{1-\theta}\|_{E(\mathcal{M},\tau)} \leq \|ab\|_{E(\mathcal{M},\tau)}$.

We shall now present some variation of the result above. For simplicity of the exposition, we shall do so for fully symmetric sequence spaces E and for symmetric operator ideals $\mathcal{E}$, although all arguments below can be repeated also for general semifinite factors.

COROLLARY 14. *Fix a fully symmetric operator ideal $\mathcal{E}$ and let $a, b_0, b_1 \in B(H)$, $b_0, b_1 \geq 0$, $\theta \in (0,1)$.*

(i) If $ab_0, b_1 a \in \mathcal{E}$, then $b_0^\theta ab_1^{1-\theta}, b_1^\theta ab_0^{1-\theta} \in \mathcal{E}$ and

$$b_1^\theta ab_0^{1-\theta} \prec\prec \max\{\mu(ab_0), \ \mu(b_1 a)\}.$$

In particular, $b_1^\theta ab_0^{1-\theta} \in \mathcal{E}$.

(ii) If $b_0 a, ab_1 \in \mathcal{E}$, then

$$\mu(b_0^\theta ab_1^{1-\theta}) \oplus \mu(b_1^\theta a^* b_0^{1-\theta}) \prec\prec \mu(ab_1) \oplus \mu(b_0 a).$$

In particular, if $a = a^*$ and $\theta = 1/2$, we have

$$\sigma_2(\mu(b_0^{1/2} ab_1^{1/2}) \prec\prec \mu(ab_1) \oplus \mu(ab_0).$$

Here $\sigma_2(a_0, a_1, \cdots) = (a_0, a_0, a_1, a_1, \cdots)$.

PROOF. (i) In $B(H \oplus H)$ consider the following operators

$$\mathbf{a} = \begin{pmatrix} 0 & 0 \\ a & 0 \end{pmatrix}, \quad \mathbf{b} = \begin{pmatrix} b_0 & 0 \\ 0 & b_1 \end{pmatrix}.$$

We obviously have $\mathbf{b} \geq 0$ and $\mu(\mathbf{ab}) = \mu(ab_0)$, $\mu(\mathbf{ba}) = \mu(b_1 a)$ and therefore $\mathbf{ab}, \mathbf{ba} \in \mathcal{E}$. Applying now Theorem 9 and Corollary 13 we arrive at

$$\begin{pmatrix} 0 & 0 \\ b_1^\theta ab_0^{1-\theta} & 0 \end{pmatrix} \prec\prec \max\left\{\mu\left(\begin{pmatrix} 0 & 0 \\ ab_0 & 0 \end{pmatrix}\right), \ \mu\left(\begin{pmatrix} 0 & 0 \\ b_1 a & 0 \end{pmatrix}\right)\right\},$$

which is the assertion.

(ii) Consider $\mathbf{b}$ as above and set

$$\mathbf{a} := \begin{pmatrix} 0 & a \\ a^* & 0 \end{pmatrix}$$

Observe that $\mathbf{a}$ is self-adjoint and that the assumption guarantees $\mathbf{ab} \in \mathcal{E}$. Thus, by Theorem 9 we have

$$\mathbf{b}^\theta \mathbf{ab}^{1-\theta} \prec\prec \mathbf{ab},$$

that is

$$\begin{pmatrix} 0 & b_0^{\theta} a b_1^{1-\theta} \\ b_1^{\theta} a^* b_0^{1-\theta} & 0 \end{pmatrix} \prec\prec \begin{pmatrix} 0 & ab_1 \\ a^* b_0 & 0 \end{pmatrix}$$

which is equivalent to the first assertion. The last assertion in (ii) is trivial. □

The following lemma extends result of [**5**, Lemma 27] and [**24**, Lemma 10].

LEMMA 15. *Let $\mathcal{E}$ be a fully symmetric operator ideal. If $a, b_0, b_1 \in B(H)$ are such that $b_0, b_1 \geq 0$ and $b_0 a, ab_1 \in \mathcal{E}$, then*

$$\|b_0^{\theta} a b_1^{1-\theta}\|_{\mathcal{E}} \leq \|b_0 a\|_{\mathcal{E}}^{\theta} \|ab_1\|_{\mathcal{E}}^{1-\theta}, \quad 0 < \theta < 1.$$

PROOF. Setting $F(z) = \tau(b_0^z a b_1^{1-z} c)$ and repeating the argument in Lemma 4, we obtain

$$\int_0^t \mu(s; b_0^{\theta} a b_1^{1-\theta}) ds \leq (\int_0^t \mu(s; b_0 a) ds)^{\theta} (\int_0^t \mu(s; ab_1) ds)^{1-\theta}.$$

Using elementary inequality $\alpha^{\theta}\beta^{1-\theta} \leq \theta\alpha + (1-\theta)\beta$, we obtain, for every $\lambda > 0$,

$$\int_0^t \mu(s; b_0^{\theta} a b_1^{1-\theta}) ds \leq (\int_0^t \mu(s; \lambda^{1-\theta} b_0 a) ds)^{\theta} (\int_0^t \mu(s; \lambda^{-\theta} ab_1) ds)^{1-\theta} \leq$$

$$\leq \theta\lambda^{1-\theta} \int_0^t \mu(s; b_0 a) ds + (1-\theta)\lambda^{-\theta} \int_0^t \mu(s; ab_1) ds.$$

Since the ideal $\mathcal{E}$ is fully symmetric, it follows that

$$\|b_0^{\theta} a b_1^{1-\theta}\|_{\mathcal{E}} \leq \theta\lambda^{1-\theta} \|b_0 a\|_{\mathcal{E}} + (1-\theta)\lambda^{-\theta} \|ab_1\|_{\mathcal{E}}.$$

The assertion follows now by setting

$$\lambda = \|ab_1\|_{\mathcal{E}} \cdot \|b_0 a\|_{\mathcal{E}}^{-1}.$$

□

Recall that the set

$$L_p(\mathcal{M}, \tau) = \{x \in S(\mathcal{M}, \tau) : \ \tau(|x|^p) < \infty\}$$

equipped with a standard norm

$$\|x\|_p := \tau(|x|^p)^{1/p}$$

is the noncommutative L_p-space associated with $(\mathcal{M}, \tau)$ for every $1 \leq p < \infty$. In the type I factor setting these are the usual Schatten-von Neumann ideals [**12, 13, 26, 28**]. The following corollary follows immediately from the above result.

COROLLARY 16. *Let $\mathcal{M}$ be a semifinite factor and $a, b_0, b_1 \in S(\mathcal{M}, \tau)$ be self-adjoint operators such that $b_0, b_1 \geq 0$ and such that $ab_0, ab_1 \in L_p(\mathcal{M}, \tau)$. We have $b_0^{1/2} a b_1^{1/2} \in L_p(\mathcal{M}, \tau)$ and*

$$2\|b_0^{1/2} a b_1^{1/2}\|_p^p \leq \|ab_1\|_p^p + \|b_0 a\|_p^p.$$

For detailed exposition of (generalized) Golden-Thompson inequality and for further references we refer to [**28**]. The following result now follows immediately from Proposition 10 and the definition of a fully symmetric space.

PROPOSITION 17. *Let E be a fully symmetric function space on $(0, \tau(\mathbf{1}))$. For any self-adjoint operators $a, b \in S(\mathcal{M}, \tau)$ and every $\theta \in (0, 1)$, we have*

$$\|e^{\theta b} e^{a} e^{(1-\theta)b}\|_{E(\mathcal{M}, \tau)} \leq \|e^{a} e^{b}\|_{E(\mathcal{M}, \tau)}.$$

We shall complete this section with a complement to [**28**, Theorem 8.3]. According to that theorem for self-adjoint operators a and b and for all $1 \le p \le \infty$ we have $\|e^{a+b}\|_p \le \|e^{a/2}e^b e^{a/2}\|_p$ and for $2 \le p \le \infty$ we have $\|e^{a+b}\|_p \le \|e^a e^b\|_p$. We claim that the latter estimate holds for all $1 \le p \le \infty$. Indeed, this follows from a combination of the former estimate and Proposition 17.

4. An application to traces

Let $\mathcal{E}$ be a (fully) symmetric operator ideal. A linear functional φ on $\mathcal{E}$ is said to be a trace if $\varphi(ab) = \varphi(ba)$ for every $a \in \mathcal{E}$ and every $b \in B(H)$. A complete characterization of symmetric operator ideals which admit a nontrivial trace has been recently given in [**31**].

For a compact operator $x \in B(H)$, the symbol $\Lambda(x)$ stands for the *set* of all sequences of eigenvalues of x counted with algebraic multiplicities and ordered by the inequality $|\lambda_{n+1}(x)| \le |\lambda_n(x)|$. The following assertion is a particular case of Theorem 3.10.3 in [**4**].

THEOREM 18. *If $a, b \in B(H)$ and if a is a compact operator, then $\Lambda(ab) = \Lambda(ba)$.*

The assertion of the previous theorem fails without the assumption of compactness.

EXAMPLE 19. *There exist bounded operators $a, b \in B(H)$ such that ab is compact while ba is not.*

PROOF. Fix an infinite projection p such that $1-p$ is also an infinite projection. Thus, projections p and $1-p$ are equivalent in $B(H)$. Select a partial isometry u such that $uu^* = p$ and $u^*u = 1-p$. We have $|up|^2 = pu^* \cdot up = p(1-p)p = 0$. Hence, $up = 0$ and $pu = (uu^*)u = u(1-p) = u - up = u$. Setting $a = u$ and $b = p$, we are done. □

The following fundamental result will first appear in [**15**], though it is essentially proved in [**14**].

THEOREM 20. *Let $\mathcal{E}$ be a symmetric operator ideal. For every $a \in \mathcal{E}$ and for every trace φ on $\mathcal{E}$, we have*

$$\varphi(a) = \varphi(\mathrm{diag}(\lambda(a))),$$

where $\mathrm{diag}(\lambda(a))$ *is a diagonal matrix corresponding to any sequence* $\lambda(a) \in \Lambda(a)$.

The following theorem is the main result of this section. It is new even in the case when φ is the standard trace Tr on $B(H)$. For the special cases of this theorem for $\theta = 1/2$, we refer to [**9, 11, 19**].

THEOREM 21. *Let $\mathcal{E}$ be a fully symmetric operator ideal and let $a, b \in B(H)$, $b \ge 0$, be such that a is compact and $ab, ba \in \mathcal{E}$. For every trace φ on $\mathcal{E}$, we have*

$$\varphi(ab) = \varphi(ba) = \varphi(b^{1-\theta}ab^{\theta}).$$

PROOF. By Theorem 9, we have $b^{1-\theta}ab^{\theta} \prec\prec \max\{\mu(ab), \mu(ba)\}$ and, therefore, $b^{1-\theta}ab^{\theta} \in \mathcal{E}$. By Theorem 18, we have

$$\Lambda(ab) = \Lambda(ba) = \Lambda(b^{1-\theta}ab^{\theta}).$$

The assertion follows now from Theorem 20. □

References

[1] Airat Bikchentaev, *Majorization for products of measurable operators*, Proceedings of the International Quantum Structures Association 1996 (Berlin), 1998, pp. 571–576, DOI 10.1023/A:1026688801188. MR1637210 (99e:46075)

[2] A. M. Bikchentaev, *On a property of L_p-spaces on semifinite von Neumann algebras*, Mat. Zametki **64** (1998), no. 2, 185–190, DOI 10.1007/BF02310299 (Russian, with Russian summary); English transl., Math. Notes **64** (1998), no. 1-2, 159–163 (1999). MR1680973 (2001a:46062)

[3] A. M. Bikchentaev, *Block projection operator on normed solid spaces of measurable operators,* (Russian) Izv. Vyssh. Uchebn. Zaved. Mat. (2012), No. 2, 86–91. English translation in Russian Math. (Iz. VUZ) **56** (2012), no. 2, 75-79.

[4] M. Sh. Birman and M. Z. Solomjak, *Spectral theory of selfadjoint operators in Hilbert space*, Mathematics and its Applications (Soviet Series), D. Reidel Publishing Co., Dordrecht, 1987. Translated from the 1980 Russian original by S. Khrushchëv and V. Peller. MR1192782 (93g:47001)

[5] Alan Carey, Denis Potapov, and Fedor Sukochev, *Spectral flow is the integral of one forms on the Banach manifold of self adjoint Fredholm operators*, Adv. Math. **222** (2009), no. 5, 1809–1849, DOI 10.1016/j.aim.2009.06.020. MR2555913 (2011a:58045)

[6] V. I. Chilin and F. A. Sukochev, *Weak convergence in non-commutative symmetric spaces*, J. Operator Theory **31** (1994), no. 1, 35–65. MR1316983 (96e:46085)

[7] J. Dixmier, *Les algebres d'operateurs dans l'espace hilbertien*, 2 edition, Gauthier - Villars, Paris, 1969.

[8] P.G. Dodds, T.K. Dodds, B. de Pagter, *Non-commutative Banach function spaces*, Math. Z. **201** (1989), 583–597.

[9] P.G. Dodds, T.K. Dodds, B. de Pagter, *Non-commutative Köthe duality*, Trans. Amer. Math. Soc. **339** (1993), 717-750.

[10] P. Dodds, B. de Pagter, F. Sukochev *Theory of noncommutative integration,* unpublished manuscript.

[11] Thierry Fack and Hideki Kosaki, *Generalized s-numbers of τ-measurable operators*, Pacific J. Math. **123** (1986), no. 2, 269–300. MR840845 (87h:46122)

[12] I. C. Gohberg and M. G. Kreĭn, *Introduction to the theory of linear nonselfadjoint operators*, Translated from the Russian by A. Feinstein. Translations of Mathematical Monographs, Vol. 18, American Mathematical Society, Providence, R.I., 1969. MR0246142 (39 #7447)

[13] I. C. Gohberg and M. G. Kreĭn, *Theory and applications of Volterra operators in Hilbert space*, Translated from the Russian by A. Feinstein. Translations of Mathematical Monographs, Vol. 24, American Mathematical Society, Providence, R.I., 1970. MR0264447 (41 #9041)

[14] N. J. Kalton, *Spectral characterization of sums of commutators. I*, J. Reine Angew. Math. **504** (1998), 115–125, DOI 10.1515/crll.1998.102. MR1656767 (99g:47078)

[15] N. J. Kalton, S. Lord, D. Potapov, and F. Sukochev, *Traces on compact operators and the non-commutative residue*, Adv. Math. **235** (2013), 1–55.

[16] N. J. Kalton and F. A. Sukochev, *Symmetric norms and spaces of operators*, J. Reine Angew. Math. **621** (2008), 81–121, DOI 10.1515/CRELLE.2008.059. MR2431251 (2009i:46118)

[17] Richard V. Kadison and John R. Ringrose, *Fundamentals of the theory of operator algebras. Vol. I*, Pure and Applied Mathematics, vol. 100, Academic Press Inc. [Harcourt Brace Jovanovich Publishers], New York, 1983. Elementary theory. MR719020 (85j:46099)

[18] Richard V. Kadison and John R. Ringrose, *Fundamentals of the theory of operator algebras. Vol. II*, Pure and Applied Mathematics, vol. 100, Academic Press Inc., Orlando, FL, 1986. Advanced theory. MR859186 (88d:46106)

[19] Hideki Kosaki, *Arithmetic-geometric mean and related inequalities for operators*, J. Funct. Anal. **156** (1998), no. 2, 429–451, DOI 10.1006/jfan.1998.3258. MR1636964 (99g:47032)

[20] S. G. Kreĭn, Yu. Ī. Petunīn, and E. M. Semënov, *Interpolation of linear operators*, Translations of Mathematical Monographs, vol. 54, American Mathematical Society, Providence, R.I., 1982. Translated from the Russian by J. Szűcs. MR649411 (84j:46103)

[21] Edward Nelson, *Notes on non-commutative integration*, J. Functional Analysis **15** (1974), 103–116. MR0355628 (50 #8102)

[22] V. I. Ovčinnikov, *The s-numbers of measurable operators*, Funkcional. Anal. i Priložen. **4** (1970), no. 3, 78–85 (Russian). MR0271763 (42 #6644)

[23] Ben de Pagter and Fyodor Sukochev, *Commutator estimates and* $\mathbb{R}$*-flows in non-commutative operator spaces*, Proc. Edinb. Math. Soc. (2) **50** (2007), no. 2, 293–324, DOI 10.1017/S0013091505000957. MR2334949 (2008f:46081)

[24] D. Potapov and F. Sukochev, *Unbounded Fredholm modules and double operator integrals*, J. Reine Angew. Math. **626** (2009), 159–185, DOI 10.1515/CRELLE.2009.006. MR2492993 (2010a:58034)

[25] M. Reed, B. Simon, *Methods of modern mathematical physics. Volume 2: Fourier analysis, self-adjointness,* Elsevier, 1975.

[26] Robert Schatten, *Norm ideals of completely continuous operators*, Second printing. Ergebnisse der Mathematik und ihrer Grenzgebiete, Band 27, Springer-Verlag, Berlin, 1970. MR0257800 (41 #2449)

[27] I. E. Segal, *A non-commutative extension of abstract integration*, Ann. of Math. (2) **57** (1953), 401–457. MR0054864 (14,991f)

[28] Barry Simon, *Trace ideals and their applications*, 2nd ed., Mathematical Surveys and Monographs, vol. 120, American Mathematical Society, Providence, RI, 2005. MR2154153 (2006f:47086)

[29] F. A. Sukochev and V. I. Chilin, *Symmetric spaces over semifinite von Neumann algebras*, Dokl. Akad. Nauk SSSR **313** (1990), no. 4, 811–815 (Russian); English transl., Soviet Math. Dokl. **42** (1991), no. 1, 97–101. MR1080637 (92a:46075)

[30] F. Sukochev, *On the A. M. Bikchentaev conjecture,* (Russian) Izv. Vyssh. Uchebn. Zaved. Mat., (2012), no.6, 67–70.

[31] F. Sukochev, D. Zanin, *Traces on symmetrically normed operator ideals,* J. Reine Angew. Math. (to appear).

[32] Masamichi Takesaki, *Theory of operator algebras. I*, Springer-Verlag, New York, 1979. MR548728 (81e:46038)

[33] M. Takesaki, *Theory of operator algebras. II*, Encyclopaedia of Mathematical Sciences, vol. 125, Springer-Verlag, Berlin, 2003. Operator Algebras and Non-commutative Geometry, 6. MR1943006 (2004g:46079)

[34] O. E. Tikhonov, *Continuity of operator functions in topologies connected with a trace on a von Neumann algebra*, Izv. Vyssh. Uchebn. Zaved. Mat. **1** (1987), 77–79 (Russian). MR892008 (88h:46120)

School of Mathematics and Statistics, University of New South Wales, Sydney, NSW 2052, Australia

E-mail address: f.sukochev@unsw.edu.au

Proceedings of Symposia in Pure Mathematics
Volume **87**, 2013
http://dx.doi.org/10.1090/pspum/087/01443

Spectral theory as influenced by Fritz Gesztesy

Gerald Teschl and Karl Unterkofler

To Fritz Gesztesy, teacher, mentor, and friend, on the occasion of his 60th birthday.

Abstract. We survey a selection of Fritz's principal contributions to the field of spectral theory and, in particular, to Schrödinger operators.

1. Introduction

The purpose of this Festschrift contribution is to highlight some of Fritz's profound contributions to spectral theory and, in particular, to Schrödinger operators. Of course, if you look at his list of publications it is clear that this is a *mission impossible* and hence the present review will only focus on a small selection. Moreover, this selection is highly subjective and biased by our personal research interests:

- Relativistic Corrections
- Singular Weyl–Titchmarsh–Kodaira Theory
- Inverse Spectral Theory and Trace Formulas
- Commutation Methods
- Oscillation Theory
- Non-self-adjoint operators

Other aspects of his work, e.g., on point interactions and integrable nonlinear wave equations are summarized in the monographs [**2**] and [**48, 49**], respectively. For some of his seminal contributions to inverse spectral theory [**22, 63–68**] we refer to Fritz's own review [**38**]. For some of his recent work on partial differential operators, Krein-type resolvent formulas, operator-valued Weyl–Titchmarsh operators (i.e., energy-dependent analogs of Dirichlet-to-Neumann maps), and Weyl-type spectral asymptotics for Krein–von Neumann extensions of the Laplacian on bounded domains, we refer to [**5, 54, 56–58**], and the detailed lists of references therein.

To be sure, this is not a survey of the state of the art on these topics but we rather focus exclusively on Fritz Gesztesy's contributions to and influence exerted on the field. Especially, the bibliography is far from being complete and only reflects the particular purpose of this survey.

2010 *Mathematics Subject Classification.* Primary 35P05, 34L40; Secondary 34B20, 34B24.

Key words and phrases. Schrödinger operators, Spectral theory, Oscillation theory.

in "Spectral Analysis, Differential Equations and Mathematical Physics", H. Holden et al. (eds), Amer. Math. Soc. (to appear).

Research supported by the Austrian Science Fund (FWF) under Grant No. Y330 and P24736-B23.

Personal note. It was 28 years ago in spring of 1985 that Fritz and Karl first met in Graz. Fritz, jointly with Wolfgang Bulla, advised his master and PhD thesis. Karl then spent two years (1990–1992) as a postdoc with Fritz at the University of Missouri, Columbia. When he returned to Austria, Karl met Gerald, who had just finished his masters with Wolfgang Bulla. Karl immediately proposed to him to pursue his PhD with Fritz. After a brief meeting with Fritz in Graz and a short visit to the University of Missouri, Columbia, Gerald enrolled in the PhD program there and a second successful cooperation began. Fritz guided us through the early stages of our careers and was much more than a mentor for us. On the one hand he was an unlimited source of ideas for new projects and on the other hand he was available for discussing mathematics close to 24 hours at 358 days a year (one week has to be subtracted for which he disappears when snorkeling in the Caribbean or Hawaii). His devotion to mathematics, his perfection when it comes to details, and his endurance has always been a role model for us. And last but not least, he is the one everybody asks for references; his collection of papers and books is legendary!

Happy Birthday, Fritz, and on to many more such anniversaries!

2. Relativistic corrections

In 1950 Foldy and Wouthuysen developed a formal perturbation scheme (which is now "regarded as obsolete", according to Kutzelnigg on page 678 in [**121**]) to obtain relativistic corrections of the nonrelativistic Pauli operator H_+. However, adding the first correction term of order c^{-2} already destroys all spectral properties of H_+. The fact, that nevertheless formal perturbation theory yields correct results in special cases has been explained in terms of spectral concentration in [**39, 43**] and [**138**].

Historically, the first rigorous treatment of the nonrelativistic limit of Dirac Hamiltonians goes back to Titchmarsh [**137**] who proved holomorphy of the Dirac eigenvalues (rest energy subtracted) with respect to c^{-2} for spherically symmetric potentials and obtained explicit formulas for relativistic bound state corrections of order $O(c^{-2})$ (formally derived in [**123**]). Holomorphy of the Dirac resolvent in three dimensions in c^{-1} for electrostatic interactions were first obtained by Veselic [**139**] and then extended to electromagnetic interactions by Hunziker [**85**]. An entirely different approach, based on an abstract set up, has been used in [**15**] to prove strong convergence of the unitary groups as $c^{-1} \to 0$. Convergence of solutions of the Dirac equation based on semigroup methods has also been obtained in [**120**]. Time-dependent electromagnetic fields are treated in [**7**].

In joint collaboration with Bernd Thaller and Harald Grosse [**41**] (see also [**40**] and [**42**]) employing an abstract framework, holomorphy of the Dirac resolvent in c^{-1} under very general conditions on the electromagnetic interaction potentials was obtained. Moreover, this approach led to the first rigorous derivation of explicit formulas for relativistic corrections of order $O(c^{-2})$ to bound state energies.

Application of these results lead to relativistic corrections for energy bands and corresponding corrections for impurity bound states for one-dimensional periodic systems [**11**] and relativistic corrections for the scattering matrix [**12**]. An explicit treatment of relativistic corrections of the scattering amplitude appeared in [**83**].

A thorough and comprehensive treatment of Dirac operators can be found in the book by Thaller [**136**] and in [**121**].

Based on the abstract approach introduced in [**85**] and [**15**], let $\mathcal{H}_\pm$ be separable, complex Hilbert spaces. One introduces self-adjoint operators α, β in $\mathcal{H} = \mathcal{H}_+ \oplus \mathcal{H}_-$ of the type

$$\alpha = \begin{pmatrix} 0 & A^* \\ A & 0 \end{pmatrix}, \quad \beta = \begin{pmatrix} 1 & 0 \\ 0 & -1 \end{pmatrix}, \tag{2.1}$$

where A is a densely defined, closed operator from $\mathcal{H}_+$ into $\mathcal{H}_-$. Next, define the abstract free Dirac operator $H^0(c)$ by

$$H^0(c) = c\,\alpha + mc^2\beta, \quad \mathrm{dom}(H^0(c)) = \mathrm{dom}(\alpha), \quad c \in \mathbb{R}\setminus\{0\}, \quad m > 0 \tag{2.2}$$

and the interaction V by

$$V = \begin{pmatrix} V_+ & 0 \\ 0 & V_- \end{pmatrix}, \tag{2.3}$$

where $V_\pm$ denote self-adjoint operators in $\mathcal{H}_\pm$, respectively. Assuming V_+ (resp.V_-) to be bounded w.r.t. A (resp. A^*), i.e.,

$$\mathrm{dom}(A) \subseteq \mathrm{dom}(V_+), \quad \mathrm{dom}(A^*) \subseteq \mathrm{dom}(V_-), \tag{2.4}$$

the abstract Dirac operator $H(c)$ reads

$$H(c) = H^0(c) + V, \quad \mathrm{dom}(H(c)) = \mathrm{dom}(\alpha). \tag{2.5}$$

Obviously, $H(c)$ is self-adjoint for $|c|$ large enough. The corresponding self-adjoint (free) Pauli operators in $\mathcal{H}_\pm$ are then defined by

$$\begin{aligned} &H_+^0 = (2m)^{-1}A^*A, \quad H_+ = H_+^0 + V_+, \quad \mathrm{dom}(H_+) = \mathrm{dom}(A^*A), \\ &H_-^0 = (2m)^{-1}AA^*, \quad H_- = H_-^0 + V_-, \quad \mathrm{dom}(H_-) = \mathrm{dom}(AA^*). \end{aligned} \tag{2.6}$$

Introducing the operator $B(c)$ (see [**85**])

$$B(c) = \begin{pmatrix} 1 & 0 \\ 0 & c \end{pmatrix}, \tag{2.7}$$

one establishes the holomorphy of the Dirac resolvent in c^{-1}.

THEOREM 2.1. *Let $H(c)$ be defined as above and fix $z \in \mathbb{C}\setminus\mathbb{R}$. Then*
(a) $(H(c) - mc^2 - z)^{-1}$ is holomorphic w.r.t. c^{-1} around $c^{-1} = 0$

$$\begin{aligned} &(H(c) - mc^2 - z)^{-1} \\ &= \left\{1 + \begin{pmatrix} 0 & (2mc)^{-1}(H_+ - z)^{-1}A^*(V_- - z) \\ (2mc)^{-1}A(H_+^0 - z)^{-1}V_+ & (2mc^2)^{-1}z(H_-^0 - z)^{-1}(V_- - z) \end{pmatrix}\right\}^{-1} \times \\ &\times \begin{pmatrix} (H_+ - z)^{-1} & (2mc)^{-1}(H_+ - z)^{-1}A^* \\ (2mc)^{-1}A(H_+^0 - z)^{-1} & (2mc^2)^{-1}z(H_-^0 - z)^{-1} \end{pmatrix}. \end{aligned} \tag{2.8}$$

(b) $B(c)(H(c) - mc^2 - z)^{-1}B(c)^{-1}$ is holomorphic w.r.t. c^{-2} around $c^{-2} = 0$

$$\begin{aligned} &B(c)(H(c) - mc^2 - z)^{-1}B(c)^{-1} \\ &= \left\{1 + \begin{pmatrix} 0 & (2mc^2)^{-1}(H_+ - z)^{-1}A^*(V_- - z) \\ 0 & (2mc^2)^{-1}[(2m)^{-1}A(H_+ - z)^{-1}A^* - 1](V_- - z) \end{pmatrix}\right\}^{-1} \times \\ &\times \begin{pmatrix} (H_+ - z)^{-1} & (2mc^2)^{-1}(H_+ - z)^{-1}A^* \\ (2m)^{-1}A(H_+ - z)^{-1} & (2mc^2)^{-1}[(2m)^{-1}A(H_+ - z)^{-1}A^* - 1] \end{pmatrix}. \end{aligned} \tag{2.9}$$

First order expansions in (2.8) and (2.9) yield

$$(H(c) - mc^2 - z)^{-1} = \begin{pmatrix} (H_+ - z)^{-1} & 0 \\ 0 & 0 \end{pmatrix}$$

$$+ c^{-1} \begin{pmatrix} 0 & (2m)^{-1}(H_+ - z)^{-1}A^* \\ (2m)^{-1}A(H_+ - z)^{-1} & 0 \end{pmatrix} + O(c^{-2}) \tag{2.10}$$

(clearly illustrating the nonrelativistic limit $|c| \to \infty$) and

$$B(c)(H(c) - mc^2 - z)^{-1}B(c)^{-1} =: R_0(z) + c^{-2}R_1(z) + O(c^{-4})$$

$$= \begin{pmatrix} (H_+ - z)^{-1} & 0 \\ (2m)^{-1}A(H_+ - z)^{-1} & 0 \end{pmatrix} + c^{-2} \begin{pmatrix} R_{11}(z) & R_{12}(z) \\ R_{21}(z) & R_{22}(z) \end{pmatrix} + O(c^{-4}),$$

where

$$\begin{aligned} R_{11}(z) &= (2m)^{-2}(H_+ - z)^{-1}A^*(z - V_-)A(H_+ - z)^{-1}, \\ R_{12}(z) &= (2m)^{-1}(H_+ - z)^{-1}A^*, \\ R_{21}(z) &= (2m)^{-2}\left((2m)^{-1}A(H_+ - z)^{-1}A^* - 1\right)(z - V_-)A(H_+ - z)^{-1}, \\ R_{22}(z) &= (2m)^{-1}\left((2m)^{-1}A(H_+ - z)^{-1}A^* - 1\right). \end{aligned} \tag{2.11}$$

To formulate and prove Theorem 2.1 Fritz et al. combined a clever decomposition of the Dirac resolvent with a skillful application of some commutations formulas by Deift [**20**] (see also Section 5).

Analyzing the relationship between the spectrum of $(H_+ - z)^{-1}$ and $R_0(z)$ now yields the following result on relativistic eigenvalue corrections.

THEOREM 2.2. *Let $H(c)$ be defined as in (2.5) and assume $E_0 \in \sigma_d(H_+)$ to be a discrete eigenvalue of H_+ of multiplicity $m_0 \in \mathbb{N}$. Then, for c^{-2} small enough, $H(c) - mc^2$ has precisely m_0 eigenvalues (counting multiplicity) near E_0 which are all holomorphic w.r.t. c^{-2}. More precisely, all eigenvalues $E_j(c^{-2})$ of $H(c) - m\,c^2$ near E_0 satisfy*

$$E_j(c^{-2}) = E_0 + \sum_{k=1}^{\infty}(c^{-2})^k E_{j,k}, \quad j = 1, \ldots, j_0, \quad j_0 \le m_0 \tag{2.12}$$

and if m_j denotes the multiplicity of $E_j(c^{-2})$ then $\sum_{j=1}^{j_0} m_j = m_0$.
In addition, there exist linearly independent vectors

$$f_{jl}(c^{-1}) = \begin{pmatrix} f_{+jl}(c^{-2}) \\ c^{-1}f_{-jl}(c^{-2}) \end{pmatrix}, \quad j = 1, \ldots j_0, \quad l = 1, \ldots, m_j \tag{2.13}$$

s.t. $f_{\pm ji}$ are holomorphic w.r.t. c^{-2} near $c^{-2} = 0$ and

$$H_+ f_{+jl}(0) = E_0 f_{+jl}(0), \quad f_{-jl}(0) = (2m)^{-1}Af_{+jl}(0) \tag{2.14}$$

and

$$(H(c) - mc^2)f_{jl}(c^{-1}) = E_j(c^{-2})f_{jl}(c^{-1}), \quad j = 1, \ldots, j_0, \quad l = 1, \ldots, m_j. \tag{2.15}$$

The eigenvectors $f_{jl}(c^{-1})$ can be chosen to be orthonormal. Finally, the first-order corrections $E_{j,1}$ in (2.12) are explicitly given as the eigenvalues of the matrix

$$(2m)^{-2}\left(Af_r, (V_- - E_0)Af_s\right), \quad r, s = 1, \ldots\ldots, m_0, \tag{2.16}$$

where $\{f_r\}_{r=1}^{m_0}$ is any orthonormal basis of the eigenspace of H_+ to the eigenvalue E_0.

Remarks: (a) The main idea behind Theorem 2.2 was to look for eigenvalues of the resolvent $(H(c) - mc^2 - z)^{-1}$ and applying the strong spectral mapping theorem ([**109**], page 109, Lemma 2) instead of looking directly for eigenvalues of the unbounded Hamiltonian $H(c) - mc^2$.

(b) Theorem 2.2 for $m_0 = 1$ is due to Fritz et al. (see [**40, 41**]). In the general case $m_0 > 1$ only holomorphy of $E_j(c^{-1})$ w.r.t. c^{-1} near $c^{-1} = 0$ and $E_j(c^{-1}) - mc^2 - E_0 =_{c\to\infty} O(c^{-2})$ has been proven in [**41**]. The above extension of this result for $m_0 > 1$ is due to [**144**].

The basic idea to prove holomorphy of $E_j(c^{-1})$ w.r.t. c^{-2} is the following: Since $(H(c)-mc^2-z)^{-1}$ is normal for $z \in \mathbb{C}\backslash\mathbb{R}$, (2.10) implies that the projection $P_j(c^{-1})$ onto the eigenspace of the eigenvalue $(E_j(c^{-1})-z)^{-1}$ is holomorphic w.r.t. c^{-1} near $c^{-1} = 0$. To prove that $E_j(c^{-1})$ is actually holomorphic w.r.t. c^{-2} near $c^{-2} = 0$ one calculates

$$\begin{aligned} \tilde{P}_j(c^{-2}) &:= B(c)P_j(c^{-1})B(c)^{-1} = \begin{pmatrix} 1 & 0 \\ 0 & c \end{pmatrix} \begin{pmatrix} p_j & 0 \\ 0 & 0 \end{pmatrix} \begin{pmatrix} 1 & 0 \\ 0 & 1/c \end{pmatrix} \\ &+ \left[\text{terms holomorphic w.r.t. } c^{-2}\right] = \left[\text{terms holomorphic w.r.t. } c^{-2}\right]. \end{aligned} \tag{2.17}$$

Here $\tilde{P}_j(c^{-2})$ and p_j are the corresponding projections associated with $B(c)(H(c)-mc^2-z)^{-1}B(c)^{-1}$ and $(H_+-z)^{-1}$ of dimension m_j, respectively. Thus, $||\tilde{P}_j(c^{-2})||$ is bounded as $c^{-2} \to 0$ and, hence, Butler's theorem ([**90**], p. 70) proves that $\tilde{P}_j(c^{-2})$ and $(E_j(c^{-1})-z)^{-1}$ are actually holomorphic w.r.t. c^{-2} near $c^{-2} = 0$.

3. Singular Weyl–Titchmarsh–Kodaira theory

The purpose of the remaining sections is to survey some highlights obtained by Fritz for one-dimensional Schrödinger operators

$$H = -\frac{d^2}{dx^2} + q, \qquad \text{in } \mathcal{H} = L^2(a,b), \quad -\infty \le a < b \le \infty. \tag{3.1}$$

In what follows we will denote by τ the underlying differential operator and by H an associated self-adjoint operator (as determined by separated boundary conditions whenever necessary, i.e., when τ is in the limit circle case at one of the endpoints). We will assume that the reader is familiar with the basic results from spectral theory for these operators and refer, e.g., to the textbooks [**102, 134, 140, 142**].

One key object in direct and inverse spectral theory is the Weyl–Titchmarsh m-function. To define it suppose that the endpoint a is regular, that is a is finite and all solutions (as well as their derivatives) extend continuously to this endpoint. For simplicity take a Dirichlet boundary condition, $f(a) = 0$, at a. Then we can choose a basis of solutions of $\tau u = zu$ satisfying the initial conditions

$$\theta(z,a) = \phi'(z,a) = 1, \qquad \theta'(z,a) = \phi(z,a) = 0. \tag{3.2}$$

In particular, $\phi(z,x)$ satisfies the Dirichlet boundary condition at a. Moreover, for $z \in \mathbb{C} \setminus \mathbb{R}$ there will also be a unique (up to scaling) solution $\psi(z,x)$ which will be square integrable near the other end point b and which satisfies a possible boundary condition at b. If we normalize this function according to $\psi(z,a) = 1$ it can be written as the following linear combination of our basis of solutions

$$\psi(z,x) = \theta(z,x) + m(z)\phi(z,x), \tag{3.3}$$

where the Weyl–Titchmarsh m-functions is given by

$$m(z) = \frac{\psi'(z,a)}{\psi(z,a)}. \tag{3.4}$$

It is a key result that $m(z)$ is a Herglotz–Nevanlinna function (i.e., an analytic function mapping the upper half plane into itself) and hence it can be represented as

$$m(z) = \mathrm{Re}(m(\mathrm{i})) + \int_{\mathbb{R}} \Big(\frac{1}{\lambda - z} - \frac{\lambda}{1+\lambda^2}\Big) d\rho(\lambda), \tag{3.5}$$

where ρ is a Borel measure with $\int_{\mathbb{R}}(1+\lambda^2)^{-1}d\rho(\lambda) < \infty$ given by the Stieltjes inversion formula

$$\frac{1}{2}\big(\rho((\lambda_0,\lambda_1)) + \rho([\lambda_0,\lambda_1])\big) = \lim_{\varepsilon\downarrow 0} \frac{1}{\pi}\int_{\lambda_0}^{\lambda_1} \mathrm{Im}\big(m(\lambda+\mathrm{i}\varepsilon)\big) d\lambda. \tag{3.6}$$

Moreover, the map

$$f(x) \mapsto \hat{f}(\lambda) = \int_a^b \phi(\lambda,x) f(x) dx, \tag{3.7}$$

initially defined on functions f with compact support in $[a,b)$, extends to a unitary map from $L^2(a,b)$ to $L^2(\mathbb{R},d\rho)$ which maps our operator H to multiplication with the independent variable λ. In particular, the spectral measure ρ contains all the spectral information of H! In his paper with Zinchenko [**82**], Fritz not only gave a particularly simple proof of this fact but also pointed out a crucial extension which is now known as singular Weyl–Titchmarsh–Kodaira theory.

In fact, the assumption that one endpoint is regular excludes many important examples which have a simple spectrum and where such a transformation should still exist by the spectral theorem. Two examples of central interest in quantum mechanics are the radial part of the Coulomb Hamiltonian, where $q(x) = \frac{l(l+1)}{x^2} - \frac{\gamma}{x}$ on $(0,\infty)$, or the harmonic oscillator, $q(x) = x^2$ on $(-\infty,\infty)$. In both cases it is still possible to define an entire (w.r.t. z) solution $\phi(z,x)$ which is square integrable near the left endpoint a and satisfies a possible boundary condition there. Moreover, a singular Weyl m-function can be defined as before once a second linearly independent solution $\theta(z,x)$ is chosen. In fact, this observation was already made by Kodaira [**91**]. However, as pointed out by Kac [**88**], this is nontrivial if $\theta(z,x)$ is required to be entire as well. In fact, we can always choose a second entire solution but the Wronskian $W(\theta(z),\phi(z)) = \theta(z,x)\phi'(z,x) - \theta'(z,x)\phi(z,x)$ might have zeros! In [**82**] this problem was overcome by showing that to define the spectral measure via (3.6) it suffices if $\theta(z,x)$ is analytic in a neighborhood of the real line. And it is easy to check that if $\theta_c(z,x)$, $\phi_c(z,x)$ are two entire solutions defined by the initial conditions $\theta_c(z,c) = \phi_c'(z,c) = 1$ and $\theta_c'(z,c) = \phi_c(z,c) = 0$, then

$$\theta(z,x) = \frac{\phi'(z,c)}{\phi(z,c)^2 + \phi'(z,c)^2}\theta_c(z,x) + \frac{\phi(z,c)}{\phi(z,c)^2+\phi'(z,c)^2}\phi_c(z,x) \tag{3.8}$$

will do the trick. In fact, by construction it is meromorphic in $\mathbb{C}$ with all poles away from the real axis and one easily checks $W(\theta(z),\phi(z)) = 1$ (evaluate the Wronskian at $x = c$ and recall that it is independent of x).

THEOREM 3.1 ([**82**]). *Suppose H has an entire solution $\phi(z,x)$ which is square integrable near the left endpoint a and satisfies a possible boundary condition there. Let $\theta(z,x)$ be defined as in* (3.8).

Then there exists a Weyl m-function which is meromorphic in $\mathbb{C}\backslash\mathbb{R}$ *(without any poles accumulating near the real line) defined via* (3.3) *and an associated spectral measure defined via* (3.6). *Moreover, the map* (3.7) *initially defined on functions* f *with compact support in* $[a, b)$, *extends to a unitary map from* $L^2(a, b)$ *to* $L^2(\mathbb{R}, d\rho)$ *which maps our operator* H *to multiplication with the independent variable* λ.

The elegant proof from [**82**] alluded to before now simply involves the following two well-known facts: First, the resolvent of H which is given by

$$(H - z)^{-1} f(x) = \int_a^b G(z, x, y) f(y) dy, \tag{3.9}$$

where

$$G(z, x, y) = \begin{cases} \phi(z, x)\psi(z, y), & y \geq x, \\ \phi(z, y)\psi(z, x), & y \leq x. \end{cases} \tag{3.10}$$

And second, Stone's formula

$$\langle f, F(H) f\rangle = \frac{1}{\pi} \lim_{\varepsilon \downarrow 0} \int_{\mathbb{R}} F(\lambda) \mathrm{Im}\langle f, (H - \lambda - \mathrm{i}\varepsilon)^{-1} f\rangle) d\lambda \tag{3.11}$$

for any bounded continuous function $F \in C_b(\mathbb{R})$. Now insert the definitions and compute the limit to obtain

$$\langle f, F(H) f\rangle = \int_{\mathbb{R}} F(\lambda) |\hat{f}(\lambda)|^2 d\rho(\lambda), \tag{3.12}$$

which in turn implies the theorem.

This seminal result initiated the development of Weyl–Titchmarsh theory at a singular endpoint and has triggered a number of results recently [**24**–**28**, **31**–**33**, **92**–**95**, **99**].

4. Inverse spectral theory and trace formulas

Next we turn to trace formulas, another area where Fritz has made profound contributions. To this end suppose that $(a, b) = \mathbb{R}$ and that the potential $q(x)$ is periodic: $q(x + 1) = q(x)$. Then it is well known by Floquet theory that the spectrum consists of an infinite number of bands

$$\sigma(H) = [E_0, E_1] \cup [E_2, E_3] \cup \cdots, \tag{4.1}$$

where $E_0 < E_1 \leq E_2 < E_3 \leq \cdots$. Moreover, if we restrict our operator to the interval $(x, x + 1)$ and impose Dirichlet boundary conditions $f(x) = f(x + 1) = 0$ at the endpoints, then we obtain a sequence of eigenvalues $\mu_1(x) < \mu_2(x) < \cdots$ depending on the base point x. Again it follows from Floquet theory that the Dirichlet eigenvalues $\mu_j(x)$ lie in the closures of the spectral gaps: $E_{2j-1} \leq \mu_j(x) \leq E_{2j}$.

Then the following trace formula

$$q(x) = E_0 + \sum_{j=1}^{\infty} [E_{2j-1} + E_{2j} - 2\mu_j(x)] \tag{4.2}$$

holds under some suitable assumptions on the potential (e.g., $q \in C^1$). This trace formula was first obtained in the case where the spectrum has only finitely many gaps (and thus the above sum is finite) and later on generalized to periodic and even some classes of almost-periodic potentials. Furthermore, trace formulas were also known in the context of scattering theory, but in this case the right-hand side contains both a sum corresponding to the eigenvalues plus an integral corresponding to the continuous spectrum. Hence, Fritz was asking for a generalization of (4.2) to more general potentials.

To this end let us briefly sketch one way of proving (4.2). Denote by $\psi_\pm(z,x)$ the Floquet solution of the underlying differential equation and recall that the diagonal Green function is given by

$$G(z,x,x) = \frac{\psi_+(z,x)\psi_-(z,x)}{W(\psi_+(z),\psi_-(z))}, \tag{4.3}$$

where

$$W_x(f,g) = f(x)g'(x) - f'(x)g(x) \tag{4.4}$$

denotes the Wronskian of two absolutely continuous functions. Recall that the Wronskian of two solutions corresponding to the same spectral parameter z is independent of x. Then, using the fact that solutions of the underlying differential equation $\tau u = zu$ corresponding to constant initial conditions are entire functions of order one half, one can obtain the following product representation

$$G(z,x,x) = \frac{1}{2\sqrt{E_0 - z}} \prod_{j=1}^{\infty} \frac{z - \mu_j(x)}{\sqrt{(E_{2j-1} - z)(E_{2j} - z)}}. \tag{4.5}$$

Comparing this product representation with the well-known asymptotics

$$G(z,x,x) = \frac{1}{2\sqrt{-z}}\left(1 + \frac{q(x)}{2z} + o(z^{-1})\right) \tag{4.6}$$

establishes (4.2).

How can one possibly generalize this argument? Since (4.6) holds for general potentials, a generalization of (4.5) needs to be found and, as observed by Fritz, the correct starting point is the following exponential Herglotz representation

$$G(z,x,x) = |G(\mathrm{i},x,x)| \exp\left(\int_{\mathbb{R}} \left(\frac{1}{\lambda - z} - \frac{\lambda}{1+\lambda^2}\right)\xi(\lambda,x)d\lambda\right). \tag{4.7}$$

Here the ξ-function is given by the Stieltjes inversion formula

$$\xi(\lambda,x) = \lim_{\varepsilon\downarrow 0} \frac{1}{\pi} \arg G(\lambda + \mathrm{i}\varepsilon, x, x), \tag{4.8}$$

where the limit exists for a.e. $\lambda \in \mathbb{R}$. Note that since $G(.,x,x)$ is a Herglotz–Nevanlinna function, we have $0 \le \xi(\lambda,x) \le 1$.

Now a trace formula which works for arbitrary potentials follows by comparing asymptotics as before. This is one of the key results obtained together with Barry Simon in [**62**].

THEOREM 4.1 ([**62**]). *Suppose q is continuous and bounded from below. Choose $E_0 \le \inf\sigma(H)$. Then*

$$V(x) = E_0 + \lim_{\varepsilon\downarrow 0} \int_{E_0}^{\infty} \mathrm{e}^{-\varepsilon\lambda}(1 - 2\xi(\lambda,x))d\lambda. \tag{4.9}$$

In fact, even more was shown in [**62**]. Namely, let $H_x^D = H_{x,-}^D \oplus H_{x,+}^D$ be the operator H restricted according to $L^2(a,b) = L^2(a,c) \oplus L^2(c,b)$ by imposing Dirichlet boundary conditions at the point $x \in (a,b)$. Then $\xi(\lambda,x)$ was identified as the Krein spectral shift function of the pair (H, H_x^D), which in turn opened the door for deriving numerous novel trace formulas for Schrödinger operators: [**37, 44–47, 50–53**].

To demonstrate the usefulness and generality of this theorem let us extract two special cases. First let us see how to obtain (4.2) by evaluating (4.8) using (4.5) (note that the limit is not needed since $G(\lambda, x, x)$ has a continuous extension to the real line away from the band edges). We begin with $\lambda < E_0$ in which case $G(\lambda, x, x)$ is positive implying $\xi(\lambda, x) = 0$ for $\lambda \in (-\infty, E_0)$. At E_0 the Green function $G(\lambda, x, x)$ will become purely imaginary implying $\xi(\lambda, x) = \frac{1}{2}$ for $\lambda \in (E_0, E_1)$. After E_1 it will become negative and change sign at its zero $\mu_1(x)$, implying $\xi(\lambda, x) = 1$ for $\lambda \in (E_1, \mu_1(x))$ and $\xi(\lambda, x) = 0$ for $\lambda \in (\mu_1(x), E_2)$. Clearly this pattern keeps repeating and we obtain (4.2) (assuming $q \in C^1$ the spectral gaps will close sufficiently fast, such that the limit in (4.9) can be taken inside the integral).

A novel application is to confining potentials, like the harmonic oscillator, satisfying $q(x) \to +\infty$ as $|x| \to \infty$. Then H has purely discrete spectrum $E_0 < E_1 < \cdots$ and the same is true for H_x^D whose eigenvalues $\mu_1(x) < \mu_2(x) < \cdots$ are known to satisfy $E_{j-1} \le \mu_j(x) \le E_j$. Then note that $G(z, x, x)$ has first order poles at the eigenvalues E_j and first order zeros at the Dirichlet eigenvalues $\mu_j(x)$ (since either $\psi_-(\lambda, x)$ or $\psi_+(\lambda, x)$ must vanish at $\lambda = \mu_j(x)$ — if both should vanish they must be linearly dependent and thus such a point must also be an eigenvalue). Hence $G(z, x, x)$ will be real-valued on the real line and change sign at every pole E_j and every zero $\mu_j(x)$ implying

$$V(x) = E_0 + \lim_{\varepsilon \downarrow 0} \frac{1}{\varepsilon} \sum_{j \in \mathbb{N}} 2\mathrm{e}^{-\varepsilon \mu_j(x)} - \mathrm{e}^{-\varepsilon E_{j-1}} - \mathrm{e}^{-\varepsilon E_j}, \tag{4.10}$$

which is just an abelianized version of (4.2).

This brings us to another item discussed in [**62**], namely the consequences of the above ideas for inverse spectral theory. While the inverse spectral theory is well understood for periodic operators, this is not the case for confining potentials. For example, one open question is to describe the isospectral class of the harmonic oscillator. To see how the ξ-function can help understanding this problem, we begin with the observation that $\xi(\lambda, x)$ determines $G(z, x, x)$ by virtue of (4.7) (the unknown constant can be determined from the asymptotics (4.6)).

Now let $\psi_\pm(z, x)$ be the solutions of $\tau u = zu$ which are square integrable near $\pm\infty$ and observe that (3.10) reads

$$G(z, x, x) = \frac{-1}{m_+(z, x) + m_-(z, x)}, \tag{4.11}$$

where

$$m_\pm(z, x) = \pm \frac{\psi_\pm'(z, x)}{\psi_\pm(z, x)} \tag{4.12}$$

are the Weyl m-functions of $H_{x,\pm}^D$. Hence if you fix x, say $x = 0$, then $\xi(\lambda, 0)$ determines the sum of the spectral measures $\rho_{0,\pm}$ corresponding to $H_{0,\pm}^D$ and it remains to split this information into its two pieces (recall that $\rho_{0,\pm}$ determine $H_{0,\pm}^D$ and thus H). Clearly this is impossible in general without additional data. However, it turns out that things get particularly simple if the following reflectionless property is assumed:

$$m_-(\lambda + \mathrm{i}0, x) = -\overline{m_-(\lambda + \mathrm{i}0, x)}, \qquad \text{a.e. } \lambda \in \sigma_{ess}(H), \tag{4.13}$$

for one (it then follows for all) x. This covers for example periodic operators, reflectionless potentials from scattering theory (which is where the name comes from), or operators with purely discrete spectrum. In fact, this condition tells us that the absolutely continuous part of the measure needs to be split equally. Moreover, assuming that there is no singularly continuous part it remains to assign a sign $\sigma_j \in \{\pm\}$ to every Dirichlet eigenvalue $\mu_j = \mu_j(0)$ which encodes if it is an eigenvalue of $H^D_{0,-}$ or $H^D_{0,+}$ (if it is an eigenvalue of both, which can only happen if it is also an eigenvalue of H as pointed out before, a number $\sigma_j \in (-1,1)$ is needed to encode how the mass should be split).

By construction we obtain that the eigenvalues $\{E_j\}_{j\in\mathbb{N}_0}$ together with the Dirichlet data $\{(\mu_j, \sigma_j)\}_{j\in\mathbb{N}}$ uniquely determine H. The remaining problem to determine the isospectral class of a given operator H_0 is to describe the set of admissible Dirichlet data. One key contribution towards answering this question is the Dirichlet deformation method to be discusses in the next section, which shows that all restrictions (apart form the obvious ones already listed above) must be of an asymptotic nature since any finite part of the Dirichlet data can be changed at will.

For further information on these circle of ideas see [**16, 17, 37, 44–47, 50–53, 55, 59, 69, 107, 111–113, 124, 132, 145**]. In particular, we mention also the review by Fritz [**38**].

5. Commutation methods

Another topic where Fritz made seminal contributions is commutation methods. These are methods of inserting (and removing) eigenvalues in spectral gaps of a given one-dimensional Schrödinger operator H. They play a prominent role in diverse fields such as the inverse scattering approach (see, e.g., [**21**], [**103**] and the references therein), supersymmetric quantum mechanics (see, e.g., [**61**] and the references therein), level comparison theorems (cf. [**6**] and the literature cited therein), and in connection with Bäcklund transformations for the KdV hierarchy (see, e.g., [**73**], [**61**] and the references therein).

Historically, these methods of inserting eigenvalues go back to Jacobi [**87**] and Darboux [**19**] with decisive later contributions by Crum [**18**], Schmincke [**118**], and, especially, Deift [**20**]. Two particular methods, shortly to be discussed in an informal manner in (5.1)–(5.5) below, turned out to be of special importance: The single commutation method, also called the Crum–Darboux method [**18**], [**19**] (actually going back at least to Jacobi [**87**]) and the double commutation method, to be found, e.g., in the seminal work of Gel'fand and Levitan [**35**].

The single commutation method relies on existence of a positive solution ψ of $H\psi = \lambda\psi$ which confines its applicability to the insertion of eigenvalues below the spectrum of H (assuming H to be bounded from below). Introducing

$$A = \frac{d}{dx} + \phi, \quad A^* = -\frac{d}{dx} + \phi, \qquad \text{where } \phi = \frac{d}{dx}\log\psi, \tag{5.1}$$

a straightforward calculation reveals

$$H = A^*A + \lambda = -\frac{d^2}{dx^2} + q, \quad \hat{H} = AA^* + \lambda = -\frac{d^2}{dx^2} + \hat{q}, \tag{5.2}$$

with

$$\hat{q} = q - 2\frac{d^2}{dx^2}\log\psi. \tag{5.3}$$

Thus (taking proper domain considerations into account) we can define two operators $H, \hat{H}$ on $\mathcal{H} = L^2(a,b)$ which turn out unitarily equivalent when restricted to the orthogonal complement of the eigenspaces corresponding to λ. In fact, using the polar decomposition $A = U|A|$, where $|A| = (A^*A)^{1/2}$ and $U : \ker(A)^\perp \to \ker(A^*)^\perp$ is unitary, one infers $\hat{H}P = (AA^* + \lambda)P = U|A||A|U^* + \lambda P = UHU^*$ with $P = UU^*$, the projection onto $\ker(A^*)^\perp$. Moreover, $H - \lambda, \hat{H} - \lambda \geq 0$ which is equivalent to the existence of the positive solution ψ [**81**]. Formulas (5.1)–(5.3) constitute the single commutation method.

The double commutation method on the other hand, allows one to insert eigenvalues into *any* spectral gap of H. To this end we assume that ψ is square integrable near a and consider two more expressions A_γ, A_γ^* as above with $\psi_\gamma = \psi/(1 + \gamma\int_a^x \psi(t)^2 dt)$. This implies

$$H = AA^* + \lambda = A_\gamma^* A_\gamma + \lambda, \quad H_\gamma = A_\gamma A_\gamma^* + \lambda = -\frac{d^2}{dx^2} + q_\gamma, \tag{5.4}$$

where

$$q_\gamma = q - 2\frac{d^2}{dx^2}\log\Big(1 + \gamma\int_a^x \psi(t)^2 dt\Big). \tag{5.5}$$

The considerations for the single commutation method show that H and H_γ are unitarily equivalent when restricted to the orthogonal complement of the eigenspaces corresponding to λ as long as ψ is positive. However, observe that q_γ is well defined even if ψ has zeros, and it is natural to conjecture that the last claim continuous to hold even in the case where all intermediate operators are ill-defined. This turned out much harder to prove and was achieved by Fritz [**36**] on the basis of Weyl–Titchmarsh m-function techniques.

In what follows we will denote by τ the underlying differential operator and by H an associated self-adjoint operator (as determined by separated boundary conditions whenever necessary, i.e., when τ is in the limit circle case at one of the endpoints).

To this end, suppose a is regular (i.e., a is finite and $q \in L^1(a,c)$ such that all solutions extend continuously to the endpoint a) and introduce the Weyl–Titchmarsh m-function as

$$m(z) = \frac{\psi'(z,a)}{\psi(z,a)}, \tag{5.6}$$

where $\psi(z,x)$ is the unique (up to scaling) solution of $\tau\psi = z\psi$ which is square integrable and satisfies a possible boundary condition at b. Then it is well known that $m(z)$ carries all the information on H and hence it suffices to find an explicit expression for the Weyl–Titchmarsh m-function of H_γ in terms of the one of H. This strategy was carried out in [**36**] and the following theorem was obtained as one of the main results.

THEOREM 5.1 ([**36, 71**]). *Let $\lambda \in \mathbb{R}$ and $\psi(\lambda,.)$ be a solution satisfying the following conditions:*

(i) *$\psi, \psi' \in AC_{loc}(a,b)$ and ψ is a real-valued solution of $\tau\psi = \lambda\psi$.*

(ii) *ψ is square integrable near a and fulfills the boundary condition (of H) at a and b if any (i.e., if τ is limit circle (l.c.) at a respectively b).*

Define

$$\psi_\gamma(\lambda, x) = \frac{\psi(\lambda, x)}{1 + \gamma \int_a^x \psi(\lambda, t)^2 dt}. \tag{5.7}$$

Let $P(\lambda)$, $P_\gamma(\lambda)$ be the projections on the subspaces spanned by $\psi(\lambda)$, $\psi_\gamma(\lambda)$, respectively (if one of these functions is not square integrable, set the corresponding projection equal to zero). Then the operator H_γ defined by

$$H_\gamma f = \tau_\gamma f, \quad \mathrm{dom}(H_\gamma) = \begin{array}{l} \{f \in \mathcal{H} | \, f, pf' \in AC_{loc}((a,b)); \tau_\gamma f \in \mathcal{H}; \\ W_a(\psi_\gamma(\lambda), f) = W_b(\psi_\gamma(\lambda), f) = 0\}, \end{array} \tag{5.8}$$

with q_γ given by (5.5) *is self-adjoint. Moreover, H_γ has the eigenfunction $\psi_\gamma(\lambda)$ associated with the eigenvalue λ. If $\psi(\lambda) \notin \mathcal{H}$ (and hence τ is limit point (l.p.) at b) we have*

$$H_\gamma(1 - P_\gamma(\lambda)) = U_\gamma H U_\gamma^{-1}(1 - P_\gamma(\lambda)) \tag{5.9}$$

for some unitary operator U_γ and thus

$$\begin{array}{rclrcl} \sigma(H_\gamma) & = & \sigma(H) \cup \{\lambda\}, & \sigma_{ac}(H_\gamma) & = & \sigma_{ac}(H), \\ \sigma_p(H_\gamma) & = & \sigma_p(H) \cup \{\lambda\}, & \sigma_{sc}(H_\gamma) & = & \sigma_{sc}(H). \end{array} \tag{5.10}$$

(Here $\sigma_{ac}(.), \sigma_{sc}(.)$ denotes the absolutely and singularly continuous spectrum, respectively.) If $\psi(\lambda) \in \mathcal{H}$ there is a unitary operator $\tilde{U}_\gamma = U_\gamma \oplus \sqrt{1 + \gamma \|\psi(\lambda)\|^2} U_\gamma$ on $(1 - P(\lambda))\mathcal{H} \oplus P(\lambda)\mathcal{H}$ such that

$$H_\gamma = \tilde{U}_\gamma H \tilde{U}_\gamma^{-1}. \tag{5.11}$$

As already mentioned, the main part of this result is due to Fritz [**36**]. In Gesztesy and Teschl [**71**] the transformation operator was identified to be

$$(U_\gamma f)(x) = f(x) - \gamma \psi_\gamma(\lambda, x) \int_a^x \psi(\lambda, t) f(t) dt, \tag{5.12}$$

and the above theorem reduces to checking that U_γ has the claimed properties. Moreover, the whole method was extended to Sturm–Liouville operators and some technical assumptions were relaxed. For the connections with singular Weyl–Titchmarsh–Kodaira theory see [**93**].

A further decisive contribution was made by Gesztesy, Simon, and Teschl [**70**] motivated by a commutation method first introduced by Finkel, Isaacson, and Trubowitz [**30**] in connection with an explicit realization of the isospectral torus of periodic potentials. This method was again used by Buys and Finkel [**13**] (see also Iwasaki [**86**]) in the context of periodic finite-gap potentials and by Pöschel and Trubowitz [**106**] and Ralston and Trubowitz [**108**] for various boundary value problems on compact intervals. As in the previous case, this method formally consists of two single commutations, but this time at different values of the spectral parameter μ and ν. The resulting operator will have the Wronskian $W(\psi(\mu), \psi(\nu))$ in its denominator and it turns out that it will be well-defined as long as both μ and ν lie in the same spectral gap of H. This is related to the fact, that $(H - \mu)(H - \nu)$ will still be nonnegative under this assumption and hence can still be factorized as $(H - \mu)(H - \nu) = B^* B$, as explained later by Schmincke [**119**]. Moreover, the fact that the zeros of the Wronskian of two solutions is related to the spectrum lead to the development of renormalized oscillation theory [**72**] to be discussed in Section 6 below.

To explain the relevance of this method recall that an important role in inverse spectral theory is played by the so called Dirichlet data as introduced at the end of the previous section. The key result of [**70**] states that the above commutation method can be used to move a given Dirichlet eigenvalue to any other admissible position within its gap as well as change its *sign*. Hence it is also known as the Dirichlet deformation method. We refrain from further details here and refer to Fritz's own review [**38**] instead. For generalizations to other operators see [**114, 115, 128, 129, 133**].

For further generalizations of commutation methods see [**3, 8, 9**]. Finally, these methods are also relevant in connection with inverse scattering theory (see, e.g., [**1, 103**] and the references cited therein) and yield a direct construction of N-soliton solutions relative to arbitrary background solutions of the (generalized) KdV hierarchy along the methods of [**73**] (see also [**79**]). Generalizations to other integrable equations can be found in [**10, 14, 60, 104, 116, 131**].

6. Oscillation theory

Another area where Fritz had an important impact is oscillation theory. As indicated at the end of the previous section, a crucial ingredient in the development of the Dirichlet deformation method was the fact that the Wronskian of two solutions $W(\psi(\mu), \psi(\nu))$ is nonzero as long as there is no part of the spectrum inside the interval (μ, ν). This observation naturally lead to the question how the number of points in the spectrum between μ and ν are related to the number of zeros of the Wronskian. This question was answered by Gesztesy, Simon, and Teschl in [**72**].

Denote by $W_0(f, g)$ be the number of zeros of the Wronskian in the open interval (a, b) not counting multiplicities of zeros. Given $E_1 < E_2$, we let $N_0(E_1, E_2) = \dim \operatorname{ran} P_{(E_1,E_2)}(H)$ be the dimension of the spectral projection $P_{(E_1,E_2)}(H)$ of H. Then the main result from [**72**] is the following:

THEOREM 6.1 ([**72**]). *Let $\psi_-(\lambda, x)$, $\psi_+(\lambda, x)$ be solutions of $\tau\psi = \lambda\psi$ which are square integrable near a, b and satisfy a possible boundary condition at a, b, respectively. (Such solutions will in general only exist inside a spectral gap). Suppose $E_1 < E_2$. Then*

$$W_0(\psi_-(E_1), \psi_+(E_2)) = N_0(E_1, E_2). \tag{6.1}$$

This result constituted an important generalization suitable for counting the number of eigenvalues inside essential gaps. In fact, there has been considerable efforts to generalize classical oscillation theory to singular operators by Hartman and others around 1950. However, while these efforts were successful below the essential spectrum, they were only partly successful for counting eigenvalues in essential spectral gaps. For example Hartmann [**84**] could show that for a regular endpoint a and a limit point endpoint b, counting the difference of zeros of the solutions $\psi_-(E_2, x)$ and $\psi_-(E_1, x)$ on (a, c) will have $N_0(E_1, E_2)$ as its lim inf when $c \to b$. As a simple example, the above theorem allows to extend this result to the case when a is non-oscillatory (thus covering important examples like radial Schrödinger equations).

Finally let us outline a proof for the above result. The first step is to prove this result in the case where both endpoints are regular. This can be done using the usual Prüfer techniques and we refer to the textbook [**135**] for a simple proof. Next one can approximate H on (a,b) by regular operators on $(c,d) \subseteq (a,b)$. Again this technique is standard (for a nice review see [**143**]) and the restricted operators will converge to H in the strong resolvent sense. Since the spectrum cannot expand for such limits, one obtains $W_0 \geq N_0$. To obtain the reverse inequality one uses a variational argument. To this end observe that at every zero of the Wronskian the two solutions $\psi_-(E_1)$ and $\psi_+(E_2)$ can be glued together (by scaling one to match the value of the other) to give a trial function in the domain of H (since the derivatives will then automatically match at this point). Hence one obtains a W_0 dimensional space of trial functions ψ satisfying

$$\|(H - \frac{E_2 - E_1}{2})\psi\| < \frac{E_2 - E_1}{2}\|\psi\| \tag{6.2}$$

and the required reverse inequality $W_0 \leq N_0$ follows from the spectral theorem. For further recent reviews of these methods we refer to the ones by Fritz [**38**] and Simon [**125**].

Again this result has triggered several extensions. In particular, it was shown by Krüger and Teschl [**96**] that one can take solutions of different operators if the right-hand side is interpreted as spectral shift between these two operators. We refer to [**96, 97**] and the references therein. The question whether eigenvalues accumulate at the boundary of an essential spectral gap based on these methods is considered in [**98, 117**]. In this respect we should also mention the beautiful result by Fritz and Ünal [**78**] which gives the most general version of Kneser's theorem. Extensions to other operators can be found in [**4, 29, 126, 127, 130**].

7. Non-self-adjoint operators

While all results so far were concerned with self-adjoint operators, Fritz always was quite active in the area of non-self-adjoint operators as well. As a prototypical example we mention his beautiful work with Tkachenko [**75, 76**] solving the long-standing open question when a non-self-adjoint Hill operator is a spectral operator of scalar type in the sense of Dunford. This problem had been open for about 40 years.

We have already encountered Hill's equation (i.e, (3.1) with periodic potential $q(x+1) = q(x)$) in Section 4. But now we will allow q to be complex valued (in addition it is assumed to be locally square integrable). Recall that one of the key objects from Floquet theory is the Floquet discriminant

$$\Delta(z) = \frac{\theta(z,1) + \phi'(z,1)}{2}, \tag{7.1}$$

where $\theta(z,x)$, $\phi(z,x)$ is a fundamental system of solutions satisfying the initial conditions (3.2) at $a = 0$. It was shown by Serov [**122**] that the spectrum of H is given by $\sigma(H) = \{z \in \mathbb{C} | \Delta(z) \in [-1,1]\}$. In particular, the spectrum consists of a number of analytic arcs which, however, might intersect in inner points [**105**].

The following version of this criterion involves the spectrum $\sigma(H)$ of H, the Dirichlet spectrum $\{\mu_k\}_{k\in\mathbb{N}}$, the periodic spectrum $\{E_k(0)\}_{k\in\mathbb{N}_0}$, and the antiperiodic spectrum $\{E_k(\pi)\}_{k\in\mathbb{N}_0}$, and is connected with the algebraic and geometric multiplicities of the eigenvalues in the sets $\sigma(H(t))$, where $H(t)$, $t \in [0, 2\pi]$, denotes the densely defined closed realization of (3.1) in $L^2(0,1)$ in terms of the t-dependent boundary conditions $f(1) = e^{it} f(0)$, $f'(1) = e^{it} f'(0)$, $t \in [0, 2\pi]$. It is known that the union of the periodic and antiperiodic spectra is formed by the numbers

$$\lambda_k^{\pm} = \left(k + \frac{\int_0^1 q(x)dx}{2k} + \frac{s_k^{\pm}}{k} \right)^2, \ k \in \mathbb{N}, \quad \sum_{k \in \mathbb{N}} |s_k^{\pm}|^2 < \infty, \tag{7.2}$$

$$\{E_k(0)\}_{k \in \mathbb{N}_0} = \{\lambda_0^+, \lambda_{2k}^+, \lambda_{2k}^-\}_{k \in \mathbb{N}}, \quad \{E_k(\pi)\}_{k \in \mathbb{N}_0} = \{\lambda_{2k+1}^+, \lambda_{2k+1}^-\}_{k \in \mathbb{N}_0}, \tag{7.3}$$

THEOREM 7.1. *A Hill operator H is a spectral operator of scalar type if and only if the following conditions (i) and (ii) are satisfied:*
(i) For all $t \in [0, 2\pi]$ and all $E_k(t) \in \sigma(H(t))$, each root function of the operator $H(t)$ associated with $E_k(t)$ is an eigenfunction of $H(t)$.
(ii) Let

$$\mathcal{Q} = \{k \in \mathbb{N} \,|\, d_k = \operatorname{dist}(\delta_k, \sigma(H)) > 0\}. \tag{7.4}$$

Then

$$\sup_{k \in \mathcal{Q}} \frac{|\lambda_k^+ - \lambda_k^-|}{\operatorname{dist}(\delta_k, \sigma(H))} < \infty, \quad \sup_{k \in \mathcal{Q}} \frac{|\mu_k - \lambda_k^-|}{\operatorname{dist}(\delta_k, \sigma(H))} < \infty, \quad \sup_{k \in \mathcal{Q}} \frac{|\mu_k - \lambda_k^+|}{\operatorname{dist}(\delta_k, \sigma(H))} < \infty. \tag{7.5}$$

Here a root function of $H(t)$ associated with the eigenvalue $E_k(t)$ denotes any element ψ satisfying $(H(t) - E_k(t))^m \psi = 0$ for some $m \in \mathbb{N}$ (i.e., any element in the algebraic eigenspace of $H(t)$ corresponding to $E_k(t)$). Of course, the conditions (i) and (ii) are always satisfied in the self-adjoint context where q is real-valued.

It can be shown that the conditions (i) and (ii) imposed in Theorem 7.1 yield the remarkable fact that the property of a Hill operator being a spectral operator is independent of smoothness (or even analyticity) properties of the potential q. In addition, also a functional model for periodic Schrödinger operators that are spectral operators of scalar type was established and the corresponding eigenfunction expansion was developed in [**75**], [**76**].

In their recent paper [**77**], under the assumption that $q \in L^2(0,1)$, necessary and sufficient conditions in terms of spectral data for (non-self-adjoint) Schrödinger operators in $L^2(0,1)$ with periodic and antiperiodic boundary conditions to possess a Riesz basis of root vectors were derived. Without entering details, we mention that this problem generated an enormous amount of interest and remained open for a long time. Their key result reads:

THEOREM 7.2. *Assume $q \in L^2(0,1)$, then the following results hold:*
(i) The system of root vectors of $H(0)$ contains a Riesz basis in $L^2(0,1)$ if and only if

$$\sup_{\substack{k \in \mathbb{N}, \\ \lambda_{2k}^+ \neq \lambda_{2k}^-}} \frac{|\mu_{2k} - \lambda_{2k}^{\pm}|}{|\lambda_{2k}^+ - \lambda_{2k}^-|} < \infty. \tag{7.6}$$

(ii) The system of root vectors of $H(\pi)$ contains a Riesz basis in $L^2(0,1)$ if and only if

$$\sup_{\substack{k \in \mathbb{N}, \\ \lambda_{2k+1}^+ \neq \lambda_{2k+1}^-}} \frac{|\mu_{2k+1} - \lambda_{2k+1}^{\pm}|}{|\lambda_{2k+1}^+ - \lambda_{2k+1}^-|} < \infty. \tag{7.7}$$

Here $\sup_{k\in\mathbb{N},\,\lambda_j^+\neq\lambda_j^-}$ signifies that all subscripts $j \in \mathbb{N}$ in (7.6) and (7.7) for which λ_j^+ and λ_j^- coincide are simply excluded from the supremum considered.

One observes that only the simple periodic (resp., antiperiodic) eigenvalues enter in the necessary and sufficient conditions (7.6) (resp., (7.7)) for the existence of a Riesz basis of root vectors of $H(0)$ (resp., $H(\pi)$). The multiple periodic (resp., antiperiodic) eigenvalues play no role in deciding whether or not the system of root vectors of $H(0)$ (resp., $H(\pi)$) constitutes a Riesz basis in $L^2(0,1)$. In addition, only every other Dirichlet eigenvalue (i.e., half the Dirichlet spectrum) enters the criterion (7.6) (resp., (7.7)).

For additional detailed results in this direction we also refer to [**23**] and the extensive literature cited therein.

Acknowledgments. We thank Barry Simon for valuable comments improving the presentation of the material.

References

[1] Tuncay Aktosun, Martin Klaus, and Cornelis van der Mee, *Scattering and inverse scattering in one-dimensional nonhomogeneous media*, J. Math. Phys. **33** (1992), no. 5, 1717–1744, DOI 10.1063/1.529650. MR1158993 (93g:81200)

[2] S. Albeverio, F. Gesztesy, R. Høegh-Krohn, and H. Holden, *Solvable models in quantum mechanics*, 2nd ed., AMS Chelsea Publishing, Providence, RI, 2005. With an appendix by Pavel Exner. MR2105735 (2005g:81001)

[3] S. Albeverio, R. Hryniv, and Ya. Mykytyuk, *Factorisation of non-negative Fredholm operators and inverse spectral problems for Bessel operators*, Integral Equations Operator Theory **64** (2009), no. 3, 301–323, DOI 10.1007/s00020-009-1696-8. MR2521239 (2010h:47027)

[4] Kerstin Ammann and Gerald Teschl, *Relative oscillation theory for Jacobi matrices*, Difference equations and applications, Uğur-Bahçeşehir Univ. Publ. Co., Istanbul, 2009, pp. 105–115. MR2664180 (2011j:47096)

[5] Mark S. Ashbaugh, Fritz Gesztesy, Marius Mitrea, and Gerald Teschl, *Spectral theory for perturbed Krein Laplacians in nonsmooth domains*, Adv. Math. **223** (2010), no. 4, 1372–1467, DOI 10.1016/j.aim.2009.10.006. MR2581375 (2011d:47052)

[6] Bernhard Baumgartner, *Level comparison theorems*, Ann. Physics **168** (1986), no. 2, 484–526, DOI 10.1016/0003-4916(86)90041-2. MR844889 (87j:81046)

[7] Philippe Bechouche, Norbert J. Mauser, and Frédéric Poupaud, *(Semi)-nonrelativistic limits of the Dirac equation with external time-dependent electromagnetic field*, Comm. Math. Phys. **197** (1998), no. 2, 405–425, DOI 10.1007/s002200050457. MR1652738 (2000g:81048)

[8] Paul A. Binding, Patrick J. Browne, and Bruce A. Watson, *Darboux transformations and the factorization of generalized Sturm-Liouville problems*, Proc. Roy. Soc. Edinburgh Sect. A **140** (2010), no. 1, 1–29, DOI 10.1017/S0308210508000905. MR2592710 (2010m:34042)

[9] Amin Boumenir and Vu Kim Tuan, *A trace formula and Schmincke inequality on the half-line*, Proc. Amer. Math. Soc. **137** (2009), no. 3, 1039–1049, DOI 10.1090/S0002-9939-08-09659-7. MR2457445 (2009g:34214)

[10] W. Bulla, F. Gesztesy, H. Holden, and G. Teschl, *Algebro-geometric quasi-periodic finite-gap solutions of the Toda and Kac-van Moerbeke hierarchies*, Mem. Amer. Math. Soc. **135** (1998), no. 641, x+79. MR1432141 (99b:58109)

[11] W. Bulla, F. Gesztesy, and K. Unterkofler, *On relativistic energy band corrections in the presence of periodic potentials*, Lett. Math. Phys. **15** (1988), no. 4, 313–324, DOI 10.1007/BF00419589. MR952454 (90c:81038)

[12] W. Bulla, F. Gesztesy, and K. Unterkofler, *Holomorphy of the scattering matrix with respect to c^{-2} for Dirac operators and an explicit treatment of relativistic corrections*, Comm. Math. Phys. **144** (1992), no. 2, 391–416. MR1152379 (93b:81302)

[13] Mutiara Buys and Allan Finkel, *The inverse periodic problem for Hill's equation with a finite-gap potential*, J. Differential Equations **55** (1984), no. 2, 257–275, DOI 10.1016/0022-0396(84)90083-4. MR764126 (86a:34052)

[14] Radu C. Cascaval, Fritz Gesztesy, Helge Holden, and Yuri Latushkin, *Spectral analysis of Darboux transformations for the focusing NLS hierarchy*, J. Anal. Math. **93** (2004), 139–197, DOI 10.1007/BF02789306. MR2110327 (2006e:37128)

[15] Ross J. Cirincione and Paul R. Chernoff, *Dirac and Klein-Gordon equations: convergence of solutions in the nonrelativistic limit*, Comm. Math. Phys. **79** (1981), no. 1, 33–46. MR609226 (84b:81023)

[16] Steve Clark and Fritz Gesztesy, *Weyl-Titchmarsh M-function asymptotics, local uniqueness results, trace formulas, and Borg-type theorems for Dirac operators*, Trans. Amer. Math. Soc. **354** (2002), no. 9, 3475–3534 (electronic), DOI 10.1090/S0002-9947-02-03025-8. MR1911509 (2003i:34191)

[17] Steve Clark, Fritz Gesztesy, and Walter Renger, *Trace formulas and Borg-type theorems for matrix-valued Jacobi and Dirac finite difference operators*, J. Differential Equations **219** (2005), no. 1, 144–182, DOI 10.1016/j.jde.2005.04.013. MR2181033 (2006e:47068)

[18] M. M. Crum, *Associated Sturm-Liouville systems*, Quart. J. Math. Oxford Ser. (2) **6** (1955), 121–127. MR0072332 (17,266g)

[19] G. Darboux, *Sur une proposition relative aux équations linéaires*, C. R. Acad. Sci. (Paris) **94**, 1456–1459 (1882).

[20] P. A. Deift, *Applications of a commutation formula*, Duke Math. J. **45** (1978), no. 2, 267–310. MR495676 (81g:47001)

[21] P. Deift and E. Trubowitz, *Inverse scattering on the line*, Comm. Pure Appl. Math. **32** (1979), no. 2, 121–251, DOI 10.1002/cpa.3160320202. MR512420 (80e:34011)

[22] Rafael del Rio, Fritz Gesztesy, and Barry Simon, *Inverse spectral analysis with partial information on the potential. III. Updating boundary conditions*, Internat. Math. Res. Notices **15** (1997), 751–758, DOI 10.1155/S1073792897000494. MR1470376 (99a:34032)

[23] Plamen Djakov and Boris Mityagin, *Criteria for existence of Riesz bases consisting of root functions of Hill and 1D Dirac operators*, J. Funct. Anal. **263** (2012), no. 8, 2300–2332, DOI 10.1016/j.jfa.2012.07.003. MR2964684

[24] J. Eckhardt, *Inverse uniqueness results for Schrödinger operators using de Branges theory*, Complex Anal. Oper. Theory (to appear).

[25] J. Eckhardt, F. Gesztesy, R. Nichols, and G. Teschl, *Weyl–Titchmarsh theory for Sturm–Liouville operators with distributional potentials*, Opuscula Math. **33** (2013), 467–563, DOI 10.7494/OpMath.2013.33.3.467.

[26] J. Eckhardt, F. Gesztesy, R. Nichols, and G. Teschl, *Inverse spectral theory for Sturm-Liouville operators with distributional potentials*, arXiv:1210.7628.

[27] J. Eckhardt and G. Teschl, *Uniqueness results for one-dimensional Schrödinger operators with purely discrete spectra*, Trans. Amer. Math. Soc. (to appear).

[28] J. Eckhardt and G. Teschl, *Sturm–Liouville operators with measure-valued coefficients*, J. Analyse Math. (to appear).

[29] Julia Elyseeva, *On relative oscillation theory for symplectic eigenvalue problems*, Appl. Math. Lett. **23** (2010), no. 10, 1231–1237, DOI 10.1016/j.aml.2010.06.004. MR2665602 (2011k:39012)

[30] Allan Finkel, Eli Isaacson, and Eugene Trubowitz, *An explicit solution of the inverse periodic problem for Hill's equation*, SIAM J. Math. Anal. **18** (1987), no. 1, 46–53, DOI 10.1137/0518003. MR871819 (88d:34037)

[31] Charles Fulton, *Titchmarsh-Weyl m-functions for second-order Sturm-Liouville problems with two singular endpoints*, Math. Nachr. **281** (2008), no. 10, 1418–1475, DOI 10.1002/mana.200410689. MR2454944 (2009f:34071)

[32] C. Fulton and H. Langer, *Sturm-Liouville operators with singularities and generalized Nevanlinna functions*, Complex Anal. Oper. Theory **4** (2010), no. 2, 179–243, DOI 10.1007/s11785-009-0026-0. MR2652523 (2011i:34053)

[33] C. Fulton, H. Langer, and A. Luger, *Mark Krein's method of directing functionals and singular potentials*, Math. Nachr. 285, No. 14–15, 1791–1798 (2012)/DOI 10.1002/mana.201100301

[34] Clifford S. Gardner, John M. Greene, Martin D. Kruskal, and Robert M. Miura, *Korteweg-deVries equation and generalization. VI. Methods for exact solution*, Comm. Pure Appl. Math. **27** (1974), 97–133. MR0336122 (49 #898)

[35] I. M. Gel′fand and B. M. Levitan, *On the determination of a differential equation from its spectral function*, Amer. Math. Soc. Transl. (2) **1** (1955), 253–304. MR0073805 (17,489c)

[36] F. Gesztesy, *A complete spectral characterization of the double commutation method*, J. Funct. Anal. **117** (1993), no. 2, 401–446, DOI 10.1006/jfan.1993.1132. MR1244942 (94m:47093)

[37] F. Gesztesy, *New trace formulas for Schrödinger operators*, Evolution equations (Baton Rouge, LA, 1992), Lecture Notes in Pure and Appl. Math., vol. 168, Dekker, New York, 1995, pp. 201–221. MR1300430 (95i:34155)

[38] Fritz Gesztesy, *Inverse spectral theory as influenced by Barry Simon*, Spectral theory and mathematical physics: a Festschrift in honor of Barry Simon's 60th birthday, Proc. Sympos. Pure Math., vol. 76, Amer. Math. Soc., Providence, RI, 2007, pp. 741–820. MR2307754 (2008d:34010)

[39] F. Gesztesy, H. Grosse, and B. Thaller, *Spectral concentration in the nonrelativistic limit*, Phys. Lett. B **116**, 155–157 (1982).

[40] F. Gesztesy, H. Grosse, and B. Thaller, *Efficient method for calculating relativistic corrections for spin-1/2 particles*, Phys. Rev. Lett. **50**, 625–628 (1983).

[41] F. Gesztesy, H. Grosse, and B. Thaller, *A rigorous approach to relativistic corrections of bound state energies for spin-$\frac{1}{2}$ particles*, Ann. Inst. H. Poincaré Phys. Théor. **40** (1984), no. 2, 159–174 (English, with French summary). MR747200 (86m:81042)

[42] F. Gesztesy, H. Grosse, and B. Thaller, *On relativistic corrections to bound state energies for two-fermion systems*, Phys. Rev. **D 30**, 2189–2193 (1984).

[43] F. Gesztesy, H. Grosse, and B. Thaller, *First-order relativistic corrections and spectral concentration*, Adv. in Appl. Math. **6** (1985), no. 2, 159–176, DOI 10.1016/0196-8858(85)90009-0. MR789851 (86j:81026)

[44] F. Gesztesy and H. Holden, *On trace formulas for Schrödinger-type operators*, and molecular physics (Minneapolis, MN, 1995), IMA Vol. Math. Appl., vol. 89, Springer, New York, 1997, pp. 121–145, DOI 10.1007/978-1-4612-1870-8_5. MR1487920 (98m:34172)

[45] F. Gesztesy and H. Holden, *On new trace formulae for Schrödinger operators*, Acta Appl. Math. **39** (1995), no. 1-3, 315–333, DOI 10.1007/BF00994640. KdV '95 (Amsterdam, 1995). MR1329568 (96f:35126)

[46] F. Gesztesy and H. Holden, *Trace formulas and conservation laws for nonlinear evolution equations*, Rev. Math. Phys. **6** (1994), no. 1, 51–95, DOI 10.1142/S0129055X94000055. MR1263198 (95h:35198a)

[47] F. Gesztesy and H. Holden, *Errata: "Trace formulas and conservation laws for nonlinear evolution equations"*, Rev. Math. Phys. **6** (1994), no. 4, 673, DOI 10.1142/S0129055X94000493. MR1290693 (95h:35198b)

[48] Fritz Gesztesy and Helge Holden, *Soliton equations and their algebro-geometric solutions. Vol. I*, Cambridge Studies in Advanced Mathematics, vol. 79, Cambridge University Press, Cambridge, 2003. $(1+1)$-dimensional continuous models. MR1992536 (2005b:37173)

[49] Fritz Gesztesy, Helge Holden, Johanna Michor, and Gerald Teschl, *Soliton equations and their algebro-geometric solutions. Vol. II*, Cambridge Studies in Advanced Mathematics, vol. 114, Cambridge University Press, Cambridge, 2008. $(1+1)$-dimensional discrete models. MR2446594 (2011b:37130)

[50] F. Gesztesy, H. Holden, and B. Simon, *Absolute summability of the trace relation for certain Schrödinger operators*, Comm. Math. Phys. **168** (1995), no. 1, 137–161. MR1324393 (96b:34110)

[51] F. Gesztesy, H. Holden, B. Simon, and Z. Zhao, *Trace formulae and inverse spectral theory for Schrödinger operators*, Bull. Amer. Math. Soc. (N.S.) **29** (1993), no. 2, 250–255, DOI 10.1090/S0273-0979-1993-00431-2. MR1215308 (94c:34127)

[52] F. Gesztesy, H. Holden, B. Simon, and Z. Zhao, *Higher order trace relations for Schrödinger operators*, Rev. Math. Phys. **7** (1995), no. 6, 893–922, DOI 10.1142/S0129055X95000347. MR1348829 (97d:34094)

[53] F. Gesztesy, H. Holden, B. Simon, and Z. Zhao, *A trace formula for multidimensional Schrödinger operators*, J. Funct. Anal. **141** (1996), no. 2, 449–465, DOI 10.1006/jfan.1996.0137. MR1418515 (97i:47098)

[54] Fritz Gesztesy, Nigel J. Kalton, Konstantin A. Makarov, and Eduard Tsekanovskii, *Some applications of operator-valued Herglotz functions*, Operator theory, system theory and related topics (Beer-Sheva/Rehovot, 1997), Oper. Theory Adv. Appl., vol. 123, Birkhäuser, Basel, 2001, pp. 271–321. MR1821917 (2002f:47049)

[55] F. Gesztesy, M. Krishna, and G. Teschl, *On isospectral sets of Jacobi operators*, Comm. Math. Phys. **181** (1996), no. 3, 631–645. MR1414303 (97i:47048)

[56] Fritz Gesztesy and Marius Mitrea, *Generalized Robin boundary conditions, Robin-to-Dirichlet maps, and Krein-type resolvent formulas for Schrödinger operators on bounded Lipschitz domains*, Perspectives in partial differential equations, harmonic analysis and applications, Proc. Sympos. Pure Math., vol. 79, Amer. Math. Soc., Providence, RI, 2008, pp. 105–173. MR2500491 (2010k:35087)

[57] Fritz Gesztesy and Marius Mitrea, *A description of all self-adjoint extensions of the Laplacian and Kreĭn-type resolvent formulas on non-smooth domains*, J. Anal. Math. **113** (2011), 53–172, DOI 10.1007/s11854-011-0002-2. MR2788354 (2012d:47126)

[58] Fritz Gesztesy, Marius Mitrea, and Maxim Zinchenko, *Variations on a theme of Jost and Pais*, J. Funct. Anal. **253** (2007), no. 2, 399–448, DOI 10.1016/j.jfa.2007.05.009. MR2370084 (2008k:35081)

[59] F. Gesztesy, R. Ratnaseelan, and G. Teschl, *The KdV hierarchy and associated trace formulas*, Recent developments in operator theory and its applications (Winnipeg, MB, 1994), Oper. Theory Adv. Appl., vol. 87, Birkhäuser, Basel, 1996, pp. 125–163. MR1399359 (97m:58095)

[60] F. Gesztesy and W. Renger, *New classes of Toda soliton solutions*, Comm. Math. Phys. **184** (1997), no. 1, 27–50, DOI 10.1007/s002200050051. MR1462498 (99f:58097)

[61] F. Gesztesy, W. Schweiger, and B. Simon, *Commutation methods applied to the mKdV-equation*, Trans. Amer. Math. Soc. **324** (1991), no. 2, 465–525, DOI 10.2307/2001730. MR1029000 (92b:35132)

[62] Fritz Gesztesy and Barry Simon, *The xi function*, Acta Math. **176** (1996), no. 1, 49–71, DOI 10.1007/BF02547335. MR1395669 (97e:47078)

[63] Fritz Gesztesy and Barry Simon, *Inverse spectral analysis with partial information on the potential. I. The case of an a.c. component in the spectrum*, Helv. Phys. Acta **70** (1997), no. 1-2, 66–71. Papers honouring the 60th birthday of Klaus Hepp and of Walter Hunziker, Part II (Zürich, 1995). MR1441597 (98f:81347)

[64] Fritz Gesztesy and Barry Simon, *On the determination of a potential from three spectra*, Differential operators and spectral theory, Amer. Math. Soc. Transl. Ser. 2, vol. 189, Amer. Math. Soc., Providence, RI, 1999, pp. 85–92. MR1730505 (2000i:34026)

[65] Fritz Gesztesy and Barry Simon, *Inverse spectral analysis with partial information on the potential. II. The case of discrete spectrum*, Trans. Amer. Math. Soc. **352** (2000), no. 6, 2765–2787, DOI 10.1090/S0002-9947-99-02544-1. MR1694291 (2000j:34019)

[66] Fritz Gesztesy and Barry Simon, *On local Borg-Marchenko uniqueness results*, Comm. Math. Phys. **211** (2000), no. 2, 273–287, DOI 10.1007/s002200050812. MR1754515 (2001b:34020)

[67] F. Gesztesy and B. Simon, *Uniqueness theorems in inverse spectral theory for one-dimensional Schrödinger operators*, Trans. Amer. Math. Soc. **348** (1996), no. 1, 349–373, DOI 10.1090/S0002-9947-96-01525-5. MR1329533 (96e:34030)

[68] Fritz Gesztesy and Barry Simon, *A new approach to inverse spectral theory. II. General real potentials and the connection to the spectral measure*, Ann. of Math. (2) **152** (2000), no. 2, 593–643, DOI 10.2307/2661393. MR1804532 (2001m:34185b)

[69] F. Gesztesy, A. Pushnitski, and B. Simon, *On the Koplienko spectral shift function. I. Basics*, Zh. Mat. Fiz. Anal. Geom. **4** (2008), no. 1, 63–107, 202 (English, with English and Ukrainian summaries). MR2404174 (2009k:47042)

[70] F. Gesztesy, B. Simon, and G. Teschl, *Spectral deformations of one-dimensional Schrödinger operators*, J. Anal. Math. **70** (1996), 267–324, DOI 10.1007/BF02820446. MR1444263 (98m:34171)

[71] F. Gesztesy and G. Teschl, *On the double commutation method*, Proc. Amer. Math. Soc. **124** (1996), no. 6, 1831–1840, DOI 10.1090/S0002-9939-96-03299-6. MR1322925 (96h:34171)

[72] F. Gesztesy, B. Simon, and G. Teschl, *Zeros of the Wronskian and renormalized oscillation theory*, Amer. J. Math. **118** (1996), no. 3, 571–594. MR1393260 (97g:34105)

[73] Fritz Gesztesy and Roman Svirsky, *(m)KdV solitons on the background of quasi-periodic finite-gap solutions*, Mem. Amer. Math. Soc. **118** (1995), no. 563, iv+88. MR1303091 (96c:35162)

[74] F. Gesztesy and G. Teschl, *Commutation methods for Jacobi operators*, J. Differential Equations **128** (1996), no. 1, 252–299, DOI 10.1006/jdeq.1996.0095. MR1392402 (97i:47079)

[75] Fritz Gesztesy and Vadim Tkachenko, *When is a non-self-adjoint Hill operator a spectral operator of scalar type?*, C. R. Math. Acad. Sci. Paris **343** (2006), no. 4, 239–242, DOI 10.1016/j.crma.2006.06.014 (English, with English and French summaries). MR2245385 (2007b:34211)

[76] Fritz Gesztesy and Vadim Tkachenko, *A criterion for Hill operators to be spectral operators of scalar type*, J. Anal. Math. **107** (2009), 287–353, DOI 10.1007/s11854-009-0012-5. MR2496408 (2010d:47063)

[77] Fritz Gesztesy and Vadim Tkachenko, *A Schauder and Riesz basis criterion for non-self-adjoint Schrödinger operators with periodic and antiperiodic boundary conditions*, J. Differential Equations **253** (2012), no. 2, 400–437, DOI 10.1016/j.jde.2012.04.002. MR2921200

[78] F. Gesztesy and M. Ünal, *Perturbative oscillation criteria and Hardy-type inequalities*, Math. Nachr. **189** (1998), 121–144, DOI 10.1002/mana.19981890108. MR1492926 (99a:34069)

[79] F. Gesztesy and K. Unterkofler, *Isospectral deformations for Sturm-Liouville and Dirac-type operators and associated nonlinear evolution equations*, Rep. Math. Phys. **31** (1992), no. 2, 113–137, DOI 10.1016/0034-4877(92)90008-O. MR1227036 (94f:35124)

[80] F. Gesztesy and R. Weikard, *Spectral deformations and soliton equations*, Differential equations with applications to mathematical physics, Math. Sci. Engrg., vol. 192, Academic Press, Boston, MA, 1993, pp. 101–139, DOI 10.1016/S0076-5392(08)62376-0. MR1207152 (93m:34138)

[81] F. Gesztesy and Z. Zhao, *On critical and subcritical Sturm-Liouville operators*, J. Funct. Anal. **98** (1991), no. 2, 311–345, DOI 10.1016/0022-1236(91)90081-F. MR1111572 (93f:34146)

[82] Fritz Gesztesy and Maxim Zinchenko, *On spectral theory for Schrödinger operators with strongly singular potentials*, Math. Nachr. **279** (2006), no. 9-10, 1041–1082, DOI 10.1002/mana.200510410. MR2242965 (2007h:47076)

[83] D. R. Grigore, G. Nenciu, and R. Purice, *On the nonrelativistic limit of the Dirac Hamiltonian*, Ann. Inst. H. Poincaré Phys. Théor. **51** (1989), no. 3, 231–263 (English, with French summary). MR1034588 (91h:81029)

[84] Philip Hartman, *A characterization of the spectra of one-dimensional wave equations*, Amer. J. Math. **71** (1949), 915–920. MR0033419 (11,438a)

[85] W. Hunziker, *On the nonrelativistic limit of the Dirac theory*, Comm. Math. Phys. **40** (1975), 215–222. MR0363275 (50 #15713)

[86] Katsunori Iwasaki, *Inverse problem for Sturm-Liouville and Hill equations*, Ann. Mat. Pura Appl. (4) **149** (1987), 185–206, DOI 10.1007/BF01773933. MR932784 (89d:34053)

[87] C. G. J. Jacobi, *Zur Theorie der Variationsrechnung und der Differentialgleichungen*, J. Reine Angew. Math. **17**, 68–82 (1837).

[88] I. S. Kac, *The existence of spectral functions of generalized second order differential systems with boundary conditions at the singular end*, AMS Translations (2) **62**, 204–262 (1967).

[89] I. Kay and H. E. Moses, *Reflectionless transmission through dielectrics and scattering potentials*, Res. Rep. No. EM-91, Div. Electromag. Res., Inst. Math. Sci., New York Univ., 1956. MR0090367 (19,805e)

[90] Tosio Kato, *Perturbation theory for linear operators*, 2nd ed., Springer-Verlag, New York, Berlin, 1980.

[91] Kunihiko Kodaira, *The eigenvalue problem for ordinary differential equations of the second order and Heisenberg's theory of S-matrices*, Amer. J. Math. **71** (1949), 921–945. MR0033421 (11,438c)

[92] Aleksey Kostenko, Alexander Sakhnovich, and Gerald Teschl, *Weyl-Titchmarsh theory for Schrödinger operators with strongly singular potentials*, Int. Math. Res. Not. IMRN **8** (2012), 1699–1747, DOI 10.1093/imrn/rnr065. MR2920828

[93] Aleksey Kostenko, Alexander Sakhnovich, and Gerald Teschl, *Commutation methods for Schrödinger operators with strongly singular potentials*, Math. Nachr. **285** (2012), no. 4, 392–410, DOI 10.1002/mana.201000108. MR2899633

[94] Aleksey Kostenko and Gerald Teschl, *On the singular Weyl-Titchmarsh function of perturbed spherical Schrödinger operators*, J. Differential Equations **250** (2011), no. 9, 3701–3739, DOI 10.1016/j.jde.2010.10.026. MR2773183 (2012b:34066)

[95] A. Kostenko and G. Teschl, *Spectral asymptotics for perturbed spherical Schrödinger operators*, will appear in Comm. Math. Phys. DOI 10.1007/s00220-013-1698-x.

[96] Helge Krüger and Gerald Teschl, *Relative oscillation theory, weighted zeros of the Wronskian, and the spectral shift function*, Comm. Math. Phys. **287** (2009), no. 2, 613–640, DOI 10.1007/s00220-008-0600-8. MR2481752 (2010g:47031)

[97] Helge Krüger and Gerald Teschl, *Relative oscillation theory for Sturm-Liouville operators extended*, J. Funct. Anal. **254** (2008), no. 6, 1702–1720, DOI 10.1016/j.jfa.2007.10.007. MR2396019 (2009b:34260)

[98] Helge Krüger and Gerald Teschl, *Effective Prüfer angles and relative oscillation criteria*, J. Differential Equations **245** (2008), no. 12, 3823–3848, DOI 10.1016/j.jde.2008.06.004. MR2462706 (2009k:34077)

[99] Pavel Kurasov and Annemarie Luger, *An operator theoretic interpretation of the generalized Titchmarsh-Weyl coefficient for a singular Sturm-Liouville problem*, Math. Phys. Anal. Geom. **14** (2011), no. 2, 115–151, DOI 10.1007/s11040-011-9090-6. MR2802457 (2012i:34031)

[100] E. A. Kuznetsov and A. V. Mikhaĭlov, *Stability of stationary waves in nonlinear weakly dispersive media*, Ž. Èksper. Teoret. Fiz. **67** (1974), no. 5, 1717–1727 (Russian, with English summary); English transl., Soviet Physics JETP **40** (1974), no. 5, 855–859. MR0387847 (52 #8685)

[101] B. M. Levitan, *Inverse Sturm-Liouville problems*, VSP, Zeist, 1987. Translated from the Russian by O. Efimov. MR933088 (89b:34001)

[102] B. M. Levitan and I. S. Sargsjan, *Sturm-Liouville and Dirac operators*, Mathematics and its Applications (Soviet Series), vol. 59, Kluwer Academic Publishers Group, Dordrecht, 1991. Translated from the Russian. MR1136037 (92i:34119)

[103] Vladimir A. Marchenko, *Sturm-Liouville operators and applications*, Operator Theory: Advances and Applications, vol. 22, Birkhäuser Verlag, Basel, 1986. Translated from the Russian by A. Iacob. MR897106 (88f:34034)

[104] Mayumi Ohmiya and Masami Urakubo, *Isospectral prosperty of double Darboux transformation*, Osaka J. Math. **43** (2006), no. 4, 877–893. MR2303554 (2008b:37118)

[105] L. A. Pastur and V. A. Tkachenko, *On the geometry of the spectrum of the one-dimensional Schrödinger operator with periodic complex-valued potential*, Mat. Zametki **50** (1991), no. 4, 88–95, 159, DOI 10.1007/BF01137736 (Russian); English transl., Math. Notes **50** (1991), no. 3-4, 1045–1050 (1992). MR1162916 (93h:34147)

[106] Jürgen Pöschel and Eugene Trubowitz, *Inverse spectral theory*, Pure and Applied Mathematics, vol. 130, Academic Press Inc., Boston, MA, 1987. MR894477 (89b:34061)

[107] Alexander Pushnitski and Ian Sorrell, *High energy asymptotics and trace formulas for the perturbed harmonic oscillator*, Ann. Henri Poincaré **7** (2006), no. 2, 381–396, DOI 10.1007/s00023-005-0253-5. MR2210237 (2007d:34157)

[108] James Ralston and Eugene Trubowitz, *Isospectral sets for boundary value problems on the unit interval*, Ergodic Theory Dynam. Systems **8*** (1988), no. Charles Conley Memorial Issue, 301–358, DOI 10.1017/S0143385700009470. MR967643 (89m:34035)

[109] Michael Reed and Barry Simon, *Methods of modern mathematical physics. IV. Analysis of operators*, Academic Press [Harcourt Brace Jovanovich Publishers], New York, 1978. MR0493421 (58 #12429c)

[110] A. V. Rybin and M. A. Sall′, *Solitons of the Korteweg-de Vries equation on the background of a known solution*, Teoret. Mat. Fiz. **63** (1985), no. 3, 333–339 (Russian, with English summary). MR805515 (86k:35145)

[111] Alexei Rybkin, *On the trace approach to the inverse scattering problem in dimension one*, SIAM J. Math. Anal. **32** (2001), no. 6, 1248–1264 (electronic), DOI 10.1137/S0036141000365620. MR1856247 (2002f:34201)

[112] Alexei Rybkin, *Some new and old asymptotic representations of the Jost solution and the Weyl m-function for Schrödinger operators on the line*, Bull. London Math. Soc. **34** (2002), no. 1, 61–72, DOI 10.1112/S0024609301008645. MR1866429 (2002j:34149)

[113] Alexei Rybkin, *Necessary and sufficient conditions for absolute summability of the trace formulas for certain one dimensional Schrödinger operators*, Proc. Amer. Math. Soc. **131** (2003), no. 1, 219–229 (electronic), DOI 10.1090/S0002-9939-02-06555-3. MR1929041 (2003g:34187)

[114] Alexander Sakhnovich, *Iterated Backlund-Darboux transform for canonical systems*, J. Funct. Anal. **144** (1997), no. 2, 359–370, DOI 10.1006/jfan.1996.3003. MR1432589 (98i:34123)

[115] Alexander Sakhnovich, *Generalized Bäcklund-Darboux transformation: spectral properties and nonlinear equations*, J. Math. Anal. Appl. **262** (2001), no. 1, 274–306, DOI 10.1006/jmaa.2001.7577. MR1857228 (2002k:35012)

[116] A. L. Sakhnovich, *Discrete canonical system and non-abelian Toda lattice: Bäcklund-Darboux transformation, Weyl functions, and explicit solutions*, Math. Nachr. **280** (2007), no. 5-6, 631–653, DOI 10.1002/mana.200410506. MR2308488 (2008k:37154)

[117] Karl Michael Schmidt, *An application of Gesztesy-Simon-Teschl oscillation theory to a problem in differential geometry*, J. Math. Anal. Appl. **261** (2001), no. 1, 61–71, DOI 10.1006/jmaa.2001.7471. MR1850956 (2002g:58016)

[118] U.-W. Schmincke, *On Schrödinger's factorization method for Sturm-Liouville operators*, Proc. Roy. Soc. Edinburgh Sect. A **80** (1978), no. 1-2, 67–84, DOI 10.1017/S0308210500010143. MR529570 (80f:34025)

[119] U.-W. Schmincke, *On a paper by F. Gesztesy, B. Simon, and G. Teschl concerning isospectral deformations of ordinary Schrödinger operators: "Spectral deformations of one-dimensional Schrödinger operators" [J. Anal. Math. 70 (1996), 267–324; MR1444263 (98m:34171)]*, J. Math. Anal. Appl. **277** (2003), no. 1, 51–78, DOI 10.1016/S0022-247X(02)00492-4. MR1954462 (2004a:34160)

[120] Andrew Y. Schoene, *On the nonrelativistic limits of the Klein-Gordon and Dirac equations*, J. Math. Anal. Appl. **71** (1979), no. 1, 36–47, DOI 10.1016/0022-247X(79)90216-6. MR545860 (81a:81024)

[121] P. Schwerdtfeger (ed.), *Relativistic electronic structure theory - part 1. Fundamentals*, Elsevier Science, 2002.

[122] M. I. Serov, *Certain properties of the spectrum of a non-selfadjoint differential operator of the second order*, Soviet Math. Dokl. **1** (1960), 190–192. MR0131016 (24 #A870)

[123] G.L. Sewell, *An appropriate relation between the energy levels of a particle in a field of given potential energy, calculated in the relativistic and non-relativistic theories*, Proc. Cambridge Phil. Soc. **45**, 631–637 (1949).

[124] Tomoyuki Shirai, *A trace formula for discrete Schrödinger operators*, Publ. Res. Inst. Math. Sci. **34** (1998), no. 1, 27–41, DOI 10.2977/prims/1195144826. MR1617621 (2000c:47067)

[125] Barry Simon, *Sturm oscillation and comparison theorems*, Sturm-Liouville theory, Birkhäuser, Basel, 2005, pp. 29–43, DOI 10.1007/3-7643-7359-8_2. MR2145076

[126] Robert Stadler and Gerald Teschl, *Relative oscillation theory for Dirac operators*, J. Math. Anal. Appl. **371** (2010), no. 2, 638–648, DOI 10.1016/j.jmaa.2010.05.069. MR2670140 (2011h:34192)

[127] Gerald Teschl, *Oscillation theory and renormalized oscillation theory for Jacobi operators*, J. Differential Equations **129** (1996), no. 2, 532–558, DOI 10.1006/jdeq.1996.0126. MR1404392 (98m:47053)

[128] Gerald Teschl, *Spectral deformations of Jacobi operators*, J. Reine Angew. Math. **491** (1997), 1–15, DOI 10.1515/crll.1997.491.1. MR1476087 (99b:39025)

[129] Gerald Teschl, *Deforming the point spectra of one-dimensional Dirac operators*, Proc. Amer. Math. Soc. **126** (1998), no. 10, 2873–2881, DOI 10.1090/S0002-9939-98-04362-7. MR1451831 (99a:34230)

[130] Gerald Teschl, *Renormalized oscillation theory for Dirac operators*, Proc. Amer. Math. Soc. **126** (1998), no. 6, 1685–1695, DOI 10.1090/S0002-9939-98-04310-X. MR1443411 (98g:34137)

[131] Gerald Teschl, *Trace formulas and inverse spectral theory for Jacobi operators*, Comm. Math. Phys. **196** (1998), no. 1, 175–202, DOI 10.1007/s002200050419. MR1643529 (99i:47065)

[132] Gerald Teschl, *On the Toda and Kac-van Moerbeke hierarchies*, Math. Z. **231** (1999), no. 2, 325–344, DOI 10.1007/PL00004732. MR1703351 (2000f:37105)

[133] Gerald Teschl, *Jacobi operators and completely integrable nonlinear lattices*, Mathematical Surveys and Monographs, vol. 72, American Mathematical Society, Providence, RI, 2000. MR1711536 (2001b:39019)

[134] Gerald Teschl, *Mathematical methods in quantum mechanics*, Graduate Studies in Mathematics, vol. 99, American Mathematical Society, Providence, RI, 2009. With applications to Schrödinger operators. MR2499016 (2010h:81002)

[135] Gerald Teschl, *Ordinary differential equations and dynamical systems*, Graduate Studies in Mathematics, vol. 140, American Mathematical Society, Providence, RI, 2012. MR2961944

[136] Bernd Thaller, *The Dirac equation*, Texts and Monographs in Physics, Springer-Verlag, Berlin, 1992. MR1219537 (94k:81056)

[137] E. C. Titchmarsh, *On the relation between the eigenvalues in relativistic and non-relativistic quantum mechanics*, Proc. Roy. Soc. A **266** (1962), 33–46. MR0131598 (24 #A1447)

[138] Krešimir Veselić, *The nonrelativistic limit of the Dirac equation and the spectral concentration*, Glasnik Mat. Ser. III **4 (24)** (1969), 231–241 (English, with Serbo-Croatian summary). MR0267294 (42 #2196)

[139] K. Veselić, *Perturbation of pseudoresolvents and analyticity in* $1/c$ *in relativistic quantum mechanics*, Comm. Math. Phys. **22** (1971), 27–43. MR0300579 (45 #9624)

[140] Joachim Weidmann, *Linear operators in Hilbert spaces*, Graduate Texts in Mathematics, vol. 68, Springer-Verlag, New York, 1980. Translated from the German by Joseph Szücs. MR566954 (81e:47001)

[141] Joachim Weidmann, *Spectral theory of ordinary differential operators*, Lecture Notes in Mathematics, vol. 1258, Springer-Verlag, Berlin, 1987. MR923320 (89b:47070)

[142] Joachim Weidmann, *Lineare Operatoren in Hilberträumen. Teil II*, Mathematische Leitfäden. [Mathematical Textbooks], B. G. Teubner, Stuttgart, 2003 (German). Anwendungen. [Applications]. MR2382320 (2008k:47002)

[143] Joachim Weidmann, *Spectral theory of Sturm-Liouville operators approximation by regular problems*, Sturm-Liouville theory, Birkhäuser, Basel, 2005, pp. 75–98, DOI 10.1007/3-7643-7359-8_4. MR2145078

[144] A. Wiegner, *Über den nichtrelativistischen Grenzwert der Eigenwerte der Diracgleichung*, Master's thesis, Fernuniversität-Gesamthochschule Hagen, Hagen, Germany, 1984.

[145] Maxim Zinchenko, *Trace formulas and a Borg-type theorem for CMV operators with matrix-valued coefficients*, Math. Nachr. **283** (2010), no. 2, 312–329, DOI 10.1002/mana.200810207. MR2604124 (2011b:47069)

Faculty of Mathematics, University of Vienna, Nordbergstrasse 15, 1090 Wien, Austria — and — International Erwin Schrödinger Institute for Mathematical Physics, Boltzmanngasse 9, 1090 Wien, Austria

E-mail address: Gerald.Teschl@univie.ac.at

URL: http://www.mat.univie.ac.at/~gerald/

Vorarlberg University of Applied Sciences, Hochschulstr. 1, 6850 Dornbirn, Austria

E-mail address: Karl.Unterkofler@fhv.at

URL: https://homepages.fhv.at/ku/

Proceedings of Symposia in Pure Mathematics
Volume **87**, 2013
http://dx.doi.org/10.1090/pspum/087/01437

Prescribed Asymptotic Behavior for Nonlinear Second-Order Dynamic Equations

Mehmet Ünal and Ağacık Zafer

Dedicated to Professor Fritz Gesztesy on the occasion of his sixtieth birthday

ABSTRACT. Let u and v be recessive and dominant solutions of

$$\left(p(t)x^{\Delta}(t)\right)^{\Delta} + q(t)x^{\sigma}(t) = 0, \quad t \in [a,\infty)_{\mathbb{T}},$$

where $\mathbb{T}$ is an arbitrary time scale which is unbounded from above. It is proved that the nonlinearly perturbed dynamic equation

$$\left(p(t)x^{\Delta}(t)\right)^{\Delta} + q(t)x^{\sigma}(t) = F(t,x(t)), \quad t \in [a,\infty)_{\mathbb{T}}$$

has solutions asymptotic to recessive and dominant solutions at infinity.

1. Introduction

We consider, on arbitrary time scales, the asymptotic integration problem for a class of nonlinear dynamic equations of the form

$$\left(p(t)x^{\Delta}\right)^{\Delta} + q(t)x^{\sigma} = F(t,x), \quad t \in [a,\infty)_{\mathbb{T}}, \tag{1.1}$$

where $\mathbb{T}$ is a time scale unbounded above, $[a,\infty)_{\mathbb{T}} := [a,\infty) \cap \mathbb{T}$ and $x^{\sigma} := x \circ \sigma$. Although it is assumed that the reader is already familiar with the notations and the subject of the theory of time scales, for completeness, we will provide some essential information about time scale calculus in the section 2.1. We should only mention here that this theory was introduced in order to unify continuous and discrete analysis, however it is not only unify the theories of differential equations and of difference equations, but also it is able to extend these classical cases to cases "in between", e.g., to so-called q-difference equations. Also note that when $\mathbb{T} = \mathbb{R}$, equation (1.1) is reduced to the second–order nonlinear differential equation

$$(p(t)x')' + q(t)x = F(t,x) \tag{1.2}$$

and when $\mathbb{T} = \mathbb{Z}$, it becomes a difference equation

$$\Delta\left(p(n)\Delta x(n)\right) + q(n)x(n+1) = F(n,x(n)). \tag{1.3}$$

2010 *Mathematics Subject Classification.* 34N05, 34D05.

Key words and phrases. Asymptotic behavior, dominant, recessive, time scale.

The first author's research was supported in part by the Scientific and Research Council of Turkey (TUBITAK) under contract 108T688.

In the case of quantum calculus which is defined as $\mathbb{T} = q^{\mathbb{N}} := \{q^m : m \in \mathbb{N}\}$, $q > 1$ is a real number, equation (1.1) leads to the q-difference equation

$$\Delta_q \left(p(t)\Delta_q x(t)\right) + q(t)x(qt) = F(t, x(t)), \tag{1.4}$$

where $\Delta_q f(t) = \frac{f(qt)-f(t)}{(q-1)t}$.

In the sequel we assume that

(a) $\sup \mathbb{T} = \infty$;
(b) $p : [a, \infty)_{\mathbb{T}} \to \mathbb{R}_+ := (0, \infty)$ and $q : [a, \infty)_{\mathbb{T}} \to \mathbb{R}$ are rd-continuous;
(c) $F : [a, \infty)_{\mathbb{T}} \times \mathbb{R} \to \mathbb{R}$ is rd-continuous;
(d) the linear equation

$$\left(p(t)x^{\Delta}\right)^{\Delta} + q(t)x^{\sigma} = 0, \quad t \in [a, \infty)_{\mathbb{T}} \tag{1.5}$$

has an eventually positive solution.

In this paper, we will show that the dynamic equation (1.1) has solutions $x_1(t)$ and $x_2(t)$ that behave asymptotically like

$$x_1(t) = u(t)(1 + o(1)), \quad t \to \infty \tag{1.6}$$

and

$$x_2(t) = v(t)(1 + o(1)), \quad t \to \infty, \tag{1.7}$$

where the functions u and v are the recessive and dominant solutions of (1.5), which will be described in the section 2.2.

Also original results are obtained for difference equation (1.3) and for q-difference equation (1.4). The results given by [**21**] for equation (1.2) are improved in a sense that the monotonicity condition is dropped. Our technique in proving the results naturally has some common features with the one employed in [**21**] but necessarily it is quite different due to difficulties peculiar to the time scale calculus. The main tools we use in our arguments are a time scale version of the Arzela-Ascoli Theorem (Theorem 2.2), Trench factorization for a second-order linear differential operator, and the Schauder fixed point theorem.

To the best of our knowledge, the first result on asymptotic integration was given in 1941 by Caligo [**10**] for second-order linear equations

$$x'' + a(t)x = 0,$$

where it was shown that all solutions possess the asymptotic representation

$$x(t) = c_1 t + c_2 + o(1), \quad t \to +\infty \tag{1.8}$$

provided that the function a satisfies a simple estimate $|a(t)| \leq l/t^{2+\rho}$ for sufficiently large t, where c_1 and c_2 are arbitrary constants, $l > 0$ and $\rho > 1$ are real numbers. Since then there has been tremendous amount of research searching for the existence of solutions satisfying (1.8) for linear and nonlinear differential equations, see [**1, 4–6, 10, 11, 13, 15, 16, 18–22, 24, 26, 28–30, 32–37**]. There is however relatively less work concerning the analogous results for difference equations [**12, 14, 17, 23, 25, 27, 38**], and only a few especially for dynamic equations on time scales [**2, 3, 7**]. In literature, a solution satisfying (1.8) is usually referred to as a linear-like solution.

It appears that there is hardly any work regarding the asymptotic integration of solutions of equations of the form (1.1). By using some well-known fixed point theorems it was shown in [**3**] that the solutions of second-order equations on time scales of the form

$$x^{\Delta\Delta} + f(t, x) = 0, \quad t \in \mathbb{T} \tag{1.9}$$

are linear-like under certain conditions imposed on the function f. The results may be considered as extensions of the ones obtained in [**2**, **7**]. The functions 1 and t appearing in (1.8) are clearly solutions of unperturbed equation corresponding to equation (1.9), that is,

$$x^{\Delta\Delta} = 0, \quad t \in \mathbb{T}. \tag{1.10}$$

It turns out that these functions are indeed recessive and dominant solutions of (1.10).

2. Preliminaries

In this section, we recall some of the notations, definitions, and theorems that we use throughout the paper.

2.1. Time scale. A time scale is an arbitrary nonempty closed subset of the real numbers $\mathbb{R}$, and is denoted by the symbol $\mathbb{T}$. The two most popular examples are $\mathbb{T} = \mathbb{R}$ and $\mathbb{T} = \mathbb{Z}$. Several other interesting time scales exist, and they give rise to plenty of applications such as the study of population dynamic models (see [**8**], pages 15 and 71).

Some elementary facts concerning the delta derivative are as follows:

(1) If f is differentiable at t, then

$$f^\sigma(t) = f(\sigma(t)) = f(t) + \mu(t)f^\Delta(t).$$

(2) If f and g are differentiable at t, then fg is differentiable at t with

$$(fg)^\Delta(t) = f^\sigma(t)g^\Delta(t) + f^\Delta(t)g(t) = f(t)g^\Delta(t) + f^\Delta(t)g^\sigma(t).$$

(3) If f and g are differentiable at t and $g(t)g(\sigma(t)) \neq 0$, then $\frac{f}{g}$ is differentiable at t with

$$\left(\frac{f}{g}\right)^\Delta(t) = \frac{f^\Delta(t)g(t) - f(t)g^\Delta(t)}{g(t)g^\sigma(t)}.$$

We say $f : \mathbb{T} \to \mathbb{R}$ is right-dense continuous ($f \in \mathrm{C_{rd}}(\mathbb{T}, \mathbb{R})$) provided f is continuous at right-dense points in $\mathbb{T}$ and its left-sided limit exists (finite) at left-dense points in $\mathbb{T}$. The importance of rd-continuous functions is that *every rd-continuous function possesses an antiderivative.* A function $F : \mathbb{T}^\kappa \to \mathbb{R}$ is called an antiderivative of $f : \mathbb{T} \to \mathbb{R}$ provided $F^\Delta(t) = f(t)$ holds for all $t \in \mathbb{T}^\kappa$. An excellent comprehensive treatment of calculus on time scales can be found in [**8**, **9**].

2.2. Recessive and dominant solutions. The condition (d) in our hypothesis ensures the existence of a real number $t_0 \in [a, \infty)_\mathbb{T}$ such that equation (1.5) has positive solutions u and v, called recessive and dominant solutions, defined on $[t_0, \infty)_\mathbb{T}$. It is well-known that u is unique up to a multiplication by a constant, and v is any other solution of equation (1.5) which is linearly independent of u. We may therefore assume without loss of generality that recessive and dominant solutions u and v are positive for sufficiently large t. These special solutions satisfy

$$\int_t^\infty \frac{1}{p(s)u(s)u^\sigma(s)}\Delta s = \infty \text{ (recessive)} \tag{2.1}$$

and

$$\int_t^\infty \frac{1}{p(s)v(s)v^\sigma(s)}\Delta s < \infty \text{ (dominant)}, \tag{2.2}$$

where $t \in \mathbb{T}$.

Trench factorization [**8**, Theorem 4.59] enables us to write

$$Lx := \left(p(t)x^\Delta\right)^\Delta + q(t)x^\sigma \tag{2.3}$$

in the form

$$Lx = \delta_1^\sigma\left(\delta_2\left(\delta_1 x\right)^\Delta\right)^\Delta, \qquad \delta_1 := \frac{1}{u}, \quad \delta_2 := puu^\sigma \tag{2.4}$$

and

$$Lx = \beta_1^\sigma\left(\beta_2\left(\beta_1 x\right)^\Delta\right)^\Delta, \qquad \beta_1 := \frac{1}{v}, \quad \beta_2 := pvv^\sigma. \tag{2.5}$$

These factorizations are useful tools in converting equation (1.1) to an equivalent integral equation, which we will employ in the present paper.

2.3. Schauder's fixed point theorem and a compactness criterion. We will need the following two theorems. The first theorem is the well-known Schauder's fixed-point theorem, and the second one is a time scale version of the Arzela-Ascoli theorem.

THEOREM 2.1. *Let X be a Banach space and Y be a nonempty closed convex subset of X. If $\mathcal{F} : Y \to Y$ is a continuous mapping such that $\mathcal{F}Y$ is a relatively compact subset of X, then $\mathcal{F}$ has at least one fixed point.*

Let w be either u or v and $t_a \in (a,\infty)_\mathbb{T}$ be sufficiently large to be specified later. We introduce the Banach space

$$\mathcal{B}_w[t_a,\infty)_\mathbb{T} = \left\{ f \in C_{rd}\left([t_a,\infty)_\mathbb{T}\right) : \sup_{t\in[t_a,\infty)_\mathbb{T}} \frac{|f(t)|}{w(t)} < +\infty \right\} \tag{2.6}$$

with

$$\|f\| = \sup_{t\in[t_a,\infty)_\mathbb{T}} \frac{|f(t)|}{w(t)}.$$

Set

$$X_{w,\gamma} = \left\{x \in \mathcal{B}_w[t_a,\infty)_\mathbb{T} : \ 0 \le x(t) \le 2\gamma\, w(t)\right\}, \tag{2.7}$$

where γ is a positive real number. It is clear that $X_{w,\gamma}$ is a bounded, convex, and closed subset of $\mathcal{B}_w[t_a,\infty)_\mathbb{T}$.

THEOREM 2.2. *Let $\mathcal{B}_w[t_a,\infty)_\mathbb{T}$ and $X_{w,\gamma}$ be given by (2.6) and (2.7), respectively. Then $X_{w,\gamma}$ is relatively compact if*

(i) *$X_{w,\gamma}$ is bounded;*
(ii) *on every compact subinterval J of $[a,\infty)_\mathbb{T}$, there exist, for any $\epsilon > 0$, $\delta > 0$, t_1, $t_2 \in J$ such that $|t_1 - t_2| < \delta$ implies*

$$\left|\frac{f(t_1)}{w(t_1)} - \frac{f(t_2)}{w(t_2)}\right| < \epsilon$$

for all $f \in X_{w,\gamma}$.(i.e., the set functions $X_{w,\gamma}$ is locally equicontinous);

(iii) *for every* $\varepsilon > 0$, *there exists* $b \in [a, \infty)_{\mathbb{T}}$ *such that* t_1, $t_2 \in [b, \infty)_{\mathbb{T}}$ *implies*

$$\left|\frac{f(t_1)}{w(t_1)} - \frac{f(t_2)}{w(t_2)}\right| < \epsilon$$

for all $f \in X_{w,\gamma}$.

PROOF. The proof of the theorem can be found in [**31**] for the special case $w(t) \equiv 1$. Since the arguments are similar, we omit the proof. □

3. The Main Results

In the first theorem we obtain an asymptotic integration of a solution $x(t)$ of equation (1.1) with respect to a recessive solution $u(t)$ as $t \to \infty$.

THEOREM 3.1. *Let (a)–(d)of our hypothesis hold, and let u be a positive recessive solution of equation (*1.5*). Suppose that there exist rd-continuous function* $\phi : [a, \infty)_{\mathbb{T}} \to \mathbb{R}_+$ *and a continuous function* $\psi : \mathbb{R}_+ \to \mathbb{R}_+$ *such that*

$$|F(t, x)| \leq \phi(t)\psi\left(\frac{|x|}{u(t)}\right). \tag{3.1}$$

If

$$\int_a^\infty \frac{1}{p(t)u(t)u^\sigma(t)} \int_t^\infty u^\sigma(s)\phi(s)\Delta s\Delta t < \infty \tag{3.2}$$

then for any given positive real number γ_1, *there exists a solution* $x_1(t)$ *of equation (*1.1*) such that*

$$x_1(t) = \gamma_1 u(t) + o(u(t)), \quad t \to \infty. \tag{3.3}$$

PROOF. Fix $\gamma_1 > 0$ and put

$$M_{\gamma_1} = \sup\{\psi(s) : \ s \in [0, 2\gamma_1]\}.$$

In view of (3.2), we may choose $t_a \in [a, \infty)_{\mathbb{T}}$ sufficiently large so that

$$\int_{t_a}^\infty \frac{1}{p(t)u(t)u^\sigma(t)} \int_t^\infty u^\sigma(s)\phi(s)\Delta s\Delta t \leq \gamma_1/M_{\gamma_1}. \tag{3.4}$$

Define an operator $\mathcal{S} : \ X_{u,\gamma_1} \to \mathcal{B}_u[t_a, \infty)_{\mathbb{T}}$ by

$$(\mathcal{S}x)(t) = \gamma_1 u(t) + u(t) \int_t^\infty \frac{1}{p(s)u(s)u^\sigma(s)} \int_s^\infty u^\sigma(r)F\left(r, x(r)\right)\Delta r\Delta s, \ t \in [t_a, \infty)_{\mathbb{T}}. \tag{3.5}$$

Taking into account (2.4), we see that if x is a fixed point of $\mathcal{S}$, then

$$Lx = F(t, x), \quad t \in [t_a, \infty)_{\mathbb{T}},$$

i.e., $x(t)$ is solution of equation (1.1) for $t \geq t_a$.

We will show that the operator $\mathcal{S}$ defined by equation (3.5) has a fixed point $x_1 \in X_{u,\gamma_1}$ satisfying the asymptotic representation (3.3).

CLAIM 1. $\mathcal{S}$ *is a self map on* X_{u,γ_1}, *i.e.*, $x \in X_{u,\gamma_1} \Rightarrow \mathcal{S}x \in X_{u,\gamma_1}$.

Let $x \in X_{u,\gamma_1}$. Using (c) of our hypothesis and (3.4), we see that

$$\begin{aligned} |(\mathcal{S}x)(t) - \gamma_1 u(t)| &\leq u(t) \int_t^\infty \frac{1}{p(s)u(s)u^\sigma(s)} \int_s^\infty u^\sigma(r) \left| F\left(r, x(r)\right)\right| \Delta r \Delta s \\ &\leq u(t) \int_t^\infty \frac{1}{p(s)u(s)u^\sigma(s)} \int_s^\infty u^\sigma(r) \phi(r) \psi\left(\frac{x(r)}{u(r)}\right) \Delta r \Delta s \\ &\leq u(t) \int_{t_a}^\infty \frac{1}{p(s)u(s)u^\sigma(s)} \int_s^\infty u^\sigma(r) \phi(r) M_{\gamma_1} \Delta r \Delta s \\ &\leq \gamma_1 u(t) \end{aligned} \tag{3.6}$$

which means that $\mathcal{S}x \in X_{u,\gamma_1}$.

Claim 2. *$\mathcal{S}$ is continuous.*

Let $\{x_j\}$ be a sequence in X_{u,γ_1} converging to $x \in X_{u,\gamma_1}$. Then

$$\frac{|\mathcal{S}x_j(t) - \mathcal{S}x(t)|}{u(t)} \leq \int_t^\infty \frac{1}{p(s)u(s)u^\sigma(s)} \int_s^\infty u^\sigma(r) |F(r, x_j(r)) - F(r, x(r))| \Delta r \Delta s. \tag{3.7}$$

Since

$$\begin{aligned} &\int_t^\infty \frac{1}{p(s)u(s)u^\sigma(s)} \int_s^\infty u^\sigma(r) |F(r, x_j(r)) - F(r, x(r))| \Delta r \Delta s \\ &\qquad \leq 2 \int_{t_a}^\infty \frac{1}{p(s)u(s)u^\sigma(s)} \int_s^\infty u^\sigma(r) \phi(r) M_{\gamma_1} \Delta r \Delta s \\ &\qquad \leq 2\gamma_1, \end{aligned}$$

by employing the Lebesgue dominated convergence theorem, it follows from (3.7) that

$$\lim_{j\to\infty} ||\mathcal{S}x_j(t) - \mathcal{S}x(t)|| = 0$$

and hence $\mathcal{S}$ is continuous.

Claim 3. *$\mathcal{S}X_{u,\gamma_1}$ is relatively compact.*

We will verify the conditions of Theorem 2.2. It is clear that (i) is trivial. Let $J = [t_a, T_1]_{\mathbb{T}}$, where T_1 is sufficiently large. Let $t_1, t_2 \in J$ with $t_2 \geq t_1$. For any given $\epsilon > 0$ we may choose $\delta > 0$ such that

$$\int_{t_1}^{t_2} \frac{1}{p(s)u(s)u^\sigma(s)} \int_s^\infty u^\sigma(r) \phi(r) \Delta r \Delta s < \epsilon / M_{\gamma_1} \tag{3.8}$$

whenever $|t_1 - t_2| < \delta$. In this case, using (3.8) we have

$$\begin{aligned}\left|\frac{Sx(t_1)}{u(t_1)} - \frac{Sx(t_2)}{u(t_2)}\right| &\leq \int_{t_1}^{t_2} \frac{1}{p(s)u(s)u^\sigma(s)} \int_s^\infty u^\sigma(r)|F(r, x(r))|\Delta r \Delta s.\\ &\leq \int_{t_1}^{t_2} \frac{1}{p(s)u(s)u^\sigma(s)} \int_s^\infty u^\sigma(r)\phi(r)M_{\gamma_1}\Delta r \Delta s\\ &< \epsilon\end{aligned}$$

for all $x \in X_{u,\gamma_1}$, meaning that (ii) of Theorem 2.2 holds. Finally, to verify (iii) of Theorem 2.2, let $\epsilon > 0$ be fixed. Then we may choose T_1 sufficiently large so that

$$\int_{T_1}^\infty \frac{1}{p(s)u(s)u^\sigma(s)} \int_s^\infty u^\sigma(r)\phi(r)\Delta r \Delta s < \epsilon/M_{\gamma_1}. \tag{3.9}$$

Let $t_1, t_2 \in [T_1, \infty)_\mathbb{T}$. In view of (3.9), one easily gets

$$\begin{aligned}\left|\frac{Sx(t_1)}{u(t_1)} - \frac{Sx(t_2)}{u(t_2)}\right| &\leq \int_{t_1}^{t_2} \frac{1}{p(s)u(s)u^\sigma(s)} \int_s^\infty u^\sigma(r)|F(r, x(r))|\Delta r \Delta s.\\ &\leq \int_{T_1}^{\infty} \frac{1}{p(s)u(s)u^\sigma(s)} \int_s^\infty u^\sigma(r)\phi(r)M_{\gamma_1}\Delta r \Delta s\\ &< \epsilon\end{aligned}$$

for all $x \in X_{u,\gamma_1}$.

By the Schauder fixed-point theorem (Theorem 2.1), we have a fixed point, say $x_1 \in X_{u,\gamma_1}$. Thus,

$$x_1(t) = \gamma_1 u(t) + u(t) \int_t^\infty \frac{1}{p(s)u(s)u^\sigma(s)} \int_s^\infty u^\sigma(r)F(r, x_1(r))\Delta r \Delta s, \quad t \in [t_a, \infty)_\mathbb{T}.$$

Finally, using (c) of our hypothesis and (3.2), we easily see that the asymptotic property (3.3) follows. This completes the proof. □

The following two corollaries are immediate consequences of above theorem and the proofs are thus omitted.

COROLLARY 3.2. *Let u be a positive recessive solution of the difference equation*

$$\Delta(p(n)\Delta x(n)) + q(n)x(n+1) = 0. \tag{3.10}$$

Suppose that (3.1) holds. If

$$\sum_{n=a}^\infty \frac{1}{p(n)u(n)u(n+1)} \sum_{j=n}^\infty u(j+1)\phi(j) < \infty,$$

then for any given positive real number γ_1, there exists a solution $x_1(n)$ of equation (1.3) such that

$$x_1(n) = \gamma_1 u(n) + o(u(n)), \quad n \to \infty.$$

COROLLARY 3.3. *Let u be a positive recessive solution of the q-difference equation*

$$\Delta_q \left(p(t)\Delta_q x(t)\right) + q(t)x(qt) = 0. \tag{3.11}$$

Suppose that (3.1) holds. If

$$\sum_{i=a}^{\infty} \frac{q^i}{p(q^i)u(q^i)u(q^{i+1})} \sum_{j=i}^{\infty} q^j u(q^{j+1})\phi(q^j) < \infty,$$

then for any given positive real number γ_1, there exists a solution $x_1(t)$ of equation (1.4) such that

$$x_1(t) = \gamma_1 u(t) + o(u(t)), \quad t \to \infty.$$

Similarly we have the following theorem on asymptotic integration of a solution of equation (1.1) with respect to a dominant solution $v(t)$ as $t \to \infty$.

THEOREM 3.4. *Let (a)–(d) hold, and v be a positive dominant solution of (1.5). Suppose that there exist rd-continuous function $\phi : [a, \infty)_{\mathbb{T}} \to \mathbb{R}_+$ and a continuous function $\psi : \mathbb{R}_+ \to \mathbb{R}_+$ such that*

$$|F(t,x)| \le \phi(t)\psi\left(\frac{|x|}{v(t)}\right). \tag{3.12}$$

If

$$\int_a^{\infty} \frac{1}{p(t)v(t)v^{\sigma}(t)} \int_a^t v^{\sigma}(s)\phi(s)\Delta s \Delta t < \infty, \tag{3.13}$$

then for any given positive real number γ_2, there exists a solution $x_2(t)$ of equation (1.1) such that

$$x_2(t) = \gamma_2 v(t) + o(v(t)), \quad t \to \infty. \tag{3.14}$$

PROOF. The proof is similar to that of Theorem 3.1. We choose $t_a \in [a, \infty)_{\mathbb{T}}$ sufficiently large so that

$$\int_{t_a}^{\infty} \frac{1}{p(t)v(t)v^{\sigma}(t)} \int_{t_a}^t v^{\sigma}(s)\phi(s)\Delta s \Delta t \le \gamma_2/M_{\gamma_2}. \tag{3.15}$$

Define $\mathcal{U} : X_{v,\gamma_2} \to \mathcal{B}_v[t_a, \infty)_{\mathbb{T}}$ by

$$(\mathcal{U}x)(t) = \gamma_2 v(t) - v(t) \int_t^{\infty} \frac{1}{p(s)v(s)v^{\sigma}(s)} \int_{t_a}^t v^{\sigma}(r)F\left(r, x(r)\right)\Delta r \Delta s, \ t \in [t_a, \infty)_{\mathbb{T}}. \tag{3.16}$$

In view of (2.5) it is easy to see that any fixed point of this operator is a solution of equation (1.1) for $t \ge t_a$. As in the proof of Theorem 3.1 one can show that the operator $\mathcal{U}$ defined by (3.16) has a fixed point $x_2 \in X_{v,\gamma_2}$ satisfying the asymptotic representation (3.14). □

The following two corollaries virtually follow from the above theorem.

COROLLARY 3.5. *Let v be a positive dominant solution of equation (3.10). Suppose that (3.12) holds. If*

$$\sum_{n=a}^{\infty} \frac{1}{p(n)v(n)v(n+1)} \sum_{j=a}^{n-1} v(j+1)\phi(j) < \infty,$$

then for any given positive real number γ_2, there exists a solution $x_2(n)$ of equation (1.3) such that

$$x_2(n) = \gamma_2 v(n) + o(v(n)), \quad n \to \infty.$$

COROLLARY 3.6. *Let v be a positive dominant solution of equation (3.11). Suppose that (3.12) holds. If*

$$\sum_{i=a}^{\infty} \frac{q^i}{p(q^i)v(q^i)v(q^{i+1})} \sum_{j=a}^{i-1} q^j v(q^{j+1})\phi(q^j) < \infty,$$

then for any given positive real number γ_2, there exists a solution $x_2(n)$ of (1.4) such that

$$x_2(t) = \gamma_2 v(t) + o(v(t)), \quad t \to \infty.$$

4. Examples

We give two examples to illustrate the results.

EXAMPLE 4.1. Consider the following second-order nonlinear dynamic equation

$$\left(t\sigma(t)x^{\Delta}\right)^{\Delta} = Q(t)\,|x|^{\alpha-1}\,x, \tag{4.1}$$

where $\alpha > 0$, and Q is rd-continuous on $[a,\infty)_{\mathbb{T}}$ with $a \geq 1$ and $p(t) > 0$.

We have $p(t) = t\sigma(t)$, $q(t) \equiv 0$ and $F(t,x) = Q(t)\,|x|^{\alpha-1}\,x$. Since the corresponding unperturbed equation is

$$\left(t\sigma(t)x^{\Delta}\right)^{\Delta} = 0,$$

we may take $u(t) = 1/t$ and $v(t) = 1$ as recessive and dominant solutions. Note that (3.1) holds with $\phi(t) = |Q(t)|/t^{\alpha}$ and $\psi(x) = x^{\alpha}$.

By Theorem 3.1 we can conclude that if

$$\int_a^{\infty}\int_t^{\infty} \frac{|Q(s)|}{s^{\alpha}\sigma(s)} \Delta s \Delta t < \infty, \tag{4.2}$$

then there is a positive solution $x_1(t)$ of equation (4.1) having the asymptotic representation

$$x_1(t) = \frac{\gamma_1}{t}(1 + o(1)) \ \text{ as } t \to \infty.$$

Notice that (3.12) holds with $\phi(t) = |Q(t)|$ and $\psi(x) = x^{\alpha}$. So, if

$$\int_a^{\infty} \frac{1}{t\sigma(t)} \int_a^{t} |Q(s)| \Delta s \Delta t < \infty, \tag{4.3}$$

then we see from Theorem 3.4 that equation (4.1) has also a positive solution $x_2(t)$ satisfying

$$x_2(t) = \gamma_2 + o(1) \ \text{ as } t \to \infty.$$

EXAMPLE 4.2. Consider the second-order nonlinear difference equation

$$x(n+2) - 6x(n+1) + 8x(n) = (-1)^n 3^{-n} 2^{3-n} x^3(n). \tag{4.4}$$

The equation can be written in the form

$$\Delta\left(2^{-3n}\Delta x(n)\right) + (3/8)2^{-3n}x(n+1) = (-1)^n 3^{-n} 2^{-4n} x^3(n).$$

It is easy to see that $u(n) = 2^n$ and $v(n) = 4^n$ are recessive and dominant solutions of the corresponding unperturbed equation

$$\Delta\left(2^{-3n}\Delta x(n)\right) + (3/8)2^{-3n}x(n+1) = 0.$$

We see that $F(n,x) = (-1)^n 3^{-n}2^{-4n}x^3$. Let $\phi(n) = 3^{-n}2^{-n}$ and $\psi(x) = x^3$. Then equation (3.1) holds and

$$\sum_{n=0}^{\infty}\frac{1}{p(n)u(n)u(n+1)}\sum_{j=n}^{\infty}u(j+1)\phi(j) = (3/2)\sum_{n=0}^{\infty}(2/3)^n < \infty. \quad (4.5)$$

Taking $\phi(n) = 3^{-n}2^{2n}$ and $\psi(x) = x^3$, we see that (3.12) holds, but

$$\sum_{n=0}^{\infty}\frac{1}{p(n)v(n)v(n+1)}\sum_{j=0}^{n-1}v(j+1)\phi(j) = (3/13)\sum_{n=0}^{\infty}[(8/3)^n - (1/2)^n] = \infty. \quad (4.6)$$

In view of (4.5), we can conclude from Corollary 3.2 that there is a positive solution $x_1(n)$ of (4.4) having the asymptotic representation

$$x_1(n) = \gamma_1 2^n(1+o(1)), \qquad n\to\infty.$$

Corollary 3.5 fails to apply for equation (4.4) due to divergence of the series in equation (4.6).

References

[1] Ravi P. Agarwal, Smail Djebali, Toufik Moussaoui, and Octavian G. Mustafa, *On the asymptotic integration of nonlinear differential equations*, J. Comput. Appl. Math. **202** (2007), no. 2, 352–376, DOI 10.1016/j.cam.2005.11.038. MR2319962 (2008k:34184)

[2] Elvan Akin, Martin Bohner, Lynn Erbe, and Allan Peterson, *Existence of bounded solutions for second order dynamic equations*, J. Difference Equ. Appl. **8** (2002), no. 4, 389–401, DOI 10.1080/1026190290017414. In honor of Professor Lynn Erbe. MR1898019 (2003c:39016)

[3] Elvan Akın-Bohner, Martin Bohner, Smaïl Djebali, and Toufik Moussaoui, *On the asymptotic integration of nonlinear dynamic equations*, Adv. Difference Equ. (2008), Art. ID 739602, 17. MR2393474 (2009b:34010)

[4] C. Avramescu, *Sur l'existence des solutions convergentes des systèmes d'équations différentielles non linéaires*, Ann. Mat. Pura Appl. (4) **81** (1969), 147–168 (French). MR0249738 (40 #2979)

[5] I. Bihari, *A generalization of a lemma of Bellman and its application to uniqueness problems of differential equations*, Acta Math. Acad. Sci. Hungar. **7** (1956), 81–94 (English, with Russian summary). MR0079154 (18,38c)

[6] Mary L. Boas, R. P. Boas Jr., and Norman Levinson, *The growth of solutions of a differential equation*, Duke Math. J. **9** (1942), 847–853. MR0007550 (4,158a)

[7] Martin Bohner and Stevo Stević, *Asymptotic behavior of second-order dynamic equations*, Appl. Math. Comput. **188** (2007), no. 2, 1503–1512, DOI 10.1016/j.amc.2006.11.016. MR2335647

[8] Martin Bohner and Allan Peterson, *Dynamic equations on time scales*, Birkhäuser Boston Inc., Boston, MA, 2001. An introduction with applications. MR1843232 (2002c:34002)

[9] *Advances in dynamic equations on time scales*, Birkhäuser Boston Inc., Boston, MA, 2003. Edited by Martin Bohner and Allan Peterson. MR1962542 (2004d:34003)

[10] D. Caligo, *Comportamento asintotico degli integrali dell'equazione $y''(x) + A(x)y(x) = 0$, nell'ipotesi $\lim_{x\to+\infty}A(x) = 0$*, Boll. Un. Mat. Ital. (2) **3** (1941), 286–295 (Italian). MR0005219 (3,119e)

[11] Donald S. Cohen, *The asymptotic behavior of a class of nonlinear differential equations*, Proc. Amer. Math. Soc. **18** (1967), 607–609. MR0212289 (35 #3163)

[12] Saber N. Elaydi, *An introduction to difference equations*, Undergraduate Texts in Mathematics, Springer-Verlag, New York, 1996. MR1410259 (97m:39001)

[13] F. Gesztesy and M. Ünal, *Perturbative oscillation criteria and Hardy-type inequalities*, Math. Nachr. **189** (1998), 121–144, DOI 10.1002/mana.19981890108. MR1492926 (99a:34069)

[14] J. R. Graef, A. Miciano, P. W. Spikes, P. Sundaram, and E. Thandapani, *On the asymptotic behavior of solutions of a nonlinear difference equation*, Equations (San Antonio, TX, 1994), Gordon and Breach, Luxembourg, 1995, pp. 223–229. MR1678694

[15] Thomas G. Hallam, *Asymptotic integrations of second order differential equations with integrable coefficients*, SIAM J. Appl. Math. **19** (1970), 430–439. MR0269935 (42 #4828)

[16] Otto Haupt, *Über das asymptotische Verhalten der Lösungen gewisser linearer gewöhnlicher Differentialgleichungen*, Math. Z. **48** (1942), 289–292 (German). MR0008296 (4,276c)

[17] I. Kubiaczyk and S. H. Saker, *Oscillation and asymptotic behavior of second-order nonlinear difference equations*, Fasc. Math. **34** (2004), 39–54. MR2099622 (2005g:39020)

[18] Takaŝi Kusano, Manabu Naito, and Hiroyuki Usami, *Asymptotic behavior of solutions of a class of second order nonlinear differential equations*, Hiroshima Math. J. **16** (1986), no. 1, 149–159. MR837319 (87f:34053)

[19] Takaŝi Kusano and William F. Trench, *Global existence theorems for solutions of nonlinear differential equations with prescribed asymptotic behaviour*, J. London Math. Soc. (2) **31** (1985), no. 3, 478–486, DOI 10.1112/jlms/s2-31.3.478. MR812777 (87a:34012)

[20] Takaŝi Kusano and William F. Trench, *Existence of global solutions with prescribed asymptotic behavior for nonlinear ordinary differential equations*, Ann. Mat. Pura Appl. (4) **142** (1985), 381–392 (1986), DOI 10.1007/BF01766602. MR839046 (87j:34060)

[21] Kusano Takaŝi, Vojislav Marić, and Tomoyuki Tanigawa, *The asymptotic behaviour of solutions of a class of nonlinear second-order differential equations*, Bull. Lond. Math. Soc. **39** (2007), no. 3, 413–418, DOI 10.1112/blms/bdm017. MR2331568 (2008e:34084)

[22] Olivia Lipovan, *On the asymptotic behaviour of the solutions to a class of second order nonlinear differential equations*, Glasg. Math. J. **45** (2003), no. 1, 179–187, DOI 10.1017/S0017089502001143. MR1973349 (2004a:34092)

[23] Rigoberto Medina and Manuel Pinto, *Asymptotic behavior of solutions of second order nonlinear difference equations*, Nonlinear Anal. **19** (1992), no. 2, 187–195, DOI 10.1016/0362-546X(92)90119-Y. MR1174467 (93i:39001)

[24] Fan Wei Meng, *A note on: "The asymptotic behavior of a class of nonlinear differential equations of second order" [Proc. Amer. Math. Soc. **84** (1982), no. 2, 235–236; MR0637175 (83d:34062)] by J. C. Tong*, Proc. Amer. Math. Soc. **108** (1990), no. 2, 383–386, DOI 10.2307/2048285. MR993758 (90d:34071)

[25] Janusz Migda, *Asymptotic behavior of solutions of nonlinear difference equations*, Math. Bohem. **129** (2004), no. 4, 349–359. MR2102609 (2005h:39026)

[26] Octavian G. Mustafa and Yuri V. Rogovchenko, *Global existence of solutions with prescribed asymptotic behavior for second-order nonlinear differential equations*, Nonlinear Anal. **51** (2002), no. 2, Ser. A: Theory Methods, 339–368, DOI 10.1016/S0362-546X(01)00834-3. MR1918348 (2003f:34069)

[27] Małgorzata Migda and Janusz Migda, *On the asymptotic behavior of solutions of higher order nonlinear difference equations*, Proceedings of the Third World Congress of Nonlinear Analysts, Part 7 (Catania, 2000), 2001, pp. 4687–4695, DOI 10.1016/S0362-546X(01)00581-8. MR1975862 (2004f:39021)

[28] Manabu Naito, *Asymptotic behavior of solutions of second order differential equations with integrable coefficients*, Trans. Amer. Math. Soc. **282** (1984), no. 2, 577–588, DOI 10.2307/1999253. MR732107 (85m:34069)

[29] Ch. G. Philos and I. K. Purnaras, *Asymptotic behavior of solutions of second order nonlinear ordinary differential equations*, Nonlinear Anal. **24** (1995), no. 1, 81–90, DOI 10.1016/0362-546X(94)E0023-A. MR1308471 (96b:34071)

[30] Ch. G. Philos, I. K. Purnaras, and P. Ch. Tsamatos, *Asymptotic to polynomials solutions for nonlinear differential equations*, Nonlinear Anal. **59** (2004), no. 7, 1157–1179, DOI 10.1016/j.na.2004.08.011. MR2098511 (2005f:34152)

[31] Pavel Řehák, *On certain comparison theorems for half-linear dynamic equations on time scales*, Abstr. Appl. Anal. **7** (2004), 551–565, DOI 10.1155/S1085337504306251. MR2084935 (2005f:39025)

[32] Svitlana P. Rogovchenko and Yuri V. Rogovchenko, *Asymptotic behavior of solutions of second order nonlinear differential equations*, Portugal. Math. **57** (2000), no. 1, 17–33. MR1752217 (2001c:34093)

[33] Svitlana P. Rogovchenko and Yuri V. Rogovchenko, *Asymptotic behavior of certain second order nonlinear differential equations*, Dynam. Systems Appl. **10** (2001), no. 2, 185–200. MR1843736 (2002f:34110)
[34] Yuri V. Rogovchenko, *On the asymptotic behavior of solutions for a class of second order nonlinear differential equations*, Collect. Math. **49** (1998), no. 1, 113–120. MR1629766 (99d:34101)
[35] William F. Trench, *On the asymptotic behavior of solutions of second order linear differential equations*, Proc. Amer. Math. Soc. **14** (1963), 12–14. MR0142844 (26 #412)
[36] Paul Waltman, *On the asymptotic behavior of solutions of a nonlinear equation*, Proc. Amer. Math. Soc. **15** (1964), 918–923. MR0176170 (31 #445)
[37] J. Ernest Wilkins Jr., *On the growth of solutions of linear differential equations*, Bull. Amer. Math. Soc. **50** (1944), 388–394. MR0010227 (5,265b)
[38] Xiaoyun Zeng and Bao Shi, *Asymptotic behavior of solutions of some nonlinear difference equations*, Ann. Differential Equations **21** (2005), no. 3, 507–513. MR2175742 (2006g:39048)

Department of Mathematics, Çanakkale Onsekiz Mart University, Terzioglu Yerleşkesi, 17020 Çanakkale, Turkey
E-mail address: `munal@comu.edu.tr`

Department of Mathematics, Middle East Technical University, 06800 Ankara, Turkey
E-mail address: `zafer@metu.edu.tr`